T0189817

Comparative Ecology of Microorganisms
and Macroorganisms

John H. Andrews

Comparative Ecology of Microorganisms and Macroorganisms

Second Edition

John H. Andrews
University of Wisconsin–Madison
Madison, WI
USA

ISBN 978-1-4939-8331-5   ISBN 978-1-4939-6897-8   (eBook)
DOI 10.1007/978-1-4939-6897-8

1st edition: © 1991 Springer-Verlag New York Inc.
2nd edition: © Springer Science+Business Media LLC 2017
Softcover reprint of the hardcover 2nd edition 2017

Printed on acid-free paper

This Springer imprint is published by Springer Nature
The registered company is Springer Science+Business Media LLC
The registered company address is: 233 Spring Street, New York, NY 10013, U.S.A.

"The great thing is to last…
and write when there is something that you know;
and not before;
and not too damned much after."

Ernest Hemingway

This book is dedicated to two of my inspiring teachers, each in his own way:

Stanley Bishop Hardacker
Robert L. Vadas

Preface to the Second Edition

There is something fascinating about science. One gets such wholesale returns of conjecture out of such a trifling investment of fact.

Mark Twain, 1874, p. 156

Notwithstanding the typically amusing sentiment expressed by Twain, my hope and intent with this book is that the conjecture and ideas emerge from and by no means overwhelm a solid basis in fact. To that end, in part, this edition is thoroughly referenced to bring to the reader the most recent authoritative information, which builds upon the historical foundation cited that underpins the many theories discussed.

This second edition follows the framework of the first but is completely revised and is generally a more detailed treatment of the topics. The rationale for and goals of the initial treatise remain the same and were presented in detail in the Preface to the first edition (1991), which follows this Preface. While science and in particular molecular biology have progressed remarkably in the intervening years, the gulf between the two major branches of ecology—that of plants and animals on one hand and of microorganisms on the other—has not narrowed appreciably.

When the first edition was being written in the late 1980s, scientists and the public were attempting to come to grips with the realities of recombinant DNA and the prospect of genetically engineered organisms. This contentious debate (still in progress) had, perhaps predictably, splintered largely along disciplinary lines.[1] It set in stark contrast the two disciplines. Microbiologists, many of whom are molecular biologists, were generally confident about the new technologies, that molecular genetic interventions could be made with surgical precision, and that the phenotype emerges predictably from the genotype. Plant and animal ecologists, used to dealing with complex systems and attendant complications such as exotic species and unintended community invasions, were much less sanguine. A common refrain, frequently unspoken but omnipresent, was that microbiologists did not really understand "ecology," to which the rebuttal was that ecologists did not appreciate the reductionist rigor of microbiology or behavior of microorganisms. Actually, a closely parallel, longstanding debate had simmered within ecology between molecular camps and organismal camps. This issue was addressed years earlier by the famous paleontologist George Gaylord Simpson (*Am. Scholar* 36:363–377, 1967).

---

[1] For examples, see P. Berg et al. (Science 185:303.1974); H.O. Halvorson et al. (*Engineered Organisms in the Environment: Scientific Issues.* Am. Soc. Microbiol. 1985); and A. Kelman (Chair, NAS Comm. Rept. *Introduction of Recombinant DNA-Engineered Organisms into the Environment: Key Issues.* Natl. Acad. Sci., Washington, DC, 1987).

The discipline of microbiology was born with the discovery of the microscope separately by Hooke and by van Leeuwenhoek in the mid-1600s. To a large extent it remains pragmatically and philosophically distinct from macroscopic biology. There are multiple reasons for the parallel but largely separate evolution of these disciplines, not the least of which is an educational system that still rapidly channels undergraduates away from a unified and conceptual biology and into specialist subdisciplines. Inroads are being made by such avenues as molecular systematics. Likewise, since the 1970s and 80s there have been numerous examples of "macroorganism ecologists" who are using microbial systems to test ecological theory (see examples in Chap. 8). Whether the manifold emergent specializations such as transcriptomics, metabolomics, metagenomics, and the like, serve to reunite biologists or further splinter biology remains to be seen. Every biologist should periodically reread Mott Greene (*Nature* 388:619–620, 1997). It remains my hope that this text will help to provide a conceptual synthesis of ecological biology.

Acknowledgements

I greatly appreciate the time and energy of numerous colleagues who have helped me in many ways. The following scientists have criticized drafts of portions of the book: K. Clay, T. Gordon, R. Guries, R. Harris, J. Irwin, M. Kabbage, M. Smith, R. Vadas, T. Whitham, and S. Woody. Peter Blenis, my friend at the University of Alberta, not only read the entire manuscript exhaustively and provided a detailed critique, but in preparation for this onerous task slogged through the entire first edition as well! To him go praise and deep appreciation. For high-resolution scanning and Photoshop® reconstruction, I am also especially indebted to Russell Spear and Richard Day, M.D., who gave unstintingly of their time and expertise. High-resolution scans were also done by David Null and Jean Phillips. L. Ballantine, campus librarians, and the IT staff of the Herrling Visual Resources Laboratory answered many questions, provided help in various ways and untangled various messes. J.E.K. Andrews contributed his skills as an electronic systems engineer. The talented artist Kandis Elliot created the beautiful charcoal drawings that accompany each chapter, as they did in the first edition. I am indebted to Elaina Mercatoris of Springer for her encouragement, patient instructions, and help in many ways.

John H. Andrews
Madison, WI, USA
October, 2016

Preface to the First Edition (1991)

*The most important feature of the modern synthetic theory of evolution
is its foundation upon a great variety of biological disciplines.*

G.L. Stebbins, 1968, p. 17

This book is written with the goal of presenting ecologically significant
analogies between the biology of microorganisms and macroorganisms.
I consider such parallels to be important for two reasons. First, they serve to
emphasize that however diverse life may be, there are common themes at the
ecological level (not to mention other levels). Second, research done with either
microbes or macroorganisms has implications that transcend a particular field of
study. Although both points may appear obvious, the fact remains that attempts
to forge a conceptual synthesis are astonishingly meager. While unifying concepts
may not necessarily be strictly correct, they enable one to draw analogies across
disciplines. New starting points are discovered as a consequence, and new ways of
looking at things emerge.

The macroscopic organisms ('macroorganisms') include most representatives of
the plant and animal kingdoms. I interpret the term 'microorganism' (microbe)
literally to mean the small or microscopic forms of life, and I include in this category the bacteria, the protists (excluding the macroscopic green, brown, and red
algae), and the fungi. Certain higher organisms, such as many of the nematodes,
fall logically within this realm, but are not discussed at any length. Most of the
microbial examples are drawn from the bacteria [defined here in the generic sense
to include archaeans] and fungi because I am most familiar with those groups, but
the concepts they illustrate are not restrictive.

Because the intended audience includes individuals who are interested in 'big'
organisms and those interested in 'small' organisms, I have tried to provide sufficient background information so that specific examples used to illustrate the
principles are intelligible to both groups without being elementary or unduly rep-
etitious to either. Evolutionary biologists in either camp may also find the work
of interest because the central premise is that organisms have been shaped by
evolution operating through differential reproductive success. Survivors would
be expected to show some analogies in 'tactics' developed through natural selection. The only background assumed is a comprehensive course in college biology.
Hence, the book is also appropriate for advanced undergraduates.

The need for a synthesis of plant, animal, and microbial ecology is apparent
from an historical overview (e.g., McIntosh 1985). Plant ecology was originally
almost exclusively descriptive. It remains so to a large degree. A strongly
predictive science has emerged, but one based largely on correlation rather than
causation (Harper 1984). Within the past few decades it has branched into three

main streams: phytosociology, population biology, and the physiological or bio-physical study of individual plants (Cody 1986).

In contrast, animal ecologists have tended to study groups of ecologically similar species (i.e., guilds; Cody 1986). Although animal ecologists have probably been the most influential in developing the field of population ecology, many of their mathematical models derive from work based on microorganisms (e.g., Gause 1932), and from medical epidemiology (again involving microorganisms). Zool-ogists have also been the main force behind ecological theory, particularly as it applies to communities. Significantly, in both plant and animal ecology, the pro-cesses underlying patterns are usually inferred: The evidence marshalled is typi-cally correlative rather than causative, and frequently the difference between the two is overlooked, if recognized.

Many microbial ecologists study systems. Ecology as practiced by microbiolo-gists, however, has been mainly autecology, is typically reductionist in the extreme (increasingly often at the level of the gene), and lacks the strong theoretical basis of macroecology. Microbiology is a diverse field and microbial ecologists may be protozoologists, bacteriologists, mycologists, parasitologists, plant or animal pathologists, epidemiologists, phycologists, or molecular biologists. Approaches and semantics differ considerably among these subdisciplines. Nevertheless, microbial ecology is in general like plant ecology in its strong descriptive ele-ment. For example, often the goal is to identify and quantify the members of a particular community. Despite increasing emphasis on field research in the last few decades, microbes are usually brought into and remain within the laboratory for study under highly controlled conditions to determine the genetic, biochemi-cal, growth rate, or sporulation characteristics, or the pathological attributes of a particular population, and how these are influenced by various factors. Thus, the major operational difference between microorganisms and macroorganisms is that the former must generally be cultured to understand their properties. It is this requirement that brings the microbial ecologist so quickly from the field to the laboratory.

To oversimplify, one could say that macroecology consists of phenomena in search of mechanistic explanation, whereas microbial ecology is experimentation in search of theory.

The foregoing picture is reinforced by observations on the parochialism evident in science. Examples include the streamlined makeup of most professional socie-ties, the narrow disciplinary affiliation of participants at a typical conference, or the content of many journals and textbooks. Microbes as illustrations of ecological phenomena are omitted from the mainstream of ecology texts and are mentioned in cursory fashion only insofar as they relate to nutrient cycling or to macroor-ganisms. Discussion is left for microbiologists to develop in books on 'microbial ecology', which in itself suggests that there must be something unique about being microscopic. This is not meant as criticism, but rather as an observation on the state of affairs and an illustration of the gulf between disciplines. Finally, if additional evidence were necessary, one need only reflect on the current furor about the release of genetically engineered organisms. To a large degree this contentious issue has polarized along disciplinary lines: Ecologists express concern that microbiologists have little real knowledge of 'ecology'; microbiologists respond that ecologists really do not understand microbial systems. While acknowledging obvious differences between the two groups of organisms, one must ask whether their biology is so distinct that such isolationism is warranted. This book addresses that question.

A synthesis of microbial ecology and macroecology could be attempted at the level of the individual, the population, or the community. There would be gains and losses with any choice. Even the use of all three hierarchical levels would not ensure absence of misleading statements. Relative to the individual, comparisons based on populations or communities offer considerably increased scope and allow for inclusion of additional properties emerging at each level of complexity (e.g., density of organisms at the population level or diversity indices at the community level). Parallels based, for example, on supposed influential (i.e., organizing) forces in representative communities would be precarious, however, because unambiguous data implicating the relative roles of particular forces such as competition are usually lacking. The data that do exist have been variously interpreted, often with intense debate. Moreover, the same type of community may be organized quite differently in various parts of the world. For example, barnacle communities in Scotland (Connell 1961) are structured differently from those in the intertidal zone of the California coastline (Roughgarden et al. 1988). Entirely different pictures could emerge depending on the communities of microbes and macroorganisms arbitrarily selected for comparison.

Thus, I have decided to focus comparisons at the level of the individual. This implies neither that the individual is the only level on which selection acts, nor that shortcomings of the sort noted above do not exist. However, I consider the individual (defined in Chapter 1) through its entire life cycle to be operationally the fundamental unit of ecology in the sense that it is the primary one on which natural selection acts. Strictly speaking, natural selection acts at many levels, and a proper analysis of this issue requires that a distinction be made between what is transmitted and what transmits (Gliddon and Gouyon 1989). Efforts to relate the former (genetic information, broadly construed to include genes and epigenetic information) to a particular organizational level in a biological hierarchy extending from a nucleotide sequence to an ecosystem have engendered lively debate (e.g., Chapter 6 in Bell 1982; Brandon 1984; Tuomi and Vuorisalo 1989a, b; Gliddon and Gouyon 1989; Dawkins 1989). Without extending this controversy here, I note that at least my choice of the individual as the relevant ecological unit in this context is consistent with the conventional ecological viewpoint expressed by naturalists from Darwin onward (e.g., Dobzhansky 1956; Williams 1966; Mayr 1970, 1982; Begon et al. 1996; Ricklefs 1990). Comparisons of traits at the level of the individual can then be carried either downward to the corresponding genes—

which have passed the screen of natural selection—or upward to assess the role of selection of traits at the group level.

This book proceeds first, by developing a common format (in Chapter 1) for comparing organisms; second, by contrasting (in subsequent chapters) the biology of macro-and microorganisms within that context; third (and also in the subsequent chapters), by drawing ecologically useful analogies, that is, parallels of interpretive value in comparisons of traits or 'strategies' between representative organisms. The first two phases can be accomplished relatively objectively; the third is necessarily subjective. There are situations where macroorganisms provide the wrong model for microbes or vice versa. I have tried to avoid forcing the parallels and have stated where close parallels simply do not exist.

All citations appear in a bibliography at the end of the work. Where books are cited in the text, I also specify, when appropriate, the specific chapters or pages concerned, in order to assist the reader in locating the relevant passages. Finally, at the conclusion of each chapter, a few suggestions are made for additional reading on each particular topic.

This is a book mainly of ideas. I find that the study of analogues is instructive and productive, not to mention interesting and frequently exciting. There is considerable speculation and a strong emphasis on generalizations, which I hope will prove stimulating and useful to the reader. Inevitably, there will be exceptions to any of the general statements. I am not particularly concerned by these oddities and think that it would be a mistake for the book to be read with the intent of finding the exception in each case in an effort to demolish the principle. However, this is not the same as advocating that it be read uncritically! MacArthur and Wilson (1967, p. 5) have said that even a crude theory, which may account for about 85% of the variation in a phenomenon, is laudable if it points to relationships otherwise hidden, thus stimulating new forms of research. Williams (1966, p. 273) points out that every one of Dalton's six postulates about the nature of atoms eventually turned out to be wrong, but because of the questioning and experimentation that resulted from them, they stand as a beacon in the advance of chemistry. Finally and most explicitly, Bonner (1965, p. 15) states,

》 It is, after all, quite accepted that in a quantitative experiment, a statistical significance is sufficient to show a correlation. The fact that there are a few points that are off the curve, even though the majority are [sic] on it, does not impel one to disregard the whole experiment. Yet when we make generalizations about trends among animals and plants, such as changes in size, it is almost automatic to point out the exceptions and throw out the baby with the bath [sic]. This is not a question of fuzzy logic or sloppy thought; it is merely a question whether the rule or the deviations from the rule are of significance in the particular discussion.

I started this book during a sabbatical leave in 1986–1987 with John L. Harper, to whom I am deeply indebted for his generosity and kindness. His enthusiasm, innumerable ideas, and critical insight have contributed much to what is presented here. I thank R. Whitbread and the faculty of the School of Plant Biology at the University College of North Wales in Bangor, Wales, UK for putting up with me for a year; in particular I acknowledge with appreciation the hospitality of N.R. and C. Sackville Hamilton, the help of Michelle Jones in many ways, and the wise counsel of Adrian Bell. Finally, I thank the many colleagues elsewhere who have helped in various ways: G. Bowen, J. Burdon, D. Ingram, and J. Parke read a draft prospectus for the book. Aspects I have discussed with P. Ahlquist, J. Handelsman, and D. Shaw. I am grateful to the following individuals who have commented on one or more of the chapters: C. Allen, A. Bell, G. Carroll, K. Clay, A. Dobson, I. Eastwood, A. Ellingboe, R. Evert, J. Farrow, R. Forge, J. Gaunt, J. Harper, M. Havey, R. Jeanne, D. Jennings, L. Kinkel, W. Pfender, G. Roberts, N. Sackville Hamilton, B. Schmid, and K. Willis. Errors of fact or interpretation remain mine.

I thank my editor, T.D. Brock, for constructive criticism and for encouraging me to write this book. I am indebted to Kandis Elliot for the outstanding artwork, Ruth Siegel for copyediting, Steve Heinemann for assistance with the computer graphics, and Steve Vicen for some of the photographs. Finally, I thank my family for their forbearance.

John H. Andrews
Bangor, Wales
and
Madison, WI, USA

Contents

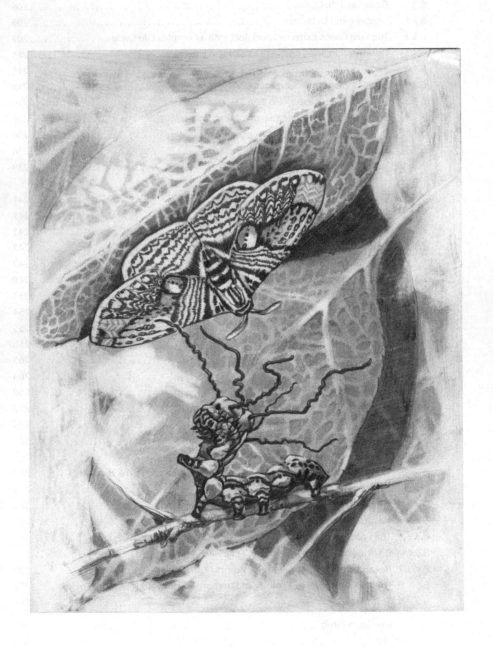

Introduction: Prospects for a Conceptual Synthesis

© Springer Science+Business Media LLC 2017
J.H. Andrews, *Comparative Ecology of Microorganisms and Macroorganisms*,
DOI 10.1007/978-1-4939-6897-8_1

When you're young, all evolution lies before you, every road is open to you, and at the same time you can enjoy the fact of being there on the rock, flat mollusk-pulp, damp and happy. If you compare yourself with the limitations that came afterwards, if you think of how having one form excludes other forms, of the monotonous routine where you finally feel trapped, well, I don't mind saying, life was beautiful in those days.

Italo Calvino, 2014, p. 138.

1.1 Organizing Life

In 1960, *The Forest and the Sea*, written by the accomplished naturalist Marston Bates, was published. It was his story of biology as told to the general public, largely by comparing the remarkable beauty and richness of life in a tropical rain forest and a tropical reef. This lovely book was an inspirational rallying cry to all biologists and aspiring biologists. But, you do not need to experience the splendor of a tropical rain forest or the Great Barrier Reef to appreciate that living forms are beautiful and astoundingly diverse. Just go for a walk in a local patch of forest, or a conservancy, or explore a trout stream. How do we interpret such a staggering array of shapes, sizes, and colors? What attributes could an invisible bacterium possibly share with a whale or a giant sequoia tree some tens of orders of magnitude its size? Do they share common patterns in behavior or in survival strategies? How are the millions of species of microorganisms and macroorganisms categorized and compared meaningfully? They will be **compared ecologically** in this book, as will be to some extent the processes underlying the apparent patterns. But, first, we should pause to examine the big picture, namely, the broad diversity of life and what relationships or order underpin this seeming chaotic assortment of creatures.

The most biologically informative analytical grouping of organisms is based on evolutionary or ancestral relationships as determined by phylogenetic systematics (cladistics). Broadly speaking, the features showing heritable variation used to develop phylogenies can either be phenotypic (i.e., physiological, anatomical, or morphological, with fossils being the classic example) or genotypic (i.e., molecular, with nucleic acids, proteins, or chromosomes as examples) (Hillis et al. 1996). Which of these two phylogenetic approaches is the 'better' repository of information for determining a true historic genealogy? The respective merits of 'morphology versus molecules' as the gold standard, along with how the biological world would be categorized (below), were actively debated for many years.[1] It became generally recognized that both types of data have inherent strengths and some weaknesses (see, e.g., Schopf et al. 1975;

---

1   See, for example, papers by Woese (1998a, b) and rebuttals by Mayr (1990, 1998). The long-running but now largely settled debate is well summarized (Pace 2009; Pace et al. 2012; Doolittle and Zhaxybayeva 2013). Until the ascendancy of molecular phylogenetics in the late 1970s, organisms were placed into as few as two or as many as 13 kingdoms (summarized by Margulis 1993). Of these, the most conventional depiction, particularly among botanists and zoologists, was and remains to varying degrees the 5-kingdom scheme *(Monera, Protista, Fungi, Plantae, Animalia)*. This is based on various traditional criteria including morphology, anatomy, nutritional mode, and cell structure. The groupings are, in general, phenotypically intuitive but, importantly, phenotypic similarity is not necessarily equivalent to evolutionary proximity. The most significant failing of the 5-kingdom organization of life is that the evolutionary relationships among, as opposed to within, the kingdoms are not clear and molecular data show that two kingdoms *(Protista, Monera)* are not clades, i.e., neither includes all descendants of an ancestral species.

◘ Table 1.1   Some contrasts among *Bacteria, Archaea, and Eukarya* (summarized from Madigan et al. 2015)

Characteristic	Bacteria	Archaea	Eukarya
Prokaryotic cell structure	Yes	Yes	No
Peptidoglycan cell wall	Yes	No	No
Membrane-enclosed nucleus	Absent	Absent	Present
Histone proteins present	No	Yes	Yes
RNA polymerases	1 (4 subunits)	1 (8–12 subunits)	3 (12–14 subunits each)
Ribosomes (mass)	70S	70S	80S
Initiator tRNA	Formylmethionine	Methionine	Methionine
Introns in most genes	No	No	Yes
Operons	Yes	Yes	No
mRNA capping and poly(A) tailing	No	No	Yes
Plasmids	Yes	Yes	Rare

Brower et al. 1996; Raff 2007). Either can be more or less suitable depending on the line of inquiry; ideally they are complementary. Hillis et al. (1996) summarized three fundamental requirements of any approach: (i) that the characters chosen should show appropriate levels of variation, together with (ii) a demonstrable and independent genetic basis, and that (iii) the data should be collected and analyzed in such a way that phylogenetic hypotheses can be tested meaningfully. In general, the modern molecular phylogenies have supported those based on morphology. Frequently, they have clarified, extended, or enabled the construction of accurate phylogenies where the morphological approach was impossible or misleading—a classic case is with the diverse truffle-like fungi (see, e.g., Bonito et al. 2013). It is also becoming increasingly common for organism phenotypic traits to be predicted from genotypic profiles (Costanzo et al. 2010; Ellison et al. 2011; Sunagawa et al. 2015).

Emerging from the 'molecules versus morphology' debate is whether cellular life is more appropriately organized into two (Prokaryota, Eukaryota; or possibly even *Archaea* and *Bacteria*; see below) or three (*Archaea, Bacteria, Eukarya*) major groupings, variously called empires, domains, or superkingdoms (◘Table 1.1). The informal division of the entire biological world broadly into prokaryotes versus eukaryotes based on physiological and morphological criteria was the traditional separation until molecular biologists discovered that prokaryotes were not monophyletic (i.e., all members did not share a common ancestor; Woese and Fox 1977). A further argument against the prokaryote/eukaryote divide was that categorization based on the *absence* of a structure (the nucleus) made little sense (Pace 2006, 2008). The counterpoint was that splitting prokaryotes into *Archaea* and *Bacteria* exaggerated their differences and placed too much emphasis on genotypic rather than phenotypic criteria (Mayr 1998; Cavalier-Smith 2006). While the prokaryotes generally are not morphologically ornate, two distinct evolutionary lineages are indeed evident and are now well accepted; the members of each are diverse genetically and metabolically. **In this text, for convenience and**

1

brevity, the term prokaryote is retained to refer informally and collectively to members of both domains *Archaea* and *Bacteria*. Such usage is in the same informal taxonomic sense that 'invertebrate' is a general term for any one of quite different organisms unified by the lack of a backbone. 'Prokaryote' is not being used in an evolutionary connotation in this book. Again, for reasons of brevity and simplicity, 'bacteria' with a small 'b' includes both bacteria and archaeons; when used with a capital 'B' and italicized, the word refers strictly to the evolutionarily distinct group *Bacteria* as opposed to *Archaea*.

Whether the so-called **universal 'tree of life'** is two or three domains, what can be said of their respective evolutionary origins? While a single-celled, bacterial-like, universal ancestor to all life commonly has been broadly assumed, how and when the more cytologically complex eukaryotes arose, and whether they are more closely related to *Bacteria* or *Archaea* remain highly contentious (▶Chap. 4 and Embley and Williams 2015). Actually, four possible scenarios are being debated (Mariscal and Doolittle 2015): (i) that the three branches—*Archaea*, *Bacteria*, and *Eukarya*—emerge more or less concurrently from a common primordial ancestor within a 'community pool' of pre-cell entities; (ii) that a **Last Universal Common Ancestor (LUCA)**, i.e., the entity common to all life, is a relatively complex eukaryote-like form from which the sister clades *Archaea* and *Bacteria* subsequently diverge and become simplified, while the eukaryote branch evolves further complexity; (iii) that the tree is rooted on the branch leading to the *Bacteria* with further complexity arising after the *Archaea* and *Eukarya* diverge; and finally (iv) that the tree is rooted on the bacterial branch as in (iii) but the LUCA is a relatively complex entity as in (ii) from which the *Archaea* diverge and become simplified as do the *Bacteria*; i.e., both lineages become convergently prokaryotic. Overall, educated opinions among these options vary, but recently the weight of evidence is shifting back to a two-domain primary clustering of life; however, those two primary domains may well be *Archaea* and *Bacteria* (Williams et al. 2013; Raymann et al. 2015). The eukaryotes arguably arose from the archaeal domain (Spang et al. 2015).

A phylogenetically based tree integrating all species would have to be based on molecular data (as supplemented by taxonomy where necessary, see below). This is because of several limitations in the applicability of morphological criteria to the *Archaea* and *Bacteria*. With some exceptions discussed in ▶Chap. 4, there is in general relatively little fossil evidence for the prokaryotes. Even if there were, morphology is not very useful because of their frequently simple, uniform external structure (though, interestingly, evidence is emerging that they are internally complex; Graumann 2007). Another limitation is that where species are compared over vast phylogenetic distances, few if any homologous morphological features exist. Differences in nucleotides within homologous (strictly, orthologous) genes among all known species provide a quantitative surrogate for conventional, morphologically based phylogenies. (Homologous, as used in the present sense, refers to a similarity derived from the same ancestral feature [phenotypic context], or to genes at the same locus in the genome [genotypic context]; orthologous refers to homologous genes that have diverged from each other because of separation of the species in which they are found [Barton et al. 2007]). The tree in ◧Fig. 1.1 shows a panoramic view of life as might be seen through the eyes of a molecular systematist. Indeed, the results of a massive phylogenic and taxonomic collaborative undertaking culminating in a comprehensive **'Open Tree of Life'** currently containing 2.3 million tips (equivalent to 'operational taxonomic units'; i.e., species where possible) have recently been published (Hinchliff et al. 2015).

Given rapid breakthroughs in molecular biology of recent decades, and in the mathematical algorithms used to create the trees, molecular phylogenies are beguiling in their apparent

5

1

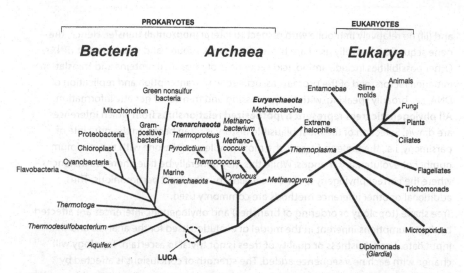

□ **Fig. 1.1** A universal phylogenetic tree of life based on sequence variation in the rRNA gene showing three domains (*Bacteria, Archaea,* and *Eukarya*). The last universal common ancestor (LUCA) to all cellular life is shown here within the very early *Bacteria* domain and the tree implies that the *Archaea* and *Eukarya* are on a separate main lineage distinct from the one leading to the *Bacteria*. Position of groups remains in flux and numerous other phylogenetic hypotheses exist (see, e.g., Raymann et al. 2015; Spang et al. 2015). Reproduced from: Madigan, Michael T.; Martinko, John M.; Stahl, David A.; Clark, David P. *Brock Biology of Microorganisms, 13*th *Ed.* ©2012. Printed and electronically reproduced by permission of Pearson Education, Inc. New York, New York

authority and simplicity. Nevertheless, it should not be forgotten that all trees, whether morphologically- or molecularly based, are basically hypotheses and interpreting them correctly is not trivial. The complexities and hidden assumptions in making phylogenetic inferences need to be understood (see **Sidebar**, below).

SIDEBAR: Some Comments on Phylogenetic Trees

Evolutionary trees, i.e., those depicting lines of descent, and particularly those based on molecular data, are now commonplace even in introductory biology. While an author's conclusions are typically stated fairly boldly, often the critical assumptions, inferences, and caveats are absent or not emphasized. Here are some things to keep in mind in interpreting phylogenies (for details see Hillis et al. 1996; Raff 1996; Lengeler et al. 1999; Baldauf 2003; Dawkins 2004; Barton et al. 2007; Baum and Smith 2013):

- The evolutionary rate among species and lineages may be assumed to be constant but this is not strictly true, especially over large evolutionary distances. Cladograms are trees that show branching pattern only, whereas phylograms depict branch lengths proportionate to the amount of evolutionary change.
- Genes selected as a universal phylogenetic marker for comparison must: (i) occur and function equivalently across all life forms; (ii) contain enough homologous nucleotides so that they are as phylogenetically informative as possible (this will be limited given the billions of years over which life has evolved), and the mutation rate should be neither too swift nor too slow over the relevant geological time span (the molecular variants must be selectively neutral or at least operate consistently with neutral theory);

and (iii) be relatively immobile with respect to lateral (horizontal) transfer. Hence, the gene sequences typically used are those of the large subunit and small subunit rRNAs. Other possibilities include amino acid sequences of ribosomal proteins and translation initiation factors, and of the proteins associated with transcription and replication of DNA, i.e., broadly speaking, with the processing and transfer of genetic information.

- **All phylogenetic trees represent hypothesized relationships** from which inferences are drawn. Selection of the most plausible history is often based on the principle of parsimony, i.e., the preferred tree is the most simple one, the one with the minimum number of evolutionary changes. While this may generally be true, there are instances where the correct phylogeny is almost certainly not the most parsimonious. Thus, additional or other inference methods are commonly used.

- Tree shape (topology or ordering of branches) and phylogenetic inferences are affected by the assumptions inherent in the model of evolution used for the analysis and the input data. The robustness or quality of trees is not trivial to ascertain. Topology will change with each new sequence added. The strength of conclusions is affected by such factors as the quality of multiple sequence alignments (occasionally ambiguous); statistical support for the groupings; choice of a particular exemplar for a group; and occurrence of homoplasies (in the context of trees, **homoplasy** refers to an inconsistency between a given tree topology and multiple occurrence of a character state). Statistical estimates only give a measure of how well the data appear to fit the tree, **not necessarily an indication whether the tree (model) is the correct one.**

- Strictly speaking, the dimension measured in molecular trees is change in nucleotide or amino acid sequence and, though correlated with time, is not time per se. For this reason and because of the different times in evolutionary origin of species representatives of a clade, trees represent the chronology in phylogenetic branching (sequential pattern of common ancestors), not age of a clade or the vintage of the organisms compared. A phylogenetic tree should be read as a set of successive convergences working backwards in time (for a dramatic demonstration of this see Dawkins 2004), not across the branch tips.

- Molecular-based trees are, for the most part, gene trees. As such, they depict relationships among the sequences of one or more **genes** in the compared species, **not the species** (organisms) themselves. Furthermore, even when compiled from summaries of multiple genes, they may or may not be close approximations of actual species relationships (see concluding remarks below). Species-based trees, still in their infancy, offer the prospect of enhanced inference of true relationships.

Now that the era of molecular systematics is firmly established, it is important to periodically remind ourselves about the distinctions between genotype and phenotype and what these mean in interpretations of species biology and species relationships, past and present. At the outset of this new era, Lewontin said in his treatise on evolutionary genetics (1974, p. 20) that "*to concentrate only on genetic change, without attempting to relate it to the kinds of physiological, morphogenetic, and behavioral evolution that are manifest in the fossil record and in the diversity of extant organisms and communities, is to forget entirely what it is we are trying to explain in the first place.*" A reasonable compromise is to recognize, as has Harold (1990) in his authoritative inquiry into the origin of form in microorganisms, that traits and especially complex traits such as morphology, are linked only indirectly to genes and that a real understanding of form can only be made at a much higher plane, orders of magnitude above the genetic. Multiple genes, multiple pathways, epigenetic factors, and the rules of biochemistry and physics are involved. This also is the thesis of Stewart (1998), who argues that **the other, largely overlooked, secret to life is mathematics**, which shapes what the genes can do: he emphasizes that genes at best are a recipe (not a blueprint) for life, a necessary but insufficient ingredient.

1.2   Microorganisms and Macroorganisms: Differences and Similarities

As noted at the outset, to even a casual observer the great diversity among organisms in size, form, locomotion, and color is evident. Mayr (1997, p. 124) says simply … *"The most impressive aspect of the living world is its diversity."* A common, informal subdivision of this diversity is at about the level of resolution of the human eye into **microorganism** or microbe (from micro = small and bios = life) and **macroorganism**. The division is also a practical one because microorganisms, if they are to be understood as organisms and not as gene sequences, have to be studied in large part by use of various kinds of microscopes. Macroscopically visible green, red, and brown algae, as well as the plants, and most of the animals, constitute the latter category. Microorganisms include most of the protists (Adl et al. 2005; e.g., the flagellates, amoebas and relatives, sporozoans, ciliates, many of the green algae), the bacteria (as noted earlier, with a small 'b' taken to be synonymous with prokaryote and to include both *Bacteria* and *Archaea*), the fungi, and certain microscopic invertebrates such as many of the nematodes. Based on their size, viruses obviously are microscopic; they are considered by some scientists to be microorganisms and are often included in texts on microbiology. They are not discussed other than in passing in this book. In practice, most biologists probably do not consider very small animals such as the nematodes and rotifers to be microorganisms, although strictly speaking, they fall within the microscopic realm and 'see' the world to some extent as do the fungi and bacteria. Thus, **obviously, microbiology includes much more than just bacteria, though this is frequently overlooked.**

Not surprisingly, exceptions to the above generalities exist, and the distinction between 'microorganism' and 'macroorganism' is clearly an arbitrary and occasionally hazy one. For example, the individual cells of some species of bacteria, such as *Epulopiscium fishelsoni*, are macroscopically visible, being about a million times the size of typical *E. coli* cells (Angert et al. 1993). Of course the eukaryotic microorganisms are much larger than the prokaryotes. Interestingly, some portion of the life cycle of *every* creature is microscopic; conversely, many microbes produce macroscopic structures and stages. Whatever the terms may mean to different people, a major subdiscipline of biology and entire university departments have become devoted to the study of bacteriology, or in some cases more broadly, '**micro**biology'. Operationally, second only to the need for microscopy to understand microbes, the major general difference between microorganism and macroorganism is that, in general, members of the former group have to be cultured in order for most of their properties to be studied. Although the culturing requirement for identification purposes has diminished with the availability of molecular methods, study of microorganisms under controlled, laboratory conditions will remain a distinctive attribute of microbial ecology.

Of course, microorganisms as a group differ from macroorganisms other than in size alone. The key characteristics that distinguish microbes quantitatively or qualitatively from macroorganisms appear below and are discussed at length in subsequent chapters (for detailed comparisons between prokaryote and eukaryote cell biology, see Neidhardt et al. 1990; Madigan et al. 2015). In keeping with the earlier comments regarding the taxa that constitute 'microorganisms', note that the following synopsis applies to microorganisms in general, of which bacteria are simply the most extreme example:

- Capacity for dormancy, or occasionally, an extended quiescent or slow growth state. Most plants and some animals such as the insects share this trait at least to a degree. Bacteria excel at negotiating a 'feast-or-famine' existence.
- Highest metabolic rates (bacteria) and potentially high population growth rates (bacteria and some fungi) and thus high numbers of separate genomes all produced with relatively little biomass. Doubling times for some bacteria as short as 12 min under optimal conditions. For unicellular microorganisms, because there can be and typically are many cells, and because each is potentially capable of forming more cells, beneficial mutations and accessory genetic elements such as plasmids (bacteria) can be rapidly established in a population by natural selection.
- Active growth (population increase) typically within a relatively narrow range of environmental conditions, which are usually distributed discontinuously, but very widely and often globally.
- High metabolic dexterity, including the capacity for rapid physiological adjustment, frequently involving entire biochemical pathways (bacteria). Fast response ('adaptability') to changing environments is thus possible, as is clonal propagation (bacteria) of effectively a single genotype, both in the laboratory and to a greater or lesser extent in nature.
- Direct exposure, frequently of individual cells, to the environment, rather than enclosure within a multicellular, homeostatic soma. Biochemical orderliness but organizational simplicity, reflected by fewer cell types and interactions; true division of labor nonexistent or rudimentary.
- The foregoing attributes imply a usually extremely high number of 'individuals' (as used in this case to mean physiologically independent, functional units, or ramets; see ▶Sect. 1.6) per genetic individual and, in terms of local population, per unit of actively occupied or colonized area.
- Predominantly haploid condition (bacteria) in the vegetative part of the life cycle. Consistent with their small size, bacteria have the smallest genomes of any cell.
- Smaller role in bacteria for conventional sexuality as a genetic recombination mechanism relative to its role in macroorganisms; gene transfer and assortment by various 'unconventional' means such as parasexuality in the fungi, or in fragmentary fashion such as by transformation, conjugation, and transduction in the bacteria. Bacteria can also exchange large blocks of genes by horizontal transmission from phylogenetically distant sources (species). The bacterial genome is unsurpassed among organisms in its plasticity.

Given these appreciable differences, one might well ask what attributes micro- and macroorganisms share. The most general **intrinsic** commonality is that every living thing is an island of order in a sea of entropy or disarray. The individual bacterial cell of a single cell type and the blue whale of about 120 cell types (Bonner 1988, p. 122) are constructed and operate in orderly fashion. Cell structure differs in detail among life forms, but the cells of essentially all organisms consist of a peripheral lipid bilayer membrane (the *Archaea* being the sole known exception) surrounding a cytoplasm. All cells take up chemicals from the environment, transform them, and release waste products. All can conserve and transfer energy, direct information flow, and differentiate to some degree. ATP is essentially *the* currency of biologically usable energy in all organisms. As a first approximation and notwithstanding exceptions, biosynthetic pathways are fundamentally the same in all cells, though organisms vary in their complement of the pathways (details in ▶Chap. 3; see also Neidhardt et al. 1990). The metabolic machinery, including the enzymes, pathways, program for cell division, and sequence of the reactions is remarkably similar (although the subcellular sites and controlling mechanisms differ). For example, the assembly of a rod-shaped virus particle in an infected cell proceeds through much the same stepwise process as does a microtubule in that cell.

The genetic code and its associated parts such as various RNAs and translation machinery, are fundamentally the same, even down to the four nucleotide building blocks and the specific triplets—which code for the same amino acids in a human and a bacterium. (Actually, the code is not quite universal, because most but not all the code words across taxa are the same; for some exceptions and mechanisms of code flexibility, see Ivanova et al. 2014; Ling et al. 2015.) For all life forms the genetic code is distributed such that, with minor exceptions, every cell of the organism has a complete copy of the DNA recipe. The flow of genetic information from DNA to RNA to protein is universal (the well-rehearsed **Central Dogma**), even though there are variations in how this information flows. All cells can communicate by chemical signals and all respond to some extent to environmental signals by switching genes on and off. Although the switching mechanisms vary, and are much more complex in eukaryotes than in prokaryotes, all organisms are able to receive and to react to stimuli. Moreover, as discussed later in the book, the general features of gene regulation are quite similar in bacteria and higher organisms.

The seemingly universal imprint of biochemical building block molecules and cell features noted above, together with a common function adhering to the same general principles, are what Lehninger (1970, pp. 3–13) referred to as *"the molecular logic of the living state."* This remarkable conformity is unlikely to be just one big coincidence; rather, logically, it can be taken as implying that all known life emerged from a common ancestor (▶Chap. 4). The fundamentally identical cellular biochemistry among organisms led Bonner (1965, pp. 129–130) to state that this characteristic is non-selectable in the sense that, being essential to all known life, it has not been altered by selection since at least Precambrian times. In a similar vein, Monod (quoted by Koch 1976, p. 47) has said that *"what is true of* E. coli *is also true of the elephant, only more so."* These molecular commonalities brought interdisciplinary research teams together at the start of the DNA era in the late 1950s. **Does a comparable ecological commonality exist? The thesis of this book is that it does.**

1.3 The Centrality of Natural Selection

The most important **extrinsic** property common to all extant organisms is that they and their ancestors have passed through the filter of natural selection. As survivors, all are 'successful' (see below) in contemporary terms, even if they are ultimately destined for extinction: the Darwinian truth is that all have been shaped by this process of differential survival. While the process of microbial evolution differs in dynamics and detail from that of macroorganisms (▶Chap. 2), all life forms share an evolutionary history extending over some 3 billion years. It could be argued that the 8.7 million or so estimated species (Mora et al. 2011) represent 8.7 million responses to natural selection. But, more interestingly, **what, if any, ecological**

commonalities are evident among the survivors? Here is where the words of Jacob (1982 pp. 14–15) are important: *"Natural selection, however, does not act merely as a sieve eliminating detrimental mutations and favoring the reproduction of beneficial ones, as is often suggested. In the long run, it integrates mutations and orders them into adaptively coherent patterns... [and] gives direction to changes, [and] orients chance."* The central question that sets the unifying theme for this book is thus the extent to which the ecology of microorganisms and macroorganisms has been analogously molded by differential reproductive success operating among individuals.

Since the premise is that natural selection is a major organizing principle of biology, it is worthwhile to diverge briefly to consider some important aspects and implications of Darwin's theory. This pertains especially to the terms 'fitness', 'success', 'optimum', and 'maximize', which, as common jargon in the ecological literature, are used to some extent in this book. According to Darwin (1859, p. 81) natural selection was... *"this preservation of favourable [sic] variations and rejection of injurious variations."* In contemporary terminology we could say that natural selection is evolutionary change in the heritable characteristics of a population from generation to generation resulting from the differential reproductive success of genotypes. So the first point is that **natural selection is thus only one of the means by which evolution can occur and it hinges critically on the proposition that different individuals leave different numbers of descendants (and <u>descendants</u> is the operative word, <u>not</u> progeny).** That, in turn, is determined both by the characteristics of the individual and those of the environment with which it interacts. Other evolutionary forces or phenomena include, but are not limited to, founder (chance) effects; archetype effects (see phylogenetic constraints, below); drift; pleiotropism and epistasis—phenomena that result in the association of multiple and apparently independent alleles or traits carried through the evolutionary process together, though not the target of selection themselves (for discussion, see Harper 1982). In passing it should be noted that many evolutionists have cautioned against the blithe application of 'adaptation' to explain any seemingly beneficial trait. Williams (1966) has stressed that adaptation *"... should be used only where it is really necessary. When it must be recognized, it should be attributed to no higher a level than is demanded by the evidence* (p. 4) and further *...it should not be invoked when less onerous principles, such as those of physics and chemistry or that of unspecified cause and effect, are sufficient for a complete explanation"* (p. 11).

Second, **the organisms we see about us today are the outcome of natural selection acting on <u>past</u> organisms in <u>past</u> environments.** Thus it has been argued by Harper (1982), somewhat facetiously, that a better term than **ad**aptation (which implies movement toward a desired state, i.e., a teleology) is '**ab**aptation' (implying movement away from a past state; how the past has influenced the present). Equivalently, the effects of current natural selection will become manifest in the nature of organisms in the *future* in *future* environments. **In other words, in actuality natural selection acts to improve fitness of descendants for the environment of their parents, not for their own environment.** (The inevitable parallel here is with generals preparing their troops to fight that last war rather than how to anticipate future conflicts.) Since the environment inevitably changes (▶Chap. 7), the truly optimum state can never be reached. In essence it is pursued through time by the organism, or as van Valen (1973), Lewontin (1978) have phrased it, environments are tracked by organisms.

Third, **natural selection at best can only operate on the best of what can be produced by mutation and recombination of existing genotypes: the choice is restricted to what is available at a given time and place in the gene pool.** Mayr (1982, p. 490) likens natural selection to a statistical concept: possessing a superior genotype does not ensure survival; it only offers a higher probability. For reproductive success it is merely sufficient to be better (or temporarily lucky!), not perfect or optimal. Every genotype is necessarily a compromise among opposing selection forces (Dobzhansky 1956; Mayr 1982). Thus, *Darwin's theory does not predict an optimum state or perfection* (although such terms are often used loosely), merely that some individuals, namely those with particular features, will leave more descendants than others (e.g., Sober 1984; Begon et al. 1996, pp. 6–7). For these reasons it is technically incorrect to say that fitness is maximized. These points are elaborated below on constraints. Nevertheless, natural selection is a potent force because, unlike the largely random process of mutation that generates the original variation (see ►Chap. 2), or random genetic drift, natural selection is *nonrandom*—on balance, the organisms with certain features are the ones that survive and leave relatively more descendants.

Fourth, **selection operates at the level of 'the individual'** (not to mention other levels; see Preface and following discussion of what is an individual in various contexts). **An important corollary is that the individual cannot be meaningfully 'atomized' into adaptive traits with a view of seeking optimization for each of the dissected parts.** In a now classic paper, Dobzhansky (1956) argued eloquently that "*traits have no adaptive significance in isolation from the whole developmental pattern of the organism which exhibits them at certain stages of its life cycle* (p. 346)... *a disadvantage in some respects may be compensated by advantages in other respects* (p. 339)... *it is the trajectory of the whole which conveys upon the genotype or the individual its fitness to survive and to reproduce*" (p. 346). While this principle has been echoed and embellished over the years (e.g., see Gould and Lewontin 1979), it bears periodic re-emphasis in an era of reductionist, genomics-oriented, molecular biology.

Finally, for the sake of brevity and simplicity, and in keeping with traditional ecological terminology, expressions such as 'choosing', 'strategy', 'tactic', and so forth are used throughout the text, albeit cautiously. They carry no teleological implications for the organisms concerned. For an excellent critique of loose thinking and loose terminology, see Harper (1982).

1.3.1 Constraints on Natural Selection: Phylogenetic, Ontogenetic, and Allometric

Natural selection does not operate on a uniform or blank slate but within the context of many interacting factors and processes such as chance, historical precedent, habitat complexity, and inherent biological constraint. In general, three major, interrelated constraints have been recognized: phylogenetic (taxa-related), ontogenetic (development-related), and allometric (size-related). Unlike the trade-off concept, which underlies most of the examples in this book and is discussed later, evolutionary constraints are absolute and cannot be moved to-and-fro by changing an opposing selection force (Stearns 1977, 1982, 1992). Thus, one way to look at any organism is "*as a mosaic of relatively new adaptations embedded in a framework of relatively old constraints*" (Stearns 1982, p. 249). Stearns (1992) has further noted that not only are certain traits fixed within lineages but they constrain within limits those that do vary (only those varying are involved in trade-offs). Moreover, since each lineage is constrained uniquely, each has its own trade-off structure, e.g., for plants, see Niklas (2000). For these reasons, organisms cannot technically be 'optimal' entities (whatever an author may mean by that expression) and optimality models in ecology have to operate within the boundaries discussed below (see discussion and caveats in Stearns 1992).

The constraints placed on a species by its evolutionary pedigree are **phylogenetic**, also termed historical. Organisms evolve in lineages of ancestral and descendant populations. No lineage begins with a blank slate; each is constrained by its legacy. A specific pattern of development for each species is both characteristic of and similar to the ancestral form. McKitrick (1993) has likened this to a game of scrabble where players are limited in spelling words by the letters they receive at the outset. There are many excellent examples in biology: Humans, unlike millipedes, have only two legs and lack the power of regenerating amputated appendages. Starfish and their relatives all show five-point symmetry. All plants are **modular** in growth form; all higher animals are **unitary**.[2] In every instance, limitations as well as opportunities are bestowed on the organism by its birthright. So, given a particular design, certain things are impossible or, though possible are strongly selected against, or both (Stearns 1983). Buss (1987) has discussed the different evolutionary directions of the plant, animal, and fungal kingdoms preordained by the different ancestral features of the three clades. These are described later (▶Chaps. 4 and 5) and in Andrews (1995).

Second, selection also can be limited in what it can do at any point in the life cycle by what has gone before in the developmental program (Bonner 1982b; see especially pp. 1–16). Such limitations are termed **ontogenetic** or developmental. One would expect that more complex organisms and more complex life cycles would be particularly vulnerable to this constraint (▶Chaps. 4 and 6). It is considerably more complicated to build the Space Shuttle than a Volkswagen Beetle. In biological terms, the complicated ontogeny of a mammal tends to preclude radical changes because early occurring mutations would likely be lethal, whereas microorganisms are relatively free from these constraints (Bonner 1982a, b). Changes in the timing of events in the life cycle (heterochrony) may occur (events speeded up, as in larvae that are sexually mature; slowed down, as in development of the large brain in humans), but critical phases or structures cannot be eliminated. This is true even when functionally useless vestigial structures (gill arches in vertebrates; appendices in humans; tails in birds and mammals) appear in the fetus or adult. Such structures persist as a consequence of engineering size-increase by the building block method (Dobzhansky 1956; Chap. 4 in Bonner 1988). What is important is not the particular item so much as the overall process and the end product. A cautionary note in interpreting such constraints in some contexts, however, is that simply because one character state precedes another (ancestral condition) in a phylogeny, it is unwarranted to conclude that the presence of the second state necessarily depends on the occurrence of the first (Herron and Michod 2008).

Third, **allometric** limitations relating to size include associated changes in chemistry, physiology, and morphology. Evolution is limited by physical and chemical laws, a key example of which is surface-to-volume relationships that affect all sorts of fundamental processes such as biomechanics and diffusion. Carroll (2001) has asked the provocative question … *"are there universal rules to the shapes of life?"* This and related issues are explored in ▶Chap. 4.

---

2   The distinction between unitary and modular organisms is a major topic in this book taken up at length in ▶Chap. 5. In brief, the body plan of modular organisms is fundamentally an indeterminate structure of iterated, multicellular units (modules) arrayed at successive levels of complexity (such as leaves, twigs, branches). Examples of modular organisms include plants and many sessile, benthic invertebrates such as bryozoans, corals, and hydroids. In contrast, the body plan of unitary organisms is a determinate structure based on a strictly defined number of parts (wings, legs) established during embryogenesis. Mobile animals are the best example. The two fundamentally different modular/unitary growth forms have distinctive suites of life history correlates.

One example is that a geometrical consequence of increasing size (volume) is decreasing surface area (S), dictated by the relationship $S \propto V^{2/3}$ (the "surface area law" or the "**2/3-power law**"). Metabolically important surfaces, however, scale not by the two-thirds power but directly in proportion to volume—we see this rule manifested as convolutions or sheets in the living world in the form of structural accommodations like villi, alveoli, capillaries, leaves, roots, and root hairs (▶Chap. 4). Metabolic and energetic ground rules set another form of constraint at the level of cell physiology (Lehninger 1970; Feldgarden et al. 2003). Size differences and the attendant allometric relationships can be examined during the course of development of an organism (ontogenetic comparisons) or across taxa at arbitrarily selected stages (phylogenetic comparisons) in an historical or contemporary context. We are considering here and later in detail (▶Chap. 4) the extent to which microorganisms and macroorganisms see the world differently and the ecological implications of this size differential. Such considerations must allow for not only differences of the conventional allometric sort applicable to organisms of similar geometry, but acknowledge also the pronounced differences in shape as well as size of the microorganisms versus macroorganisms. There are no spherical cows to compare with coccoid bacteria! Even if there were, the types of environments and the scale on which the organisms would interact with these environments would be entirely different.

1.4 Analogies, Homologies, and Homoplasies

As noted in the Preface, a principal focus of this book is to develop analogies between the ecology of microorganisms and macroorganisms. In this context, '**analogy**' is broadly construed and used in its popular connotation to refer to parallels, similarities, or resemblances of interpretive value in contrasting strategies or life histories, *without* implying common ancestry. The word 'analogy' also can be used more specifically in a scientific context (though rarely so in this book) for comparisons of similar traits of independent evolutionary origin, i.e., those that have arisen by parallel evolution or convergence. Thus, the wing of an insect and the wing of a bird are analogous. Such traits are said to be **homoplasies** (Glvnish and Sytsma 1997). In contrast, **homologies** imply similarity due to common ancestry, and features can be homologous even if they differ functionally, as in the forelimbs of bats, porpoises, and humans. Homologies can be further subdivided as primitive or derived in ancestral origin. For instance, since all mammals are vertebrates, they have a backbone, which is considered a primitive characteristic manifested by the ancestor common to the entire vertebrate lineage, including animals such as the lampreys, fish, and reptiles. Possession of hair, however, is unique to the mammal clade. Since hair is a trait that arose in an ancestor more recent than the ancestral vertebrate, it is a derived character.

1.5 A Framework for Comparisons

1.5.1 Competing Demands, Trade-Offs, and Resource Allocation

All organisms can be viewed simply as input/output systems (▣Fig. 1.2; Pianka 1976, 2000). Each acquires some kind of resource (energy and materials) as input. Progeny are the output. Under natural selection, each creature will tend to allocate limited inputs optimally among the competing demands of growth, maintenance, and reproduction (Gadgil and Bossert 1970;

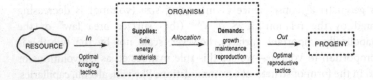

☐ **Fig. 1.2** The organism as an input/output system. Redrawn and revised figure based on Pianka (1976) originally from *American Zoologist* by permission of Eric Pianka and Oxford University Press ©1976

Abrahamson and Caswell 1982; Karasov and Martinez del Rio 2007). To the extent to which each of these activities is clearly an alternative (a debatable point, below), an increase in one necessarily results in a decrease in one or both of the others. As developed below and in later chapters, this allocation principle entails trade-offs because there are finite resources and finite time to meet these competing needs.

As is the case for all models, the input/output concept is of course a simplification. The evidence for allocation patterns in particular taxa is mixed and 'alternative' demands when examined in detail are not as clear-cut as appears intuitively and superficially. Nevertheless, the generality holds and our attention here is not as to whether it is invariably correct but that it provides a good vehicle for comparing how microorganisms and macroorganisms live. Clearly, the focus in the model is on the major life history attributes. It is worthwhile to consider briefly what is not represented explicitly. The most glaringly obvious yet perhaps the most easily overlooked point is that it centers on the individual operating as a singular unit, whereas probably all organisms are functionally collaborative entities through various mutualisms. A classic case is the reliance of macroorganisms on microorganisms for services (e.g., ►Chap. 3 and McFall-Ngai et al. 2013). A 'hidden' output in the model is metabolic products, typically invisible (e.g., oxygen in the case of plants; various secretions in the case of animals; extracellular polysaccharides and organic acids for microbes), but are not unimportant, especially in the microbial world! Responses of the individual to predation or competition, while not recognized explicitly, would be reflected in the model by altered biomass allocation patterns and in the output term as altered number of progeny. Foraging tactics involve the obvious maneuvers such as predators hunting in packs rather than alone, with the likely microbial equivalent being the primitively multicellular microbes such as *Myxococcus*, which *"feeds as a pack of microbial wolves, with each cell benefiting from the enzymes secreted by its mates"* (Kaiser 1986). Other foraging devices may include protecting a resource by various secondary compounds such as antibiotics, mycotoxins, or staling products (e.g., Janzen 1977a; Burkepile et al. 2006), or directly poisoning competitors. In the same vein, attracting mates or pollinators, or defending a breeding territory from competitors or predators, are a part of optimal reproductive tactics. In other words, fitness can be increased by high acquisition and efficient allocation of one's own resources, but also by interfering with those same activities of a competitor.

Fitness or 'success' is measured by the progeny or output over time, but it is more than just progeny numbers: By definition, the fittest individuals leave the most descendants as alluded to above. In population genetics terms, the number of descendants amounts to the proportion of the individual's genes left in the population gene pool (Pianka 1976). Given the diversity of life forms, what constitutes an individual is not clear and need not be in the context of the overall model. Some connotations of 'individual' are discussed below in ►Sect. 1.6 and at length in ►Chap. 5. In addition to being transient, success is thus assessed in a relative way, against other members of the population. The issue of the fitness or competency of the offspring, that is, the likelihood that they will go on to reproduce and perpetuate the lineage, and thereby become descendants, is not specified in the general model.

Acquisition (input) Three general points will be made here concerning resources as a prelude to subsequent chapters (◘Fig. 1.2). **First**, a basis for grouping all life forms is how they meet requirements for organic carbon compounds and energy for growth, maintenance, and reproduction (►Chap. 3). With respect to carbon, organisms are either autotrophic (literally, 'self-feeders', or CO_2 fixers) or heterotrophic ('fed from others', i.e., they use organic forms of carbon). With respect to energy source, organisms are either phototrophic (source is light) or chemotrophic (source is chemical substances). If inorganic chemicals are oxidized for energy, the organisms are said to be lithotrophic, whereas organic chemicals are oxidized by organotrophs. These characteristics set very broad limits on what organisms can do and consequently where they are found, or at least where they can grow actively to competitive advantage. One example pertinent to each requirement will suffice to make the point. The need for sunlight restricts most aquatic plants to relatively shallow depths (photic zone) of oceans and most lakes. The distribution of those bacteria that obtain their energy by oxidizing specific inorganic compounds such as sulfur or iron (lithotrophs) tends to mirror that of the particular deposit. Members of the genus *Sulfolobus*, for example, live mainly in hot springs and in other geothermal habitats rich in sulfur.

Assignment of all organisms to broad resource categories also establishes the well-known trophic structure of communities, which is informative in terms of energy flow and nutrient cycling. Since a high percentage of energy is dissipated by organisms in maintenance (and as heat and motion), not only is it probable that the number of links is set ultimately by energy losses at each, but the density of individuals and total biomass typically decrease at each successive trophic level (for caveats and details, see ►Chap. 3).

The **second** generality concerns how resources are presented to the organism. All organisms must contend with variation in the distribution and abundance of resources. Environments can be characterized with respect to resource availability (and other attributes) from the organism's viewpoint as relatively heterogeneous (discontinuous, patchy) or comparatively homogeneous (uniform). Patches are dynamic in that they vary in characteristics such as content, size, time, and spatial orientation.

The idea of **environmental grain** (►Chap. 7; see also Levins 1968; Jasmin and Kassen 2007a) relates the size of a resource patch to the size of the individual and to the space within which the organism is active. **Coarse-grained** environments are sufficiently large that the individual either comes by fate to spend its entire life in a patch (for instance, when deposited there as a seed or spore) or chooses among them (for example, a millipede living only within a rotten log). Where the patches are so small that they appear uniform to the individual, or when the individual encounters many states of the heterogeneity during its lifetime, the environment is **fine-grained**. The larger relative to the patch size, the more mobile, or the longer lived the organism, the more likely it is to 'see' its surroundings as fine-grained. Plants in a field would appear fine-grained to humans or a grazing deer, but coarse-grained to an insect larva and immensely so, even at the level of the individual leaf, to a bacterium. A butterfly flits across several hundred square meters of a field in fine-grained fashion, while a slug, sliding slowly on its mucus trail across only a few meters of that same field, experiences its surroundings as coarse-grained. Over its life span of several hours or days, a bacterial or yeast cell on the skin of an animal or in a plant exudate confronts a coarse-grained environment, whereas typical changes on this order of time would appear relatively fine-grained to the host on which that microbe finds itself. In contrast to unitary organisms such as deer or humans, in modular organisms (see below), the genetic individual commonly samples many environments concurrently. Sporadic and cyclic changes in resource availability (and other environmental characteristics) have immense influence on life history features such as dispersal, migration, and dormancy (►Chaps. 6 and 7).

Third is the issue of efficiency of resource utilization. This has been expressed in optimal foraging theory (OFT) and appears explicitly as part of ◘Fig. 1.2. OFT was developed originally in a particular context by animal ecologists (Chap. 5 in Pianka 2000). When the concept is broadly construed and used in conjunction with optimal digestion theory (►Chap. 3), however, it is easy to see analogies and implications for all organisms. The basis for the theory is that there are both benefits, such as matter and energy, and costs, such as exposure to predators or parasites, or time and energy diverted from other activities such as reproduction, associated with foraging. Under natural selection, organisms presumably have evolved some sort of efficient 'foraging strategy' that recognizes both the benefits from and costs directly associated with resource acquisition, as well as implications to other competing demands. The use of optimal foraging models as one case in point for optimization models in general should be employed not as an attempt to assume or prove that organisms are 'optimal', but to understand how particular adaptations, trade-offs, and constraints shape evolution (Maynard Smith 1978a; Parker and Maynard Smith 1990; for broader insights on the utility of optimality models, see Stearns 1992).

To a species of bird, for example, optimal foraging might concern net energy gained or lost in the catching of insects of various sizes at various distances from its perch. How small a prey item is too small, or how far away is too far, to make a trip worthwhile? Under what circumstances would it be better to adopt a 'sit-and-wait' tactic as opposed to actively searching for prey? A microbial analogue might be whether it is better to synthesize a key metabolite, such as ATP, quickly if wastefully (lower yield), or implement pathways that produce substantially more product but much more slowly (►Chap. 3 and Pfeiffer et al. 2001). Plants 'forage' from a more or less fixed location in the sense that they display leaf canopies for photosynthesis and root systems to collect water and nutrients. How branch, leaf, and root architecture has evolved to meet these needs and in the face of competitive and predation pressure is of much interest (►Chap. 5; see also Horn 1971; Bell 1984; Givnish 1986). Analogously, sedentary filter-feeding invertebrates (e.g., bivalves, brachiopods, ectoprocts, and phoronids), and both attached and motile microbes 'forage' (e.g., van Gestel et al. 2015), although in many cases foraging may entail a sit-and-wait strategy (quiescent spores of a root-infecting fungus in soil; sessile, stalked bacteria of the genus *Caulobacter*).

The form of living things as it pertains to resource utilization can be generalized further. Sessile modular organisms such as plants, corals, bryozoans, fungi, and clonal ascidians (as opposed to unitary organisms; ►Chap. 5) capture space and other resources by a branching habit of growth. In so doing they create 'resource depletion zones' (Harper et al. 1986). These zones might represent, for instance, a shaded area on understory leaves resulting from interception of light by a tree, or a volume of soil from which nutrients had been partially removed. Many biological systems exhibit dichotomous branching. Why? A major challenge to branching, modular organisms regardless of size is development of a branching pattern to effectively capture resources. As detailed in ►Chap. 5, a continuum can be visualized between two extreme growth forms, phalanx (closely packed branches; resource site densely occupied) and guerrilla (infrequent branching; rapid extension; much unoccupied intervening space) (Lovett-Doust 1981a, b). Thus, there is a correspondence between optimal foraging theory and optimal branching strategy. The issue is not straightforward because selection for one activity or function cannot proceed independently of other selection pressures. The challenge is visually dramatic with respect to morphology and hence has been emphasized in the case of branching and resource harvesting.

Reproduction (output) Having acquired resources, the organism must then allocate them among the competing needs of growth, maintenance, and reproduction. To delay reproduction will be advantageous only if more descendants are ultimately contributed to future generations. For example, formation of large, perennial fruiting structures by certain of the basidiomycete fungi is undoubtedly expensive in diverted materials, time, and energy, but offers the multiple advantages of an enhanced platform for sexual recombination and widespread aerial dispersal of millions of propagules.

Use of time and energy by organisms tends to vary seasonally (▶Chap. 6) as is apparent to anyone who has collected mushrooms in the autumn or followed the varying activities of birds over a few months. Reproductive activities are typically well synchronized to periods of the year. For many animals the breeding season coincides with environmental conditions favorable for survival of parents and offspring. Birds become highly territorial and nest in the spring in temperate regions; parasites are remarkably well coordinated to the behavior or phenological development of their hosts. The fact that sexuality in algae and fungi is frequently triggered by environmental adversity is not inconsistent with the above generality, because the resulting zygote is typically enclosed in a thick-walled, resistant, dormant structure functionally analogous to seeds in plants. While maintenance and repair activities are usually ongoing at the cellular level (Kirkwood 1981, 2005), they are especially apparent at the level of the individual as stages of rest. In one form, maintenance is recognizable in the circadian sleep cycle typical of higher animals; in another, as seasonal inactivity (winter hibernation; plant dormancy). Although obviously essential for survival, and the preparations necessary for resuming activity, maintenance represents lost time, energy, and materials to the extent that limited resources are diverted from reproduction.

So-called 'optimal reproductive tactics' (◙Fig. 1.2) seemingly have evolved by natural selection acting on organisms, which as a result have improved long-term reproductive success. As for foraging, the issue is again one of relative allocation: resources and time assigned to reproduction are diverted from other activities. And, what is allocated to reproduction may be invested in various ways. For instance, a partial suite of either/or reproductive 'choices' includes whether to: (i) engage in only one round of reproduction (semelparity) or several (iteroparity) during a lifetime; in the latter case a further question is how many episodes to have; (ii) have one mate or several or none (clonal or asexual reproduction); (iii) reproduce early or late in the life cycle (or season); (iv) produce few large, highly endowed or 'competent' progeny or many small, relatively 'incompetent' progeny; related to this frequently is the length of gestation; (v) reproduce locally or risk migration to new breeding grounds; (vi) maintain separate sexes (sexual dimorphism; dioecy; heterothallism) or unite sexual function within one individual (hermaphrodism; monoecy; homothallism); related to this is the capacity to vary the sex ratio and the even larger question of why there should be two, rather than some other number of sexes; (vii) disperse progeny widely or concentrate offspring near the parents. Regardless of the countless variations on the theme of reproduction, the ultimate test of a strategy is whether it results *not* in more progeny, but more *descendants*.

Longevity is tied closely to reproduction. In evolutionary terms, senescence is expected to occur wherever the reproductive value of the individual declines with time (for terminology and details, see ►Chap. 6). Under relatively stable environmental conditions and where the reproductive value of the individual increases with age, as it typically does for modular (unlike unitary) organisms, senescence should be delayed or not occur (Harper et al. 1986). Under these circumstances, the genetic individual (►Sect. 1.6) lives for an indefinite time and there is the potential for exponential increase in offspring, especially where the individual is also clonal (►Chap. 6).

The ecological issues noted above and summarized in ◨Fig. 1.2 are developed in subsequent chapters as follows: Since variation provides the raw material for evolution, the stage is set in ►Chap. 2 with an overview of the genetics and mechanisms of genetic variation of the larger as opposed to the minute organisms. In ►Chap. 3 we take up the supply side of ◨Fig. 1.2 and consider the kinds of resources different organisms use, and how they acquire and allocate them in the face of competing demands. This leads to a consideration of how size (►Chap. 4) and growth form (►Chap. 5) influence foraging and reproduction. ►Chapter 6 concerns the origins and molding of the life cycle, in essence how the developmental play is acted out under the direction of the genes. Senescence is considered at some length, including the controversial issue of whether it is inevitable. The role of the environment in shaping the plot (life history features) is discussed in ►Chap. 7. Some general conclusions are summarized and perspectives are advanced in ►Chap. 8.

1.6   What Is an Individual?

The individual organism figures prominently in this book. As discussed in the Preface, this is largely because natural selection acts primarily at the level of the individual. But what is an individual? The word is ambiguous because it may be used by ecologists in at least three contexts. These are not mutually exclusive and are considered below (◨Fig. 1.3). In passing, we should note that the related matter of what constitutes 'individuality' is not explored in this book, other than briefly with respect to size increase in the volvocine algae in ►Chap. 4. The evolutionary point(s) at which a mass of cells becomes sufficiently differentiated, organized, and interdependent to be declared 'an individual' is contentious and at the extreme becomes a metaphysical issue. Definitions are subject to various debatable criteria, such as whether or not germ–soma differentiation is a necessary condition. Individuality is considered elsewhere (e.g., Buss 1987; Maynard Smith and Szathmáry 1995; Michod 2006, 2007; Herron and Michod 2008; Folse and Roughgarden 2010; Szathmáry 2015). Several of the ideas may lend themselves relatively well to animals but not to plants or microbes, for example because of the totipotency of their cells.

With respect to the term 'individual', **first**, it is often used operationally in a **numerical or quantitative sense** to mean a representative of a particular population or species, something that can be counted. For some organisms this is intuitively clear: we can all visualize one rabbit or one maple tree. Individuals by this definition are supposedly discrete and functionally independent units. For colonial or clonal organisms such as bacteria, fungi, various algae, bryozoans, and coelenterates, however, it is not at all clear what level of cellular aggregation fulfills these criteria (e.g., Larwood and Rosen 1979). For example, the reason the term CFU (colony-forming unit) is used in bacteriology and to some extent in mycology is that in practice it is generally unknown whether growth (as on solid medium in a petri dish) is produced from one cell or a cluster of cells, or whether all potentially living cells seen microscopically in a sample will multiply in the appropriate medium.

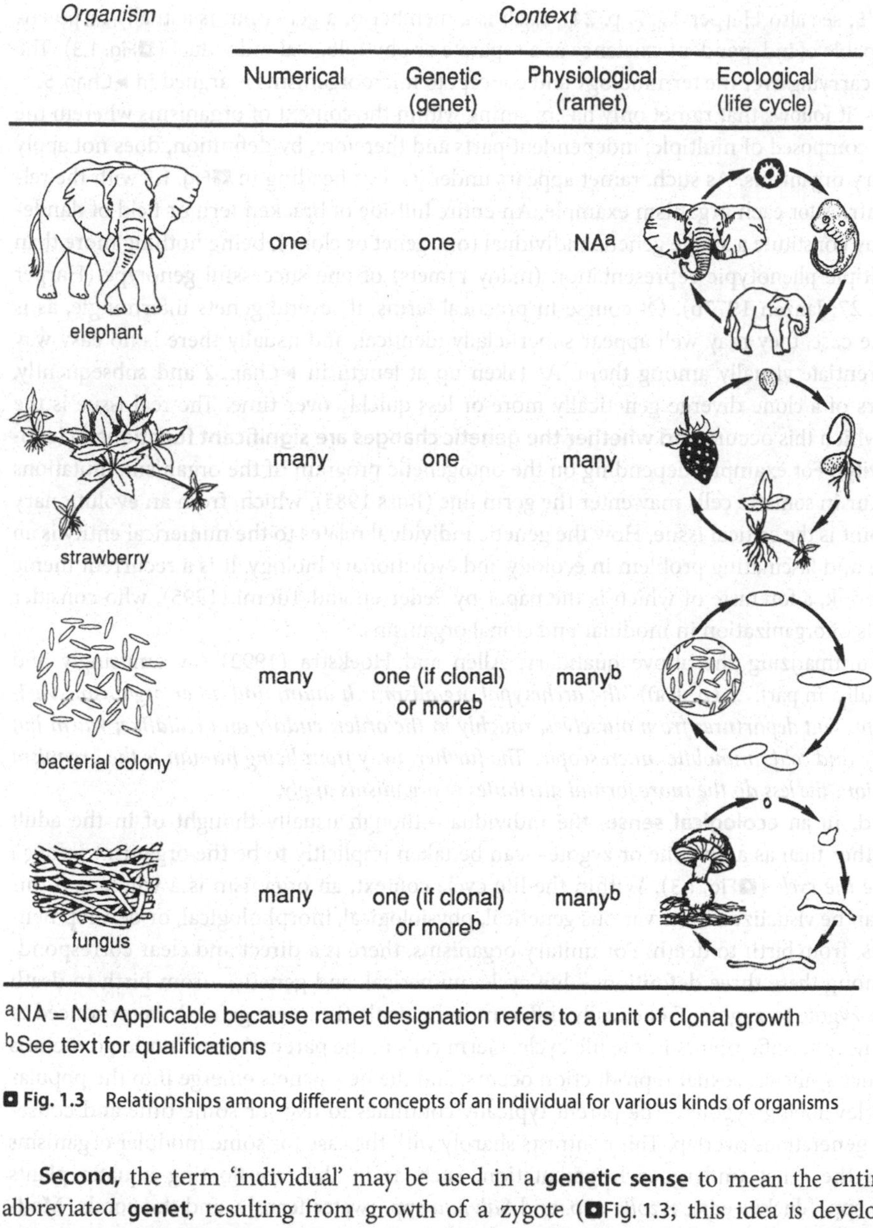

Organism	Context			
	Numerical	Genetic (genet)	Physiological (ramet)	Ecological (life cycle)
elephant	one	one	NA[a]	
strawberry	many	one	many	
bacterial colony	many	one (if clonal) or more[b]	many[b]	
fungus	many	one (if clonal) or more[b]	many[b]	

[a]NA = Not Applicable because ramet designation refers to a unit of clonal growth
[b]See text for qualifications

□ Fig. 1.3 Relationships among different concepts of an individual for various kinds of organisms

Second, the term 'individual' may be used in a **genetic sense** to mean the entire unit, abbreviated **genet**, resulting from growth of a zygote (□Fig. 1.3; this idea is developed in ►Chap. 5; see also Harper 1977, p. 26; Anderson and Kohn 1998). This usage is equivalent to the numerical definition when applied to unitary organisms such as most mobile animals (►Chap. 5), that is, those in which the zygote develops into a determinate body repeated only when a new life cycle starts. A single deer is both a numerical individual and also technically a genet (though rarely referred to as such). **The genet concept is meaningful in evolutionary terms because it focuses attention on the genotype through time.** It came into vogue because, for modular organisms such as corals and those plants with **clonal (asexual) growth,** the number of discrete, countable 'individuals' is not the same as the number of genets (Kays and Harper 1974). The term **ramet** is given to each of these countable units

(▶Chap. 5; see also Harper 1977, p. 24). Each is a member of a genet but is actually or potentially capable of independent existence as a separate or physiological individual (◘Fig. 1.3). The case for carrying over the terminology and concept to microorganisms is argued in ▶Chap. 5.

Thus, it follows that ramet only has meaning within the context of organisms wherein the genet is composed of multiple, independent parts and therefore, by definition, does not apply to unitary organisms. As such, ramet appears under its own heading in ◘Fig. 1.3 with the relevant entries for each organism example. An entire hillside of bracken fern or field of dandelions may constitute a single genetic individual (one genet or clone), being nothing more than the multiple phenotypic representation (many ramets) of one successful genotype (Harper 1977, p. 27; Janzen 1977b). Of course in practical terms, if several genets intermingle, as is often the case, they may well appear superficially identical, and usually there is no easy way to differentiate visually among them. As taken up at length in ▶Chap. 2 and subsequently, members of a clone diverge genetically more or less quickly over time. The real issue is the rate at which this occurs **and whether the genetic changes are significant functionally (evolutionarily).** For example, depending on the ontogenetic program of the organism, mutations that occur in somatic cells may enter the germ line (Buss 1985), which, from an evolutionary standpoint is the critical issue. How the genetic individual relates to the numerical entity is an ongoing and fascinating problem in ecology and evolutionary biology. It is a recurrent theme in this book, a foretaste of which is the paper by Pedersen and Tuomi (1995), who consider the levels of organization in modular and clonal organisms.

In summarizing the above quandary, Allen and Hoekstra (1992) say amusingly and insightfully, in part … (p. 160) "*the archetypal organism is human, and other organisms variously represent departures from ourselves, roughly in the order: cuddly and childlike; warm but big; scaly and cold; immobile; microscopic. The further away from being human is the organism in question, the less do the more formal attributes of organisms apply.*"

Third, in an **ecological sense**, the individual—though usually thought of in the adult form rather than as a juvenile or zygote—can be taken implicitly to be the organism *through its entire life cycle* (◘Fig. 1.3). Within the life cycle context, an organism is a whole and this entity can be visualized in its various genetical, physiological, morphological, or developmental states, from birth to death. For unitary organisms, there is a direct and clear correspondence among these three definitions—life cycle, numerical, and genetic—from birth to death or from zygote to zygote. Genetically different individuals (i.e., new genets) are produced at one or more specific points in the life cycle. Germ cells in the parental genet undergo meiosis to produce gametes, sexual reproduction occurs, and the new genets emerge into the population as developing zygotes. The parent typically continues to live for some time and consequently generations overlap. This contrasts sharply with the case for some modular organisms wherein the genet undergoes fragmentation, such as in the free-floating aquatic plants (*Lemna* spp. [duckweeds], *Azolla* spp. and *Salvinia* spp. [water ferns]), and the corals. Modules may be sloughed and are free to move about passively. Kariba weed (*Salvinia molesta*), which propagates clonally, has come to occupy vast areas and may weigh millions of tons (Barrett 1989). Thus, it considerably exceeds the mass of the blue whale (the largest extant unitary organism on earth), but may be challenged for this record by the fungi (Smith et al. 1992)! Of course the numerical as well as genetic individual in the case of microbes tends not to be discrete in either time or space; indeed the microbial clone is often widely and occasionally globally distributed.

In summary, there is no simple, unambiguous concept of an individual that can be used across the spectrum of life as a benchmark in the comparative ecology of micro- and macroorganisms. Comparisons can be made conceptually to a greater or lesser extent by use of any of the three definitions depending on context, but in a strict sense not functionally. Definition one, the numerical usage, is rejected here as a useful common denominator because it means little in evolutionary terms and operationally in microbiology is quite contentious, notwithstanding the application of molecular methods, which in the past two decades have largely overcome limitations such as the 'viable but uncultivable' problem discussed in later chapters (Brock 1971; Colwell and Grimes 2000; Walker and Pace 2007). Thus, to arbitrarily choose individual bacterial cells, fungal colonies, and slime mold amoebae, and attempt to compare them with individual trees, insects, and birds would be futile.

To the extent that the genetic individual, definition two, implies simply that a genet is the product of a zygote, it can be applied widely, in many cases literally and in others conceptually where a true zygote is not strictly produced. But it should not be mistaken to imply genetic homogeneity, i.e. absence of sub-organismal variation. A key difficulty posed by microbes is that an individual cell or clone can change genetically in a manner different from the conventional process of reproduction with meiosis, so it may be impossible to know when the genetic individual starts to exist or when it ends. As seen above, for unitary organisms it begins at the time of genetic fusion, in other words, when genetic recombination occurs at zygote formation and the contribution from each parental genome is diluted by exactly one half. Moreover, the point in the life cycle when karyogamy can occur is fairly standard and usually easily recognized and associated with particular terminology such as reproductive phase; sexual maturity; seed production. This is not the case for bacteria and fungi, which display indeterminate, highly plastic development and experience novel genetic events. A bacterial cell may divide repeatedly, yielding descendants that may be relatively sequestered as an initially genetically homogeneous, clonal microcolony, albeit one that changes progressively by mutation (Milkman and McKane 1995). Moreover, horizontal (lateral) gene transfer may occur through various atypical sexual processes (▶Chap. 2). How many genes have to mutate, be acquired, or lost in a haploid organism for our conceptual genet to effectively cease to exist?

A roughly analogous, ambiguous phenomenon occurs in the fungi that have a dikaryotic (n + n) phase or exhibit parasexuality (▶Chaps. 2 and 6). For instance, *Puccinia graminis tritici*, the parasite causing stem rust of wheat, has five spore stages with associated mycelial growth involving both barberries and wheat as the alternate host. The life cycle includes haploid, dikaryotic, and diploid phases. Shortly after meiosis, the fungus becomes partially or fully dikaryotic by various mechanisms (▶Chap. 6), but karyogamy is delayed and true diploidy is a transient stage. The dikaryotic phase, technically haploid but functionally diploid, is of variable duration. Furthermore, in absence of the barberry as an alternate host but in the presence of a favorable environment, the fungus may cycle indefinitely in the dikaryotic condition. What, then, is the genet for the wheat rust fungus? Whatever the debatable answer, the genet corresponds neither to a single numerical entity, nor is the ploidy static throughout the full cycle. This is a much more complicated situation than occurs with, say, a clonal buttercup or an aclonal maple tree, and may be most closely approximated by plants that have an extended gametophyte stage such as the fern.

The genet concept could be brought into line with what intuitively seems to physically represent the organism by allowing for a certain amount of genetic flux after recombination in the zygote, or at the start of the life cycle in the case of haploid and asexual organisms. This would recognize that, while the genetic basis for the individual is established by genetic recombination in some form at the outset, the genotype could be molded by additional genetic events that might periodically occur during the life cycle. As noted above, in bacteria this can take the form of mutation or HGT (horizontal gene transfer, ▶Chap. 2). For plants and benthic invertebrates it can take the form of somatic mutations, some of which may enter the germ line. However, the full evolutionary consequences remain to be seen.

Definition three, the individual depicted as an organism passing in more or less organized fashion through stages in its life cycle, provides a reasonable, universal basis for ecological comparisons. Unless noted otherwise, this is the frame of reference that is used throughout this text. The scheme is applicable to all of the various phases of diverse organisms, and accommodates changes in ploidy, morphology (phenotype) including size, and variation in the time and extent of genetic change. Bonner (1965) has interpreted the life cycle as proceeding sequentially from a state of minimum size (generally at the point of fertilization) to maximum size (generally the point of reproductive maturity). The transition involves growth over a relatively long period to adulthood, followed by return to minimal size over a relatively short period, accomplished by separation of buds or gametes from the adult. For single-celled organisms, the cell cycle is the life cycle of the physiological but not of the genetic individual. These matters are taken up at length in ▶Chap. 6 for which the stage is set in the next chapter.

1.7   Summary

Although every organism is unique in detail, all share fundamental properties. The cellular chemistry and metabolic machinery of life forms is basically the same across taxa. All represent order as opposed to disorder. All display cellular organization, growth, metabolism, reproduction, differentiation to some degree, the ability to communicate by chemical signals, and a hereditary (replication/transcription/translation) mechanism based on transfer of information encoded in DNA. Of more significance ecologically is the fact that all have been shaped by evolution operating through differential reproductive success. All are survivors of lineages shaped by lengthy evolution, in most cases over geological timescales. As such they would be expected to show some analogies in strategies resulting from natural selection.

The term 'individual' has various connotations: numerical, genetical, physiological, and ecological. Perhaps the most informative way to view an individual is in the ecological context as the entity through its entire life cycle. Based on this concept, a comparative ecology of microorganisms and macroorganisms is developed in the ensuing chapters within the general conceptual framework of the individual as an input/output system. All individuals acquire resources as input and allocate them among the functions of growth, maintenance, and reproduction. The output is progeny. In practice, resources and time are limiting and the alternative demands on them are competing. Thus, selection over time should generally favor organisms that optimally partition, thereby increasing fitness as measured by the number of descendants. The life history attributes bearing on this model are the subject of the following chapters and are presented in a way to highlight the analogies and distinctions between the ecology of micro- versus macro-organisms: genetic variation and the means by which different organisms recombine and transfer genetic information (▶Chap. 2); nutritional mode; size; growth dynamics and growth form of the individual (▶Chaps. 3–5, respectively), life cycle (▶Chap. 6), and interaction between the individual and its environment (▶Chap. 7). ▶Chapter 8 consolidates and recapitulates the major themes.

Suggested Additional Reading

Bonner, J.T. 1965. Size and Cycle: An Essay on the Structure of Biology. Princeton Univ. Press, Princeton, NJ. *The biology of organisms from the life cycle perspective. An outstanding synthesis; stimulating; full of ideas.*

Corner, E.J. H. 1964. The Life of Plants. Univ. Chicago Press, Chicago, IL. *A classic; one of the best biology books of all time.*

Dawkins, R. 1989. The Selfish Gene, 2nd Ed. Oxford University Press, Oxford, U.K. *A well written, speculative, stimulating thesis that the fundamental unit of natural selection is the gene.*

Dobzhansky, T. 1956. What is an adaptive trait? Am. Nat. 90: 337–347. *An eloquent and classic paper emphasizing that traits have no adaptive significance in isolation from the entire organism that expresses them at various developmental stages throughout its life cycle.*

Jackson, J.B.C., L.W. Buss, and R.E. Cook (Eds.). 1985. Population Biology and Evolution of Clonal Organisms. Yale Univ. Press, New Haven, CT. *Collected papers on representative organisms posing a challenge to the definition of 'the individual'.*

Pianka, E.R. 1976. Natural selection of optimal reproductive tactics. Am. Zool. 16: 775–784. *Optimal foraging theory, optimal reproductive tactics, and the model of the organism as an input/output system.*

Rosenberg, E., E.F. DeLong, S. Lory et al. 2013 (Eds.). The Prokaryotes: Prokaryotic Biology and Symbiotic Associations. Springer-Verlag, NY. *A contemporary, comprehensive overview of the biology of bacteria, including numerous chapters relevant to bacterial ecology and a synopsis of the prokaryote/eukaryote terminology controversy.*

Stewart, I. 1998. Life's Other Secret: The New Mathematics of the Living World. Wiley, NY. *An engaging investigation into mathematical patterns of life and the argument that life is not prescribed by genes but that genes and mathematical laws are partners in shaping life.*

Genetic Variation

© Springer Science+Business Media LLC 2017
J.H. Andrews, *Comparative Ecology of Microorganisms and Macroorganisms*,
DOI 10.1007/978-1-4939-6897-8_2

Diversity is a way of coping with the possible.

François Jacob, 1982 p. 66.

Genomic components have significance only in terms of phenotypic expression.

Steven M. Stanley, 1979 p. 56.

2.1   Introduction

What processes account for the remarkable variation in form and function that we see in the living world and how might they be similar or different between microorganisms and macroorganisms? Because natural variation provides the raw material on which evolution acts, this chapter sets the stage for later comparisons in the book.

Here we review briefly the generation, maintenance, and transmission of genetic variation in micro- and macroorganisms. Microbes generally have plastic, dynamic genomes and all the capability of generating genetic variability that do macroorganisms, though adaptive evolution proceeds rather differently in the two groups. We discuss clones and what constitutes a 'genetic individual' in various taxa and show that the concept is more-or-less abstract, with major evolutionary implications. The important role that the cellular and external environments plays in reciprocal interactions with the genome and its phenotypic manifestation is developed in ►Chap. 7. The major kinds of change that affect the central information (DNA) repository for cells are mutations; rearrangements, and recombinations; and epigenetic and other gene expression controls. Changes in sequence information typically manifest themselves on a scale of multiple organism generations, whereas at least some epigenetic and related gene regulatory changes, such as prions, can be gained or lost quickly and at higher frequencies than DNA mutations, i.e., within the cycle of a single cell or involving a few cell generations (►Chap. 7 and Shapiro 2005; Jarosz et al. 2014 a, b).

Arguably the biggest revolution in genetics of the past half-century has been the shift in analytical focus from gene to genome. The term 'genome' has been used in many contexts; Shapiro (2002, p. 112) defines it as "all the DNA sequence information of a particular cell, organism, or species." Along with this paradigm change is recognition that the latter is fluid and internally dynamic, rather than static and subject to change only through background random point mutation. A computer storage system metaphor for genome function is described by Shapiro (2002, 2011), summarized as follows: The genome contains many categories of information, each associated with a certain sequence code. Among these are the well-known coding sequence for RNA and protein; sites for initiation and termination of transcription; sites for covalent modification of DNA; and binding sites for the spatial organization of the genome. So the genome is in effect formatted for various functions it carries out in conjunction with other cellular systems designed to read, replicate, transmit, package, or reorganize the DNA sequence information. Such formatting is analogous to the formatting of computer programs wherein various generic signals recognize particular files regardless of their specific content. Among species there are differences in how control systems operate. Bacteria handle regulatory decisions somewhat differently from eukaryotes, just as different computer systems recognize different codes. New data and new programs are written when the genome is restructured by natural genetic engineering that can occur in many ways (◘Tables 2.1 and 2.2).

☐ Table 2.1 Changing concepts and metaphors related to genetic information and variation (from Shapiro 2002 with minor rewording and abbreviation)

Process or concept	Classical genetics era	Molecular genetics era
Hereditary theory	Genes as the unit	Genomes as interactive systems
Genome organization	Beads on a string	Computer operating system
Sources of inherited novelty	Localized mutations due to base changes; one gene affected at a time	Epigenetic modifications and rearrangement of genomic components due to internal genetic engineering
Evolutionary processes	Natural selection acting on independent background random mutations; cells passive	Non-random, genome-wide rearrangements; cells active in restructuring genome

☐ Table 2.2 Some mechanisms and consequences of DNA restructuring (natural genetic engineering; abbreviated and slightly revised from Shapiro 2002)

DNA reorganization process	DNA rearrangements involved (examples)
Homologous recombination	Reciprocal exchange; unequal crossing-over; gene conversion
Site-specific recombination	Insertion, deletion, or inversion of DNA carrying specific sites
Site-specific DNA cleavage	Direct localized gene conversion by homologous recombination; create substrates for gene fusions
Nonhomologous end-joining systems	Precise and imprecise joining of broken DNA ends; genetic fusions; localized hypermutation
DNA transposition	Self-insertion; transposons carry signals for transcriptional control, RNA splicing and DNA bending; amplifications
Retroviruses and other terminally repeated retrotransposons	Self-insertion and amplification; carry signals for transcriptional control, RNA splicing and chromatin formatting; mobilization of sequences acquired from other cellular RNAs

2.2 Mechanisms

2.2.1 Mutation

Mutations are the ultimate source of most variation in all organisms. By definition (e.g., Barton et al. 2007), they consist of changes in the genetic message (either at the level of the gene or the chromosome) that are heritable and, implicitly, detectable. Mutation generates the variation; recombination amplifies this by mixing it, in essence by repackaging it, not by creating brand new packages. We note in passing that molecular biologists frequently speak of inherent mutation rates (usually referring to replacement rates of one DNA base by another, as may be read, for example, from a genomic sequence), whether or not there is a change in the wild-type phenotype. This fundamental level of mutation may go undetected if the protein is not functionally altered, if two or more mutations compensate for each other giving a pseudo wild-type, or if another gene product takes over the function of the altered protein. Thus, there are really three mutation rates—the inherent rate of uncorrected change in DNA sequences; the rate at which the mutants survive; and the rate at which survivors are detected in a population by virtue of their phenotypic differences. Although there are many types of mutation, fundamentally most involve errors introduced into the genome. These may occur spontaneously or be induced by endogenous or exogenous mutagens, typically occurring during replication; and errors in recombination, repair, or in large-scale chromosomal rearrangements (Alberts et al. 2015; Griffiths et al. 2015). To these long-established sources of failure in cellular machinery must now be added the more recently recognized category of genomic change caused by mobile genetic elements such as transposons (below).

Mutations are mechanically inevitable—damage to DNA occurs and copying, proof-reading, and segregating mechanisms are not and cannot be perfect. For this reason Williams (1966, p. 12) has argued that mutations are not adaptive, i.e., they should not be considered to be a means for ensuring evolutionary plasticity. That they still occur is largely in spite of natural selection rather than because of it (Williams 1966, p. 139); However, this does not mean that there cannot be selection for a particular mutation *rate* and there are clear differences in rates among species (below). Overall, because many of these alterations are deleterious, there has been strong selection pressure (reduced reproductive fitness) over millennia to reduce rates in organisms. Incremental benefit in increased fidelity below a certain point may be offset by greater physiological costs in correction (Kimura 1967; Drake 1991). It has also been argued (Lynch 2010) that the lower limit is set by genetic drift. Regardless, there is some value in having background genetic 'noise' (Drake 1974; Drake et al. 1998). This is illustrated by the essential function of mutation (somatic hypermutation) in antibody genes of the B lymphocyte cells, which contributes to antibody diversity in vertebrates (French et al. 1989). Analogous genetic rearrangements provide the operon structure needed for expression of *nif* genes in a nitrogen-fixing cyanobacterium, *Anabaena* (Golden et al. 1985). The terminal step in heterocyst differentiation involves excision of DNA in response to an environmentally triggered, site-specific DNA recombinase (Haselkorn et al. 1987). Thus, much the same rearrangement process is used in a prokaryote and an advanced eukaryote.

With the mechanistic characterization of transposable elements in both prokaryotes and eukaryotes (for an overview, see Griffiths et al. 2015), the distinction at the molecular level between mutation and recombination is often one of semantics. Classification schemes are more-or-less subjective because the degree of genetic change occurs as a continuum. For instance, where relatively large segments of DNA (hundreds or thousands of nucleotides) are rearranged, such as by inversion or exchange between chromosomes, the event is often called mutation by molecular biologists (Chaps. 9 and 11 in Watson et al. 2008). Yet it falls within the domain of recombination (in the sense of mixing of nucleotides) as defined by eukaryote geneticists. For example, site-specific recombination (see Recombination, later) is a mechanism that can lead to mutation. Probably the best-known form of recombination is the even exchange of sequences between aligned, homologous chromosomes at meiosis. However, recombination can occur between any two regions of sufficient DNA similarity on the same chromosome or between paired chromosomes, an event known as ectopic (nonallelic) recombination. This potentially leads to mutation due, for example, to deletions or inversions (in the case of intrachromosomal recombination) or, if two chromosomes are involved, to unequal crossing-over where one offspring receives too many copies of the sequence and the other too few (Chap. 12, Barton et al. 2007).

Before proceeding into the details, some important general points about mutations need to be made. First, they occur in somatic as well as in germline cells, the implications of which will be discussed in ▶Sect. 2.5 (see Shendure and Akey 2015). Suffice it to say here that, with few exceptions (such as among the somatic cell lines that produce antibodies), high rates must be avoided in *both* lineages. The implications of mutation in somatic cells tend to be overlooked because of the focus by evolutionary biologists on the germline.

Second, the impact of a mutation will depend on the environment, broadly construed (▶Chap. 7). A mutation deleterious in one set of circumstances could be beneficial in another. Also, as discussed later, while mutation rates can vary with environmental conditions, the environment does not induce mutations that are specifically adaptive; i.e., adaptively directed mutation does not occur (Barton et al. 2007; Futuyma 2009). Current evolutionary thinking is that mutation and selection are separate processes. It was well established for bacteria by Luria and Debruck in the 1940s, confirmed by the Lederbergs in the 1950s, and reconfirmed with elaborations since then, that mutations conferring resistance to phage or antibiotics happen both before and after the selective agent is applied and are random with respect to their adaptive value (Barton et al. 2007). (For commentary on the controversy that some mutations in bacteria are adaptive, allegedly being 'directed' by the environment, see Sniegowski and Lenski 1995; Foster 2000; Sniegowski 2005.)

Third, the impact of a mutation can depend on the stage of development of an organism. Mutations occurring in early ontogeny, for example, in animals at the blastula stage, are much more likely to be lethal because of their far-reaching impact on subsequent differentiation. In contrast, those occurring later may affect relatively superficial properties such as eye color (Bonner 1965, pp. 123–128; 1988, p. 168). As Bonner notes, it follows from this that any mutation at *any* stage in the life of a unicellular organism is more likely to be lethal than would a similar mutation during most of the life of a macroorganism.

Extent of mutations If we categorize mutations by the degree of the base changes involved, the first kind consists of simple base position (typically called point) substitutions such as a C to a T in the nucleotide sequence, or by addition or deletion of a single base (Graur and Li 2000; Griffiths et al. 2015). Because of the degeneracy feature of the genetic code ('wobble' in the third codon position), often there is no change in the amino acid sequence ('synonymous' or 'silent' mutation). Mutations may also be silent if a related amino acid is inserted (**missense conservative** mutation) or if the affected region of the gene product is unimportant. If the altered codon specifies a different amino acid, the change is a **missense nonconservative** mutation. This is manifested either as an altered but functional ('leaky') protein product or as a defective protein. The latter category is very common, as in sickle cell anemia, Tay-Sachs disease, and phenylketonuria. One form of retinitis pigmentosa is apparently caused by a C to A transversion in codon 23, corresponding to a proline to histidine substitution (Dryja et al. 1990). Finally, if translation is terminated (**nonsense** mutation), a shortened and likely defective protein results.

It should be emphasized that point and other small changes in the **noncoding** regions of a gene, such as in regulatory sequences, can potentially be very significant in evolutionary terms. For example, such changes may create or interrupt a binding site thereby changing or even obliterating the expression of the gene. In other words, alteration in gene regulation, accomplished in multiple ways, has arguably had an even greater impact than actual mutation in the polypeptide coding regions of those same genes (see later section and ▶Chap. 7; also Doebley and Lukens 1998; Carroll et al. 2005).

Second in extent of nucleotides involved are small insertions or deletion (**indel**) mutations involving one or a few base pairs. When this happens within regions coding for polypeptides, the mutations are called **frameshift** if the reading frame of the codons is shifted out of phase (Barton et al. 2007; Alberts et al. 2015). This can lead to premature termination or obliteration of a stop codon, so a frameshift mutation may result in a truncated, possibly unstable protein, or it may generate a completely nonfunctional product, depending on its position in the structural gene. In general, the foregoing changes, except deletions, are revertible to wild-type. Reversion of deletions depends on the nature of the deletion and the encoded gene product.

The third category of mutation comprises the potentially large-scale chromosomal changes such as bigger deletions, inversions, duplications, and insertions (transpositions) that intergrade with recombination, discussed later. Thus, the size or number of genes in chromosomes may be increased or decreased; genes may change in location (by inversion or translocation); and there may be wholesale changes in the number of chromosomes. Where a deletion is large enough to remove or disrupt a gene coding for a critical enzyme, the result is likely to be death of the cell or organism. At the other extreme, while polyploidy (more than two chromosome sets) is generally considered to be an abnormal condition, it appears to be well tolerated by some organisms, especially plants. Here, it is estimated to have resulted in 7% of the speciation events in ferns and 2–4% in angiosperms (Otto and Whitton 2000), as well as in the associated evolution of plant gene families and structural complexity (De Bodt et al. 2005). Most such polyploidizations are followed by gene deletions over time resulting in eventual diploidy. Plants, however, again are distinctive in being able to tolerate imperfectly matched chromosomes that result from loss or rearrangement (Walbot and Cullis 1983). Stanley (1979) has emphasized the role of chromosomal rearrangement together with gene regulation in 'quantum speciation'.

Gene expansion, contraction, duplication, or loss may contribute significantly to the evolution of complexity, functional diversity, and adaptation to the environment over evolutionary or contemporary time (see ▶Chap. 4 and also: Verstrepen et al. 2005; Hittinger and Carroll 2007; Pränting and Andersson 2011; Andersson et al. 2015). A classic example is the evolution of the globin gene family in multicellular animals and how duplication was instrumental in creating new proteins (Hoffmann et al. 2012; Alberts et al. 2015). Analogously, the evolutionary expansion of enzyme families in plants associated with specialized metabolism for which plants are famous is attributable to various kinds of gene duplication events (Moore and Purugganan 2005). In contrast, genes can be inactivated by insertion of mobile DNA elements into the chromosome and subsequently eroded or lost. These insertions may be fragments (generally 1,000–2,000 nucleotides) that do not code for any characters beyond those needed for their own transposition (simple transposon or insertion sequence; discussed later in Recombination section). Alternatively, they may be longer elements (complex transposon or transposable element) that code for a product such as antibiotic resistance in bacteria. The rates of gene loss and gain are particularly dynamic in prokaryotes due in large part to a process of horizontal or lateral gene transfer (discussed later under Sex and Adaptive Evolution in Prokaryotes), offset by inactivation and deletion (Mira et al. 2001; Lerat et al. 2005). Genome plasticity is also notable in plant (Ibarra-Laclette et al. 2013) and fungal (Raffaele and Kamoun 2012) evolution.

Rates As alluded to above, mutation rates can be estimated by recording spontaneous phenotypic changes over time in a population of organisms in nature (animals or plants) or in the laboratory (tissue culture or microbial culture). Because the frequencies of most muta-

◘ **Table 2.3** Mutation rates per nucleotide site (×10⁻⁹) for selected species and tissues (abbreviated from Lynch 2010)

Species	Tissue	Cell divisions per generation	Mutation rates	
			Per generation	Per cell division
Homo sapiens	Germline	216	12.85	0.06
	Retina	55	54.45	0.99
	Intestinal epithelium	600	162.00	0.27
Mus musculus	Male germline	39	38.00	0.97
	Liver		237.88	
Rattus norvegicus	Prostate		448.90	
Drosophila melanogaster	Germline	36	4.65	0.13
	Whole body		380.92	
Caenorhabditis elegans	Germline	9	5.60	0.62
Arabidopsis thaliana	Germline	40	6.50	0.16
Saccharomyces cerevisiae		1	0.33	0.33
Escherichia coli		1	0.26	0.26

tions are very low in natural populations and can only be detected in large sample sizes, it is common to examine specific proteins electrophoretically or DNA sequences directly (e.g., in restriction fragment length polymorphisms, or as single nucleotide variants on a genome-wide basis). The mutations tallied and mutational rates published typically pertain to single nucleotide substitutions and, in some cases, indels or copy number variants (e.g., Lynch 2010; Shendure and Akey 2015), but rarely thus far on large structural (>20 bp) variants (Campbell and Eichler 2013; Kloosterman et al. 2015).

Rates are expressed in numerous ways, e.g., on the basis of per base pair per cell replication; per genome per replication; or per genome per sexual generation (unless otherwise stated, typically on a haploid basis). For an example of such data see ◘Table 2.3 and Lynch (2010). Different eukaryotes (or even sexes within a species) have different numbers of cell divisions involved in gamete production per sexual generation. In humans there are about 30 germline cell divisions in females, but more than 100 for spermatogenesis in males (Crow 2000; Barton et al. 2007), which thus have a higher mutation rate on a generational basis. Higher eukaryotes have a genome up to several orders of magnitude larger than viruses or prokaryotes, but most of this is in introns and inter-genic regions where most mutations are neutral rather in than functional genes. Thus, rates for macrorganisms are often expressed as **per effective genome** (i.e., excluding the genomic regions where most mutations are neutral) per replication (cell division) or per sexual generation (Drake et al. 1998). There are further complications or caveats with eukaryotes, among them age- or sex-related effects alluded to above, as well as substantially higher rates in somatic versus germline cells, discussed later. Nevertheless, as a generality, when expressed as number of mutations per base pair per replication, there are appreciably higher rates for smaller than larger genomes. When expressed on a per genome basis, there is less variation across genome sizes or species. Rates in microbes are typically about 1/300 per genome per replication (versus about 1 per genome per replication event in the lytic RNA viruses; Drake et al. 1998). **For the higher macroorganisms, on an effective genome basis, rates per sexual generation are in the range of 0.1–100, but on a cell division basis per effective genome are approximately comparable to the 1/300 for microorganisms** (Drake et al. 1998; Hershberg 2015). The remarkable comparability across taxa remains to this day a source of animated debate as to, for example, whether it reflects an inherent rate of change or a common level of error correction. The implication of various determinants on the ultimate mutation rate, including factors such as sexual versus asexual reproduction and population size, has been discussed at length (see e.g., Drake et al. 1998; Lynch 2010; Sung et al. 2012).

Beyond the differences noted above, the incidence of mutation in a local sense varies with numerous other factors. Among these are the specific genomic region (so-called 'hotspots' vs. 'coldspots,' in part affected by degree of DNA superhelicity; see Foster et al. 2013), organelle (nucleus vs. mitochondrion), and type of base (AT-biased rather than GC). Thus, whereas mutation is generally described as a random phenomenon, where it occurs in the genome is far from random. The consequences of mutation are inextricably linked to life cycle (►Chap. 6). For instance, mutations are masked by diploidy and even more so by polyploidy. In relatively simple organisms (such as algae with a prominent haploid gametophyte phase, see ►Chap. 4) the individual experiences selection mostly in the haploid stage of the life cycle. Under such circumstances mutations will not be masked and there will be strong selection pressure against mutant cells. Possibly this furthered the maintenance of haploidy, as well as the correlation of diploidy with more complex development (Otto and Orive 1995; this point is also developed in ►Chap. 6). In tetraploids like the agronomic crop alfalfa, mutations are hidden, but selfing results in inbreeding depression. This may be due to loss of third-order interactions or homozygous deleterious recessive alleles (Jones and Bingham 1995).

The molecular clock Notwithstanding the substantial variation in mutation patterns among species, it was recognized several decades ago that the average rates of amino acid or nucleotide substitution are approximately constant *for a particular protein or gene* among taxa (Kimura 1987). This fairly steady accumulation in sequence divergence over time has been called the 'molecular clock' (Zuckerandl and Pauling 1965; Wilson et al. 1977; Kumar 2005). Presumably, it reflects the fact that the same functional constraints exist for a given gene or gene product in different organisms. Clock analysis involves comparing the amino acid or gene sequences in a particular highly conserved protein (such as hemoglobin), or base sequences in a given gene, in several species. For protein comparisons, each discrepancy is assumed to represent a stable change in a codon corresponding to the altered amino acid. In this analysis, the original (ancestral) state, being extinct, is of course unavailable for comparison. So the versions of the sequence that appear in two or more living relatives of the phyla of interest are compared and expressed as units of accumulated amino acid or nucleotide changes. The organisms are then arranged in a branching diagram drawn to minimize the number of changes needed to describe all the permutations from the ancestral sequence. The number of stable genetic changes can then be related to chronological estimates of evolutionary time, obtained from fossil records, since the species diverged from the common ancestor (recall ►Chap. 1). At a detailed level, molecular clocks have proven in general to give more resolution and reliability than the fossil record (cf. ►Chap. 1).

How and why the rates have stabilized where they have is unknown. An inference is often made that the clock is centrally standardized, ticking away uniformly for all species, much like a radioactive decay process, but this is not strictly the case (Jukes 1987). Substitutions are nonlinear such that irregularities occur and these may or may not reconcile over time. Also, for species of both microorganisms and macroorganisms, the clock can tick at different rates among genes, even when only silent substitution rates are considered (Sharp et al. 1989). Nevertheless, among other interesting comparative observations are the following by Ochman and Wilson (1987): (i) the silent mutation rate for protein-coding genes in *Salmonella typhimurium* and *Escherichia coli* is comparable to that in the nuclear genes of invertebrates, mammals, and flowering plants; (ii) the average substitution rate for 16S rRNA of bacteria is similar to that of 18S rRNA in vertebrates and flowering plants; (iii) the rate for 5S rRNA is about the same for bacteria and eukaryotes.

To summarize, the present indications are that the rates of DNA divergence for a gene encoding a given function are approximately the same regardless of the organism. From this generality we derive not only very useful phylogenetic information but also insight into the origin of sequence variations and how gene families diverge (Kumar 2005).

Finally, it should be noted that the rate of phylogenetic change evidently is not controlled by the mutation rate. Rather, the evolutionary history of the major groups of organisms appears to depend on ecological opportunities afforded the simple or complex genetic variants (Wright 1978, pp. 491–511). Evolution, overall, has been for increased size and complexity, although there are several exceptions (►Chap. 4). Evolutionary rates, unlike mutation rates, vary greatly even along single phyletic lines. For some existing genera (e.g., the lungfishes and the opossum among animals; horsetails [*Equisetum*] and club mosses [*Lycopodium*] among plants), evolution has been virtually at a standstill for hundreds of millions of years (Wright op. cit.). Where genetic variation has provided for a major and entirely new way of life, swift adaptive radiation follows. Examples include the development of feathers, wings, and temperature regulation in the case of the birds; efficient limbs, hair, temperature regulation, and mammary glands in the mammals. Among prokaryotes, genes can be transmitted horizontally and in blocks among distant relatives with major evolutionary implications; see later discussion with respect to bacteria in ►Sect. 2.3. Other ecological opportunities are presented when new forms colonize an area where either the niches are unoccupied or are better filled by the mutant than the wild-type (Wright 1978).

2.2.2 Recombination

Recombination is a process leading to the rearrangement of nucleotides by the breaking and rejoining of DNA molecules. There are three general categories (Alberts et al. 2015): homologous; conservative site-specific; and transpositional. The term 'homologous' in this context pertains to regions of sequence similarity (homology) between the DNA molecules involved. As noted earlier, modifications induced by processes such as transposition, or in aberrant homologous recombination, are also considered to be forms of mutation.

General or homologous recombination involves genetic exchange between members of a pair of homologous DNA sequences (i.e., those of extensive sequence similarity). Its most widespread role is in the accurate repair of double-strand breaks but it also is involved in genetic exchange between two paired chromosomes in eukaryotes (or DNA strands in the case of prokaryotes; Lengeler et al. 1999). Recombination or crossing-over does not occur with every chromosome at every meiosis.

Where recombination is part of the formal meiotic process in eukaryotes, it is frequently referred to as **meiotic recombination** and includes (i) independent or Mendelian assortment of entire maternal and paternal homologs during metaphase I and anaphase I of meiosis; and (ii) reciprocal recombination of chromosome segments (crossing-over) that occurs between the nonsister chromatids of paired homologs during prophase I of meiosis (Alberts et al. 2015). In this role, recombination is both a mechanical process to insure the equivalent segregation of chromosomes to the two daughter germ cells, as well as a mixing process that reassorts genes on those individual chromosomes. Although genes may be reassorted, gene order on the chromosomes involved usually remains the same, since the recombining sequences are quite similar (cf. transposition, below). As alluded to earlier, an analogous phenomenon also occurs in bacterial transformation and conjugation, where exogenous chromosome fragments are integrated into the genome of the recipient cell by homologous recombination (see Sex, and Adaptive Evolution in Prokaryotes in ►Sect. 2.3). Thus, features of homologous recombination are common to all organisms. For eukaryotes, assortment is quantitatively more significant than crossing-over in all organisms with a haploid chromosome number exceeding two (Crow 1988). Chromosomes in all organisms can break spontaneously. An additional key role (nonmeiotic) of homologous recombination is to accurately repair such single- or double-strand breaks. (Double-strand breaks can also be corrected by a cruder mechanism known as nonhomologous end-joining, which reseals the DNA but results in a mutation where the strands broke; Alberts et al. 2015.)

Transposition and conservative site-specific recombination differ from general recombination in that diverse, specialized segments of DNA are moved about the genome (Alberts et al. 2015). The pieces moved, which vary considerably from a few hundreds to many thousands of nucleotide base pairs, go by various colloquial ('jumping genes'; 'selfish DNA'), as well as specific names. For example, in bacteria, there are two general types of transposable elements: (i) **insertion-sequence or IS elements** that can move themselves but do not carry genes other than those related to the movement; and (ii) **transposons**—carry the movement genes as well as others (Griffiths et al. 2015). Transposition and conservative site-specific recombination differ in the reaction mechanisms involved and because the conservative process requires that there be specialized DNA sequences on both donor and recipient DNA, whereas transposition generally requires these specialized sequences only on the transposon.

Transposition is further divided into three classes involving (i) DNA-only transposons; (ii) retroviral-like retrotransposons; and (iii) nonretroviral retrotransposons. Transposons are typically 'cut' from one place and 'pasted' into another place in the genome without duplication, whereas the retrotransposons (retroposons) are duplicated because they are transcribed into RNA, reverse-transcribed into DNA, and then reintegrated. Different kinds of TEs predominate in different kinds of organisms: bacterial transposons tend to be the DNA type, whereas **it has been estimated that about half of mammalian genomes originate from TEs primarily of the retroelement group** (Van de Lagemaat et al. 2003). Similarly, 64% of the genome of the powdery mildew pathogen *Blumeria* consists of TEs (Spanu et al. 2010). At least four identical transposon families occur in invertebrates and vertebrates, where they have moved horizontally from the former to the latter evidently by parasite-host interactions (Gilbert et al. 2010).

TEs have broad evolutionary implications because they can mutate existing genes, create new genes, and affect gene regulation at both the transcriptional and post-transcriptional levels when they insert nearby (see Epigenetics below; also Slotkin and Martienssen 2007; Freschotte 2008; Elbarbary et al. 2016). The resulting variation can be adaptive and, for example, IS have been shown to promote the evolution of specialists (as opposed to generalists; see ►Chap. 3) in controlled growth experiments with bacteria (Zhong et al. 2004, 2009). Frequently, they have deleterious consequences because the insertion and rearrangement can lead to disease (e.g., the Alu elements in humans; Kazazian 2004). For this reason and because the elements replicate themselves independently of host chromosomes, they have been called 'intra-genomic parasites' or 'selfish DNA' (Charlesworth 1985; also see ►Sect. 2.3), though such terms are simplistic and potentially misleading (Kidwell and Lisch 2001). Moreover, repetitive DNA may well have a functional role in the physical ordering of the genome (Shapiro and von Sternberg 2005). McClintock (1956) discovered TEs in the late 1940s and 1950s in maize, and called them 'controlling elements' because, although distinct from genes, they could modify gene expression. They have since been well documented in many other taxa including bacteria (phage Mu; insertion sequences; transposons conferring antibiotic/metal resistance or surface antigen variation); yeasts (Ty and mating type elements of *Saccharomyces*); and animals (*Drosophila* transposable elements and hybrid dysgenesis determinants; vertebrate and invertebrate retroviruses). At least in maize, and presumably in other macroorganisms, transposition occurs at predictable times and frequencies in the ontogeny of the individual. In maize, a controlling element can have a similar effect on genes governing different biochemical pathways and at different places in the genome (McClintock 1956; Fedoroff 1983, 1989). Moreover, a single element can control more than one gene concurrently.

Both transposition and site-specific recombination are complex processes in detail and a molecular biology text such as Alberts et al. (2015) should be consulted for specifics. **The key conceptual point is that these phenomena occur broadly if not universally and add considerable genetic versatility or plasticity to organisms beyond the conventional mechanism of recombination normally associated with sexuality.** For additional comments, see Genomic Plasticity and Epigenetics sections, below.

An analogous process pertains to integration of some **plasmids** (small, ancillary, self-replicating extrachromosomal elements) into the bacterial chromosome and 'promiscuous' (organellar) DNA into the nuclear chromosomes. To date, plasmids (see Sect. 2.3) are known to occur ubiquitously in bacteria. Though uncommon in eukaryotes, they are found in many fungi and in some higher eukaryotes, often in association with mitochondria (Funnell and Phillips 2004). Promiscuous DNA has been detected in most eukaryote species examined, including plants, filamentous fungi, yeasts, and invertebrates (Timmis et al. 2004). The term originated with Ellis (1982) for DNA that appeared to move from chloroplasts to mitochondria. Subsequently, evidence has accrued for a broader process, including the insertion of mitochondrial and chloroplast DNA sequences into nuclear DNA (Herrmann et al. 2003; Matsuo et al. 2005; Bock and Timmis 2008). The extent to which such transpositions produce functional transcripts remains unclear; for example, most of the plastid DNA engulfed by the nucleus may be eliminated by genome shuffling (Matsuo et al. 2005). If the genes are expressed, there are potentially significant evolutionary implications because of the different modes of inheritance of a nuclear as opposed to an organellar gene. Presumably such transpositions also can interrupt nuclear gene function, depending on where they insert.

To what extent is mobile DNA favored by natural selection? At the level of the 'selfish gene' (Dawkins 1989) selection is presumably for these mobile elements, especially in eukaryotes with excess DNA, or in bacteria where they add unique (useful but generally described as nonessential) features as plasmids, discussed later under prokaryote recombination. However, once essential gene functions are disrupted, selection at the level of the gene will be offset by counter-selection at the level of the physiological individual. So the tendency should be toward some balance in opposing forces. Note, however, that deleterious genes can still spread in a population by over-replication or if they alter reproductive mechanisms to favor themselves (Campbell 1981; Chap. 6 in Bell 1982). Certain TEs (retroviruses, below) provide an independent mechanism for moving genetic material horizontally.

Perhaps the most intriguing subcategory of site-specific recombination involves the retroviruses. They are unique in having an RNA genome that replicates by reverse transcription through a DNA intermediate, which can then integrate as provirus into host chromosomal DNA. Retroviruses are considered with transposons because of similar structure and functional properties. They do not transpose in the same way that bacterial transposons do, but are analogous in that they can be viewed as intermediates in the transposition of viral genes from proviral integration sites in the host chromosomes (Varmus 1983; Varmus and Brown 1989). Retroviruses and viral-like elements have been described from diverse genomes, including those of mammals, the slime mold *Dictyostelium*, yeast, fish, reptiles, birds, and plants (McDonald et al. 1988). The most information is on mammalian retroviruses and because of the interesting evolutionary implications, this is summarized briefly below (see also Doolittle et al. 1989).

There are two retroviral categories (Benveniste 1985; Varmus 1988): *Infectious* or *exogenous* retroviruses occur as a few copies of proviral DNA per cell, only in the genome of infected cells; they are infectious and often pathogenic (as in HIV); and they are transmitted horizontally, i.e., among individuals rather than from mother to daughter. *Endogenous* retroviruses occur as multigene families in the host DNA of somatic cells (and occasionally germline cells, in which case they are transmitted vertically) of all animals of the species of origin (Benveniste 1985; Jern and Coffin 2008). They have been viewed as fossil representatives of retroviruses extant in the geological era when they entered the germline (Jern and Coffin 2008). About 7–8% of the human genome is of retroviral origin (Jern and Coffin 2008). Endogenous retroviruses are usually not infectious to cells of the species of origin, but are often so to those of other species. In fact, it is this property of being able to replicate in heterologous cells that sets them apart from conventional cellular genes.

Both types of retroviruses can cause host genes to mutate, or can carry host genes with them. This is significant because of the spreading of somatic variation through the soma, and the possibility of introducing variation directly to the germline (▶Sect. 2.5). **From an evolutionary standpoint, the endogenous group is particularly intriguing because some members have been transmitted horizontally and, once established, may subsequently have been incorporated into the germ line and transmitted vertically** (i.e., from mother cell-to-daughter cell and from parent to offspring). Benveniste (1985, p. 362) reviews evidence that retrovirus transfers have included those "*from ancestors of primates to ancestors of carnivores, from rodents to carnivores, from rodents to primates, from rodents to artiodactyls, from primates to primates, and from primates to birds.*" A specific example is the baboon type C viruses, which are transmitted vertically in primates and which were transferred millions of years ago to ancestors of the domestic cat where they were incorporated into germ cells and inherited thereafter in conventional Mendelian fashion (Benveniste and Todaro 1974, 1975). Benveniste (1985) proposes that retroviruses may promote genetic interaction above the species level, much as do plasmids in bacteria (discussed later).

Are retroviruses a major force in evolution? They do seem to play a major role by influencing gene regulation (McDonald 1990) and by their phjylogenetic implications noted above. The analogue that comes closest to this is the transfer and incorporation of bacterial DNA into plant chromosomes. Crown gall and hairy root diseases of plants involve transfer of a plasmid (tumor-inducing or Ti plasmid) from a pathogen, *Agrobacterium tumefaciens*, to the plant, where a fragment (T-DNA) is covalently integrated into the host nuclear genome. In essence, the agrobacteria use genetic engineering methods to force the infected plant to synthesize nutrients (opines), which the bacteria utilize (discussed in ►Chap. 3; Zambryski 1989; Platt et al. 2014). (Although the process of *Agrobacterium* oncogenesis is often considered with processes involving transposable elements, strictly speaking the T-DNA is not a transposon because it does not jump about the chromosome. The fascinating *Agrobacterium/plant* tumor story is summarized in ►Chap. 3.)

2.2.3 Genomic Plasticity

The sorts of nucleotide changes reviewed above emphasize that both protein-coding and non-coding regions are subject to dynamic evolutionary change. Cells can read multiple messages from the same DNA sequence and these do not just pertain to protein structure (Shapiro 1999, 2002). Mutations are not limited to nucleotide substitution but can be genome-wide rearrangements involving potentially large blocks of nucleotides. As Shapiro states (2002, p. 9) "... *living cells can rearrange their genomes in any way that is compatible with the rules of DNA biochemistry.*" Such rearrangements allow rapid genotypic and phenotypic changes by organisms, as in the response of microorganisms to antibiotic selection pressure discussed later, or by the vertebrate immune system to novel antigens. Physical changes in the genome accomplished by the rearrangements are amplified by DNA interactions with cellular complexes that do not alter the sequences (❑Table 2.2 and Shapiro 2002; see Epigenetics below). The bacterial genome in particular is highly plastic, with multiple mobile components of the genome interacting, being transferred, gained and lost in a dynamic equilibrium (Touchon and Rocha 2016; developed in Sect. 2.3).

2.2.4 Epigenetics and Gene Regulation

Epigenetic controls include nongenetic, enzyme-mediated chemical modifications of DNA structure (methylation of DNA residues after replication) and changes to the associated protein (mostly histones) (Feng et al. 2010; Griffiths et al. 2015). Both processes affect transcription and thereby gene activity. Some alterations to histone as well as DNA methylation marks can be inherited stably and such instances are referred to as 'epigenetic inheritance' (although terminology varies; see Eichten et al. 2014). Epigenetic changes such as erasure of DNA methylation or 'reprogramming' can also occur in both plants and animals where they play an important role in development (Feng et al. 2010). Classic cases of epigenetic inheritance include gender-specific gene silencing even though both the maternal and paternal copies are functional ('genomic imprinting'), and even the silencing of an entire chromosome (random inactivation of one of the two copies of X chromosomes in female mammals).

It is perhaps in the realm of gene regulation more so than any other that Shapiro's (2002) computational metaphor of the genome, noted above, is most apt. The inert DNA storage medium (hard drive) interacts with cell complexes to format the information in a readable, transmissible manner. Though complex in detail, regulation of the message takes many forms

◘ Table 2.4   Six points at which eukaryotic gene expression can be regulated (Alberts et al. 2015). Bacteria generally employ similar mechanisms (see Chaps. 3 and 7)

Type of control	Overview of mechanism
1. Transcriptional	When and how often a gene is transcribed[a]
2. RNA processing	How the RNA transcript is spliced/processed
3. RNA transport/localization	Which completed mRNAs in cell nucleus are transported to cytosol and where localized
4. Translational	Which mRNAs are translated by ribosomes
5. mRNA degradation control	Selective destabilization of certain mRNAs
6. Protein activity control	Selective activating, deactivating, degrading or compartmentalizing certain proteins

[a]There are multiple controls at this level. One form, chemical changes to DNA (e.g., methylation) or histones altering gene function but not DNA sequence, is termed 'epigenetic'

and is exerted at numerous points broadly directed at the transcription and post-transcription levels (◘Table 2.4). In bacteria, one method of control involves proteins that inhibit or activate transcription of specific genes (see discussion of the *lac* operon in ▶Chap. 3). In eukaryotes, transcription is controlled in part by multiple *cis*-regulatory sequences, so-named because they typically are located on the same DNA molecule as the gene affected. In plants, chromatin modifications and other forms of genetic regulation influence development and reaction to environmental stimuli (Feng et al. 2010; Eichten et al. 2014). In animals, there may be 5–10 times as many such regulatory modules as there are genes (Davidson 2006). Sequence-specific DNA-binding proteins (transcription regulators), which are themselves variably active by time in the life cycle and place in the organism, read this information and determine the time and location of genes to be transcribed (Davidson 2006; Tuch et al. 2008; Gilbert and Epel 2009).

It appears from the model organisms studied to date that most if not all the eukaryote genome is transcribed and among the products are several classes of small, noncoding RNAs (ncRNAs) (Amaral et al. 2008). Two such classes of ncRNAs are termed microRNAs (miRNAs) and small interfering RNAs (siRNAs) (Amaral et al. 2008; Ghildiyal and Zamore 2009). They appear to be primarily regulatory, achieving their effect by interacting with transcription factors, RNA polymerase, or directly with DNA (Amaral et al. 2008). For example, transcription can be affected in various ways, including through various modifications of chromatin structure (Slotkin and Martienssen 2007; Figueiredo et al. 2009). The miRNAs direct mRNA degradation or repress translation (Amaral et al. 2008) and may function at multiple hierarchical levels in regulatory networks (Makeyev and Maniatis 2008; Dekker 2008). Among the siRNAs, the so-called exo- and endo-forms (Ghildiyal and Zamore 2009) are derived from dsRNA and are associated with Argonaute (Dicer) proteins that execute the regulation. This specific form of silencing, differing in details but broadly represented among eukaryotes, is known as **RNAi (interference)** (Ghildiyal and Zamore 2009). At least in plants, it serves also as a form of antiviral defense (Baulcombe 2004). As a regulating mechanism on gene expression, RNA interference (RNAi) is a topic of intense research on gene silencing and has practical implications, for example, in disease and pest control (Zhang et al. 2015). Morphological evolution may depend largely on changes in gene expression accomplished by mutations in regulatory networks (Prud'homme et al. 2006, 2007; Davidson 2006; Carroll 2008), though the extent to which such changes drive evolution is controversial.

The far-reaching impact that seemingly minor or innocuous changes to the genome can have is evident in the following example: There is a class of genes (proto-oncogenes) that appears to have a normal housekeeping function within the cell but which, if altered by mutation in a coding or noncoding region, can lead to malignant transformation. In the human Ha-*ras* gene, a single point mutation within the fourth intron can cause a tenfold increase in gene expression and transforming activity (Cohen and Levinson 1988). This is only one example of many that show *there are several ways to change gene expression without changing the message itself* (see also McDonald 1990). Undoubtedly, numerous ways of altering expression are important in evolution and may explain why humans have so many genes in common with other organisms. It was pointed out insightfully by King and Wilson in 1975 that probably humans differ from chimpanzees largely because of differences in gene expression, rather than in gene structure. As the human genome continues to be studied intensively in the years since being sequenced in 2001, it has been a surprise to find that the protein-coding regions, some 3 billion bases in all, account for only a trivial amount (about 1.5%) of the total. At least some and perhaps much (still actively debated) of the vast noncoding portion—often formerly derided as 'junk DNA'—now appears to have a critical regulatory function.

Finally, that there are many gene copies in the more complex life forms provides the opportunity for alterations in the genome by changing introns, exons, or both. New gene products can be exposed to natural selection while the organism is buffered through continued function of the unaltered product encoded at another site(s). Genes that have been rendered silent can be retrieved, and other genes turned on or off, all with the phenotypic expression of a point mutation in a coding region, but accomplished simply by changing patterns of regulation. Controlling the timing or extent of gene expression has important ecological implications because this mechanism in effect increases the phenotypic plasticity of the organism as discussed in some detail in ▸Chap. 7. Cells of more complex organisms in particular have a large repertoire of mechanisms to generate genetic variability!

2.3 Sex and Meiotic Recombination

2.3.1 Definitions, Origin, and Maintenance

Definitions of sex vary considerably (Michod and Levin 1988). Here it will be considered to be the bringing together in a single cell of genes from two (or rarely more) genetically different genotypes or individuals (Maynard Smith 1978b). In most but not all organisms, sex is tied to reproduction, i.e., to the production of a new individual that arises from the zygote or its functional equivalent. The most important consequence of sex is the acquisition by the zygote of new genes (mutations in the germ cells of one or both parents) and new gene combinations. The latter arise from reciprocal exchange (crossing-over) of genes between homologous chromosomes, and reciprocal exchange of chromosomes (independent assortment) during meiotic recombination as part of gametogenesis.

Processes that are sexual are general (though maybe not universal; see ▸Sect. 2.4) among the prokaryotes and eukaryotes. Among extant prokaryotes, genetic exchange can occur in three ways (discussed in detail later in the next section): (i) by direct uptake of DNA from the environment under specific conditions (transformation); and as mediated by (ii) phage infection (transduction) or (iii) plasmids (conjugation). Transformation is likely very ancient. Arguably, it can be traced as far back as the emergence of protocells in primordial communities of progenotes that existed ca. 4.0–3.6 billion years ago, prior to the divergence of the Last Common Ancestor of the three lineages Archaea, Bacteria, and Eukaryotes (see ►Chap. 4). Woese (2002), among others, has postulated an era dominated by the lateral transmission of information with communities as a whole varying in descent and selection operating largely at the level of community optimization.

Meiotic sex, traceable to the last eukaryotic common ancestor (Speijer et al. 2015), is postulated to have arisen from mitosis (Wilkins and Holliday 2009). It may have begun by the interposition of one new step (homolog synapsis) followed by 'parameiosis' in the occasional diploid protocells within an otherwise haploid cell population of early protists. It is perhaps vestiges of this innovation that remain today in 'parasexual' cell cycles, most notably among the fungi, as discussed later. Over evolutionary time, sex cells, and sexual fusion arose with enhanced inter-genic recombination during the pairing step, as well as related meiotic properties such as synaptonemal complexes. With meiosis thus began the classic alternation of generations, haploid/diploid phases of the eukaryotic life cycle (►Chap. 6). Wilkins and Holliday (2009), however, reason that mitosis originated before meiosis because it occurs universally among the eukaryotes, whereas meiosis is both more complex and, though very widely represented as noted, is not universal.

But what was the major selection pressure for the evolution of sex and the origin of meiosis? While increased recombination would imply the creation of new, potentially favorable gene combinations and disruption of unfavorable ones (see below), this feature alone is not generally regarded as conveying sufficiently immediate benefit to constitute a potent evolutionary force. It conveys instead future benefits to the population or lineage rather than to the individual. An alternative, interesting hypothesis (one of many) is that meiosis facilitates repair of DNA damage (Bernstein et al. 1985). If damage occurs to only one DNA strand, it can be repaired by using the other strand as template. However, when both strands are damaged, correction requires proximity of the homologous chromosome for a template and a process akin to recombination. A variation of the Bernstein argument is that sex arose in prokaryotes as a side-effect of processes to promote DNA replication and repair. Indeed, much of the biochemical machinery and the underlying genes are homologous, including the RecA family of so-called recombination enzymes and their eukaryotic homologs (Cox 1999; Marcon and Moens 2005). Notwithstanding the name ('Rec' for recombination), DNA replication and repair appear to be their primary function (Redfield 2001). In the broader context of the implications of sex for endogenous and exogenous repair, Stearns (1992, p. 183) says that "… *the evolution of sex can be viewed as the evolution of the mechanisms preventing the ageing of the germ line.*"

For eukaryotes, Wilkins and Holliday (2009) have modified the Bernstein repair hypothesis by arguing that the benefit of meiosis was not in restoration of the original wild-type DNA message but prevention of recombination-induced injury. Consequently, meiosis improved recombinational accuracy and confined the process to a localized period in the cell cycle (while also likely increasing the frequency of genetic recombination mainly among the 'right' sequences; Wilkins and Holliday 2009). Recombination is error-prone because the 'wrong' or ectopic pairing may occur leading to various irregularities in the message including deletions, duplications, or aneuploidy. The invention of homolog synapsis in meiosis would have enforced accurate alignment so that only identical regions were in register, not diverged homologous sequences elsewhere on the chromosome.

The foregoing may explain why sex arose but, having arisen, why is it maintained in some form in virtually all taxa? The fact that sexual reproduction is ubiquitous yet carries significant costs is commonly referred to as the paradox of sex and has been called by Bell (1982, p. 19) *"the queen of problems in evolutionary biology."* This controversial matter has been debated in countless papers, review articles, and books (e.g., Williams 1975; Maynard Smith 1978b; Bell 1982; Michod and Levin 1988; Otto and Lenormand 2002; Rice 2002; Otto 2009). At the risk of trivializing a very complex issue, the following general points can be made in passing. In eukaryotes the costs, relative to an alternative of asexual reproduction, are various but principally of three sorts: (i) in anisogamous species (i.e., those in which the male and female sex cells contribute unequally in terms of gamete characteristics to the production of progeny) the twofold 'cost of producing males', since only females produce offspring; (ii) in sexual eukaryotes, the twofold 'cost of meiosis', or more accurately the cost of genome dilution, since each parent's genes are diluted by one-half in their progeny; and (iii) the disruption of favorable gene combinations resulting from past selection, analogous to deciding to reshuffle your hand of cards when you already have a good hand in a game of poker (Otto 2009). With respect to (i), a significant general distinction between eukaryotic microorganisms—many species of which are single celled—as opposed to macroorganisms, is that the former typically are isogamous. In such cases there is no 'cost of producing males'.

Against these handicaps of sexual reproduction is set the traditionally acclaimed advantage, namely the ability to combine beneficial alleles from different individuals, restoring variation that would otherwise become dissipated in asexual reproduction (Otto 2009). Simultaneously, in changing environments, sex also would eliminate genetic associations that may have been favorable in a previous selective environment but are no longer so (Otto 2009). An example of such oscillating conditions as they influence coevolving species is that sex can produce novel genotypes that enable lineages of macroorganisms to survive attack by much shorter lived (hence more rapidly evolving) microbial parasites. Conversely, asexually reproducing lines would be vulnerable both because the parasite quickly evolves virulence to overcome host resistance genes, and because as the size of the host clone increases from generation to generation it presents a progressively larger target. This is one form of the 'Red Queen Hypothesis' (Hamilton 1980; Clay and Kover 1996; Lively and Morran 2014).

Through sex, deleterious mutations that would otherwise accumulate in a finite population in the absence of recombination (Muller's 'ratchet' 1964; Bell 1988a) also are purged. These advantages accrue over time (Rice 2002). Crow (1994) has shown conceptually how sexual species can in effect clump harmful mutations and eliminate several at once by a mechanism such as truncation selection (i.e., selection eliminating all individuals beyond a certain phenotypic state or value). In contrast, asexual species can only eliminate them in the original genotype in which they occur. Also, in outbreeding sexual species, genes influencing the mutation rate will become separated from the corresponding mutations, whereas they will not in asexual species (Drake et al. 1998), so evolutionary processes may be quite different in the two situations. A recent test of the longstanding dogma that sex accelerates adaptation, executed by comparing evolutionary events in sexual and asexual populations of *Saccharomyces cerevisiae*, confirms that sex acts by providing a sorting mechanism to separate the beneficial from the deleterious mutations: advantageous mutations are combined into the same background, whereas deleterious mutations are separated from advantageous backgrounds that would otherwise carry them to fixation (McDonald et al. 2016).

The above arguments do not imply that sex necessarily increases variation or that such variation necessarily increases fitness (Otto and Lenormand 2002; Otto 2009). So while competitively superior genotypes can be produced, sexual recombination overall may not have a net advantage. While an earlier generation of models suggested fairly constrained conditions wherein sexuality would be maintained, recent evolutionary models (Otto 2009) run under more realistic conditions imply the evolution of sexuality when (i) selection varies over time (past genetic associations no longer favorable); (ii) selection varies over space (where migration-driven genetic associations are locally disadvantageous); (iii) rates of sex as opposed to asexual reproduction are relatively high for less fit individuals and relatively low for fit individuals, i.e., when facultatively sexual individuals in poorer condition allocate more resources to sexual reproduction; and (iv) populations are finite. With respect to (iv), most evolution-of-sex models are deterministic and assume infinitely large populations. This is not realistic and can lead to the wrong conclusions because the best genotype can be lost by drift in finite populations. Sexual recombination would allow it to be regenerated relatively quickly whereas asexuality would not. For discussion of this and the other conditions, see Otto (2009).

2.3.2 Sex and Adaptive Evolution in Prokaryotes

Not only do bacteria engage in sex, but if one means by the term that genes from different sources are recombined in a single entity, then "*bacteria are particularly sexy organisms*," to use the words of Levin (1988), and subsequently, with colleagues (Johnsen et al. 2009) that "... *Bacteria may not have sex often, but when they do it can be really good, at least evolutionarily speaking*"! A distinctive attribute of bacterial sexuality is that sex is not formally linked to reproduction, whereas in eukaryotes there is a linkage. Prokaryotes reproduce asexually (clonally and in most species, though with notable exceptions [Angert 2005] by binary fission, see later section and Fraser et al. 2007) and the current consensus is that they undergo recombination sporadically. An extension of Levin's wry comments would be that, the degree of bacterial sex apparently varies considerably among populations and species, ranging from the arguably relatively promiscuous *Neisseria* to the relatively asexual *Pseudomonas syringae* (Sarkar and Guttman 2004) or *Salmonella* (Maynard Smith et al. 1991, 1993; Feil and Spratt 2001; however, see Tibayrenc and Ayala 2015). Overall, recombination now appears to be the norm rather than the exception, at least among pathogens (Maynard Smith et al. 2000; Touchon et al. 2009; Bobay et al. 2015).

Unlike recombination in macroorganisms that characteristically involves two complete genomes, recombination in prokaryotes is asymmetrical, typically involving a relatively large and a small donation, respectively, from the two partners. This entails replacement of small nucleotide regions in the recipient cell by corresponding regions moving almost always in unidirectional fashion from the donor bacterium. Furthermore, prokaryotic sex involves, in addition to chromosomal genes, various accessory genetic elements differing in their degree of mobility and autonomy, ranging from phages, plasmids, and transposons at the high end to genomic islands and integrons[1] at the other (Levin and Bergstrom 2000; Touchon and Rocha

---

1   Genomic islands are discrete DNA inserts presumptively acquired horizontally that encode various functions such as those involved in symbiosis or pathogenicity (Hacker and Kaper 2000). Integrons are gene expression elements that capture promotorless genes by site-specific recombination from external sources, thereby converting them to functional genes. All consist of three parts: (i) an attachment site; (ii) a gene encoding an integrase; (iii) and a promoter directing transcription of the captured genes (Mazel 2006).

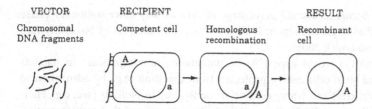

VECTOR   RECIPIENT   RESULT

Chromosomal   Competent cell   Homologous   Recombinant
DNA fragments   recombination   cell

□ Fig. 2.1 Gene transfer in bacteria by acquisition of free DNA (transformation). Incoming chromosomal fragments from the environment (*dark lines*) bind to bacterial cell (*rectangle*); one enters the bacterium and is incorporated into the genome (*light circle*) by homologous recombination. From Levin (1988); reproduced by permission of Sinauer Associates, Inc., Sunderland, MA ©1988

2016). These and related mobile elements have been called a *"motley riff-raff of DNA and RNA fragments"* (Dawkins 1982, p. 159). The ability of bacteria to routinely accept DNA from other species and even entire genes and gene clusters appears to far exceed the capability of eukaryotes in doing so.

While the three processes involved in prokaryotic sex—transformation, transduction, and conjugation—are distinct from each other and from eukaryotic sex, all produce effectively the same end result: acquisition of and usually recombination of DNA from genetically different individuals (cells). Each mechanism is quite complicated in detail and beyond the scope of this discussion. The synopsis here is intended to provide a basis for comparisons between prokaryotic and eukaryotic sex; specifics are available in general microbiology texts and advanced treatises (e.g., Levin 1988; Neidhardt et al. 1990; Bushman 2002; Madigan et al. 2015). Regardless of the process, the entering DNA may either (i) become degraded by restriction enzymes; (ii) replicate by itself (if it has its own origin of replication, as in the case of phage or plasmids) or; (iii) recombine with the recipient's chromosome by homologous recombination. Occasionally, it may recombine as mediated by phage integrases or mobile element transposases, or by various 'illegitimate' or nonhomologous means such as by double-strand break repair (Ochman et al. 2000). These latter mechanisms pertain particularly to incorporation of sequences by horizontal gene transfer. In practice, because of the limitations of detection methods, it may not be known which of the three processes is responsible for recombination in a given situation. The presence of synteny (gene blocks similarly arranged in the species compared) within and surrounding the genome break points, as well as absence of viral- (phage) related sequences, is usually sufficient to eliminate transduction. If the bacterium is naturally transformable (below), it is very difficult to separate transformation from conjugation.

In **transformation**, a bacterial or archaeal cell takes up naked DNA from the surrounding medium (originating usually from a lysed or decomposing cell) which is then integrated into and replicates with the recipient's genome (□Fig. 2.1). The amounts of genome transferred vary over an order of <1–100 kilobases. The mechanism hinges on several conditions, among them development of a transient ability (competent state) by the recipient to be transformed. Competency is a complex trait and the selection pressures for the genes involved are unclear (Levin and Cornejo 2009; Johnston et al. 2014) though recognition and uptake of foreign DNA are highly evolved processes. Transformation occurs naturally and has been documented in many genera, including *Streptococcus, Staphylococcus, Hemophilus, Neisseria,* and *Pseudomonas* (Levin 1988; Madigan et al. 2015). Even within such genera only certain species and strains are transformable and under specific conditions. For instance, in

B. subtilis, competency occurs in a small percentage of cells as they enter stationary phase of growth and is a stochastic phenomenon. It is one of several examples of bistability (see ►Chap. 7 and Dubnau and Losick 2006).

In nature, transformation would appear to be potentially most significant in habitats where DNA in dead and lysed cells can be protected from digestion (e.g., by adsorption to a matrix such as clay particles) and where cells occur densely, as in biofilms (see, e.g., Hall-Stoodley et al. 2004; Hiller et al. 2010). Some authors have speculated that transformation may not even be primarily a sexual process but rather has evolved to provide the cell with nutrients (Redfield 2001), though this opinion has been challenged (Johnston et al. 2014). Others postulate that its most important function is to acquire genes from without as a source of variation (Levin and Bergstrom 2000) and specifically to restore ('reload') genes lost or degraded in a local population though still present in the overall population at large or metapopulation (Szollosi et al. 2006). Under simulation conditions, Redfield (1988) showed that even when the acquired DNA is from dead cells, transformation can reduce mutational load and transformed populations had a higher mean fitness than asexual populations. It is also noteworthy that competent, nongrowing cells may have a transient selective advantage over noncompetent, dividing cells under episodic conditions that kill growing cells (Johnsen et al. 2009). The striking evolutionary dexterity of the human pathogen *Streptococcus pneumoniae* has been attributed to gene transfer among strains by transformation (Hiller et al. 2010; Croucher et al. 2011). This is not only interesting from the standpoint of basic microbial ecology but has important practical implications for understanding the pathogenesis and epidemiology of diseases caused by *S. pneumoniae*. The genomic analysis shows that this lineage evidently acquired both drug resistance and evolved adaptations (antigen switches) to counter vaccine pressure multiple times (Croucher et al. 2011). This bacterium is a natural resident commensal of the human nasopharynx and also exists as a potentially invasive pathogen. It habitually causes ear infections of children as well as frequently fatal infections such as meningitis, bacteremia, and pneumonia.

Transduction occurs when bacterial DNA is packaged within the protein capsid of a bacteriophage particle and injected into a recipient cell during the viral infection process (Levin 1988; Madigan et al. 2015; Salmond and Fineran 2015). In generalized transduction, chromosomal genes from the donor bacterium are transferred when a small proportion of progeny phage carry some random portions of bacterial DNA instead of phage DNA. The donor's genes must recombine with homologous sequences in the recipient's chromosome, otherwise they will be lost. In specialized transduction, temperate phage move specific, adjacent bacterial genes when the occasional phage genome excises imprecisely from its latent or prophage state in the bacterial chromosome at the onset of the lytic cycle (◨Fig. 2.2). Transduction has been shown to occur in numerous genera of soil and aquatic bacteria under nonsterile experimental conditions, though not all bacteria are transducible and not all phages can transduce. Nevertheless, since many phage can infect diverse bacterial species, DNA can be moved across significant evolutionary distances. Transduction is a fortuitous process, essentially resulting from mistakes in phage growth.

Conjugation involves cell-to-cell contact and is controlled by genes carried on certain so-called 'conjugative plasmids'. The result is transmission from donor to recipient of plasmid (extrachromosomal) DNA alone or, occasionally, both plasmid and various lengths of chromosomal DNA (◨Fig. 2.3). Furthermore, there is a diverse group of mobile genetic elements maintained largely as part of the chromosome that can also be excised and transferred to another cell during conjugation but which, unlike plasmids, cannot replicate autonomously.

(a) NORMAL PHAGE INFECTION

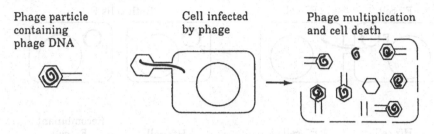

Phage particle
containing
phage DNA

Cell infected
by phage

Phage multiplication
and cell death

(b) GENERALIZED TRANSDUCTION

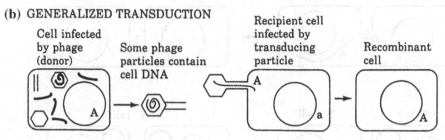

Cell infected
by phage
(donor)

Some phage
particles contain
cell DNA

Recipient cell
infected by
transducing
particle

Recombinant
cell

(c) SPECIALIZED TRANSDUCTION

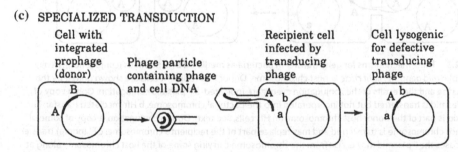

Cell with
integrated
prophage
(donor)

Phage particle
containing phage
and cell DNA

Recipient cell
infected by
transducing
phage

Cell lysogenic
for defective
transducing
phage

▫ **Fig. 2.2** Three possible results of infection of bacteria by bacteriophage (transduction). *Heavy lines* = phage genome; *light lines* = host genome. **a** In normal phage infection by lytic phages, replication of the phage leads to packaging of phage DNA in phage particles. **b** In generalized transduction, a few of the progeny phage contain random portions of bacterial DNA instead of phage DNA. These progeny phage then transfer the host DNA into new cells where it replaces the recipient's genes. **c** In specialized transduction, part of the phage genome is replaced by adjacent host genes when the phage excises from the bacterial chromosome; these genes are inserted by site-specific mechanisms when the virus enters the genome of its new host bacterium. From Levin (1988); reproduced by permission of Sinauer Associates, Inc., Sunderland, MA ©1988

These are collectively referred to as 'integrative and conjugative elements' (Wozniak and Waldor 2010). The best known of the conjugative plasmids is the F (for fertility) plasmid of *E. coli* and closely related enteric bacteria. One of the proteins specified by a cluster of F genes is for the sex pilus, a temporary projection that joins the F$^+$ cells and F$^-$ cells and through which plasmid DNA moves. The result of mating is two F$^+$ cells. In rare F$^+$ cells (known as Hfr = <u>h</u>igh <u>f</u>requency of <u>r</u>ecombination), where the F particle is integrated into the bacterial chromosome, chromosomal genes also are transferred. The F particle can also mobilize a class of nonconjugal plasmids when both occur in the same donor cell. However, to put these events in perspective, Hfr formation, even under laboratory conditions, is a comparatively rare event. Such Hfr's, once formed, are relatively unstable because the F factor excises

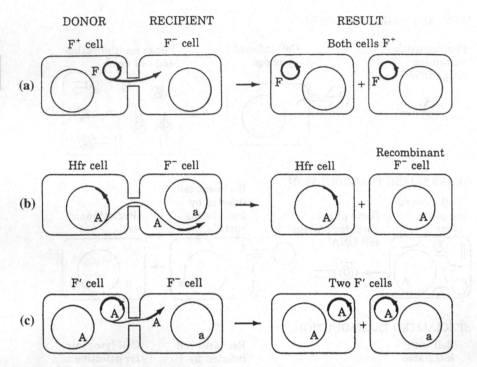

Fig. 2.3 Three mechanisms for gene transfer in bacteria as mediated by F plasmids (conjugation). *Heavy lines* = plasmid genome; *light circle* = host chromosome. Donor and recipient cells are shown united by the pilus bridge and the results of the conjugation events are indicated. **a** F plasmid conjugation: Only a copy of the F plasmid is transferred but not incorporated into the bacterial chromosome. **b** Hfr-mediated transfer: The F plasmid is part of the donor host chromosome (in Hfr cells, see text). During conjugation, a copy of some of the donor chromosome is transferred and may replace part of the recipient's chromosome. **c** F' (prime) transfer: In rare cases the F plasmid may excise from the chromosome carrying some of the host chromosome along at which point it becomes known as an F' plasmid. When conjugation subsequently occurs, it can transmit the host genes it acquired as well as itself. In this case, the plasmid may not integrate into the host chromosome and the recipient becomes diploid for the newly acquired chromosomal genes. From Levin (1988); reproduced by permission of Sinauer Associates, Inc., Sunderland, MA ©1988

at a high frequency. Therefore, the contribution of this type of genetic exchange to variation of bacterial populations in nature is unclear. While the Hfr's may be a relatively rare phenomenon, transfer of the F plasmid is not, the process occurring rapidly and efficiently and rapidly spreading the plasmid infectiously within a population (see **Sidebar**). In another rare phenomenon, recombination may occur between a site on the F+ plasmid and a site on the host chromosome resulting in what becomes known as a F' (F prime) plasmid containing host genes. When an F' mates with an F− cell usually the entire F plasmid is transferred and the result is a partial diploid, i.e., for the genes carried on the F plasmid as well as on the recipient's chromosome. Plasmids can also be transferred by transduction and transformation as well as by conjugation.

As alluded to earlier, **plasmids** are ubiquitous among bacteria (Funnell and Phillips 2004; Touchon and Rocha 2016) and constitute the most common form of semi-autonomous replicating pieces of DNA (so-called replicons). Bacterial cells may carry more than 20 of these elements. Some may be very large (on the order of Mb DNA) and differ little from secondary chromosomes (see footnote under summary below and Touchon and Rocha 2016). In the older literature those plasmids that could integrate into the chromosome were called episomes. Bacteria lacking them generally multiply normally under laboratory conditions; hence, plasmid DNA seemingly does not encode essential functions and may be best regarded as a desirable albeit expendable source of accessory traits. However, 'essential' should be qualified as what appears to be nonessential under laboratory conditions, where such assessments are made, may well be essential in nature. Moreover, exchange of key genes among replicons and the chromosome may well lead to acquisition of essential genes by such elements and thereby persistence in the bacterial lineage.

To the extent to which plasmids contribute to bacterial competitiveness, they also benefit themselves indirectly by enhancing their representation in the population of bacterial carriers. Plasmid DNA contributes to the genetic plasticity of the carrier and confers many characteristics of adaptive value in diverse environments. The best known of these is antibiotic or heavy metal resistance (R factor plasmids; see **Sidebar**); others include the ability to induce tumors in plants (*Agrobacterium tumefaciens*), nitrogen-fixing capability (*Rhizobium* spp.), increased virulence (*Yersinia enterocolitica*), and antibiotic synthesis (*Streptomyces* spp.). Transfer of plasmids has been observed between bacterial strains, species, and, in some instances, even between unrelated genera (Funnell and Phillips 2004). Remarkably, in the case of plant tumors (the crown gall disease, above), the mobilization function of a bacterial plasmid promotes its transfer to *plant* hosts: Not only can this plasmid move among bacteria, but plants have access to the gene pool of at least some bacteria (Buchanan-Wollaston et al. 1987; McCullen and Binns 2006). While plasmids may be the key means by which bacterial genes are transferred in nature, it is not yet clear that conjugation is the mechanism involved, although this is generally inferred to be the case.

SIDEBAR: A Case Study: Transferable Drug Resistance in Bacteria

A central message of this chapter is that although all living things have generally analogous means of generating and transmitting genetic variation, the potential evolutionary rates of microorganisms are much higher than those of macroorganisms. This is a function of their short generation times, hence large population sizes (▶Chap. 4), widespread dissemination and mixing, efficient means of genetic exchange, numerous accessory genetic elements, and the relatively broader phylogenetic range over which gene exchange occurs.

A classic example of natural selection in action (evolution in changing environments) is the phenomenon of antibiotic resistance in bacteria. Such resistance is either encoded by chromosomal genes or plasmid-borne (genes typically on one class of conjugative plasmids known as resistance or R factors) (Nikaido 2009). Plasmids are especially significant ecologically because resistance to several different antibiotics can be combined in a single element, which can also serve a role as an efficient vector as well as mediating genetic rearrangement (Koch 1981; Sandegren and Andersson 2009). In the presence of an antibiotic, the plasmids may increase in size due to gene duplication and/or in plasmid copy number per cell; i.e., the antibiotic resistance genes can be amplified when necessary and deamplified when not needed (Sandegren and Andersson 2009). Early stages in the evolution of resistance of *E. coli* to ciprofloxacin involve extensive cytological changes, including the production of multi-chromosome-containing filaments (Bos et al. 2015).

In the presence of an antibiotic(s), the drug-resistant phenotype obviously has a selective advantage. In absence of the drug, there are typically fitness costs associated with resistance manifested as a decline in growth rate or virulence, but bacteria often respond by compensatory mutations at other loci or amplification of the affected gene (Gagneux et al. 2006; Andersson and Hughes 2010). Amelioration of fitness costs in the absence of the drug commonly result in a bacterial population that is fitter in drug-free culture than the uncompensated resistant population, but less fit that the original wild-type. Levin et al. (2000) speculate that the common ascent of intermediate-fitness compensated mutants, rather than high-fitness revertants, may be attributable to higher rates of compensatory mutation relative to reversion and to bottlenecks in culture associated with serial passage. From a microbial ecology standpoint, as well as with respect to the strategy of antibiotic administration, the fact that reversion to sensitivity is difficult has important implications (e.g., Tanaka and Valckenborgh 2011).

Resistance to many if not most antibiotics is known, and frequently the genes responsible are carried within transposons. Conjugative plasmids typically replicate during transfer; hence, acquisition by the recipient is not at the expense of loss from the donor. The resistance genes spread so quickly (often exponentially) that the phenomenon has been called *infectious* drug resistance. For instance, following the introduction of antibiotic therapy with streptomycin, chloramphenicol, and tetracycline from 1950 to 1965 in Japan, the proportion of drug-resistant *Shigella* (the bacterium that causes bacillary dysentery) increased from about 1–80% of the isolates (Mitsuhashi 1971). The history of the use of various antibiotics to control *Staphylococcus aureus* is analogous. Waves of resistance dated from the introduction of penicillin in the 1940s, through other ß-lactam antibiotics, in particular methicillin. *S. aureus* can cause rapidly progressing, potentially fatal skin and soft-tissue infections. This situation has culminated with worldwide epidemics of methicillin-resistant *Staphylococcus aureus* (MRSA) clones that spread rapidly among healthy individuals, as opposed to earlier outbreaks associated mainly with hospitals and similar settings (Chambers and DeLeo 2009) (◻Fig. 2.4).

In a parallel situation, the genomic plasticity and hence evolutionary versatility of *Streptococcus pneumoniae* to multiple antibiotic and vaccine pressure has been thoroughly documented. This was approached by large-scale genomic analysis of a single ancestral strain (clone) by sequencing 240 isolates of the lineage PMEN1 from 22 countries (Croucher et al. 2011). Their approach allowed relatively easy determination of the number and size of base replacements, and thereby in separating variation due to recombinations from point mutations. The clone has diversified rapidly since its estimated origin in 1970. Base substitutions occurred about once every 15 weeks; though recombination events occurred at about one-tenth that rate, they introduced an average of 72 single nucleotide-polymorphisms each, with 5% of the replacements involving more than 30 kb of genome. Overall, 74% of the genome length had received a recombination event in at least one isolate. On average, some 74,000 bp of sequence were affected by recombination in each strain. A related intensive study of recombination events among four strains in the nasopharynx in a single patient repeatedly sampled over a much shorter period (7 months) confirms the rapid evolutionary potential of *S. pneumoniae* (Hiller et al. 2010). At least 156 kb corresponding to about 7.8% of the genome was exchanged, probably by transformation, during the multiple recombination events. The authors suggest that this case supports the 'distributed genome hypothesis,' which proposes that: (i) bacterial species consist of multiple, co-occurring, complementary strains among which the fluid 'pangenome' or species-level genome is distributed (Ehrlich et al. 2010); and (ii) pathogen genetic diversity accomplished by concurrent infection with multiple strains (so-called polyclonal infection) can overwhelm a host's immune response (Hiller et al. 2010).

The activity of diverse mobile genetic elements in the origin and spread of resistance genes provides an interesting case study in the plasticity of bacterial evolutionary processes.

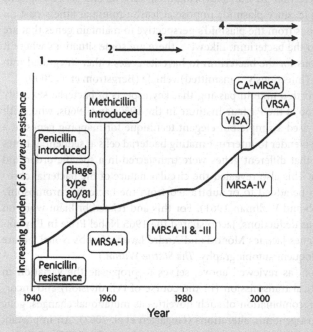

☐ **Fig. 2.4** Four major epidemic waves of antibiotic resistance in *Staphylococcus aureus* with the respective eras shown above the graph. Wave 1 began in the mid-1940s shortly after the introduction of penicillin, driven by strains producing a plasmid-encoded penicillinase. The second wave following within 2 years after the introduction of methicillin in 1959 was caused by a strain designated here as MRSA-I (for methicillin-resistant *S. aureus*) that carried a gene encoding a low-affinity penicillin-binding protein that confers resistance to the entire class of β-lactam antibiotics. Wave 3 began in the mid-late 1970s with clones derived from earlier infections as well as new lineages; the use of vancomycin to treat MRSA infections led to vancomycin-intermediate *S. aureus* (VISA). Wave 4 began in the mid-late 1990s and marked the emergence of MRSA strains with distinct attributes in the community at large (CA-MRSA) as opposed to clinical settings. Strains tolerating even higher doses of vancomycin are called vancomycin-resistant *S. aureus* (VRSA). From Chambers and DeLeo (2009); reproduced from *Nature Reviews Microbiology* by permission of Macmillan Publishers Ltd. ©2009

To the role of conventional point mutation in molding the genome must now be added such dynamic processes and factors as phage transduction, transposons, plasmids, horizontal gene transfer, pathogenicity islands, and probably others awaiting discovery. Transfer of drug resistance is not only an important phenomenon within the context of basic microbial ecology, but obviously has profound implications in practical terms of how antibiotics can be over-prescribed in medicine. Furthermore, because drug-resistant bacteria of animal origin can cause serious diseases in humans (Wegener 2003), the routine use of antibiotics as animal feed supplements should be restricted. For additional reading on the ecology of transferable drug resistance, see Pränting and Andersson 2011; Jackson et al. 2011).

The evidence has been reviewed (Levin 1988; Touchon and Rocha 2016) for plasmid- and phage-mediated bacterial sex being simply coincidental to the infectious (parasitic) transfer of the elements involved and the availability of recombination repair systems in the bacterial hosts. The population biology of the highly mobile elements such as plasmids prompts many interesting questions as to the selection pressures favoring their survival and genetic options for the host bacteria. For example, since plasmids impose at least a minimal fitness cost on the host, it should be advantageous from the plasmid's perspective to maintain genes that are at least occasionally beneficial to the bacterium. Likewise, there are some situations where it may be more or less advantageous for the bacterium to have its genes either integrated into the chromosome or on a mobile (infectiously transmitted) vehicle (Bergstrom et al. 2000).

In an historical context it worth a note in passing that key aspects of bacterial sexuality were discovered by François Jacob at the Pasteur Institute in the 1950s and 1960s, who, with his colleague, Elie Wollman, devised a simple but elegant technique for mapping genes in a linear sequence. Using a Waring blender to interrupt mating bacterial cells at sequential times during the process, they found that different genes were transferred in a specific order and could be mapped as to position. This also revealed the circular nature of the bacterial chromosome and that episomes can be added to or subtracted from the bacterial chromosome (see especially Chap. IX in Jacob and Wollman 1961). For this and related brilliant work on gene regulation and his ingenious deductions, Jacob shared the 1965 Nobel Prize in Physiology or Medicine with his colleagues Jacques Monod and André Lwoff. (Jacob's Nobel lecture is well worth reading, as is his eloquent autobiography, *The Statue Within*.)

Bacterial sexual reproduction, as reviewed above, serves to propagate the variation in existing genes by so-called vertical transmission (in the course of cell division) and incorporation through homologous recombination of such novelties as mutational changes, gene rearrangements, and related intra-genomic alterations (Gogarten et al. 2002). An important extension of the sexual process called **lateral or horizontal gene transfer (HGT)** (Ochman et al. 2000), alluded to in various earlier contexts, involves the integration within a recipient cell of entire genes or gene clusters that are fundamentally new (exogenous) to the genome. **Unlike vertical transmission, in HGT the donor and recipient may be distantly related prokaryotic species or genera, at the extreme even in different domains** (for transfers from prokaryotes to eukaryotes or among eukaryotes see Dunning Hotopp et al. 2007; Keeling and Palmer 2008; Andersson 2005, 2009; Gilbert et al. 2010). The sequences transferred tend to retain characteristics of the donor and so can be distinguished from ancestral DNA (for a discussion of how such inferences are made, see Ochman et al. 2000). Traits introduced by HGT frequently are complex phenotypic features including antibiotic resistance, virulence, and metabolic properties.

HGT has profound implications for organism taxonomy and phylogeny as well as evolutionary ecology. With respect to the former, prokaryotic species boundaries (Rossello-Mora and Amann 2001; Rosen et al. 2015) tend to become blurred (Ochman et al. 2005; Shapiro et al. 2012). Thus, bacterial species might better be thought of less rigidly as "*distinctive arrays of varying but co-adapted gene complexes which are periodically reshuffled*" (Duncan et al. 1989, p. 1586). Perhaps more significantly, the evolutionary history of a gene does not necessarily equate with that of the organism. This is most dramatically seen in the contentious debates surrounding attempts to root the universal tree of life (Brown 2003; Chaps. 4 and 27 in Barton et al. 2007; ▶ Chaps. 1 and 4 of this text). A quantitative measure of the prevalence of lateral transfer is provided by Lawrence and Ochman (1997), who examined a sequenced region of about 30% of the *E. coli* chromosome. Based on atypical base composition and

	Utilize lactose	Utilize citrate	Produce H₂S	Produce indole	Produce urease	Lysine decarboxylase	Lifestyle
Escherichia coli	+	-	-	+	-	+	Mammalian commensal
Shigella flexneri	-	-	-	+	-	-	Primate pathogen
Salmonella enterica	-	+	+	-	-	+	Mammalian pathogen
Klebsiella pneumoniae	+	+	-	-	+	+	Soil
Serratia marcescens	-	+	-	-	+	+	Soil
Yersinia pestis	-	-	-	-	-	-	Mammalian pathogen

◻ **Fig. 2.5** Representative phenotypic properties of, and evolutionary relationships among, certain enteric bacteria based on nucleotide sequence data (+ = presence; − = absence of the trait). Many such species-specific traits are conferred to bacteria, not by conventional mutation or the reshuffling of existing genetic information, but by the acquisition of sequences through horizontal gene transfer (HGT) from distant relatives. Note in particular the differences between *E. coli* (mammalian commensal) and its sister species *S. enterica* (mammalian pathogen). These traits and others are inferred to have arisen from HGT. From Ochman et al. (2000); reproduced from *Nature* by permission of Macmillan Publishers, Ltd. ©2000

codon usage patterns, they estimated that at least 17% of the protein-coding sequences resulted from HGT since the divergence of *Escherichia* from *Salmonella* (or more than 600 kb of transferred DNA accumulated at the rate of about 31 kb per million years; smaller estimates have been made in other systems). This was viewed as being quantitatively similar to the amount of variation introduced through mutation. However, qualitatively, as alluded to above, the impact of the two processes is very different, with HGT potentially providing novel functions to the recipient, i.e., a significantly changed phenotype, immediately, and ultimately being a process that enhances genome dynamics and diversifies lineages (Nowell et al. 2014). Indeed, as an extreme example, it has been proposed recently that adaptation of *Archaea*, which were originally hyperthermophiles, to a mesophilic lifestyle is attributable to HGT from the *Bacteria* (López-García et al. 2015).

The relationships among several enteric bacteria are shown in ◻Fig. 2.5. Frequently, the key traits defining those relationships have resulted from genes transferred by HGT. For example, *E. coli* acquired the lactose operon, hence the ability to utilize lactose and thereby colonize the intestinal tract as a commensal, while genes conferring pathogenicity islands in *Salmonella* and *Shigella* also are conveyed horizontally (Ochman et al. 2000; Gogarten et al. 2002). So lineages can separate as a result of HGT and entirely new niches are created as opposed to being simply refined. The implications can be visualized in terms of Wright's classical model (1932) portraying a population at successive peaks on an adaptive landscape. Having climbed a peak, an evolving population would tend to stay there because to descend would mean to decline in fitness. As Gogarten et al. (2002) point out, peaks may never be explored if they can be reached only by changing one gene at a time (i.e., by mutational processes). HGT can overcome this constraint by introducing multiple changes simultaneously.

To summarize, sexual reproduction and adaptive evolution in prokaryotes is loosely analogous to that in the sexual eukaryotes with some important qualifications. Unlike most eukaryotes, the prokaryotes reproduce for the most part asexually (clonally) as discussed further under 2.4 The Asexual Lifestyle, later. Clones tend to progressively diverge genetically by the sequential accumulation of mutations. Of course they also diverge as a result of recombination or acquisition of foreign DNA from distant phylogenetic sources by horizontal transfer. Whether at the point of acquiring foreign DNA from close or distant sources a cell is still considered to be a clonal member is a matter of definition; see later section and Tibayrenc and Ayala (2012). Bacteria are considered to be haploid because generally there is only one copy usually of one chromosome present.[2] Any genetic change is thus immediately expressed and exposed to natural selection. Genes are exchanged over a substantially wider phylogenetic range and routinely in the prokaryotes unlike the case among eukaryotes. This provides both for the variation in existing genes typical of eukaryotes (by homologous recombination) but, more significantly, the wholesale introduction of unique traits by HGT and nonhomologous recombination allowing major changes in organism habitat or niche. Moreover, the sexual process is mediated by a diverse array of semi-autonomous accessory genetic elements that do not appear to play a significant role in eukaryotes.

Selection pressure on microorganisms, and prokaryotes in particular, typically favors a small genome and rapidity of reproduction. The evolution of prokaryotic chromosome structure and evolution of genome organization are separate but closely related and interdependent processes (Touchon and Rocha 2016). The genome is a streamlined, plastic, dynamic one where there is tension between high plasticity at one extreme and high organization at the other. One gets the impression of a busy railroad station, a scene of frenetic activity with the continuous comings and goings trains and throngs of passengers who intermingle but in an organized manner. Though the population size of prokaryotes typically is vast, dwarfing that of eukaryotes, the effective genetic size can be lower because of bottlenecks or the recurrent selective sweeps of mutants through a population (see comments on asexuality, later, and Reeves 1992; Maynard Smith et al. 2000; Levin and Bergstrom 2000).

2.3.3 Sex and Adaptive Evolution in Some Simple Eukaryotes: The Fungi

For the fungi, sex followed by meiosis is generally similar to that in the higher eukaryotes. Three distinctions with broad ramifications, however, need to be emphasized: (i) Variation among taxa in rapidity of completion of the stages of sexual reproduction—plasmogamy, karyogamy, and meiosis—establishes four basic life cycles and their associated nuclear conditions, in addition to a category putatively ascribed as nominally asexual. These patterns are summarized below along with some relevant terminology. (ii) Fungi show extreme phenotypic plasticity and can exhibit alternative or supplementary genetic systems, most notably

---

2   While this is correct as a generality, bacteria may be polyploid and some species even have >100 copies of the chromosome per cell. Additionally, some bacterial species have more than one *type* of chromosome, typically of different sizes. In such cases the smaller chromosomes are called secondary chromosomes or chromids. Some bacterial lineages (e.g., *Burkholderia* and closely related genera) are characterized by variable numbers of chromosomes, while others (*Vibrio* spp.) consistently (i.e., over geological time) carry two distinct chromosomes. For details and implications, see Touchon and Rocha (2016).

heterokaryosis and parasexuality (also reviewed briefly below). Thus, for the fungi, recombination need not be meiotic or sexual (for details see Taylor et al. 1999a,b, 2015; Billiard et al. 2012). A given fungal species or individual may display different modes of sexuality based on seasons of the year or geographic locations (Taylor et al. 1999a). Furthermore, both sexual and vegetative fusions enable information molecules such as plasmids and mitochondrial DNA to be exchanged among fungal thalli. The septa (cross-walls) transversing the hyphae are rarely complete in areas other than where reproductive structures are borne. Apart from the obvious physiological implications pertaining to cytoplasmic streaming, solute movement, and so forth, this means that multiple, potentially genetically distinct nuclei exist within a common cytoplasm. (iii) Sex is associated often with resting spore formation and is frequently triggered by adverse environmental conditions (see ▶Chap. 7). For these reasons and because the fungi display the genetic characteristics of a transitional group between the prokaryotes and the more complex eukaryotes, the following comments provide background informative for discussions in later chapters.

Life cycles of the fungi and associated nuclear states, mating systems, and degree of genetic variation are numerous and typically complex (Billiard et al. 2012). Simplistically, the options might be categorized as follows (Carlile et al. 2001): (i) **Asexual**—an artificial assemblage of species (formerly, Fungi Imperfecti or Deuteromycota) united historically by their apparent absence (based on morphology) of conventional sexuality. Existence of a sexual cycle has recently been documented or very strongly implied from genomic or population genetics data for the human pathogens *Cryptococcus neoformans, Candida albicans,* and *Aspergillus fumigatus,* all traditionally believed to be strictly clonal (Heitman 2006, 2010; Taylor et al. 2015). Mycologists abandoned this formal classification in 2012; the current concept is that fungi exhibit both episodes of recombination and clonality in their life cycles. More is said later about this category in the later section The Asexual Lifestyle. (ii) **Haploid (haploid/monokaryotic)**—the life cycle is predominately haploid and the hyphal cells generally uninucleate (monokaryotic). Karyogamy follows soon after plasmogamy. Meiosis and compartmentation of the meiotic products by septa in the hyphae follow soon after that (Ascomycota), or, if delayed, the zygote remains dormant (Zygomycota, e.g., the bread mold *Rhizopus stolonifer*) (iii) **Haploid (haploid/dikaryotic)**—the cycle is similar to (ii) except that each cell of a dikaryotic mycelium "… *contains paired, synchronously dividing nuclei, one of each given by the original gametic genotypes*" (Anderson and Kohn 2007, p. 345; see below). Classic examples are the rust fungi such as *Puccinia graminis,* discussed in ▶Chap. 6. The dikaryotic phase may be transient (typical of the Ascomycota) or exist for much of the vegetative phase where karyogamy is delayed after plasmogamy (common in the Basidiomycota). For example, typically in the mushroom-producing (agaric) fungi, the perennial dikaryotic mycelium grows indefinitely and hidden from view as a saprobe in the soil or thatch layer. It may give rise annually to a short-lived (days or weeks) flush of mushrooms in which the life cycle stages of karyogamy followed immediately by haploidy (meiosis and sporulation) occur. The dikaryotic, vegetative, mycelial phase of the well-known fairy ring mushrooms common in pastures may exist for several centuries, with the rings expanding progressively outwards (Dix and Webster 1995). (iv) **Haploid/diploid**—the cycle alternates regularly or irregularly between these two nuclear conditions as in many yeasts. (v) **Diploid**—this group overlaps with (iv) and is analogous to most higher organisms where the haploid phase is relatively inconspicuous and may be relegated to the gametes. It includes members of the Oomycota (fungus-like organisms now considered to be a monophyletic group within the kingdom Straminipila; Webster and Weber 2007). The vegetative cells and much of the life cycle of many yeasts are diploid (Chaps. 10 and 24 in Webster and Weber 2007) or preponderantly so. For example, the remarkable morphogenetic gymnastics of certain strains of the human pathogen *Candida albicans,* formerly thought to be a bland and well behaved, 'obligate diploid', classic yeast, illustrate how dynamic are the reproduction options of the fungi (Hickman et al. 2013). It is the variations on the theme that are informative.

Most fungi belong either to category (ii) or (iii), that is, they reproduce both sexually and asexually and are haploid for most of their life cycle. The haploid, asexual phase of the cycle is generally repeated numerous times annually, typically by rounds of sporulation but in some taxa by other asexual methods such as by budding or fragmentation of the soma. The sexual phase normally occurs only once a year, and may more or less overlap the asexual state. Although a cycle comprised of haploid and to greater or lesser extent diploid structures is thus conventional in the fungi, the alternation of generations is not distinct or regimented, unlike the case in many higher organisms.

Some fungi that engage in sex are hermaphroditic in that a single thallus can function simultaneously as both 'male' and 'female'; such organisms are thus self-fertile, nonoutcrossing, and are said to be **homothallic**. Others are self-sterile, requiring the union of two compatible thalli and are said to be **heterothallic**. Across the fungal world and including the related Oomycetes, the range of sexuality includes haploid selfing, diploid selfing, and outcrossing (for details see Billiard et al. 2012). Fungi have at least two mating types and in the mushrooms there are as many as several thousand (Brown and Casselton 2001).

Dikaryosis, heterokaryosis As noted in the life cycle overview above, the dominant vegetative phase characteristic of the phylum Basidiomycota is generally a dikaryotic $(n + n)$ mycelium. This prolonged, balanced nuclear phase is unique in the living world to certain fungi. As such it warrants some discussion.

In the basidiomycetes, typically the post-meiotic, haploid sexual spore (basidiospore) germinates to produce a filament (hypha) containing genetically identical (therefore, homokaryotic) nuclei in uninucleate or monokaryotic cellular compartments. Hyphae tend to fuse constitutively as they grow; such fusions serve several physiological purposes as well as setting the stage for nuclear transfer in sexually or vegetatively compatible colonies (Glass et al. 2004). When such anastomoses involve different but genetically very closely related individuals of sexually compatible mating type, a developmental program is triggered that results in the dikaryotic mycelium on which the fruiting bodies later develop (Anderson and Kohn 2007; Webster and Weber 2007). This process is dictated, at least in the mushrooms, by different allelic versions of multiallelic genes at two unlinked loci, *A* and *B*; Casselton and Economou 1985; Brown and Casselton 2001). The *B* genes encode pheromones and pheromone receptors; the *A* genes encode proteins involved in transcriptional regulation and synchronized division and cellular distribution of the conjugate nuclei described below. In the 'dikaryotization' process, each homokaryon acts simultaneously as male and female, both donating and receiving nuclei, which divide and migrate quickly and generally more or less widely throughout the recipient mycelium under control of the *B* genes. However, the mitochondria typically do not migrate, so the resultant dikaryon has a consistent nuclear background but is a spatial mosaic for cytoplasmic content, including mitochondrial DNA.

The classic dikaryon appears in the form of sexually compatible nuclei representing the original gametic genotypes, physically associated and dividing synchronously for an indeterminate period as growth ensues (☐Fig. 2.6). Cell divisions in the extending hyphal apex typically are associated with construction of a cytological feature known as the clamp connection, which ensures that the daughter compartments receive exactly two nuclei, one of each mat-

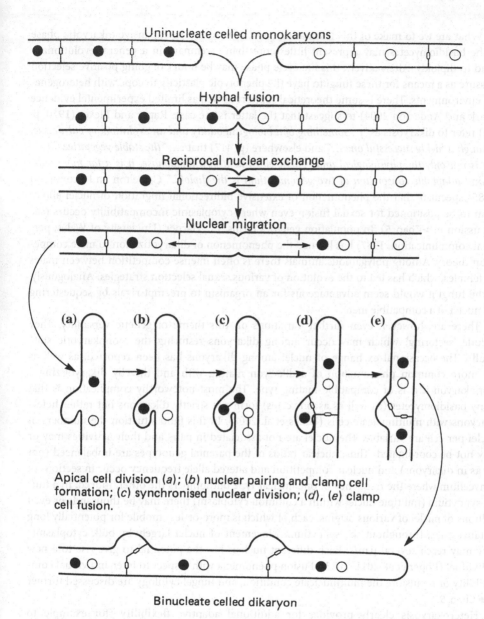

Uninucleate celled monokaryons

Hyphal fusion

Reciprocal nuclear exchange

Nuclear migration

(a)   (b)   (c)   (d)   (e)

Apical cell division (a); (b) nuclear pairing and clamp cell formation; (c) synchronised nuclear division; (d), (e) clamp cell fusion.

Binucleate celled dikaryon

Fig. 2.6 The sequence of events leading to a fungal dikaryon from two compatible monokaryons. From Casselton and Economou (1985); reproduced by permission of Cambridge University Press, ©1985

ing type. Throughout this protracted dikaryotic phase, the conjugate nuclei remain separate but physically associated and in close molecular communication—intriguingly, the distance between them has an impact on gene expression (Anderson and Kohn 2007). The two nuclei may exchange genetic material and undergo somatic recombination (Clark and Anderson 2004; Gladfelter and Berman 2009). When dikaryotization is complete, the new genetic entity functions as a unique genetic individual and tends to rebuff through somatic incompatibility further fusions and genetic invasion by other dikaryons.

What are we to make of this fascinating quirk of nature? The expansive dikaryotic phase in the basidiomycetes may represent little more than a remnant in a general evolutionary trend to diploidy. Alternatively, this life cycle phase may be under ongoing positive selection pressure as a means for these fungi to have the phenotypic plasticity to cope with heterogeneous environments. There is some theoretical support as well as limited experimental evidence (Clark and Anderson 2004) to suggest that the latter is the case. Raper and Flexer (1970, p. 419) refer to dikaryosis as "... *something of a biological oddity and an evolutionary cul-de-sac, although a highly successful one ...*" and elsewhere (p. 417) that .... "*the stable vegetative dikaryon is not only the physiological and genetic equivalent of a diplophase, it is a far more plastic and adaptable consortium of two genomes than is the diploid.*" Casselton and Economou (1985) speculate that the phenomenon of extensive bidirectional migration of nuclei allows them to be positioned for sexual fusion even where cytoplasmic incompatibility occurs (see discussion in ►Chap. 5). In population genetics terms, J.L. Harper (University of Wales; personal communication, 1987) has likened the phenomenon of dikaryotization to mate competition theory. Among polygamous animals there is often intense competition between males for females, which has led to the evolution of various sexual selection strategies. Analogously, in the fungi it would seem advantageous for an organism to preempt rivals by sequestering the nuclei of a compatible mate.

There are, however, even further variations on this theme of genetic versatility. They include 'sectoring', which may occur among dikaryons restoring the monokaryotic state locally. The occasional exchange of nuclei among dikaryons has been reported, as well as the more common phenomenon of a dikaryon mating with and thereby dikaryotizing a monokaryon if it is of compatible mating type. The most noteworthy complication is that many basidiomycetes (as well as ascomycetes) form not strictly dikaryons but rather heterokaryons with multinucleate cells (James et al. 2008). In this fluid situation the numbers of nuclei per cell are variable. The nuclei are not associated in pairs and their activities may or may not be coordinated. Thus, nuclear ratios of the parental genotypes are imbalanced (not 1:1 as in dikaryons) and nuclear competition and altered allele frequency occur in sections of a mycelium where the frequency of one nuclear type outnumbers the other. Indeed, in a fungal syncytium (multiple nuclei within a common cytoplasm) there may be thousands or even millions of nuclei of various origins, each of which is more-or-less mobile for potentially long distances, i.e., throughout the syncytium. Movement of nuclei largely by bulk cytoplasmic flow may reach several μm/s. Each different nucleus has the potential to give rise to a new individual (Roper et al. 2011, 2013). Fusion phenomena with respect to inter-individual compatibility or repulsion, the multinucleate condition, and fungal cytology are discussed further in ►Chap. 5.

Heterokaryosis clearly provides for additional adaptive flexibility (for example to changing environments) beyond that afforded by the more regulated conditions of dikaryosis or diploidy, and it has other implications. Different cells or nuclei within the mycelium may differ in ploidy or by having undergone mitotic recombination (see parasexual cycle, below). Genomic 'conflict' occurs where selection acts in opposition at different levels. For instance, at the organelle level, selection could favor a particular nuclear type, yet disfavor the resultant heterokaryon nuclear ratio at the level of the mycelium as a whole (James et al. 2008). Different nuclear ratios have been reported for conidia versus mycelium of the basidiomycete *Heterobasidion parviporum*, indicating that nuclei may compete to be included in these asexual propagules (James et al. 2008; Roper et al. 2011). Conflict also arises because of the different degrees and modes of somatic transmission of the nuclear and mitochondrial genomes—the nuclei being relatively freely exchanged between the homokaryons whereas the mitochondria in the resultant dikaryon being contributed only from the 'female' parent (Anderson and Kohn 2007). More on these interesting points is discussed in the context of the genetic individual later in this chapter and in ▶Chap. 5). For now, the important concluding point is that we see in heterokaryosis the ability of an organism to adjust the proportion of different sets of genes in response to environmental variation (e.g., available substrates). This is distinct from the formal mitotic-meiotic system of macroorganisms where the genotype (apart from somatic mutation) is continuous throughout the soma. The heterokaryotic fungus adapts genetically and physiologically literally as it grows. Successful heterokaryons, manifested by vegetative fusion and regulated nuclear exchange, as opposed to growth inhibition and cell lysis, are also evidence of nonself recognition systems in fungi, discussed later (Worrall 1997; Saupe 2000; Glass and Kaneko 2003). Finally, there are close parallels between this fungal system and cell/individual compatibility in colonial benthic invertebrates (e.g., Rosengarten and Nicotra 2011), a theme developed in ▶Chap. 5.

Parasexuality Another distinctive attribute of the sexual process in fungi involves genetic recombination outside the usual sexual mechanisms. Some fungi, such as the opportunistically pathogenic yeast *Candida albicans* that were once thought to be asexual, have subsequently been shown to have a nonmeiotic parasexual cycle (Heitman 2006; Forche et al. 2008; Hickman et al. 2013). In the classic parasexual cycle there are four unrelated phases, each of which occurs relatively rarely (Pontecorvo 1946, 1956): (i) a heterokaryotic condition is established, as described above; (ii) diploidization (nuclear fusion) occurs giving a somatic, heterozygous, diploid nucleus; (iii) as growth ensues, the numbers of all nuclei increase by mitosis; in the diploid nuclei, mitotic crossing-over occurs between homologous chromosomes and occasional mitotic error can produce aneuploids; (iv) haploidization follows eventually as chromosomes are lost randomly in successive mitoses. Recombination results both from mitotic crossing-over as well as from the haploidization process. Parasexuality, in being a consequence of these uncoordinated, fortuitous events, is a process distinct from standardized sexual recombination. It cannot replace conventional meiotic sex as a means for recombining genes, and in any *one* generation contributes insignificantly to variation (Caten 1987). The irregular karyotic variation, however, may be advantageous in regulating physiologically important genes in *C. albicans*; and the parasexual cycle, in bypassing the conventional sporulation cycle of this pathogen, may contribute to its ability to live in prolonged association with its hosts as a commensal (Forche et al. 2008). Considerably more information is needed on the extent and significance of parasexuality in nature. Despite its apparent rarity, the process provides yet one more means for genetic recombination, especially in certain supposedly asexual organisms, and it illustrates how mitosis can play a role in genetic variability (Schoustra et al. 2007).

In overview, it is evident that the fungi are a genetically versatile transitional group that spans the gamut in means of transmitting genetic variability. They exhibit some of the orderliness (meiotic mechanisms) of the macroorganisms, together with haphazard variation mechanisms akin to those of the bacteria. Their idiosyncratic life cycles and strange nuclear processes and arrangements may reflect mechanisms to control access to the germline comparable to the evolutionary forces that led to historecognition systems in animals (Buss 1987) and are discussed in detail in ▶Chap. 6.

2.4   The Asexual Lifestyle

While most if not all extant organisms reproduce sexually, some are facultatively and others apparently obligately asexual. Asexuality is discussed here at some length because of its significant evolutionary implications and to set the stage for our later discussion of what constitutes a genetic individual. There are numerous variations on the theme of asexuality, each with its own terminology that is frequently inconsistent across disciplines (Normark et al. 2003; Jackson et al. 1985; Schon et al. 2009; Tibayrenc and Ayala 2012). The prevalent modes are: (i) **apomixis**, which strictly means that female progeny arise mitotically from unfertilized eggs. Its usage, however, is varied and is often equated with agamospermy or parthenogenesis, or alternatively equated with all forms of asexual reproduction; (ii) **vegetative growth**, wherein new individuals arise by fission or budding—as is exemplified by bacteria and yeasts, respectively; or fragmentation as in some plants and animals; and (iii) **automixis**, where eggs are produced meiotically but the meiotic products refuse. Clearly, the various modes can have somewhat different genetic consequences.

The entire set of individuals that descends exclusively (i.e., asexually) from a common ancestor is called a clone. As Milkman points out (1996) 'exclusive' means that *all* the genetic material in *all* the descendants originates with the common ancestor. The exact meaning of 'clone' varies by discipline and microbiologists usually have a more specific and practical context in mind than do botanists or zoologists; also, usage in microbiology is complicated by inconsistent semantics of multiple terms such as serotype, ecotype, sequence type, lineage, strain, and clonal cluster (for microbial examples, see Ørskov and Ørskov 1983; Anderson and Kohn 1995; Spratt and Maiden 1999; Henriques-Normark 2008; Tibayrenc and Ayala 2012, 2015; for macroorganisms, see Jackson et al. 1985; Hughes 1989). As one operational example from medical microbiology pertaining to *Staphylococcus aureus* genotyping, isolates that have identical nucleotide sequences at all of seven housekeeping genes are considered to belong to the same 'clone' and receive a unique 'sequence type'. Those that are identical at five or more of the loci are known as a 'clonal complex' (Chambers and DeLeo 2009).

The word 'clone', used as a noun or more commonly today as a verb, has undergone considerable variation in meaning since its first application apparently in plant breeding in the early 1900s.[3] Regardless of its current application in many subdivisions of biology, each with its own semantics, **one should not infer that all members of a clone are necessarily genetically identical**, despite that implication being drawn by some authors. While in the early stages of growth and depending on the mode of asexual reproduction (see above), all clonal members may be essentially identical, they diverge over time due to somatic mutation. The nested relationship of progressively diverging clones, most easily visualized for bacteria, is nicely illustrated by Milkman (1996; see also Spratt and Maiden 1999) (❑Fig. 2.7). However, some authors have defined clones more broadly. Writing mainly with respect to microorganisms, Tibayrenc and Ayala (2012, p. E3305; see also Spratt et al. 2001) argue that ... "*clonality does not mean the total absence of recombination, but that it is too rare to break the prevalent pattern of clonal population structure.*" Their key defining criteria for clonality are: (i) strong

---

3   Ostensibly, the first use of the word clone was in a horticultural context by H.J. Webber (1903, p. 502) where he defined clones (in that era spelled 'clons') as one subdivision of 'variety' specifically to mean "... *groups of plants that are propagated by the use of any form of vegetative parts such as bulbs, tubers, cuttings, grafts, buds, etc., and which are simply parts of the same individual seedling*".

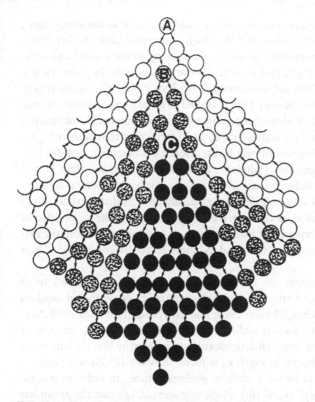

◧ **Fig. 2.7** Clonality as a hierarchical or nested relationship with clones originating exclusively from a single common ancestor. This is most easily visualized as a progressively diverging series of bacterial cells dividing by fission, but is in principle applicable to all clonal organisms. Clone A is the oldest and progenitor of sub-clone B and eventually and indirectly of sub-clone C; B arises from A by (somatic) mutation and C diverges from B by further mutation. All three clones are closely related and ultimately traceable to cell A, but each has its own age and genetic structure. The process continues indefinitely. See text for complications and caveats. From Milkman (1996); reproduced by permission of ASM Press, Washington, DC ©1996

linkage disequilibrium together with (ii) clear phylogenetic signal. This would constitute "predominant clonal evolution" (Tibayrenc and Ayala 2015). Unfortunately, it is often not a simple matter to differentiate between episodes of mutation and recombination, either in microorganisms (Bobay et al. 2015) or macroorganisms (Ally et al. 2008). In prokaryotes, genotypic and phenotypic clonal divergence can occur rapidly (within hours to days) in culture, whether in heterogeneous (Rainey and Travisano 1998) or constant (chemostat; Maharjan et al. 2006) conditions (for general remarks, see Bobay et al. 2015). Thus it is the case both in vitro and in vivo that multiple clonal sub-lineages occur, each carrying at least one and likely many mutations. These subpopulations or 'mutational cohorts' compete in a phenomenon commonly referred to as 'clonal interference' (e.g., Williams 1975), and the population is not necessarily purged of variation by selective sweeps of a clearly superior line as once thought (Lang et al. 2013). Because of their shorter generation times and relatively much larger population sizes, microbial clones diversify faster in absolute time than clones of macroorganisms such as aphids or aspen trees.

In phylogenetic terms, lineages of asexual eukaryotes tend to be scattered among branches of their sexual relatives near the tips of phylogenetic trees. This implies that, in general, asexuals have arisen relatively recently and are short-lived in evolutionary time, without having had the opportunity to diversify to a high taxonomic rank (Butlin 2002). Phylogenetic evidence, along with molecular genetics data attesting to the ancient and complex nature of the sexual process, suggest that asexuals have arisen sporadically from sexuals, rather than the other way round (Otto and Lenormand 2002; Rice 2002). Why asexuals generally do not persist is hotly debated. Stanley (1975) argued that most evolutionary change arises from speciation events and that asexual macroorganisms generally could not speciate rapidly enough over evolutionary and geological time to avoid extinction, thus accounting for the paucity of asexual clades. Another reason may be the relentless accumulation of deleterious mutations in a finite asexual population ('Muller's ratchet' noted earlier; Bell 1988; Barton et al. 2007). The rate at which the ratchet turns and its impact are subject to several assumptions. Depending on which ones are invoked, the theory actually predicts that asexuals would be eliminated quickly, slowly, or not at all (Normark et al. 2003). The common condition of polyploidy in asexuals has been interpreted as possibly insulating the organism from the deleterious effect of accumulating mutations (Otto and Whitton 2000; Pawlowska and Taylor 2004).

There have been numerous claims for ancient asexuals, occasionally referred to as "ancient asexual scandals" because, if true, the examples contradict conventional wisdom that asexual lineages cannot persist long (Muller 1964; Maynard Smith 1986; Bell 1988; Normark et al. 2003). As emphasized by Judson and Normark (1996), such records must meet all three components inherent in the term 'ancient asexual group,' namely that the lineage is: (i) descended from a common ancestral group (i.e., is monophyletic); (ii) 'ancient,' defined subjectively but commonly taken to be on a scale of geological time, an order of magnitude of millions or tens of millions of years; (iii) primitively asexual, i.e., that the group has remained asexual from its inception without interludes of sex. This latter attribute is the most difficult to establish and several taxa once assumed to be asexual have been shown to engage in cryptic sex. This includes fungi classified as arbuscular mycorrhizae (Glomeromycota), originally believed to be the eldest asexuals at ca. 400 million years (Kuhn et al. 2001; Croll and Sanders 2009; however, see Taylor et al. 2015). Males, hermaphrodites, and meiosis are unknown in a large metazoan taxon (Class Bdelloidea of the Phylum Rotifera), believed to be at least 35 million years old (Welch and Meselsen 2000; Flot et al. 2013). Examples of other kinds of evidence used to infer not only asexuality but in some instances ancient asexuality are several, including: (i) the independent evolution of two alleles at any given locus. This increasing allelic divergence due to the accumulation of neutral or possibly adaptive (Pouchkina-Stantcheva et al. 2007) mutations is called the 'Meselson effect' (Welch and Meselson 2000; Butlin 2002); (ii) high taxonomic rank with abundance of species; (iii) phylogenetic congruence of gene genealogies; (iv) strong correlation among alleles at multiple, polymorphic loci (linkage disequilibrium) leading to the recovery of the same multilocus genotype through time and often over great distances; and (v) decay of sex- and recombination-specific genes (Tibayrenc et al. 1991; Taylor et al. 1999a,b, 2015; Normark et al. 2003). These and other tests vary in rigor, are subject to caveats, and the outcome may be subject to interpretations other than a conclusion of asexuality; for insightful discussion, see Taylor et al. (1999b) (☐Fig. 2.8).

At least in contemporary time and quite likely also in geological time, many lineages have effectively integrated alternating rounds of sexual and asexual reproduction into their life cycles. At the level of the individual or species, whether sexual or asexual reproduction dominates the cycle, or indeed occurs exclusively, depends on factors such as the local envi-

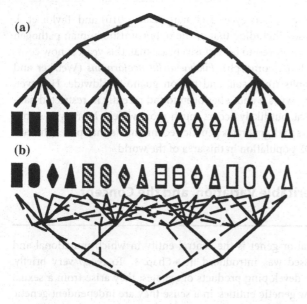

□ Fig. 2.8 A simplified example of strictly clonal versus recombination reproductive modes and resulting phylogenies based on two hypothetical characters, shape and pigmentation (*shape* = *rectangles, triangles*, etc. symbols representing the four structures shown; pigmentation = degree of blackness or whiteness within the structures). In the clonal organism (**a**), recombination is absent; the entire genome remains intact through generations and occurs In few combinations as represented by the sub-clones showing constant association between degree of pigmentation and specific shape. In the recombining organism (**b**), pigmentation and shape are found in all possible combinations. Note, however, that genetic regions may be constantly associated for reasons other than clonality; likewise, lack of association of loci may occur for reasons other than recombination. From Taylor et al. (1999b); reproduced from *Clinical Microbiology Reviews* by permission of the American Society of Microbiology, ©1999

ronment and whether a compatible mating type also occurs. In an evolutionary context, this is perhaps particularly true of pathogens or parasites (Price 1980) and fungal pathogens especially (Andrews 1984; Heitman 2006). Sexual reproduction is often associated with adverse environments and its occurrence in some facultatively sexual fungi is probably a specific instance of the broader phenomenon known as 'condition-dependent' or 'fitness-associated' sex reported in various taxa (see Hadany and Otto 2009). Tsai et al. (2008) quantified the occurrence of sexual and asexual rounds in populations of the wild yeast *Saccharomyces paradoxus* and found that a sexual cycle occurs about once in every 1,000 asexual generations.

Generally speaking, trade-offs are evident in a sexual/asexual cycle, where production of the sexual spore form typically takes significantly longer (weeks vs. days) but may be more resistant to desiccation, whereas the asexual form is produced in much greater abundance. In the case of plant pathogens, timing of the life cycle phases is typically exquisitely linked to susceptible phenological stages of the host. Repeated rounds of asexual reproduction allow the pathogen to 'track' the host in time and space; regular episodes of sex (for pathogens of plants, these typically occur during the over-wintering or dormant phase) allow for generation of novel genotypes to respond to selection pressure of the evolving host. Taylor et al. (1999a) suggest that fungi also may be thought of as mosaics of recombining and clonal populations: the sexual populations to be found on wild, heterogeneous hosts where the recombined fungal genotypes are generated and then move to genetically uniform agricultural hosts, where asexual populations cycle. Heitman (2006, 2010) and Taylor et al. (2015) review several interesting cases, including that of the opportunistic human pathogen *Cryptococcus neoformans*. Originally believed to be exclusively asexual, this yeast is now commonly referred to by its sexual state (teleomorph), *Filobasidiella neoformans* (Webster and Weber 2007, pp. 660–665). It inhabits trees, soil, and pigeon guano worldwide. However, opposite-sex mating between mating types *a* and α has been found recently in restricted habitats in India and sub-Saharan Africa, but likely occurs much more widely. Moreover, same-sex mating (self-fertility) also occurs and generates genetic recombinants that pose a hazard for the immuno-compromised AIDS population in this area of the world.

2.5   Somatic Variation, Heritable Variation, and the Concept of the Genet

The idea that the genetic individual or genet is the central entity in which mutational and recombinational events are expressed was introduced in ►Chap. 1. To recap very briefly, genets are classically viewed as the developing products of zygotes; they arise from a sexual process and therefore represent new genetic entities; in a sense they are independent genetic colonizations of a landscape (Harper 1977). Following our review in this chapter of sexuality/asexuality and the mechanisms of genetic variation, let us now return to the genet concept in a more detailed fashion. This is important, not only because 'the genetic individual' is referred to repeatedly in this book, but also because of its major influence in evolutionary biology.

The utility of the genet concept hinges on the premise that although genetic variation can occur in somatic cells, such variation cannot be transmitted to progeny. This view, propounded most forcefully by August Weismann in the late 1800s, can be stated more formally as: (i) the zygote produces somatic cells mitotically and germ cells meiotically; (ii) genetic variation developing during ontogeny cannot be inherited; and hence (iii) heritable variation is expressed only in the zygote or during meiosis in the formation of gametes (Weismann 1892; summarized by Buss 1987, p. 13). **If true, this so-called Weismannian doctrine has enormous implications: it restricts evolutionary change to a matter of selection among individuals,** as Buss (1987) insightfully develops in his book. So then, to what extent is the genet a valid and useful common denominator in phylogenetic comparisons?

First of all, it should be reemphasized that somatic variation occurs in all organisms and rates are at least as high and typically substantially higher than germline rates (earlier discussion and Lynch 2010). They are presumed to be a significant source of phenotypic and genotypic variation in plant populations (Silander 1985; Klekowski 1988) and in clonal, modular animals (Hughes 1989; more on this in ►Chap. 5).

Secondly, there is no doubt that somatic variation can affect the life of the organism. Probably the most dramatic evidence of this is that the main changes that lead to various cancers involve somatic mutations (Griffiths et al. 2015). The cells in most forms of cancer have aberrant chromosomes (e.g., deletions, inversions, translocations, aneuploidy) as well as numerous point mutations (ranging from a few to more than 1,000; Vogelstein et al. 2013). Accumulating somatic mutations also form the basis of one of the theories of aging (see ►Chap. 6). Other sorts of somatic change may be neutral or beneficial, either in allowing the organism to adapt to biotic or abiotic challenges (e.g., generation of antibody diversity, acquisition of acquired immunity), or by concomitant adjustment to specific genotypic alterations. Recently, the history of somatic mutation accumulation within individual neurons of the human brain has been traced by single-cell sequencing (Lodato et al. 2015). Remarkably, each neuron (which lives and remains transcriptionally active for decades) has its own unique genome as a result of as many as ~1,580 single nucleotide variants (SNVs), among other genetic changes. These mutations appear to arise during transcription, unlike the standard case of errors being introduced during DNA replication. Highly expressed genes were enriched for the SNVs.

The extent to which somatic variation can enter the germline, which is the real issue in evolutionary terms, depends on the ontogenetic program of the organism. It is 'the real issue' because, as seen above, while somatic genetic changes can be devastating to the individual, possibly even causing death, they are limited to that individual, not the lineage. For dipterans, as illustrated by *Drosophila*, it would be highly improbable for somatic variants to enter germ cells. The totipotent lineage in *Drosophila* is restricted to only the first 13 nuclear divisions per generation (Buss 1987, pp. 13–25)—a fleeting opportunity for the origin of a somatic variant. Similarly, in humans, germ cells established in the 56-day-old embryo remain sequestered for up to about three decades (Buss, op. cit. p. 100). Based on current evidence, both dipterans and humans come as close as any organism does to being a homogeneous genetic entity. That the period of accessibility to the germline is short for vertebrates has been confirmed elegantly in studies where foreign genes are introduced during early embryogenesis (Robertson et al. 1986; Jaenisch 1988). For instance, early embryonic cells can be infected with retroviruses in vitro and reintroduced to the embryo at the blastocyst stage of ontogeny. The infected cells contain integrated provirus that con tributes to both the somatic and germ cell lineages, as confirmed biochemically and by the chimeric phenotype of the transgenic animal and its progeny. Infection of pre-implantation stage mouse embryos results in transmission to the germline, whereas infection at the post-implantation stage (between days 8 and 14 of gestation) results in transmission to the somatic but generally not to the germline (Soriano and Jaenisch 1986). Thus dipterans, humans, and mice are examples of a type of ontogeny where all cell lineages are determined early in ontogeny, known as **preformistic development** (Buss 1987). A correlate of this developmental mode is the absence of ramet production. (Ramets, also discussed in ►Sect. 1.3 of Chap. 1, are the asexual counterpart to genets and will be taken up further in ►Chap. 5.)

At the other extreme, an ontogenetic program known as **somatic embryogenesis** is characterized by absence of a distinct germline and the ability to regenerate a new individual from some tissues at any life stage (Buss 1983, 1987). Somatic variants can be transmitted to progeny either by the mutated cell lineage passing directly to the 'new' individual during the process of asexual fragmentation, fission, etc. or by the lineage entering the gametes (Otto and Orive 1995; Orive 2001). For instance, in the fungi, mutations arising in any tissue can be transmitted sexually or asexually. Because asexual reproductive rates are so high, favorable mutants, such as those containing virulence alleles (Clay and Kover 1996), can be rapidly increased through natural selection (Caten 1987). Thus, by clonal growth, fungi can evolve significantly by mutation in absence of recombination. Among some simple animals such as *Hydra*, the zygote divides to produce an interstitial and a somatic cell lineage. The former remains totipotent and mitotically active. By the time gametes are differentiated it is highly likely that somatic variation will have arisen in the forerunners of those cells. Likewise, corals are totipotent and in Buss's words (1987, p. 107) ... *"a 20,000-year-old reef coral had passed uncounted millions of fruit fly generations."*

Plants are a particularly interesting example of the somatic embryogenesis mode of development. As we discuss in ▶Chap. 5, they grow by virtue of the activity of meristematic cells in their shoot and root apices, and additionally in some cases by lateral meristems encircling the axis. The meristems accumulate mutations over repeated cell divisions as the cell lineage increases. A practical consequence of such mutations is that horticulturists have exploited them for centuries to develop most varieties of fruit trees, potatoes, sugar cane, and bananas, not to mention countless vegetatively propagated ornamental and floricultural plants (Silander 1985). The spread and impact of a somatic mutation depends on many factors, including the strength of selection at the level of cell lineage; when the mutation occurs in the timing of the mitotic lineage; and the number of cell generations per individual generation (Otto and Orive 1995; Otto and Hastings 1998; Orive 2001). Long-lived and particularly large-statured plant species have higher mutation rates per individual generation than do short-lived species because of the greater number of cell divisions before gamete formation (Klekowski and Godfrey 1989; Schultz and Scofield 2009). A biological consequence of somatic mutation is that plants can develop as mosaics where one component, say a shoot, is genetically quite different from another. This implies that beneficial somatic mutations (e.g., resistance to parasites or insect grazers) could potentially spread easily, whereas at least some kinds of deleterious mutations would be inconsequential because the affected part could be shed (hence, it is argued, no increase in mutational load would occur) (Whitham and Slobodchikoff 1981; Gill et al. 1995). This variation has been hypothesized as being one way by which long-lived plants could contend with rapidly evolving pests and pathogens. The evidence is mixed (see ▶Chap. 5 and Whitham and Schweitzer 2002; Folse and Roughgarden 2011).

Because of the totipotency of plant meristematic cells and the clonal aspect of development, somatic mutations in precursors of a floral lineage can be transmitted to gametes. As alluded to earlier in this chapter, these events happen occasionally and have been documented in the groundbreaking work with transposable elements of maize pioneered by Barbara McClintock (McClintock 1956; Fedoroff 1983, 1989). For example, if a genetic change occurs during the first embryotic cell division, a plant with genotypically and phenotypically distinct halves is created. Each half will go on to produce different gametes. If a similar change is delayed until ears form, two different sectors with correspondingly distinct kernels will develop. Indeed, the order of genetic events can be surmised from the timing in appearance of the sectors. **The important point here, however, is that somatic changes can be reflected in the gametes and ultimately zygotes.** Hence, somatic variation can not only alter the fitness of the carrier in which they arise, but they can, at least in some instances, most notably with modular organisms, be passed on to offspring produced sexually (Otto and Orive 1995; Pineda-Krch and Lehtila 2004; see ▶Chap. 5). This mechanism extends the conventional forms of genetic variation discussed previously (▶Sect. 2.3).

Microorganisms have always posed a challenge for the genet idea. As noted above (The Asexual Lifestyle), although members of a clone are essentially identical initially, they diverge genetically over time due to somatic mutation and other processes. For the genet idea to hold, it is not necessary for daughter cells in a mitotic lineage to be genetically identical. Rather, the complication for microbes is mainly that the nature, occurrence, and transmission of a genetic change is often haphazard and may occur outside the conventional sexual cycle. Formation of the bacterial recombinant is not tied to a particular divisional event, a morphological structure, or a characteristic life cycle stage such as reproduction, dormancy, or dispersal. The recombinant cells are frequently not even evident in a mixed population unless identifiable phenotypic traits such as auxotrophic markers are involved. Some eukaryotic microbes (e.g., the ascomycete and zygomycete fungi) typically are diploid only very briefly, so the nature of the event triggering a new genet is not clear, unlike the usual case among plants and animals (for elaboration of this point see Anderson and Kohn 1995). Fungi, perhaps uniquely among organisms, tend to fuse upon contact. While such comingling tends to be restricted by vegetative incompatibility systems to close relatives, a single fungal thallus may exist as a genetic mosaic, with genetically different nuclei operating within a common hyphal cytoplasm (Peabody et al. 2000; James et al. 2008; Roper et al. 2013) or spore (Kuhn et al. 2001). As we have seen, by way of heterokaryosis and parasexuality, mutational and recombinational events can be expressed, transferred clonally, and exposed to natural selection independently of fertilization. The evolutionary implications of migration of new genes through an existing genet, followed by change in phenotype and outgrowth of a new genet, present complications to conventional modular theory, a topic taken up in detail in ▶Chap. 5. This fungal situation does not arise with unitary organisms because the germ cells are segregated from the soma, and within other modular life forms this kind of gene migration would be rare, if not unique. Thus, **while genets can be visualized clearly for most macroorganisms, the concept must be applied somewhat abstractly for some and perhaps most microorganisms.**

The preceding foray into potentially heritable somatic variation is necessary because it documents that genetic variation occurs at many levels, including the cellular, as well as those of the so-called 'physiological' and 'genetic' individual. This challenges the dogma that the developing product of the zygote (by which is implied a single entity arising from gametic fusion) is **the unit** of variation. It means, on balance, that the concept of a genet is an ideal that is more or less approximated in various phyla. The unitary organisms, most clearly illustrated by the vertebrates, come closer than do modular life forms (many invertebrates, plants, fungi) in behaving as genetic individuals. As reviewed earlier in this chapter, molecular biology is showing that mobile genetic elements can move among chromosomes of a cell, among cells, and between the somatic and germ lines. One consequence of this fluidity is increased somatic variation and potentially a direct route from soma to gametes. Even in the case of unitary macroorganisms, the concept of the genetic individual must now be revised to reflect more flexibility and fluidity.

2.6 Summary

The principal general categories or agents of genetic variation are mutation, recombination, and regulatory (expression) controls. Mutations are the ultimate source of most variation in all organisms. Broadly speaking, the sources include errors in replication of the genome; errors in segregation of the replicated genome to the daughter cells; and modification of the genome by events such as transposition or errors in recombination. Mutation rates per base pair per replication event vary over orders of magnitude from viruses to eukaryotes, but mutations are roughly comparable across taxa when expressed as per genome per replication event. The inherent rate of sequence divergence over time for a given gene or protein among taxa tends to be approximately constant (the so-called 'molecular clock'). Nevertheless, rates of phylogenetic change are far from constant, apparently governed largely by the ecological opportunity afforded the genetic variants.

Recombination, the mixis or rearrangement of nucleotides, is accomplished by general (meiotic assortment; crossing-over; segmental interchanges) and site-specific (e.g., retroviruses; plasmids; transposons; promiscuous DNA) mechanisms. The potential of site-specific mechanisms to dynamically restructure the genome and to affect genetic expression is only beginning to be fully appreciated. Sexual recombination, the bringing together in a cell of genes from two genetically different sources, is one of the means by which recombination can occur. Sex is essentially ubiquitous among organisms and is usually but not always associated with reproduction, that is, the generation of offspring. For example, sex in bacteria is never an obligatory aspect of reproduction. Most if not virtually all extant preponderantly asexual creatures can also reproduce at least sporadically sexually, although the sexual stage may occur infrequently and in some locations may be absent. Apparent absence of sex in certain living things may simply be a consequence of insufficient observation or an overly restrictive concept of sex, e.g., where it is construed to exclude events (e.g., transduction, transformation, plasmid-mediated conjugation, anastomosis and resulting heterokaryosis, parasexuality) that are in effect sexual but do not involve the fusion of gametes. Retroviral gene transport in mammals is analogous to transduction in bacteria. Meiotic recombination in eukaryotes has an obvious parallel in conjugal transfer in prokaryotes.

Sexual reproduction arguably predated the asexual process. It may have arisen in a process akin to bacterial transformation or as a mechanism to repair DNA from which the recombination function then evolved. In conventional diploid organisms, sexual reproduction entails a dilution of 50% of the genome per generation. Given this, as well as other disadvantages relative to an asexual alternative, why sex has been consistently maintained is unclear. In evolutionary terms, it has probably been retained for several reasons (e.g., resetting genetic variation; avoidance of Muller's ratchet; host-parasite coevolution), which may well be different or of differing importance in the various phyla. If this is the case, efforts to find a singular role for sex are unlikely to succeed.

The most obvious evidence for gene regulation at multiple levels from transcription to protein activity is that cells of the same genetic constitution differentiate into various tissue types. Transcription of bacterial genes is controlled by various mechanisms, among them repressor or activator proteins. In eukaryotes, controls include the sequestering of portions of the genome into transcriptionally inactive regions (heterochromatin); existence of cooperative groups of regulatory proteins; methylation of DNA; RNA processing, transport, and degradation; and many other factors. Because of numerous controls on gene expression, it is not necessary to disrupt the coding region of a gene in order to render a change in phenotype. The importance of changes in regulatory as opposed to coding DNA is increasingly being recognized for its seminal role in plant and animal developmental biology, as well as in the phylogenetic diversification of eukaryotes.

Overall, microorganisms appear to have essentially all of the capability of macroorganisms in generating genetic variability, though adaptive evolution proceeds rather differently in the two groups. The major distinction seems to be that variation specifically in the prokaryotes is transmitted in relatively dynamic, unordered fashion, as opposed to the orchestrated manner, characterized by meiosis and gametogenesis typical of the eukaryotic microorganisms and of higher organisms in particular. Much of the genetic machinery and reservoir of variation for bacteria is nonchromosomal, residing instead in numerous kinds of accessory elements, classically the plasmids. Perhaps most significant is that the gene pool of distant relatives is tapped by horizontal transmission that introduces fundamentally new traits in a manner that apparently occurs rarely in eukaryotes. The fungi, as eukaryotic microorganisms, have a fairly fluid genome as exhibited by such processes or conditions as parasexuality, dikaryosis, and heterokaryosis. Thus, they can be viewed to be an intermediate group, spanning the gamut in variability-generating mechanisms, dexterity, and orderliness of transmission.

The concept of a genetic individual or genet originated in Weismann's doctrine, which viewed the zygote as the seat of all heritable variation and the germline as being insulated from the soma. It has long been recognized that somatic variation occurs and that this can markedly influence the carrier and potentially its clonal descendants. Whether somatic variation can be transmitted to offspring produced sexually depends on the ontogenetic program of the organism. In taxa where the totipotent lineage is strictly limited, as in higher animals, the chance that somatic variation can occur and be passed on to the developing germ cells is remote. Where cells remain totipotent and mitotically active, as in plants, fungi, and the less complex metazoans, transmissible somatic variation is probable (see modular organisms in ►Chap. 5). This has been demonstrated strikingly, for example, in the case of mobile genetic elements of maize. Thus, the genet is a much less discrete unit in some taxa than in others. Nevertheless, the evolutionary significance of somatic variation remains sketchily documented and controversial. Even more contentious is the extent to which there may be direct genetic feedback from the soma to the germline by a mobile gene (e.g., retrovirus) mechanism. What is clear and perhaps most important in evolution is that the *ability* to transmit somatic variability *is* to varying degrees heritable. Obviously, the genet must be a more fluid entity than was conceived originally, in part because of increasing awareness of the role of mobile elements in genetic rearrangement and expression.

The main conclusions from this chapter, which sets the stage for the rest of the book, are that: (i) All organisms possess several and in principle analogous means of generating and transmitting genetic variation on which evolutionary processes, in particular natural selection, can act. The principal differences are the relatively unordered (microorganisms) versus ordered (macroorganisms) manner in which the variation is transmitted. (ii) Because of their short generation times and large population sizes, microorganisms have higher evolutionary rates as species than do macroorganisms. (iii) With respect to broad comparisons of taxa, the occurrence of different ontogenetic programs means that in some cases somatic variation can be transmitted to the germline; different and in some cases multiple nuclear conditions occur during the life cycle; and the ubiquity of mobile DNA, all mean that the concept of the genetic individual (genet) must be more fluid than as originally conceived. As such it should be used guardedly in some circumstances (such as for the bacteria, fungi, and simple metazoans).

Suggested Additional Reading

Buss, L.W. 1987. The Evolution of Individuality. Princeton Univ. Press, Princeton, N.J. *The history of life as a transition between different units of selection.*

Dawkins, R. 1982. The Extended Phenotype: The Gene as the Unit of Selection. Oxford Univ. Press, Oxford, U.K. *A continuation of Dawkins' stimulating examination, begun in "The Selfish Gene", of the levels in the hierarchy of life at which natural selection acts. The focus is on "selfish genes" and not "selfish organisms".*

Grodzicker, T., D. Stewart, and B. Stillman (Eds.). 2016. 21st Century Genetics: Genes at Work. 80th Cold Spring Harbor Symposium on Quantitative Biology. *Collected papers on genetic mechanisms and molecular biology in the areas of gene control, chromatin, genome stability, developmental biology, and evolution.*

Otto, S.P. 2009. The evolution of sex: recent resolutions and remaining riddles. *A symposium organized by Sarah P. Otto. American Naturalist, vol. 174 pp. S1–S94. July 2009.*

Worrall, J.J. (Ed.) 1999. Structure and Dynamics of Fungal Populations. Kluwer Academic Publishers, Boston, MA. *An excellent synthesis of fungal biology by multiple authors.*

Nutritional Mode

© Springer Science+Business Media LLC 2017
J.H. Andrews, *Comparative Ecology of Microorganisms and Macroorganisms*,
DOI 10.1007/978-1-4939-6897-8_3

It's a very odd thing.
As odd as can be -.
That whatever Miss T eats.
Turns into Miss T.

Walter De La Mare, 1913.

3.1   Introduction

The quest for food and energy is universal. We consider in this chapter how creatures can be grouped depending on the nutrients they use and how they go about harvesting them. Not much can be said about the biology of an organism until one knows how it provides for itself. Every living thing must acquire elements in some chemical form from its environment and then process and allocate them among the competing demands of growth, maintenance, and reproduction. Inextricably linked to nutrients is energy needed in the biochemical polymerizations, depolymerizations, and maintenance of molecular complexity and orderliness.Though the quest is universal, the way it is conducted is not. Mode of nutrition/ energy scavenging is one of the key criteria for the first-order grouping of species at the kingdom level. Indeed, if you knew nothing else about an organism other than the way in which it accomplished these tasks, you could still make reasonable first approximations about some of its biological attributes. These characteristics might include the nature of a good part of its design and cell machinery, where it probably lives (or could not live), it's likely association with other organisms, etc.

3.1.1 Some General Principles and Ground Rules

Foraging, broadly construed, is essentially the 'input' or resource acquisition function of the organism-as-system model presented in ◘Fig. 1.2 of Chap. 1. **Metabolic rules** governing how energy sources and matter are acquired and processed underpin all organism functions, however, beginning with resource acquisition and ending with output of progeny. **Metabolic rates** largely determine the rates of most biological activities (Brown et al. 2004). Thus, before looking in more detail into the diverse lifestyles with respect to nutrition, we need to consider some common underlying processes and principles.

In an abiotic environment that is energetically disordered, living things maintain structural and functional order by converting energy from one form to another. The work done by cells operating as isothermal, homeostatic chemical engines includes such activities as biosynthesis, catabolism, osmosis and active transport, maintenance of concentration gradients, excretion, and mechanical work in the form of locomotion or contraction. In all organisms, the biochemicals and their reactions conform to chemical and physical ground rules pertaining to inanimate matter, as well as to rules governing how metabolites interact with each other in living cells. Lehninger (1970) has referred to these protocols collectively as the "molecular logic of the living state." The most important among these are the principles of chemical thermodynamics and the physical and chemical laws pertaining to energy balance and mass. Stoichiometry is the branch of chemistry that deals with such laws. It is the rules of stoichiometry that specify allowable states (Sterner and Elser 2002; for a summary of the key axioms and theorems see their Chapter 1).

■ Table 3.1 The 12 universal precursor metabolites[a] (Neidhardt et al. 1990)	
Glucose 6-phosphate	Phosphoenolpyruvate
Fructose 6-phosphate	Pyruvate
Ribose 5-phosphate	Acetyl CoA
Erythrose 4-phosphate	α-Ketoglutarate
Triose phosphate	Succinyl coA
3-Phosphoglycerate	Oxaloacetate

[a]These metabolites are the basis for the synthesis of all known building blocks (e.g., amino acids, fatty acids, purines and pyrimidines), coenzymes, and prosthetic groups in all organisms. The precursor metabolites are synthesized by fueling reactions, which are collectively known as central metabolism, and which also act to conserve energy and reducing power. Compare with ■ Fig. 3.1

Organisms from the smallest to the largest are unified in another way: They operate according to a very similar biochemical master plan, an attribute sometimes referred to as the "universality of cellular biochemistry" (the implications of this are discussed in ►Chap. 4) (Neidhardt et al. 1990; Stephanopoulos et al. 1998; Lengeler et al. 1999). For example, there are core cellular features such as the conventional lipoprotein cellular envelope; DNA as the fundamental genetic material together with its transcriptional and translational machinery; and universal intermediary metabolism. Though organisms may differ with respect to specific proteins and certain other macromolecules, all cell components are built from the same building blocks, such as amino acids and nucleotides. In turn, these 75 or so building blocks originate from 12 central precursor metabolites (■Table 3.1). Overall metabolic coordination and the precursor metabolites are universal. The reactions by which they are produced from whatever substrates are available to the cell are very similar in all organisms. With minor exception, the central key pathways and reactions, such as glycolysis, the pentose phosphate pathway, and the citric acid (Krebs) cycle are common to both prokaryotes and eukaryotes. This is generally true even though the enzymes participating in the comparable reactions may differ among organisms, or be regulated differently, or where the same substrate is processed differently in different organisms.

Metabolically speaking, reactions can be clustered with respect to their role in growth into four general groups common to all living things, even though the specifics may differ among species: fueling, biosynthesis, polymerization, and assembly (■Fig. 3.1 and Neidhardt et al. 1990). For these processes to function, all organisms must have raw materials to feed the fueling reactions and an energy source to make them run. In the broad category of reactions comprising biosynthesis or anabolism, ATP is consumed in all organisms; conversely, ATP is generated from ADP and P_i universally in catabolism. The biosynthetic pathways from precursor metabolites to building blocks are, with minor exceptions, the same in all organisms. Though the fueling reactions differ, for example, depending on whether the organism is a phototroph or chemotroph (discussed below), the common thread is that they serve to generate the precursor metabolites noted above, as well as to conserve energy and reducing power. Those metabolites then enter the biosynthetic pathways. A significant contribution emerging from this era of genomics and metabolomics is that metabolites and metabolic pathways of organisms can be inferred and compared from their sequenced genomes and protein databases.

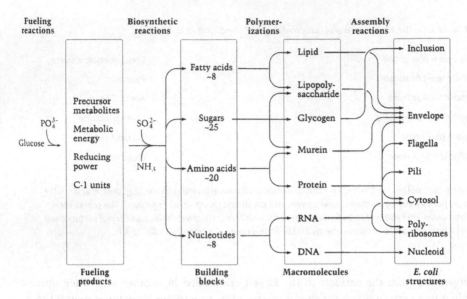

◘ Fig. 3.1 The four universal categories of biochemical reactions in cellular metabolism. *Fueling reactions* produce the universal precursor metabolites (see ◘ Table 3.1), and energy and reducing power for biosynthesis. *Biosynthetic reactions*, which begin with one or more of the precursor metabolites, produce building blocks for polymerization and related metabolites. *Polymerizations* link the building blocks into large polymers. *Assembly reactions* use the macromolecules to construct cellular features. The example shown is for *E. coli* but the scheme is applicable also to eukaryotes. From Neidhardt et al. (1990). Reproduced by permission of Sinauer Associates, Inc., Sunderland, MA ©1990

These analyses show there to be a basic, common core of interconnected, conserved enzymes across the three domains of life. This suggests, among other things, that 'enzyme recruitment' plays a major role in metabolic evolution (Caspi et al. 2006; Peregrin-Alvarez et al. 2009), i.e., new pathways emerge by recruiting enzymes and their metabolites from existing pathways.

3.2   Carbon and Energy as Resources

Regardless of their complexity, nutrients ultimately are resolvable into the essential atomic elements and from a functional standpoint they have energetic, structural, electrochemical, mechanical, or catalytic roles. Of the 94 naturally occurring elements in the Periodic Table, all organisms consist of some 30, of which C, N, P assume particular importance (Sterner and Elser 2002). Why this particular nonrandom sampling from the Table and of the elements available on Earth has been made by natural selection is itself an interesting question, explored elsewhere (Fraùsto de Silva and Williams 1991; Sterner and Elser 2002). Carbon is included as a major dimension in the categorization of life forms because, apart from water, molecules based on C are the most abundant in all organisms and it is unique in its properties, including a proclivity to form polymers. In organisms C tends to occur in reduced (hydrogenated or energy-rich) organic forms as opposed to environmental sources that are preponderantly gaseous or present as bicarbonates or carbonates (oxidized or energy-poor). The energy cycle is thus closely coupled specifically to the carbon cycle, both because of what the chemical states of carbon imply energetically and because light energy is converted to chemical energy for carbon fixation in photosynthesis.

All living organisms can be categorized into broad groupings based on their energy sources and the chemical form of carbon they obtain. Energy as ATP is probably *the* common currency of all living things, because it is required for almost every activity. Energy can be harvested in three ways: as light energy directly from the sun, or as chemical energy from inorganic compounds or organic compounds. In the great diversity of life, there are still only two basic mechanisms for generating ATP: The first, electron transport phosphorylation (oxidative phosphorylation or respiratory-chain phosphorylation), involves the flow of electrons 'downhill' from inorganic or organic donors with a relatively negative redox potential (higher energy) to those of a relatively positive (lower energy) potential, tied to the synthesis of ATP from ADP and inorganic phosphate. This may occur in cyclic and noncyclic photophosphorylation (phototrophic organisms only) and in respiratory chains (most organisms). The second mechanism, substrate-level phosphorylation, is not associated with the process of electron transport, and occurs when organic substrates containing high-energy level phosphoryl bonds are degraded. For example, metabolism of some intermediates in glycolysis (Embden-Meyerhof Pathway), such as the catabolism of phosphoenolpyruvate to pyruvate, is associated with the transfer of the phosphate group to ADP to generate ATP.

A fairly detailed and inconsistently defined terminology applies to energy and carbon dynamics. Nevertheless, it can be a useful guide provided that its limitations are recognized. Definitions appearing in ecology and biology texts are more general than those used by most microbiologists. Bacteriologists in particular use the term 'substrate' to refer to the nutrient component of a resource. Substrates used in laboratory culture are often well defined chemically and their physiological function is usually known. Accordingly, an organic or inorganic chemical is referred to as a carbon source or an energy source or a nitrogen source, and so forth. Depending on the particular microbe, specific components might be O_2 or NO_3^- (electron acceptors), NH_4^+ (nitrogen source), H_2 (inorganic electron donor), or glucose or acetate (which are both organic carbon and energy/electron donor sources).

Another source of confusion is that the terms 'autotroph' and 'heterotroph' have been used variously across the biological disciplines to refer either to the energy source, or to the carbon source, or both (that is, unspecified). This may be because a large group of organisms, the animals (not to mention most microbes), use organic carbon both as a carbon source and as an electron donor (energy source). The scheme set out below and summarized in ◧Table 3.2 follows largely that within bacteriology and is used here because it is both specific and universally applicable.

With respect to energy, biologically usable energy can be in the form of photons or chemical bonds. Organisms deriving their energy from light, which induces a flow of electrons for photophosphorylation, are **phototrophs**; those deriving electrons from a chemical energy source are **chemotrophs**. In either instance, the electron donor may be an inorganic or an organic substance. **Lithotrophs** use electrons from inorganic sources such as H_2O, H_2S, H_2, Fe^{+2}, S^0, or NH_4^+. **Organotrophs** use organic substrates such as microbial, plant, and animal biomass that may be living or dead.

With respect to carbon source, some organisms can fix CO_2 as their sole carbon source, in which case they are **autotrophs**. Organisms using mainly organic compounds rather than CO_2 to supply cell carbon are **heterotrophs**. (As noted above, authors differ in their use of these terms; occasionally, for example, heterotroph and organotroph are used synonymously.) The options resulting from the energy/carbon source permutations form the basic categories

▪ Table 3.2 Classification of organisms, with representative examples, based on energy source, electron donor, and carbon source

| Energy source | Substrates by electron donor | | Examples by carbon source | |
	Inorganic	Organic	Carbon dioxide	Organic compounds
Light	Photolithotrophs use H_2O, H_2S, S, H_2	Photoorganotrophs use succinate, acetate	Photolithoautotrophs: plants, most algae, cyanobacteria, some purple and green bacteria	Photoorganoheterotrophs: some bacteria (Rhodospirillaceae)
Chemicals	Chemolithotrophs use H_2, H_2S, NH_4^+, Fe^{2+}, NO_2^-	Chemoorganotrophs use many organic substrates	Chemolithoautotrophs: hydrogen bacteria, colorless sulfur bacteria, nitrifying bacteria, iron bacteria, methanogenic bacteria, methylotrophs	Chemoorganoheterotrophs: animals, most bacteria, fungi, many protists

Based mainly on Gottschalk (1986), Carlile (1980), Schlegel and Bowien (1989)

The demarcations are not absolute and terms are defined differently by various authors. For example, the criterion that all organisms that can utilize CO_2 as a sole C source are autotrophs begs the question as to the status of those that grow better upon the addition of an organic substrate for the functional categorization of creatures. Technically, and leaving aside for now various complications, any organism can thus be described by a 3-part prefix designating energy source/electron donor/carbon source (◻Table 3.2). Plants are photolithoautotrophs and bears are chemoorganoheterotrophs. This terminology is cumbersome, so it is frequently abbreviated. Because most photolithotrophs also use inorganic carbon, they are usually simply called photoautotrophs. Likewise, most chemoorganotrophs use organic substrates both as a source of carbon and electrons and so are simply called chemoheterotrophs. As described later, certain nonphotosynthetic bacteria use light as a proton pump and in photophosphorylation, so phototrophy is not restricted to the process of photosynthesis alone; also, some organisms known as **mixotrophs**, discussed later, have combined the advantages of photoautotrophy and chemoheterotrophy.

Two observations on the information in ◻Table 3.2 are particularly noteworthy. First, the apparent wide spectrum in patterns of energy and carbon source is really a result of metabolic diversity, and ultimately niche diversity, of the prokaryotes. Second, the resource patterns cross size boundaries. What biological properties belong exclusively to all members of a class, and what are the ecological implications? Is there a fundamental reason why the ability to form complex, multicellular macroorganisms is restricted to the photoautotrophs and the chemoheterotrophs? Why are there apparently no eukaryotic chemolithotrophs, either small or large?

A reflection of nutrient relationships is the well-known trophic structure of ecosystems (Bascompte 2009) based on food chains where a particular species, or more commonly group of species, occupies a **trophic level**. Organisms obtaining their food by the equivalent number of steps are by definition on the same trophic level. As historically depicted, the base or start of **grazer chains** (level 1) is formed by primary producers of organic matter, until recently assumed to be exclusively phototrophs that use sunlight as an energy source and consume inorganic nutrients and fix C autotrophically. Producers include plants, algae, photosynthetic bacteria and, remarkably, in some habitats (see below), chemosynthetic bacteria. All successive trophic tiers ordered basically by 'who eats whom' consist of consumer species, for example, herbivores (level 2) → first carnivores (level 3) …. → top carnivores (level n), i.e., they are energetic chemotrophs that use an organic source of carbon. The parallel in marine ecosystems is classically taken to be phytoplankton (collective term for the photoautotrophic microorganisms) and heterotrophic zooplankton grazers and their larger predators. In contrast to grazer chains, **detritus chains** are based on dead organic matter, which supplies saprophytic microorganisms and detritivores that in turn are grazed by predators. (As discussed in the following sections, microorganisms and the lifestyle known as mixotrophy pose several complications to the conventional depiction.) The distinction between grazer and detritus chains is not as clear as it may seem and there is a close interaction between them. Trophic structure is really a functional, not a taxonomic, classification because many species (bears and raccoons are good examples) operate on more than one level. Chains interconnect into webs. Extensive destruction, as in massive deforestation (Laurance et al. 2006), or even the loss of selected species (Anderson et al. 2011; Estes et al. 2011), especially if they are critical or so-called keystone species (Paine 1969, 1995), can disrupt community dynamics across multiple trophic levels.

Food chains are short, generally 3–4 links measured vertically from the base to the top, though the webs of which they are a part vary substantially in their complexity (Post 2002; Brown et al. 2004; Pascual and Dunne 2006; Bascompte 2009). Of various hypotheses to explain the limitation on links, the most generally accepted ultimate constraint is attributed to energy loss within and between stages, coupled with energy requirements of animals at the top (an observation made famous by Hutchinson in 1959). Only about 1% of the light energy intercepted by phototrophs is converted to usable energy and of that only about 10% on average is passed through each step in the sequence. This is another area of complexity and debate, since the 10% level of efficiency is generally an educated guess. Transformation from level 2 to 3 has not been measured because what constitutes level 2 is often unclear and level 3 and above cannot be unambiguously defined in any real ecosystem. As one progresses up through a trophic sequence, the amount of usable energy entrapped at each level (kilocalories/square meter/year), the number of individuals (counts/square meter), and the amount of biomass (grams/square meter) accordingly usually decrease (energy does so invariably). Trophic structure can be used to define particular communities or ecosystems, such as a lake or a forest, and can be depicted graphically as a pyramid based on energy or numbers (or biomass) of organisms at each level. The pyramidal shape for biomass or numbers is a measure of the standing crop of organisms at a point in time and reflects the fact that each species at level $n + 1$ sees only that portion of the resource at level n available to it. The shape for energy is more meaningful functionally than biomass or numbers as it is a measure of *rate* or flow of food production through the whole chain, not just the amount fixed at the level below (Odum and Barrett 2005).

The mechanistic basis of the energetic hypothesis for the pyramids is interesting, especially in an evolutionary context. The glycolytic reaction used to generate ATP is one of the oldest biochemical processes and is energetically very inefficient (2 molecules of ATP are produced per molecule of glucose catabolized). This energetic constraint means that the earliest heterotrophs arguably lived only as consumers of the chemo- and photoautotrophs, rather than of other heterotrophs. In other words, for some geological time the trophic chain would have consisted of a single link, with a large base of photoautotrophs or chemolithoautotrophs supporting a relatively small number of heterotrophs. It was the evolution of oxidative respiration, energetically an almost 20-fold improvement over glycolysis (36 molecules of ATP per molecule of glucose), that made successive tiers feasible. A possible evolutionary sequence that emphasizes the early role of microorganisms and relates changing food relationships to selection for increase in size is developed in ▶Chap. 4.

3.2.1 Microorganisms and Trophic Structure

What role do microorganisms specifically play in energy flow and biogeochemical cycling? They have not been conspicuous in typical food web diagrams and were largely overlooked until recently. In part this is because they are invisible and appropriate methods for assessing standing stocks and turnover in nature have been difficult to devise (DeLong and Karl 2005; Zinger et al. 2012). Actually, despite the focus on macroorganism-based grazer chains, most—and in some cases almost all—of the net primary production (NPP) in most natural ecosystems is utilized in the detritus webs. Microbes are important in both grazer and detritus chains (Steffan et al. 2015), but especially in the latter, where they play the seminal role as decomposers of plant and animal carcasses, then are in turn eaten. They grow as parasites or commensals on the surface of or within plants and animals (Andrews and Harris 2000; Dethlefsen et al. 2006; Dobson et al. 2006; Lafferty et al. 2006) in both terrestrial and aquatic ecosystems. Many people associate aquatic primary production with macrophytes (submerged or emergent plants), yet most of it is generated by phytoplankton. For instance,

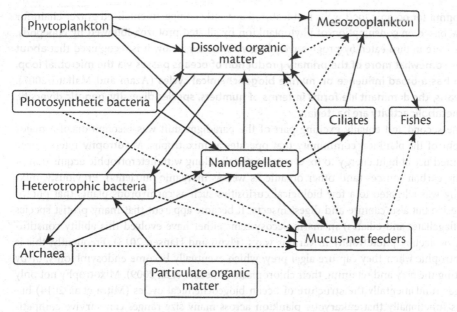

◘ Fig. 3.2 Oceanic food web emphasizing the so-called "microbial loop". All organisms shown except the fishes and mesozooplankton (*top right side* and including mucus-net feeders, *bottom*) are members of the loop. The *continuous lines* show major fluxes of energy and carbon; *dotted lines* show those of lesser magnitude. Note the central role of dissolved organic matter (*box*, top center of figure), which is utilized by *Bacteria* and *Archaea*, with the flux then moving sequentially through nanoflagellates, ciliates, mucus-net feeders or mesoplankton, and ultimately fish. A significant and in some cases the dominant portion of primary production is now interpreted to occur by mixotrophic protists (nanoflagellates and ciliates; see text and Mitra et al. 2014). From Pomeroy et al. (2007). Reproduced from *Oceanography* by permission of The Oceanographic Society © 2007

a genus of photosynthetic cyanobacterium, *Prochlorococcus*, is the most abundant (global estimates ~10^{27} cells) photosynthetic organism not only in the oceans but on Earth and may account for up to 50% of NPP (Biller et al. 2015; for planktonic eukaryotes, see de Vargas et al. 2015). It is also the smallest free-living photosynthetic cell and contains the smallest genome. In his famous book on plant biology, Corner (1964, p. 3) has described the vast communities of oceanic phytoplankton imaginatively as "... *the sea meadows of which the oceanographer now writes consist not of blades of grass but of separate cells, corresponding with those that build the blade of grass. They are not disintegrated plants, but plants too primitive for integration*".

Beginning in the 1970s, a paradigm shift occurred with respect to energy flow and nutrient cycles in the ocean. Initially, this was the recognition of the importance of a **'heterotrophic microbial loop'** (◘Fig. 3.2) based on bacterial consumption of dissolved organic matter entering the water column from primary producers (Pomeroy 1974; Azam and Malfatti 2007; DeLong and Karl 2005; Karl 2007). A large portion of the organic matter synthesized by phytoplankton, photosynthetic bacteria, and seaweeds is exuded; heterotrophic microorganisms convert the dissolved organics to biomass that re-enters the conventional nanoflagellate → ciliate → mesoplankton → herbivore/carnivore levels (shorter routes are possible; Pomeroy et al. 2007). Bacteria also play a major role themselves in releasing organics, much of which is recalcitrant to further breakdown; these products are sequestered as particulate fixed C and sink into deeper waters. These discoveries are remarkable because the dogma for oceanic food webs had featured a strictly compartmentalized grazer chain for C flow based on consumption of phytoplankton by ciliated protozoa followed by copepods, which were in turn eaten by progressively larger consumers. Now it is recognized that **about half or somewhat more of the primary production of oceans passes via the microbial loop, which has a broad influence on marine biogeochemical cycles** (Azam and Malfatti 2007). **In oceans, the dominant life forms in terms of numbers, species diversity, genetic diversity, and metabolic activity are bacteria.**

The second and equally exciting part of the paradigm shift was recognition of a major segment of the plankton community that operates as mixotrophs. **Mixotrophy** refers to the combined use of light energy to fix C autotrophically along with heterotrophic acquisition of organic carbon sources (and other organic as well as inorganic nutrients). Previously, mixotrophy was relegated to a few biological curiosities such as carnivorous plants that photosynthesize but also capture and digest insects. It became apparent that many protist species (dinoflagellates and ciliates) in marine ecosystems either have evolved this ability constitutively, or acquire phototrophy in various ways (Flynn and Hansen 2013). Frequently, this is phagotrophic when they capture algal prey, which eventually become endosymbionts, or by digesting the prey and retaining their chloroplasts (Stoecker et al. 2009). Mixotrophy not only changes fundamentally the structure of ocean biogeochemical cycles (Mitra et al. 2014) but means functionally that eukaryotic plankton across many size ranges can survive competitively in nutrient-deprived (N, P, Fe; Moore et al. 2013) surface waters. This flexible nutrition strategy appears to enable mixotrophs to dominate the plankton community under certain conditions by outcompeting strict phototrophs (Hartmann et al. 2012; Ward and Follows 2016).

Another component of this ongoing trophic reinterpretation occurred in 2000, prompted by a genomic analysis of the bacterial component of the plankton community. Among the marine bacterioplankton are forms that can obtain their energy as *nonphotosynthetic* phototrophs (Béjà et al. 2000; see also Venter et al. 2004). (Strictly speaking, these microbes are not photosynthetic because they lack chlorophyll and do not engage in the classic photosynthetic reaction typical of photolithoautotrophs.) They capture solar energy by utilizing a membrane protein called proteorhodopsin, a form of the visual pigment rhodopsin that functions as a light-driven proton pump and coupled photophosphorylation. Interestingly, these bacteria also grow as chemotrophs as has been demonstrated by growing them in culture in darkness, where they use organic carbon as an energy source. One genus of bacterium that lives this way is *Pelagibacter*, but ongoing analysis of pelagic (water column in the open ocean) waters by environmental genomics and direct gene probes indicates that as many as half of the upper ocean's bacteria may harbor this photoprotein. Laboratory experiments under light/dark and high/low nutrient conditions with marine *Vibrio* suggest that they have a fitness advantage over competitors in the frequently occurring low nutrient (oligotrophic; see terminology below) oceanic environment by supplementing conventional heterotrophy with light energy (DeLong and Béjà 2010; Gómez-Consarnau et al. 2010). To date none have been found that fix carbon, so technically they are not autotrophs but rather photoorganoheterotrophs.

A final climax in this remarkable series of discoveries is that work beginning in the 1970s documented novel deep-sea ecosystems (reviewed by Van Dover 2000; Kelley et al. 2002; Cavanaugh et al. 2013). These diverse marine invertebrate communities, both in relatively cold water and surrounding deep-sea thermal vents, are based on the productivity of lithotrophic bacteria, which oxidize H_2S or other inorganic substrates such as H_2 and CH_4. These substrates, acting as electron donors (as well as the electron acceptor CO_2), originate in the hydrothermal fluid. Thus, **we now know that there are not one but two major systems for primary production on Earth: While solar radiation provides energy for phototrophy, terrestrial thermal energy and pressure provide the energy source on venting water for anaerobic or aerobic chemosynthesis by the lithotrophic reduction of CO_2.** To put the chemosynthetic system in perspective, however, it should be noted that current estimates indicate (Van Dover 2000; McClain et al. 2012) that sinking particulate matter representing carbon originally fixed by photosynthesis in the upper reaches of the ocean contributes by far the greatest percentage of chemical energy for deep-inhabitants; chemosynthesis contributes only about 0.02–0.03% total ocean production and about 3% of the carbon flux to nonchemosynthetic organisms. These numbers may well be revised upward as further such communities become discovered and explored.

3.2.2 Resources and Biogeography

The sources of energy and carbon also set broad limits on the distribution of living things. Phototrophic micro- and macroorganisms illustrate this point well, being distributed with respect to light gradients (described later). The distribution of many species of lithotrophic microbes can be said to mirror environmental geochemistry: they reside in quantity only where their requisite inorganic electron donors reside. For example, *Sulfolobus acidocaldarius* lives primarily in sulfur-rich geothermal domains such as hot springs (Brock et al. 1972). The organism is a thermophilic (temperature range 60–85 °C), aerobic, obligate acidophile (pH range 1–5). A member of the *Archaea*, it can thrive under hot, acid conditions in part because of an unusual cellular membrane, which is an exception to the ubiquitous phospholipid bilayer type. In these high-temperature habitats, H_2S oxidizes spontaneously to elemental sulfur (S), which is in turn oxidized to H_2SO_4 by *Sulfolobus*, further acidifying the environment. The bacteria adhere to crystalline S thereby acquiring lithotrophically the few atoms going into solution that they need as an electron source (◘Fig. 3.3; Brock et al. 1972). This, incidentally, is an excellent example of the importance of scale in ecosystems and of the unique ability of microbes, especially bacteria, to finely partition a resource: One microscopic S crystal, measuring on the order of cubic micrometers, can provide an energy source for many thousands of bacterial cells, whereas an entire plant or animal may be only one of many in the diet of a herbivore or carnivore. *Sulfolobus* also illustrates how specialization tends to impose limitations (e.g., a restricted range of resources) as well as opportunities (efficient harvesting mechanisms)!

The distribution of certain heterotrophs can also be strictly limited by that of their energy and carbon sources. Several examples are developed later (see section, Generalists and Specialists). For instance, the giant panda eats only bamboo and hence cannot exist in otherwise favorable regions of the world lacking its specific food source. Obligate parasites are restricted by the spatial and temporal availability of their hosts. Heteroecious parasites are those that need two or more hosts to complete their complex life cycles (▶Chap. 6). The fungus causing black stem rust of wheat (*Puccinia graminis tritici*) requires both wheat (in fact in agroecosystems it requires specific, susceptible cultivars of wheat) and common barberry (*Berberis vulgaris*) or other species of wild native barberry or mahonia. In the case of the *Schistosoma*

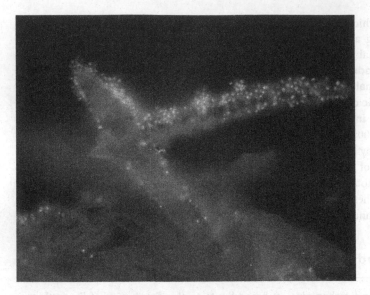

◘ Fig. 3.3   A member of the *Archaea*, the sulfur-oxidizing *Sulfolobus acidocaldarius* attached to crystals of elemental sulfur. The archaeal cells appear as bright spots because they have been stained with the fluorescent dye acridine orange. Photo courtesy of T.D. Brock, University of Wisconsin-Madison

flatworms, there is an obligatory alternation of a sexual generation in humans (occasionally other mammals) and an asexual generation in particular snails.

Terrestrial and aquatic plants, protists, animals, and algae (living or dead), provide locally high reservoirs of organic carbon as well as other nutrients, and accordingly act like nutritive islands for colonization by heterotrophic bacteria and fungi. Similarly, the pelagic zone, while impoverished as a whole, contains microenvironments enriched in nutrients. High microbial activity is associated, for example, with the relatively large fecal pellets of zooplankton and with 'marine snow', which consists of heterogeneous flocculent aggregates containing phytoplankton, detritus, bacteria, and fecal pellets embedded in mucus. Bacterial species known as **oligotrophs** predominate in nutrient-depleted waters, whereas **copiotrophs** tend to dominate in marine snow[1] (Azam and Malfatti 2007; Lauro et al. 2009). Since many of these microorganisms, especially the oligotrophs, cannot be cultured, their lifestyle properties as well as predicted niches and trophic relationships are being inferred from their genomes (de Vargas et al. 2015; Lima-Mendez et al. 2015).

Unlike mobile animals and chemotrophic microbes, obligate phototrophs cannot choose their energy diets, but are of course affected by the intensity, temporal distribution, and spectral composition of light. Although plants vary in growth form, their energy gathering mech-

---

1   Oligotrophy refers to a condition characterized by low concentrations of inorganic and organic nutrients and oligotrophs to the organisms (usually used with respect to bacteria or microscopic plankton) adapted for growth in such environments (such as soil or the open oceans or deep seabed). In contrast, copiotrophy (copiotrophs) refers to conditions (organisms) characterized by relatively high concentrations of nutrients or the ability to grow competitively in such environments. Nutrients can constrain both growth *rate* (new biomass per unit time) and ultimate growth *yield* (total biomass, often expressed relative to some input function such as resource or reactant). The rate/yield issue returns later in this chapter under optimal digestion theory and in subsequent chapters in this book. For background and terminology, see Poindexter (1981b), Moore et al. (2013), Hoehler and Jørgensen (2013), Kirchman (2016).

anisms for the common commodity, light, are essentially the same. Thus, they are unified by their need for a common energy source. This is in stark contrast to animals as a whole, which display extreme variation in diet and in anatomical features for harvesting their respective food sources. Such morphological differences among animals are analogous to the numerous architectures of plants. Once gathered and converted to precursor metabolites, however, the nutrient elements are handled biochemically and energy extracted (e.g., via glycolysis and the Krebs cycle) in much the same way by both groups of organisms, as alluded to in the Introduction.

Plants, algae, and phototrophic microbes are limited to areas where light occurs and hence do not grow in such habitats as caves, deep subterranean layers, or the intestines of animals. Remarkably, however, certain cyanobacteria and various green algae do grow inside porous rocks (endolithic phototrophs), where infiltrating water provides moisture as well as channels that facilitate access of light. In aquatic environments, the growth of attached macrophytes and benthic (bottom-dwelling) phototrophs, as well as phytoplankton, is restricted to the euphotic zone, which extends from the surface to a depth where photosynthetically active radiation is 1% of that at the surface. In general, this is sufficient for net photosynthesis to occur (i.e., photosynthesis above the compensation point, where photosynthesis equals respiration). The depth of this zone varies with water clarity, but for coastal waters is typically about the uppermost 80 m; in oligotrophic waters, it may extend to about 200 m (Biller et al. 2015). In other words, within this zone energy is generally not the limitation to growth rate or yield, but rather nutrients frequently are limiting (Moore et al. 2013; Saito et al. 2014). The energy for phototrophs also arrives essentially in a continuous stream (albeit in quantum units), and does not entail hunting or trapping discrete packages separated in space and time, as the case for animals. Similarly, their energy supply is not subject to the laws governing population biology of prey in the sense that the energy source of chemoheterotrophs is influenced by the population dynamics of the plants and animals on which they live.

Within the euphotic zone, the distribution pattern of phototrophic organisms is related to light intensity and spectral quality that change with depth. Water absorbs proportionately more of the longer (red) than shorter (blue) wavelengths. Algae and cyanobacteria growing aerobically in the upper reaches of the water column further remove the red and some of the blue portions because these are the energy quanta preferentially absorbed by their chlorophylls and accessory pigments. Anaerobic, phototrophic purple and green sulfur bacteria commonly develop in a zone or on the muddy bottom (◻Fig. 3.4; Pfennig 1989) if the water is sufficiently clear to allow light penetration to depths that are anoxic, and in the presence of H_2S, which is used as an electron donor. Thus, these organisms thrive under relatively specific, ecologically restricted conditions in stagnant water or on mud surfaces where they may form spectacular mats ranging in color from pink to purple-red or green. The phototrophic sulfur bacteria absorb primarily in the far red part of the spectrum and so can exploit wavelengths for anoxygenic photosynthesis not utilized by the algae. Their striking colors attributable to bacteriochlorophylls and carotenoid pigments are visible evidence of their resource-harvesting equipment. Adaptive divergence in pigment composition among phototrophic phytoplankters allows for efficient partitioning of light energy and can favor coexistence (Stomp et al. 2004); alteration in pigment composition to maximize efficiency under prevailing conditions is also a well-known phenomenon ('complementary chromatic adaptation'; Stowe et al. 2011). Similarly, the role of preferential light absorption on the differential distribution of the cyanobacteria *Prochlorococcus* and *Synechococcus* in the photic zone is reviewed by Ting et al. (2002). The chemical equivalent of photic zonation is vertical zonation

3

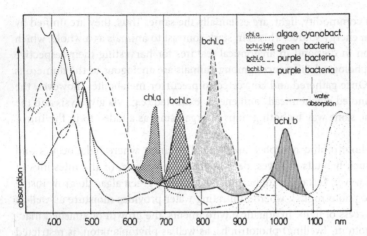

◘ Fig. 3.4 Absorption spectra of representative phototrophic microorganisms as measured on intact cells. Note that light absorption by the algae and cyanobacteria as a whole is generally mid-spectrum (stippled area as shown for chl. a represented by the green alga *Chlorella*), while that of the green and purple sulfur bacteria is at longer wavelengths (bchl. *c* for *Chlorobium*; bchl. *a* for *Chromatium*, and bchl. *b* for *Thiocapsa*). Thus, light absorbed by the algae and cyanobacteria in shallow water or in floating mats does not appreciably reduce the wavelengths available for the sulfur bacteria. From Pfennig (1989). Reproduced by permission of Springer-Verlag, Heidelberg, ©1989

of prokaryote communities in marine sediments due to the successive depletion of electron acceptors for metabolism (Fenchel 2002). Photic stratification attributable to the effect of plant morphology on competition, as well as the ability to maintain positive energy income, is also evident in the depth zonation of aquatic plants in lakes and slow-flowing streams (Spence 1982).

In practical terms, knowing the energy and carbon sources of a particular organism is of great value to bacteriologists and mycologists in their attempts to isolate it in pure culture from nature. Isolation generally proceeds by enrichment, a process that provides favorable conditions for growth of the desired organism or, conversely, counter-selects against extraneous organisms. For example, use of a simple mineral salts medium with bicarbonate as a carbon source, together with appropriate light conditions, provide the basis for selecting photosynthetic bacteria. If incubation conditions are aerobic, cyanobacteria will be selected; if anaerobic (and in the presence of certain other growth factors and H_2S as an electron donor), purple and green sulfur bacteria will be selected. Likewise, cellulose is commonly used as a carbon and energy source in isolating cellulolytic fungi.

3.3   Resource Acquisition

3.3.1   Optimal Foraging Theory

How and to what extent have organisms been shaped by natural selection with respect to the efficient acquisition of resources? And are such modifications analogous between macroorganisms and microorganisms? The evolution of search strategies has fascinated ecologists for decades (Hein et al. 2016). In papers published concurrently in the *American Naturalist* that explored costs and benefits in the economical search for food, Emlen (1966) and, separately, MacArthur and Pianka (1966) launched what became known as 'optimal foraging theory' (OFT). Emlen focused on a mathematical explanation of food preference or selectivity of predators with respect to calories gained and expended, and time invested in the search, capture, and consumption of an item. In their mathematical model addressing the optimal diet of a predator, MacArthur and Pianka addressed where (kinds of habitat patches) and on what items (extent of diet breadth; i.e., specialization) a species would feed if it did so following conventional economic principles of marginal return on investment, thereby optimizing its time or energy budget. MacArthur later expanded his remarks (1972) as has Pianka (2000). The concept has since been applied in numerous contexts (e.g., Pyke 1984; Stephens and Krebs 1986; Stephens et al. 2007). The underlying premise is that natural selection acts to shape energetic costs against gains by foragers: adaptations in attributes such as behavior, physiology, or form that result in a net increase in terms of energy gained should be favored and increase in a population; those that do not should decrease.

Before we turn to the details, brief comment is needed with respect to terminology because the literature on this subject is vast and authors have approached 'foraging' either broadly or narrowly. Many papers, especially since the 1990s, consider foraging strictly with respect to the search strategy to find an item only. Much of the foraging literature, particularly prior to the 1990s as illustrated above, considers both the search and the subsequent exploitation of the resource (pursuit, capture, processing). Here we discuss both aspects and begin by focusing on some attributes of the search process. The broader aspects of OFT are discussed subsequently.

The search for resources In *The Physics of Foraging*, Viswanathan and colleagues (2011) argue that, in their words, *"biological foraging is a special case of random searches."* In fact, at the extremes, the search process can be considered to be either random or systematic. **Random searches** occur only where there is uncertainty (the searcher is uninformed) as to the location of the food source. Here, the search rules involve stochastic processes as depicted in standard random-walk theory (Berg 1993, 2004; Bartumeus et al. 2005) centered on a diffusion process characterized by probabilistic, discrete-step statistical models where displacement events (lengths of movements or steps) are separated by reorientation events (turns). In the Brownian walk model (named after Brownian motion), step lengths are drawn from an exponential distribution; in the Lévy walk or Lévy flight model they follow a power-law distribution (de Jager et al. 2011). Simulations show that the latter type results in faster dispersion (hence Lévy motion has been called 'super-diffusive'), less intraspecific competition, and more exploration of new sites than does the Brownian. Lévy walks appear to be common in nature and are particularly effective when the resources are scarce and patchily distributed, such as in the open ocean (Humphries et al. 2012; Humphries and Sims 2014). Brownian walks are efficient when prey is abundant and/or relatively homogeneously distributed. Examples of both walks are discussed below.

At the other end of the search spectrum are **systematic or deterministic searches**. These are effective where the food source distribution is predictable and the location known by the forager. This could occur when sources emit cues, such as a scent trail or nutrient signal, or the forager remembers landscape features and past successes. Such searches include Archimedean spirals or other types of area-restricted patterns (see examples below and ▶Chap. 5 regarding phalanx growth patterns of sessile, modular organisms). Interestingly, simulations show that a Lévy walk can become systematized (spiral) by imposing various search restrictions on the walk, such as avoidance of self or self-trails (Sims et al. 2014 and ▶Chap. 5).

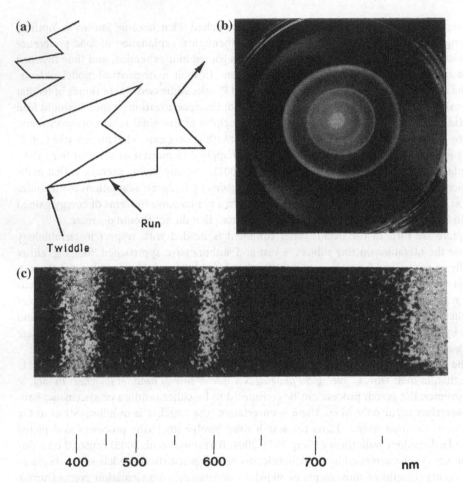

◻ Fig. 3.5   a Idealized random walk of the motile bacterium *Escherichia coli* in a uniform chemical gradient. Movement consists of brief 'runs' in a particular direction, followed by 'twiddles' during which reorientation occurs. This is only one of several forms of bacterial movement patterns. **b** If the gradient is nonuniform, directed movement can take place either towards (attractant) or away (repellent) from the source. In this example, bacteria inoculated as a drop in the center of a petri dish have moved outwards as three successive concentric rings, each following a specific nutrient that has become locally depleted. Photo courtesy of Julius Adler, University of Wisconsin-Madison. **c** Aggregation of individual cells of the phototrophic bacterium *Thiospirillum jejense* at wavelengths of light at which its pigments absorb. Technically, this particular light-mediated response is an aversion to darkness called scotophobotaxis. Photo courtesy of Norbert Pfennig, University of Konstanz

With respect to random searches, Lévy walks are shared by organisms as disparate as motile bacteria (Reynolds 2015) and albatrosses in their hunt for prey over the ocean (Humphries et al. 2012). Here we focus briefly on bacteria as an example. Motile single cells (typified by most bacteria) can move short distances called 'runs' in bacteriology, followed by directional change during 'tumbles'; the phenomenon usually is illustrated by the model bacterium *E. coli* (◻Fig. 3.5a). The biophysics of such patterns has been described in detail by Berg and colleagues in several papers and two books (e.g., Berg and Brown 1972; Berg 1993, 2004).

The actual pattern of swimming varies considerably, however, depending on many factors, including the degree of flagellation and motility adaptations of the species, swimming speed, and characteristics of the habitat (Mitchell 2002; Young 2006; Son et al. 2015). Stocker and Seymour (2012) discuss how some marine bacteria move (in contrast to the enteric bacterium *E. coli*) and the importance and distinctive attributes of **chemotaxis** (movement towards or away from a chemical) in the ocean (◘Fig. 3.5b; see also Mitchell 2002; Mitchell and Kogure 2006; Guasto et al. 2012; for movement in sediments see Fenchel 2008).

Bacterial motility is energetically expensive in terms of the construction and function of the flagellar system (Macnab 1996) suggesting that it has survival value, probably for several reasons (for one example related to resource capture, see Wei et al. 2011). Chemotactic (Armitage 1999; Eisenbach 2004) as well as two different light-mediated responses exist among bacteria (phototaxis and scotophobotaxis; see ◘Fig. 3.5c and Madigan et al. 2015). In the presence of a chemical, such as a nutrient, the runs become longer and the tumbles less common, so the organism moves up the chemical gradient; similar movements occur away from a source if it is a toxicant. Bacteria can alter their chemotactic search strategies as a function of the energetic status and other properties of the cell and the spatial resource gradient (Mitchell 2002; Mitchell and Kogure 2006).

With respect to systematic searches, and continuing the bacterial theme momentarily, certain bacteria (e.g., some plant- or animal-associated pathogens and the root-nodulating mutualists) also have evolved the ability to respond to signals produced by their specific hosts, which in turn react to their presence, so there is a two-way communication (Brencic and Winans 2005). In at least one extraordinary case, a plant pathogenic bacterium, *Agrobacterium tumefaciens*, upon gaining entry genetically engineers its specific host to produce new compounds metabolized only by the pathogen.[2] There is a recently described counterpart in medical microbiology where commensal gut bacteria induce formation of fucosylated carbohydrate moieties on intestinal epithelial cells, the fucose then being utilized by the resident beneficial bacteria that confer protection from pathogens (Goto et al. 2014).

Most if not all filamentous fungi can alter their mycelial networks in the search for, or subsequent exploitation of, food resources. Some, such as certain basidiomycetes, can play both a 'sit and wait' strategy, analogous to animal predators, by exposing an expansive mycelial network over or into substrate and capturing resources that appear. Alternatively, they can concentrate hyphae into linear explorative and transport organs to move concertedly into new terrain (Boddy et al. 2009; Darrah and Fricker 2014). The energetic trade-offs among

---

2   *A. tumefaciens* causes an important bacterial disease known as crown gall affecting the lower stems and roots of about 150 genera of herbaceous and woody plants worldwide. Compounds released from wounded plants (including opines, below) attract the pathogen cells residing in soil, which move chemotactically, attach to the damaged plant tissue, and genetically transform the host by injecting portions (~40 kb) of its large a tumor-inducing (Ti) plasmid. These plasmid-borne genes are replicated, delivered into the plant cell, integrated into the host genome in the nucleus, and expressed by the host machinery only under specific environmental conditions arising in wounded tissue. Among the T-DNA genes are those coding for the plant hormones auxin and cytokinin, resulting in the characteristic galls. But additionally are those coding for synthesis of unique metabolites, opines, as well as the catabolic genes allowing the pathogen to utilize them (some avirulent bacteria also carry these catabolic genes and can thereby benefit from association as 'freeloaders' with *A. tumefaciens*). There are chemical families of opines in multiple molecular configurations but basically all are sugar and/or amino acid derivatives. This is an extremely brief synopsis of a fascinating and very complex story. For pathology, see Agrios (2005); for ecology and molecular biology, see Brencic and Winans (2005), Platt et al. (2014).

reproduction, growth, and foraging have been modeled by Heaton et al. (2016). The mycorrhizal fungi develop elaborate morphological and physiological relationships within the roots of their plant hosts, to the benefit of both partners (Smith and Read 2008). Biotrophic fungi such as the rusts, powdery mildews, and downy mildews form similar intricate relationships as highly specialized plant parasites (see section Generalists and Specialists, below).

Perhaps even more remarkable than the fungi is the foraging behavior of their distant relatives, the slime molds (Mycetozoa of the phylum Amoebozoa; Fiore-Donno et al. 2010; Yip et al. 2014). Some of these, the myxogastrids or plasmodial slime molds, feed as a single, massive, multinucleate amoeba that tends to be web-like in shape and may be several centimeters in size. *Physarum polycephalum* forms such feeding networks, adapting its shape to the terrain, moving around obstacles, and responding locally to food patches (Takamatsu et al. 2009; Fricker et al. 2009). It advances generally with an expansive margin that maximizes the area explored but behind the advancing perimeter its conformation is dominated by an array of tubes, through which the cytoplasm streams. Tero et al. (2010) compare the characteristics of this extraordinary biological network to abiotic networks as modeled by the Tokyo rail system. They set up 36 food sources in an array that represented cities in the Tokyo area and compared connections established by foraging *Physarum* with those of the railway. **Remarkably, the exploration and transportation solutions found by the slime mold were mathematically very similar to the rail network in topology, transport efficiency, and tolerance to faults caused by disconnection.**

Finally, the strategy of foraging animals after locating a profitable habitat patch often involves modifying their tactics to capitalize on abundant prey. (For discussion of what constitutes a 'profitable' patch and how long it may remain so from the consumer's perspective, see comments on the marginal value theorem in Maynard Smith 1978a; Wajnberg et al. 2000.) Thrushes explore for worms on a lawn by running a short distance, pausing to look about, turning, and making another run. Once food is located, the birds turn more sharply and frequently (Smith 1974; see also Sulikowski and Burke 2011). Gophers search similarly (Benedix 1993). The result of this strategy is concentrated exploration over the zone of high food density, so-called **area-restricted searching**. Such patterns are not limited to animals but occur among some clonal plants and are a common response among filamentous organisms and organs, from the fungi noted above to the dynamics of root architectures (▶Chap. 5).

OFT as an economics model Classic OFT, broadly construed, is one example of optimization concepts and models developed originally in engineering and economics (Maynard Smith 1978a). Despite terminology that can be circular or misconstrued when used simplistically, optimization does not mean that selection produces optimal solutions, but rather optimization concepts help to identify specific adaptations and the operating constraints (Parker and Maynard Smith 1990; see also Pyke 1983; Stephens and Krebs 1986). Ultimately, the ability to understand the criteria and the extent to which species forage efficiently could help predict their distribution. Nevertheless, it is useful here to recall several caveats noted in ▶Chap. 1: (i) 'what is an adaptive trait' (Dobzhansky 1956) in the context of attempts to 'atomize' an organism; (ii) that natural selection does not start with a blank slate; and (iii) natural selection can only act on what is available in the gene pool.

All optimization models, of which OFT is an example, share implicitly or explicitly several features (Parker and Maynard Smith 1990). In brief these are (i) operational limits or constraints such as the range of phenotypes on which selection can act; (ii) specification of the optimization criterion, in other words what attribute is being maximized; and (iii) the assumption that the population is responding to the *contemporary* environment (i.e., not a historical legacy) and that its behavior will be reflected in its contribution to the next generation. In OFT, a typical situation would be briefly as follows (see Box 1 in Parker and Maynard Smith, 1990): (i) benefits and costs, expressed in a common currency such as calories, of adopting a particular action (strategy) are identified; (ii) an indirect optimization criterion typically is selected, usually implicitly being to maximize the net rate of energy intake per unit time spent foraging. In the case of foraging distance to catch prey, this would be as a function of movement distance, with movement continuing until the marginal value of taking the next step becomes equal to or less than the marginal costs of that step; the optimal move being the one that maximizes caloric benefits minus costs; (iii) the indirect criterion of energy gain is translated implicitly or explicitly into direct fitness (number of progeny) associated with the various strategies; (iv) predictions are tested against observations. Pyke (1984) summarized the assumptions inherent specifically in OFT, the most important being that: (i) an individual's fitness depends on its behavior while foraging; (ii) foraging behavior is heritable; (iii) the relationship between foraging behavior and fitness is known; and (iv) the animals forage 'optimally', i.e., their behavior changes in pace with or more rapidly than the presumed relevant conditions change. Pyke's and Maynard Smith's notation that foraging behavior must be or is assumed to be directly related to fitness is important, especially since, in routine tests of the theory with macroorganisms, fitness (numbers of descendants) is generally impossible to measure. In the case of microorganisms, attributes such as growth rate, competitive success, total population size, i.e., direct measures of fitness are routinely determined, as discussed later.

While much of OFT has dealt with the foraging behavior of birds, there are other interesting situations where foraging efficiency and energy economy have been studied (see Table 9.1 in Stephens and Krebs 1986 for synopsis and tests). Some are direct tests of the theory but many are not. For example, consider the giant panda bear and its food, almost exclusively bamboo (a good example of specialization; see later section). Actually, the animal appears *not* to forage efficiently for a couple of reasons. First, pandas are not even anatomically streamlined for holding the stalks of bamboo and stripping off the leaves, as recounted wittily by Steven Jay Gould in his 1980 book 'The Panda's Thumb.'[3] Second, its alimentary tract is poorly adapted to this relatively low-energy, nonnutritious diet. So the animal spends most of its time eating; it also adapts by living passively, along with having evolved physiological adjustments reflected in reduced sizes of typically high-metabolic organs (brain, liver, kidneys) (Nie et al. 2015). In contrast, many predators have episodes of extremely high-energy expenditure. Hunting styles vary considerably among felids (e.g., lions, cheetahs, pumas) consistent with their different sizes, body designs, prey species, and habitat. Pumas hunt in rugged terrain and exemplify a strategy of stalk/ambush/pounce. The energetics of each phase has been continuously monitored in the wild from animals with radio tracking collars (Williams et al. 2014). These cats have an efficient energy budget overall because most of their time awake is associated with low-energy expenditure (sitting and stalking), while the attack phase, although demanding high-energy expenditure, is of short duration and the animal's pouncing

---

3   In brief, pandas grasp the shoots between what appears to be a thumb (but is actually an enlarged wrist bone) and the other digits. The true thumb is used in another role, so to compensate, the animal has enlisted through evolution the use of the wrist bone as an integral part of the harvesting apparatus. Gould (p. 24) describes this as "a somewhat clumsy, but quite workable, solution" and uses the panda example among others to make the important point that such evolutionary tinkering, such *imperfections* rather than perfections, are the proof of evolution (p. 13).

force is calibrated to prey size. Of course, these two examples, while interesting and relevant to foraging, are not direct tests of the theory.

In formal tests of OFT there are various complications, notwithstanding its conceptual elegance and intuitive appeal. Apart from semantic difficulties alluded to above and noted by Perry and Pianka (1997), examples of some of the complexities include the following situations. (These do not mean that OFT is wrong. Rather, optimal foraging can mean different things to different species and accordingly such cases require modifications to the conventional model [see Chap. 9 in Stephens and Krebs 1986].) The standard criterion of maximizing energy return from the resource may not be the overriding factor in foraging, but rather something else limiting in the diet, such as salt intake or protein acquisition. An interesting case pertains to sodium, which is an essential element often in limiting supply in the diet of moose (Belovsky 1978, 1984). The primary source of the mineral is aquatic plants, which are available only in the summer as they are under ice in the winter, so the animals must fulfill seasonally their entire annual requirement. However, they supply less energy per unit dry mass than do terrestrial plants and, if consumed exclusively, the aquatics are too bulky and quickly fill the rumen. Thus, theory predicts and Belovsky's experiments support the postulate that moose forage on an 'appropriate' mix of both deciduous leaves and aquatic plants. This turns out to be about 18% aquatics. Analogously, herbivores and detritivores (particularly wood-decomposing fungi) typically are faced with a shortage of nitrogen needed for protein, not carbon. Microbial population size and thus decomposition rate of many forms of plant debris—wood especially because the C:N ratio is high—is driven by nitrogen availability.

Even where energy acquisition is the appropriate currency, organisms may be viewed (simplistically) as foraging either to maximize **efficiency of energy intake** (energy gain per unit expended) or **net rate**, i.e., per unit time (see microbial example in the following section). Honeybees in visiting flowers often do not fill their crops with nectar, which runs counter to optimal foraging based on a criterion of net rate of return. When metabolic costs of transporting nectar are accounted for, however, the better foraging prediction accords with crops being only partially filled (Parker and Maynard Smith 1990). With respect to rate, a further consideration is whether the organism is maximizing its *instantaneous* intake rate of digestible energy or maximizing over some longer period, such as its *daily* rate. Babin et al. (2011) studied the distribution of bison with respect to a plant community in Saskatchewan, Canada. They found that the animals were grazing on a species distribution of plants that allowed the animals to maximize instantaneous rate, even though foraging on smaller plants (which required longer cropping times and ultimately provided slower satiation) provided more digestible energy. This strategy is apparently not universal because shorebirds forage consistently with longer rather than shorter time energetic predictions (Quaintenne et al. 2010). Beyond these more global considerations is the complication of animals foraging differently depending on whether they are doing so alone or among competitors or when predators are in the vicinity.

In overview, Stephens and Krebs (1986) summarize the reaction of ecologists to classic OFT into five categories ranging from its being trivial and tautological at one extreme, to often having been verified and useful at the other. That spectrum of opinion was the status in the theory's heyday of the late 1980s and would apply now as well. The data from their extensive review of the literature (Stephens and Krebs Tables 9.1–9.5) support a conclusion that optimal foraging models generally have performed well. Enthusiasm for econometrics-centered OFT, like all theories in ecology, has peaked and waned over the ensuing decades, and

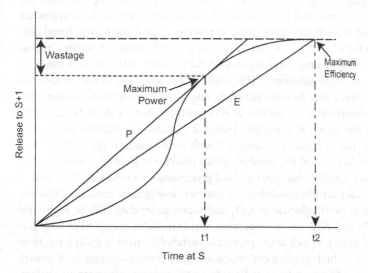

Fig. 3.6 Food processing in the animal gut showing the trade-off between maximum average rate versus maximum efficiency of digestion. Plot shows amount of nutrient released for post-digestive absorption as a function of time of digestion. The transfer function is shown by the sigmoid line. *Upper dashed horizontal line* shows maximum nutrient available. Nutrient transferred increases with time of processing in the gut with the greatest extent of transfer (efficiency) occurring at t2. The average rate of transfer for food held over time t2 is given by the line with slope E. However, the rate of transfer is maximized by holding food for a shorter period, t1, shown by the slope of line P. This occurs with a cost of 'wastage' In amount potentially transferred. From *Physiological Ecology: How Animals Process Energy, Nutrients, and Toxins* by W.H. Karasov and C. Martinez del Rio (2007). Reproduced by permission of Princeton University Press ©2007 via Copyright Clearance Center

attention shifted to subsequent refinements such as Lévy walks previously described (James et al. 2011; Pyke 2015). Like most if not all major theories in ecology, OFT has provoked intense debate and controversy (Perry and Pianka 1997). Nevertheless, both as a formal model and as a framework for empirical tests, it remains a fundamental guide to investigations (see, e.g., Plante et al. 2014).

Optimal digestion: trade-offs in rate versus efficiency To be broadly applicable to resource acquisition, foraging should include optimal digestion theory because an important consideration is the dynamics of energy and nutrient extraction after the food source is acquired. This aspect concerns issues such as gut design, retention times, and absorption efficiencies in classic predators or analogous criteria for metabolic streamlining in plants and microorganisms. For example, broadly speaking, animals can maximize either digestion *rate* or digestion *efficiency* as discussed above in regard to search strategies. In principle, the longer the time that food is processed in the gut, the more nutrient or energy is extracted, up to the theoretical limit of what is available for processing. Efficiency can be expressed as digestive efficiency (amount of usable resource extracted per unit food ingested) or in energetic terms as net energy obtained per unit of food ingested. Both generally increase over time up to a maximum. However, the rate of extraction can be visualized as rising to a maximum and then declining. A strategy of maximizing rate comes at the expense of digestion efficiency; organisms may maximize either rate or efficiency, or compromise at some intermediate level (Sibly and Calow 1986; Karasov and Martínez del Rio 2007) (**Fig. 3.6).

Microbial behavior follows similar principles. There is considerable experimental and theoretical evidence that bacteria and other unicellular microorganisms such as yeasts are selected for **either a rapid growth rate (cells per unit time) or high growth yield (total cells per unit substrate**; Pfeiffer et al. 2001; Frank 2010). In parallel fashion, microbes appear generally able to harvest energy either quickly if wastefully (moles ATP per unit time), or efficiently (moles ATP per mole substrate). This leads to the situation where competitive outcome for a shared resource may be determined by rate of ATP production, whereas ultimate population size is determined not by rate of ATP synthesis but on ultimate ATP yield (Molenaar et al. 2009). The inherent rate/yield trade-off is based on thermodynamics of chemical reactions where rate is negatively correlated with yield. Maximal rates of ATP production have been shown to result at intermediate yields (Pfeiffer et al. 2001; Molenaar et al. 2009). These dynamics are directly analogous to food processing in animals as illustrated in ◘Fig. 3.6. For example, under aerobic conditions, at relatively low glucose concentrations and low growth rates, *E. coli* converts glucose to CO_2 and water, generating ATP in high-yield fashion by respiration (38 mol ATP per mole glucose, which includes operation of both glycolysis and the citric acid cycle), as well as the precursor metabolites used in fueling reactions for biosynthesis. However, at high glucose concentrations and correspondingly high growth rates, also under aerobic conditions, cells partially redirect the glucose into energetically low-energy yielding (2 mol ATP net per mole glucose), fermentative pathways resulting in incomplete oxidation and the accumulation of by-products such as acetate ('overflow metabolism'; Molenaar et al. 2009). Why microorganisms switch from high to low-yield pathways has been explained in several ways (among them, saturation of the respiratory system; maintenance of redox balance; Molenaar et al. 2009; van Hoek and Merks 2012) and remains actively debated (Molenaar et al. 2009; Bachmann et al. 2013). The trade-off between rate and yield is a specific instance of a larger category of multi-objective decision-making models and as such has been called 'an apparent Pareto front' (Bachmann et al. 2013; for **Pareto fronts** and optimization, see, e.g., Shoval et al. 2012); this issue is closely related to *r*- and *K*-selection and will be discussed further in that context in ▸Chap. 4.

As implied above, when resources become limiting, the relative advantage among competitors tends to change in favor of ability to extract energy. In a particularly interesting case, Brown et al. (1998) compared the growth dynamics of a parental strain (P) of baker's yeast with an evolved (E) population derived from it that had undergone 450 generations of glucose-limited growth. In chemostat monoculture,[4] the amount of residual glucose was 10-fold lower for E than P, indicative of the former's enhanced scavenging ability. E out-competed P, attributed to multiple tandem mutations of high-affinity glucose transport genes that enabled it to achieve twofold greater yield biomass under substrate-limitation. In chemostat dual culture, the frequency of E continued to increase over time until P could no longer be detected. Interestingly, parallel experiments under *non*limiting conditions showed that this adaptation did *not* come at a cost in either yield or doubling time, i.e., there was *no* apparent trade-off. The advantage of E was attributable to its ability to transport limiting glucose 2–8 times faster than P and also an ability to produce more cells per mole glucose. In other words, it had both increased flux of limiting substrate into its cells, as well as a higher cell yield (for background theory and apparatus relevant to these experiments, see Gresham and Hong 2015).

---

4   The role and establishment of different environments in microbial experiments, including the use of chemostats and turbidostats, is discussed by Dykhuizen and Hartl (1983), Dykhuizen (1990).

Before we conclude this consideration of bacterial growth in culture, a perhaps obvious point nevertheless bears emphasis: Studying bacterial growth dynamics in culture can demonstrate capability but does not necessarily reflect behavior in nature. *E. coli* can grow very rapidly under favorable conditions and, like prokaryotes in general, is streamlined to do so. But this is still not as fast as theoretically possible. Why this is so is an old question (see, e.g., Koch 1988), still unanswered (Maitra and Dill 2015). The cell apparently trades-off growth speed and associated high costs of energy generation and biosynthesis for the ability to maintain seemingly wasteful ribosomal apparatus under slow-growth (maintenance) conditions. The reserve, however, ensures a fast start when transient, favorable times return. This strategy is entirely consistent with the feast-and-famine existence of *E. coli* in its enteric habitat (Koch 1971), and presumably it reflects as well the portion of the bacterium's cryptic life history in the environment at large (Savageau 1983).

Finally, the rulebook for resource utilization by microorganisms differs in part from macroorganisms because for the latter starvation leads to death, whereas microbes under nutrient stress can shift easily to a dormant phase (▶Chap. 7). The evolutionary advantage of spore formation is clear in those habitats where drought or heat kills nonspore-formers. While this observation is true for the fungi and to a lesser extent for the bacteria (few of which form spores), it needs to be added that dormancy per se is not a productive venture for a microbe and can at best lead only to survival in place or distribution in space (▶Chap. 7). However, 'opting-out' in suspended animation is time lost to growth and reproduction and hence to spreading one's genes in the gene pool. Largely quiescent individuals would be selected out of the microbial milieu unless they counterbalanced prolonged dormant states with rapid growth, reproduction, and effective dispersal characteristics when transient favorable conditions arise. How different kinds of organisms handle the downshift to dormancy and the upshift to germination are correlates of *r*- and *K*-selection (Andrews and Harris 1986). These and other microbial strategies are discussed when the implications of size are considered (▶Chap. 4).

Integration of OFT and digestion To summarize the foregoing, provided that the limitations of energy as a common currency are recognized, optimal foraging and digestion theory can serve as a useful conceptual framework for both microorganisms and macroorganisms. Nevertheless, it must be remembered that **the architecture of an organism and its behavior are compromise responses to many selection pressures, of which energy acquisition and allocation is only one.** The means by which very different organisms forage are not homologous but they are analogous (◨Table 3.3). For instance, animal ecologists refer to 'foraging,' to the 'catchability of prey,' to 'handling time,' to 'feeding efficiency' and so forth because the creatures that they study usually move about. In spinning a tensile web to trap its prey, the orb-weaving spider is behaving fundamentally like an aquatic filter-feeder that must expose a prey-capturing surface to water currents. In essence, plant biologists deal with the same issue but in such terms as canopy architecture, photosynthetic pathways, leaf orientation, and optimal leaf area indices. Analogs in microbial ecology include such phenomena as enzyme kinetics and catabolite repression. For the bacterium or fungus that secretes extracellular enzymes to degrade organic matter there are chemical trade-offs—e.g., in producing the enzyme and in diffusional losses that can limit foraging distance. Pathologists and epidemiologists deal with host range, age or sex of suscept, or the specific tissues invaded. Semantics and detail aside, the fundamental process of resource acquisition is universal, inevitably there will be trade-offs in resource capture, and an optimization approach can illuminate the analysis of adaptations. Alternative actions or strategies should be particularly analogous when

◻ **Table 3.3** Optimal foraging theory and optimal digestion theory broadly construed: some analogous components for animals, plants, and microbes		
Organism	**Terminology**	**Measures of foraging efficiency**
Animals	Prey size: digestibility; gut length; wait or pursuit; territory size; hunt individually or in packs; foraging time; filtering rate	Prey caught per unit time; net Kcal acquired per unit time
Plants	Chromatic adaptation; photosynthetic pathways; growth habit; shoot and root architecture; leaf area indices; substitutable ions	Percent solar radiation trapped
Microbes	Metabolic versatility; partial or complete oxidation; growth habit; photo- or chemotaxis; host range and tissue specificity; proto- or auxotrophy	Maximum specific rate of food acquisition (q_s^{max}); efficiency of food conversion to biomass (Y_s); ability to acquire food at very low food densities (K_s)

compared within the group of organisms that is motile, or within the group that is collectively sedentary, regardless whether the organisms are microscopic or macroscopic.

3.4   Nutritional Versatility

The dogma in microbiology is that microbes are 'versatile'. But what does this mean and are they really more versatile than macroorganisms? Certainly bacteria, for instance, can be found almost everywhere—from hot springs, to the ocean floor, to Antarctic ice sheets. Although this has been taken to mean that the *Bacteria* and *Archaea* collectively are versatile (in this case with regard to extremes of physical habitat), it says nothing about versatility of the individual species or clone, which is the relevant context here.

From the standpoint of the organism, 'versatility' can be adopted in scientific parlance from its common usage in two contexts. First, we say that someone who can do *many things*—wrestle, play the saxophone, speak several languages, etc.—is versatile. Second, a person who can master different situations or tasks quickly and with apparent ease is said to be versatile. Unlike the first definition, the emphasis in this latter case is on *speed of response* to new conditions and less on the spectrum of accomplishments. The constraint is time. After all, it can be argued, many things can be accomplished given sufficient time; the truly versatile are those who adjust quickly. Based on these two criteria and the issue of nutritional versatility let us now compare microorganisms with plants and animals. 'Metabolic' is used with reference to anabolic and catabolic pathways (in terms of their number and regulation and the metabolites involved) and to the range of substances synthesized or degraded.

3.4.1   Versatility as the Ability to Do Many Things

Evaluated by the first criterion of being able to do many things, both microbes and macroorganisms are highly versatile metabolically with respect to biosyntheses. Despite remarkable advances in genomics and proteomics, all the metabolites, end products, and metabolic pathways for even a single organism are not yet known. The potential metabolic complexity of even a small organism is illustrated by *E. coli* (about one five-hundredth the size of a plant or animal cell). This is probably the most studied and best understood of all organisms. Its genome has been sequenced, so the number of genes and known or predicted gene functions (4453) are established (Riley et al. 2006; Orth et al. 2011). Approximately 2200 metabolic reactions and more than 1100 unique metabolites have been documented. However, direct evidence exists to date for the function of slightly more than half of the protein-coding gene products, and about one-third of the proteome is functionally not annotated. Metabolic reconstructions for multicellular eukaryotes to date have been relatively few, with a predicted metabolome generally higher than for prokaryotes (de Oliveira Dal'Molin and Nielsen 2013). These data are fluid benchmarks, changing with time as knowledge of the metabolic network becomes increasingly comprehensive. Though small and often misleadingly dismissed as 'simple,' the bacterial cell is in fact complex in many ways.

Many, perhaps most, bacteria and to a lesser extent fungi can synthesize basically all of the organic compounds they need for cellular macromolecules. In essence they are specialized synthetic chemists, possessing the hundreds of enzymes required to make all their building blocks and polymerizations. As so-called **prototrophs** they will grow if almost any relatively simple carbon source is available and other conditions are not limiting. *E. coli*, a facultative inhabitant of the gut, has extraordinary biosynthetic versatility, being able to produce all its cellular components from a simple medium of glucose and mineral salts (including a nitrogen source). This means that a pathway, separate at least in part, exists for each of the 20 amino acids synthesized. In the absence of a sugar, any amino acid can serve as a sole carbon source. The bacterium's extensive synthetic capability likely reflects characteristics of its habitat—the highly oscillating nutrient realm of the gut, as well as its life elsewhere (see, e.g., Koch 1971; Savageau 1983).

At the other extreme, some bacteria and fungi called **auxotrophs** lack one or several biosynthetic steps or pathways; as such they are fastidious and require growth factors such as vitamins, amino acids, or purines and pyrimidines in small amounts. Auxotrophs tend to occur in environments of relatively rich media, such as other parts of the human body, where building blocks can consistently be found preformed, obviating the need for synthesis pathways. Such close associates of humans include species of *Staphylococcus* and *Haemophilus*. Indeed, they have become auxotrophs because, by living in nutrient-rich environments and under selection pressure, such bacteria have dispensed with redundant biochemical pathways. Instead of a nutritional strategy based on maintaining pathways of biosynthesis for all the building blocks, they have maintained active transport mechanisms capable of efficiently importing end products into their cells (Koch 1995). Another example of a fastidious microbe is the lactic acid bacterium *Leuconostoc mesenteroides*, which, unlike *E. coli*, requires a complex culture medium containing yeast extract and peptone as well as glucose. In nature, species of *Leuconostoc* are found on plant surfaces and in dairy and other food products.

Since degradative pathways are distinct from (not simply the reverse of) biosynthetic pathways, this means that there are also numerous pathways for metabolite degradation. Microbes excel in their degradative capabilities, outstripping those of either plants or animals. Indeed, bacteria collectively can grow in virtually any natural environment and, with few exceptions, can metabolize effectively any organic compound. As such, they have been called 'metabolically infallible.' Degradations are frequently accomplished by microbial consortia with intimate and occasionally obligatory interdependencies (such as cross-feeding relationships or syntrophy) among member species (Embree et al. 2015). Degradation pathways, most of which are fairly specific, exist not only for proteins but also for lipids, carbohydrates, and other biopolymers and compounds. As noted at the outset, these degradative reactions produce key precursor metabolites (plus ATP and reducing power in the form of NADH) through central pathways that are common to all species of bacteria. *E. coli* and other prokaryotes also have peripheral metabolic pathways, which may be unique to the species or function only where the bacterium grows on compounds that are not part of the central pathway. *E. coli* has more than 75 such pathways consisting of at least 200–300 different enzymatic reactions (Neidhardt et al. 1990; Lengeler et al. 1999). The number could be expected to exceed this in certain bacteria such as the pseudomonads, which grow on diverse molecules (see Generalists and Specialists, below). Such peripheral pathways illustrate the true nutritional diversity of bacteria.

Plants are even more versatile than heterotrophic bacteria and fungi (note, however, that there are photolithoautotrophic bacteria; recall earlier discussion and ◻Table 3.2). Starting with only CO_2 rather than a sugar as a carbon source, and using water as an electron or hydrogen donor and light instead of chemical energy, plants manufacture an immense array of primary and secondary compounds. Many of these groups, such as the terpenoids, flavonoids, and alkaloids, are biochemically complex (Chae et al. 2014). Finally, animals, unlike plants or many microorganisms, lack the biochemical machinery to synthesize all their requirements from a few simple molecules such as CO_2 or glucose. Vertebrates, for example, can make only half of the basic 20 amino acids; the others must be supplied in the diet. The complex nutritional requirements of animals are evident in culture, where 13 amino acids, 8 vitamins, and various undefined growth factors supplied by dialyzed serum, in addition to glucose and inorganic salts, are needed in order to grow human (HeLa) cells (Eagle 1955; this basic medium and variations such as that by Dulbecco, persist to this day).

To summarize, by the first definition, nutritional versatility in the case of plants and the bacteria or fungi entails the capacity to synthesize essentially all macromolecules from a few relatively simple, inorganic or organic compounds. In contrast, nutrition for animals generally involves degrading complex food sources, the component units of which are rearranged into new metabolites. While in terms of the ability to do many things this may seem to imply that animals are less versatile than microbes, it does not necessarily: Few microbes synthesize specialized cell types or complex molecules akin to antibodies, eye pigment, neurotransmitters, or any of the vast assortment of other chemicals and structures characteristic of at least the complex metazoans. These metazoan gene products are often encoded by many spatially separate genes. Such products interact with others in a precisely orchestrated sequence of events to coordinate assembly and function of cells, tissues, organs, and ultimately the entire organism. Microbial biochemical coordination is focused largely at the gene expression and pathway level; for macroorganisms the focus is more at the levels of interaction among gene products and timing of gene activity.

3.4.2 Versatility as the Ability to Adjust Rapidly

The second interpretation of versatility, speed of response to new (nutrient) conditions, is more amenable to analysis. Prokaryotes as a group have unrivaled metabolic rates (substrate consumed per unit cell mass per unit time) coupled with the ability to react to changing nutrient conditions by rapidly altering metabolic pathways. Thus, what is impressive about these microorganisms is not just the numbers of alternative central and peripheral pathways within a single cell, but that entire peripheral pathways can be rapidly switched on and off. The set of enzymes needed for a particular route is often present or absent as a unit or operon (see below). Details of how these pathways and intermediates change can be found in Neidhardt et al. (1990), Lengeler et al. (1999), Fraenkel (2011). We now look in more detail at some of the mechanistic bases and ecological implications.

Prompt response to ambient nutrient conditions undoubtedly reflects the rapidly fluctuating environmental conditions to which microbes are exposed. These range from excess of a particular substrate to growth-limiting levels or, commonly, to situations where the organism 'sees' different but functionally similar substrates in varying concentrations (e.g., both glucose and acetate supply energy and carbon). Enzymes useful at one moment could be useless or detrimental (wasteful) at the next. From the microbe's standpoint, the premium is on being able to avoid synthesizing catabolic enzymes if it is currently well nourished, while being able to adjust quickly to new resources if the need arises. Typically, when resources are in high concentration (not growth-limiting), bacteria respond by sequential utilization, with the nutrient supporting the highest growth rate being used preferentially from the mixture (Neidhardt et al. 1990; Lengeler et al. 1999). This phenomenon is known as **diauxie** and supports the general argument that in bacteria the enzyme machinery adjusts to permit as rapid growth as possible. If substrate concentrations drop to growth-limiting levels, the various nutrients are used simultaneously. Diauxie is one manifestation of the broader regulatory mechanism known as catabolite repression.

Despite some similarities, there are important differences between environmental sensing and gene regulation in prokaryotes and eukaryotes. This topic is discussed in ►Chap. 7, but it is pertinent to emphasize here some specific aspects that contribute to the nutrient versatility of bacteria. These pertain in part to the direct exposure to the environment in the case of microorganisms, especially those that are unicellular, as opposed to the multicellular and often relatively homeostatic existence of most eukaryotic cells. Thus, the individual cell in a plant or animal body responds to stimuli as mediated systemically by hormones, neurotransmitters, etc., but is buffered from the external environment and also a part of a centrally orchestrated ontogenetic program. The eukaryote cell is constrained by the past conditions, including the developmental program of the organism of which it is a part—an exceptional situation for bacteria (the closest approximations being the few differentiated, multicellular species, or in quorum sensing communications, see Miller and Bassler 2001; and biofilms, see Vlamakis et al. 2013).

In bacteria, fast changes in gene expression are possible because of the kinds of control mechanisms operating at both the transcriptional and translational levels (▢Fig. 3.7). With respect to the former, new mRNA molecules are produced continuously, with half-lives on average of 0.5–2 min (Lengeler et al. 1999). Even faster control is possible post-transcriptionally to modify enzymes by allosteric and covalent changes, as well as by compartmentation within the cytoplasm. While it may appear wasteful to produce a transcript only to have to modify it later, post-transcriptional regulation allows for fine-tuning. Some attributes of prokaryotes that facilitate rapid response to changing environments are summarized in ▢Table 3.4. These include that their DNA is aggregated as an unenclosed nucleoid within the cytoplasm whereas the chromosomes of eukaryotes are encased within a membrane-bound nucleus. Therefore, external signals affecting nuclear gene expression in eukaryotes must pass through the cytoplasm and then into the nucleus, and likewise, for translation to occur the transcript must move, and be protected from degradation in the process, from the nucleus to ribosomes in the cytoplasm. Bacteria combine transcription and translation, thus expediting the process and doing away with processes involved in transporting the mRNA. Events are not compartmentalized as they are in eukaryotic

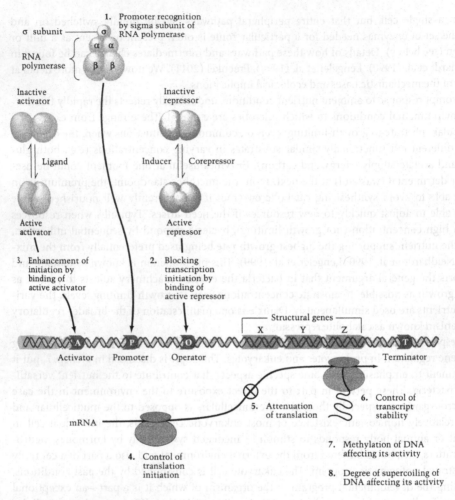

◻ Fig. 3.7 The eight major sites of control for operon expression in bacteria. See also Chaps. 2 and 7. From Neidhardt et al. (1990). Reproduced by permission of Sinauer Associates, Inc., Sunderland, MA © 1990

cells, but rather occur directly in the cytosol. Finally, bacteria exhibit gene clustering, whereby enzymes in a particular pathway may be encoded by adjacent genes. This allows for coordinated regulation of the genes, which are transcribed as one polycistronic mRNA and sequentially translated by ribosomes into each of the proteins.

Probably the classic example of coordinated gene expression is the architecture of the *lac* region in *E. coli*, which encodes three genes involved in the metabolism of lactose (Madigan et al. 2015). As alluded to above, the three structural genes are grouped as an operational unit or operon, which is transcribed into a single mRNA molecule. The essence of the mechanism was proposed formally by Jacob and Monod in 1961 (the discovery for which, in part, they received the Nobel Prize in 1965 along with André Lwoff, noted in ►Chap. 2). They postulated that the availability of external food molecules, lactose in this case, in conjunction with the energy status of the cell as mediated through cyclic AMP, control the rate of synthesis of the small inducible enzymes by regulating the synthesis of particular mRNA templates.

□ Table 3.4 Some attributes of prokaryotes (cf. eukaryotes) that facilitate rapid response to environmental changes[a] (Lengeler et al. 1999; Madigan et al. 2015)

Prokaryotes	Eukaryotes
DNA aggregated in cytoplasm as unbounded nucleoid	DNA in membrane-bound nucleus
Transcription and translation coupled (in cytoplasm)	Spatially separate (in nucleus and cytoplasm)
Introns rare	Introns and repetitive sequences
Characteristically polycistronic mRNA	Monocistronic mRNA
Unstable mRNA	Relatively stable mRNA
no mRNA cap or tail	Cap and poly(A) tail
Operons	No operons

[a]There are many molecular features that are different between the prokaryotic *Archaea* and *Bacteria*, on one hand, and the *Eukarya*, on the other. The above are some of the main attributes affecting primarily metabolic versatility

Though the basic Jacob-Monod operon model is still valid, transcriptional control has subsequently been shown to be more complex (to involve, e.g., more than just initiation of transcription) and the *lac* paradigm is not universal. The picture is also different for more complex organisms, where enzymes are coded individually and extensive RNA splicing is involved in the formation of messenger. As an example of comparative speed, translation proceeds about eight times faster in bacteria than in mammals (reviewed in Koch 1971). Transcription is also rapid: the average rate is 48 nucleotides per second, or threefold the step time for incorporation of an amino acid.

In overview, ecological comparisons based on versatility defined as the ability to do many things are ambiguous. While many microorganisms, and plants in general, have the ability to synthesize all their building blocks from a simple carbon source and inorganic ions, animals rely on sophisticated absorption systems for their nutrients from which they synthesize complex macromolecules. When versatility is defined as speed of response to changing nutrient conditions, bacteria are more versatile than macroorganisms because they can rapidly alter protein synthesis and entire metabolic pathways. Bacteria can grow at a high rate without excess enzyme 'baggage' (▶Chap. 7), on a preferred nutrient source, under nonlimiting substrate conditions, while retaining the ability to adjust quickly under nutrient limitation by using multiple substrates concurrently or sequentially.

3.5 Generalists and Specialists

Diet breadth is a specific case in foraging theory and relates also to the broader issue of niche width (Futuyma and Moreno 1988). In nature we see populations or species that, relative to others, are extreme specialists or generalist feeders, with the bulk of organisms falling somewhere in the middle of the spectrum. For example, Koala bears (which technically are not bears but rather marsupials) eat *Eucalyptus* leaves and even show clear preferences among species of *Eucalyptus*. In contrast, their closest living relatives, wombats, are also herbivores but have a relatively broad diet including grasses, bark, roots, and sedges. Rabbits and Virginia opossums are relative omnivores.[5] An analogous dietary spectrum applies among free-living microbes. For instance, *Sulfolobus*, discussed previously, is a specialist at oxidizing elemental sulfur. *Methylococcus* lives on methane. In contrast, various species of *Pseudomonas* can grow on any one of many dozens of carbon sources (Clarke 1982; MacLean and Bell 2003). *Sulfolobus* is the Koala bear of the microbial world, and *Pseudomonas* is the opossum.

Since generalists have the obvious advantage of a greater range of food sources from which to choose than do specialists, the interesting question is why do populations respond to heterogeneity by evolving specialists? In early work that shaped the field, MacArthur and Connell (1966; see their Chap. 3) examined theoretically the foraging efficiency of generalists versus specialists with respect to the extent of environmental heterogeneity or 'grain.' Environments can be classified as **'fine-grained'** or **'coarse-grained'** relative to organism foraging depending on whether the resources are consumed in the proportion in which they occur (fine-grained) or selectively, e.g., where the organism selects one because of its prevalence in preference to others (coarse-grained). Grain is discussed in detail in ►Chap. 7 where we consider the environment. The short answer from their model is that in a coarse-grained environment specialists will have an advantage over generalists at harvesting the resource of their choice. In a fine-grained environment, the outcome may favor either but, where the resources are quite similar, the generalist will theoretically win (see related work by Levins 1968; and by MacArthur and Levins 1964; MacArthur and Pianka 1966).

The above results were interpreted by MacArthur, Levins, and Connell, among many others subsequently, within the context of the well-known adage 'a jack-of-all-trades is master of none.' While this maxim has intuitive appeal, at least in regard to human activities, it does not necessarily apply in ecology. The limitations on generalism usually are ascribed to either the unavoidable interference in performing different tasks equally well (inability to maximize competing functions simultaneously) or costs of maintaining characters specifically for reducing environmental variation in fitness (Kassen 2002). For example, the genetic trade-offs in a parasite to have many hosts, or alternatively to become highly adapted to a particular host (see **Sidebar** and later section, Phylogeny), may drive the evolution of specialists (Chappell and Rausher 2016). However, there is some contradictory evidence to the assumption that a generalized phenotype embodies costs constraining optimal performance (e.g., in tests of performance as a function of body temperature, a 'jack-of-*all-temperatures* may also be a master of *all*'; Huey and Hertz 1984; see also Reboud and Bell 1997; Remold 2012). Thus, apparently, traits promoting performance in one environment can do so in others as well. Tests of the prediction of costs or trade-offs generally have produced ambiguous or at best nuanced results (recall earlier discussion under Optimal Digestion of experiment by Brown et al. 1998; for caveats and updated semantics and status of generalist/specialist theory, see Kassen 2002; Jasmin and Kassen 2007a, b; Remold 2012). The widespread assumption that

---

5   Arguably the most extreme omnivore among macroorganisms is the grizzly bear. A special issue of *National Geographic* magazine on Yellowstone National Park (v.229 [5] May 2016; see Into the Backcountry, pp. 92–123), which includes a list of the following species numbers eaten by grizzlies in Greater Yellowstone: 162 plants; 36 invertebrates; 26 mammals; 26 cultivated plants and domestic animals; 7 mushrooms; 4 fish; 3 birds; 1 alga; 1 amphibian. The bears alter their diet seasonally based on availability of a food item and their energetic needs; these data are consistent with numerous grizzly bear diet studies more formally reported (e.g., Munro et al. 2006).

being a generalist entails costs has been examined in Kassen's review (2002) as well as by Remold (2012). Kassen noted that a trade-off between fitness (performance) and breadth of adaptation in only one of the four cases considered. It should also be recognized that species may come to be associated with particular resources for reasons other than competition or feeding preference (for amplification and examples, see Futuyma and Moreno 1988).

SIDEBAR: Case Study: The Gene-for-Gene Interaction and Plant Parasites as Specialists

Nowhere are specialized interactions better illustrated than in the coevolution of parasites and their hosts (Price 1980; Thompson 1994). With respect to some plant pathogen-host relationships in particular, a precise correspondence between genes governing host resistance and those controlling parasite virulence has been well documented, beginning in the 1930s with the classic work that Flor conducted over several decades (e.g., 1956, 1971). He studied the rust disease on flax caused by the biotrophic fungal pathogen *Melampsora lini* and came to propose the now famous 'gene-for-gene' (GFG) hypothesis that for each gene conferring resistance in the host there is a specific gene conferring avirulence in the pathogen (e.g., Flor 1956; Clay and Kover 1996; Barrett et al. 2009). Inheritance studies involving pathogen genotypes (races) and plant genotypes (differential varieties of flax) led Flor to conclude that both resistance (R) in the host and avirulence (Avr) in the pathogen are inherited in dominant fashion; specificity is evident in that a resistance reaction only occurs where the pathogen is avirulent and the host is resistant *at corresponding loci* (Keen 1982; ◘Fig. 3.8). In a strict GFG model, the relevant locus in both host and pathogen is diallelic, with a resistant and susceptible host allele corresponding to a virulence and avirulence allele in the pathogen. If the host is homozygous for the susceptible allele, it will be overcome by all pathogen races, regardless of their virulence genotype; equivalently, a race homozygous for the recessive avirulence allele will infect all host cultivars regardless of their genotype.

A hallmark of GFG interaction is that under parasite pressure, host genotype frequencies oscillate through time in response to corresponding changes in the parasite population.

(a)	Host genotype		(b)	Host genotype	
Pathogen genotype	Rx	rr	Pathogen genotype	$R_1R_1r_2r_2$	$r_1r_1R_2R_2$
Ax	I	C	$A_{R1}A_{R1}, a_{R2}a_{R2}$	I	C
aa	C	C	$a_{R1}a_{R1}, A_{R2}A_{R2}$	C	I

◘ **Fig. 3.8** The gene-for-gene relationship as an example of host-parasite specialization. a Quadratic check shown for one diallelic locus in diploid host and diploid or dikaryotic pathogen. The pattern shown here, while consistent with a gene-for-gene relationship, is not definitive because it could also result from general resistance mechanisms. b Reciprocal reactions shown for two loci in host and pathogen. This pattern is diagnostic for gene-for-gene specificity. *I* incompatible (resistant or no disease); *C* compatible (susceptible or disease); $x =$ allele unspecified. Each gene of the corresponding pair R_1-A_1 or R_2-A_2 is dominant over its defective alleles r_1 and a_1; r_2 and a_2. Based on Keen (1982)

However, Clay and Kover (1996) have argued that a strict gene-for-gene interaction does not lead to cycling of gene frequencies in host and pathogen but rather that typical GFG models actually assume instead a 'matching allele' system. Some GFG models are restrictive with respect to operating assumptions; others do produce cycling in gene frequencies (frequency-dependent selection) if fitness costs to resistance or virulence are assumed (as has been demonstrated in some cases, e.g., Tian et al. 2003; Jones and Dangl 2006). The genetics of resistance and virulence are occasionally more complicated than as assumed in a strict GFG manner but the basic concept has been validated in multiple tests in both agronomic and natural ecosystems (Thompson and Burdon 1992; Lawrence et al. 2007), as well as in some insect-host relationships (Thompson 1994). If ecological, demographic, and epidemiological considerations are superimposed on the purely genetic GFG model to add realism, various outcomes are possible. For example, Thompson and Burdon (1992) showed that at the local level of a plant population and its immediately associated pathogen population there was little correlation between resistant lines and pathogen races. Instead, locally, the main influence on genetic dynamics was from drift, extinction, and gene flow. At such smaller levels of scale, frequency-dependent selection may still occur but may not be responsible for the pattern in gene frequencies. Most of the GFG models assume panmixis, no drift, and no gene flow, with only mutation and natural selection driving the coevolution of host and parasite population.

The question before us in terms of specialization is how have plant and certain pathogen populations come to be so tightly associated? What is the mechanistic basis for the GFG interaction? In brief, plant resistance to pathogens involves multiple layers of both preformed and inducible responses. The nonspecific, preformed (passive) physical and chemical barriers take such forms as a waxy cuticle overlying the epidermis or toxic peptides, proteins, or other metabolites that deter herbivory and infection. Through such mechanisms most plants are resistant to most potential pathogens. However, a second line of defense is a relatively specific, generalized basal immunity triggered when plant receptors at the cell surface recognize conserved molecular signals, so-called microbe-associated molecular patterns or MAMPs, characteristic of most microbes. (Interestingly, much of plant-based immunity including the triggering mechanisms of recognition and defense are similar in plants and vertebrates (Staskawicz et al. 2001; Jones and Dangl 2006). In response to this selection pressure, successful pathogens have evolved various virulence factors such as effector proteins that are translocated into cells to suppress signal detection or host response. These effectors not only influence the virulence of pathogens but are thought to influence host range—some parasites with a narrow range produce host-specific toxins, while those with a broad range have a broad spectrum of phytotoxic molecules, only some of which are operative on any particular host (Barrett et al. 2009). In certain plant-parasite interactions, the plant responds to virulence factors of the pathogen in a highly specific fashion at the genotype level. Here, the products of resistance (R) genes recognize the structure or activity of pathogen effectors and elicit a classical hypersensitive reaction characterized by localized necrosis and programmed cell death (PCD) of host cells. This GFG response (frequently termed 'effector-triggered immunity,' ETI) provides a specific, qualitative form of resistance to those pathogens that rely on living tissue to grow (obligate biotrophs) because the pathogen also dies and further ingress is halted. It is a form of resistance that allows rapid coevolution of host and parasite because in yet a further turn of the interaction spiral, the pathogen population can respond simply by changing the nature of the effectors being detected or by deploying yet other effectors that suppress the host's ETI. Remarkably, these suppressors themselves have occasionally then become overcome by new R genes. Both the virulence system in the pathogen and the defense system in the host are based on multiple redundancies and so components can be rapidly switched by replacing effectors and modifying resistance genes, respectively. Thus, the cycle of surveillance, detection, and evasion continues indefinitely in a coevolutionary arms race, each member driving the other ever deeper into the realm of specialization (Staskawicz et al. 1995; Schneider and Collmer 2010).

The GFG concept, so thoroughly established by Flor's insightful studies in classical genetics, has since been confirmed and is being increasingly explained by molecular genetics and biochemistry. Several R genes and corresponding avirulence genes have been cloned, sequenced, and their functions inferred (Keen 1990; Bent and Mackey 2007). Most R genes encode proteins of the nucleotide binding, leucine-rich repeat (NB-LRR) type, while avirulence genes typically encode small secreted proteins of known characteristics (Jones and Dangl 2006; Lawrence et al. 2007) that interact with the R proteins. The exact nature of the interactions and how they relate to a PCD response are among the details that remain to be elucidated.

These host-parasite systems not only provide a remarkable example of host-parasite specialization but have important practical implications. A common disease control strategy has been the sequential release of crop varieties carrying single R genes to counter newly evolved races of pathogens (Thompson and Burdon 1992; Hulbert et al. 2001). Molecular approaches will facilitate both the identification of R genes and their potential utilization.

With respect to specialization, the common assumption is that species or populations showing differential performance across environments do so because of trade-offs. These in turn are usually explained mechanistically as arising by antagonistic pleiotropy whereby a mutation providing for better performance in one environment is deleterious in another (Cooper and Lenski 2000; Kassen 2002; MacLean and Bell 2002). However, specialization can result from mutation accumulation in which, by genetic drift, various mutations arising in the environment where they are neutral prove harmful in another. Two populations may also independently adapt to their alternative environments by accumulating alleles that are advantageous in one while being neutral in the other (Elena and Lenski 2003; Zhong et al. 2009). Thus, there are three mechanisms for producing specialists. By analyzing data from a long term field study of aphid parasitoids, Straub et al. (2011) found that specialists were more abundant than generalists on their shared hosts (as predicted by the 'jack-of-all-trades is master of none' postulate) but that the generalist fitness cost depended not on the number of utilized host species per se, but rather on their taxonomic breadth.

The generalist/specialist issue can be approached in rigorous, if somewhat artificial, experiments with microbial systems in vitro. These offer control of both the genetic makeup of the populations and the environmental variables. Dykhuizen and colleagues have done such studies for many years (e.g., Dykhuizen and Davies 1980; Dykhuizen 1990, 2016; Zhong et al. 2004, 2009) and their work is illustrative of the approach. In the 1980 study they defined the generalist as a strain of E. coli that could use both of the disaccharides lactose and maltose (lac$^+$, mal$^+$), whereas the specialist could metabolize maltose only (lac$^-$, mal$^+$) because the genes for uptake and catabolism of lactose had been deleted. As a technical point, the construction of the strains involved genetic engineering of an initial (specialist) strain containing a deletion of the lactose operon that removed all of the proteins involved with lactose breakdown but no others. The generalist was then created by transferring through phage transduction (see ▶Chap. 2) only the genes for lactose metabolism into the specialist from a donor strain, thus constituting the lac$^+$, mal$^+$ generalist phenotype in a genetic background otherwise identical to the specialist. While various other strains were constructed and tested in these experiments, a key result is that when cultured separately on maltose, the specialist grew about 7% faster than the generalist, consistent with predictions. In direct competition experiments in chemostat culture under energy-limited conditions on mixtures of lactose and maltose, the strains generally coexisted, with the final ratio depending on the percent lactose in the medium. As argued by the authors, this implies that coexistence in nature would occur under certain conditions, e.g., where the specialist uses the abundant resource in common while the resource unique to the generalist is rare, or if the selection differential on the shared resource is appreciable. Extrapolation from this system to nature is constrained by the experimental conditions, wherein chemostat culture presents to *E. coli* a constant, homogeneous (fine-grained) energy-limited environment, whereas in nature *E. coli* in the gut faces a markedly heterogeneous feast-or-famine life and an even more heterogeneous survival stage outside the body (see, e.g., Koch 1971). Nevertheless, these and similar experiments are a powerful and useful intermediary step for hypothesis-testing between theoretical models and experiments in nature.

In the older literature there are numerous studies similar to the experiments above. Typically these were conducted by comparing progeny carrying introduced point mutations (auxotrophs) against fully functional (prototrophic) parents. The underlying hypotheses usually were that the metabolically simpler auxotroph 'specialist' would have a selective advantage because it would not carry the additional biosynthetic steps of what is in effect the prototroph 'generalist'. Data from numerous experimental systems involving mutants support the assertion. Zamenhof and Eichhorn (1967) compared various nutritionally deficient mutants and fully functional (back mutant) strains of *Bacillus subtilis* in continuous (chemostat) culture in liquid media. One experiment demonstrated that a histidine requiring mutant (his⁻) was competitively superior in dual culture to its histidine nonrequiring, spontaneous backmutant (his⁺). In experiments with other mutants the authors went on to show, first, that dispensing with an earlier rather than a later biosynthetic step gave an auxotrophic carrier a selective advantage and, second, that a de-repressed strain producing the final metabolite (tryptophan) in quantity was at a selective disadvantage compared with the normal repressed strain. Both outcomes are consistent with a conclusion of advantageous metabolic streamlining in specialists.

The most obvious interpretation of the foregoing is that the prototroph continues to synthesize at least some of the metabolite in question, thereby incurring energy costs, despite the availability of the chemical in the medium. These functions are completely blocked in the auxotroph. Evidently the feedback inhibition and gene repression mechanisms are incomplete in the prototroph. Zamenhof and Eichhorn went on to speculate that it may have been through energetic savings that parasitism evolved, once nutritious 'media' in the form of macroorganisms became available to support fastidious variants of fully functional, free-living microbes. Interestingly, the issue of operon repression was taken up in detail in subsequent experiments by Dykhuizen and Davies (1980). They showed that on the lactose-limited medium a lac constitutive mutant (producing the lactose enzyme whether or not lactose is present; operating as essentially a generalist) replaced the lac⁺ specialist by enabling the constitutive to subsist on lower levels of lactose than that required for derepression of the operon in the lac⁺. They further demonstrated that equilibrium of lac⁺ and lac constitutive cells was established in the presence of both maltose and lactose, implying that the lac constitutive cells were less efficient in using maltose than the lac$^+$ cells. This was interpreted not in terms of energy conservation or a diauxie, but rather 'resource interference'—a circumstance when organisms are not as efficient at using a resource when it is combined with others as when alone. Here, lactose use appeared to interfere with maltose use, which they attribute likely to lack of space in the bacterial membrane for permeases of both sugars.

Work extending the above findings and incorporating genomic techniques to study *E. coli* adaptation to sugars over hundreds of generations has been done by Zhong et al. (2004, 2009). Here, growth-limiting concentrations of either lactulose (rather than lactose), methylgalactoside, or a 72:28 mixture of the two were used. Across multiple experiments and isolates they describe eight unique gene duplications and 16 unique deletions in a genomic pattern that was consistent. The *lac* duplications and *mgl* mutations usually occurred in different backgrounds, producing specialists for the respective sugar, but not to the other. Interestingly, although growth in the mixed sugars provides conditions for generalists to evolve, they rarely did, rather growth was dominated by specialists. These experiments did not test the mutation accumulation hypothesis for specialist evolution (such mutants would have been eliminated by the experimental design) and of the remaining two (antagonistic pleiotropy and independent specialization) antagonistic pleiotropy was inferred to be the mechanism most consistent with the results.

Finally, similar studies with nutrient generalist or specialist populations considered the ability of *E. coli* to evolve simultaneously in response to distinct selection pressures over extensive periods, i.e., 6000 generations of culture. Satterwhite and Cooper (2015) tested whether replicate populations (generalists) conditioned in an environment of two resources (varying presentations of lactose and glucose) have essentially the same fitness as replicate populations grown either in glucose alone or lactose alone (specialists). They found that for the first 4000 generations, generalists were usually as fit in the individual resources as were the specialists in those resources, but this period of cost-free adaptation was followed subsequently by a cost of adaptation for all generalists. Whether such costs might eventually diminish is an open question.

3.5.1 The Phylogeny of Specialization

As noted at the outset, pathogenic microorganisms can be highly specialized as parasites of a single host species or, conversely, may have evolved as relative generalists to derive nutrition from hosts in many species or even families (Woolhouse et al. 2001; Barrett et al. 2009) (◘Fig. 3.9). To use fungi as an example, fungal phylogenies reveal that plant parasitic species frequently are closely related to species that live exclusively as saprobes or mutualists (Berbee 2001; James et al. 2006). Apart from having usually very narrow plant host ranges, the pathogens that are extreme specialists share several correlates. Among these are (i) a typically intimate association with their hosts that tends to result in minimal damage to the plant cells (indeed, some fungal parasites—the obligate biotrophs—are so nutritionally dependent on living host tissues that they cannot be cultivated on artificial media); (ii) close synchrony between host and parasite life cycle stages; and ultimately (iii) a tight coevolutionary relationship marked at the population level by coordination in genes for host resistance and parasite virulence (see the gene-for-gene relationship in earlier Sidebar). The endless spiral

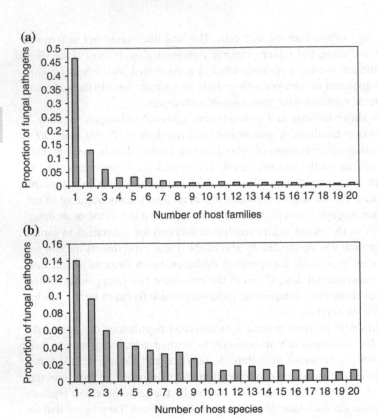

■ **Fig. 3.9**   A continuum of specialization as illustrated by fungal host plant range. **a** Proportion of fungal pathogens (total number, 1252) utilizing the number of host *families* indicated. **b** Proportion of fungal pathogens (total number, 1252) utilizing the number of host *species* indicated. From Barrett et al. (2009). Reproduced from *New Phytologist* by permission of John Wiley and Sons, ©2009

of specialization is most dramatically seen in agroecosystems, where the repeated introduction of resistant cultivars has been matched by corresponding evolution of new strains of the pathogen. Probably the most sophisticated specialists are those parasites that must alternate between two taxonomically distinct hosts to complete their life cycle. This is because they have to synchronize their developmental stages not only with one host but two quite different hosts (recall the rusts noted earlier and see ▶Chap. 6).

Does the intense specialization evident in many coevolutionary relationships such as between herbivorous insects or plant parasites and their hosts reflect optimization in the sense of a precise fit between organism and its physical and biotic environment (as many adaptationists might argue)? Or are such examples an evolutionary dead-end, an endpoint in an ongoing spiral where the partners *"each drive the other into an ever-deepening rut of specialization"* (Harper 1982; see also Moran 1988, 1989)? Broadly speaking, phylogenies show that specialists are not always on the terminal branches, i.e., they are not necessarily a derived condition and such relationships may open successful new evolutionary avenues (Thompson 1994). This interesting issue will be explored in ▶Chap. 6 where we take up the evolution of complex life cycles.

3.6 Summary

All creatures must acquire nutrient resources from their environment and then allocate them among the competing demands of growth, maintenance, and reproduction. Additionally, they must extract and transform energy needed in anabolism and catabolism, as well as maintenance of molecular complexity and orderliness. In doing so, they operate according to a very similar biochemical master plan and obey the same fundamental laws of thermodynamics, energy balance, and mass. Metabolic and stoichiometric rules are universal and metabolic rates largely determine the rates of most biological activities. However, the ways in which organisms obtain resources are multiple and a defining characteristic, so much so that mode of nutrition is one of the key criteria for the categorization of species at the kingdom level.

Resource categories can be illustrated by defined with respect to sources of energy and of carbon building blocks for an organism. ATP is the major carrier of biologically usable energy in all organisms and there are only two basic mechanisms for its generation: electron transport phosphorylation, or substrate-level phosphorylation. In phototrophs (phototrophic bacteria, algae, plants), energy is obtained directly from the sun, in which case light energy is converted by electron transport phosphorylation into the high-energy phosphate bonds of ATP. In chemotrophs (most organisms), ATP is generated from reduced inorganic compounds or from organic compounds by electron transport phosphorylation or substrate-level phosphorylation. Carbon is acquired either directly from CO_2 (autotrophs) or from organic compounds (heterotrophs). Although the resulting energy/carbon permutations form several potential resource categories, most living forms, in terms of species or biomass, use either light energy to fix CO_2 for their biosynthetic needs (photoautotrophs, e.g., plants), or derive both their energy (i.e., in this case electron donors) and carbon from organic molecules (chemoheterotrophs, e.g., animals and most microbes). Chemolithotrophs (chemotrophs using an inorganic electron donor), which evidently exist only as microorganisms, and though apparently fewer in number of species, play an essential global role in biogeochemical cycles.

A manifestation of energy/nutrient relationships within the living world is the trophic structure of ecosystems. This is usually depicted as a grazer food chain or network based on phototrophs associated with a decomposer chain based on dead organic matter. Microbes play a key role in both, particularly the latter. Apart from the microscopic plankton in aquatic systems, however, their role in grazer systems has been underestimated until recently. The central role of an explicit 'heterotrophic microbial loop,' which converts dissolved organic matter to biomass that reenters the grazer chain, is now generally recognized in aquatic systems, as is a significant class of mixotrophic plankton that combine the functions of photoautotrophy and chemoheterotrophy. Likewise, historic depictions have not generally accommodated food chains based on other sources of energy input, most notably that fixed chemosynthetically by lithoautotrophic bacteria. These microbes can, for example, support substantial oases of benthic invertebrates around deep-sea hydrothermal vents, representing a novel, second major source for primary production on Earth.

The sources of energy and carbon also set broad limits on the distribution of living things, as in the restriction of aquatic plants, algae, and photosynthetic microbes to the euphotic zone, of obligate parasites to their hosts, and of many lithotrophic bacteria to their requisite geochemicals.

As originally devised, optimal foraging theory was essentially a cost/benefit analysis in energetic terms developed primarily as an economic optimization model to interpret the foraging behavior of certain animals. It can be construed broadly, merged with optimal digestion theory, and applied informally, conceptually, and empirically to all organisms. Formal, rigorous tests of the theory can and have been undertaken successfully but are relatively few. Broadly speaking, in terms of foraging, bacteria appear to do largely by metabolic versatility what animals accomplish by mobility and behavior, and plants, fungi, and other sessile organisms by morphology. Architecture and behavior are, however, compromise responses to many selection pressures, of which energy acquisition and allocation is only one. In the past two decades classic OFT has given way to increased focus on the search process per se, the extremes of which are random (as represented by Brownian walks or Lévy walks) or systematic exploration. Both types of searches occur among microorganisms as well as macroorganisms.

From the standpoint of the individual and with respect to nutrition, versatility can be defined either as the ability to do many things, or to respond rapidly to new conditions. As judged by the first definition it cannot be said definitively whether microbes are more or less versatile than macroorganisms because all of the metabolites for even a single life form are as yet unknown. Metabolic reconstructions are being projected based mainly on genomic annotations and exist for several prokaryotes and yeasts, and a few multicellular eukaryotes. Although the metabolic pathways and metabolites in some cell types of a multicellular eukaryotic organism may be comparatively few, in aggregate for the individual they likely exceed those for a microorganism, given the range of cell types, the subcellular compartmentalization, and the diversity of complex chemicals produced. When versatility is construed as speed of response, bacteria (and possibly some other microbes such as the fungi) appear to be more versatile metabolically than macroorganisms because they can rapidly switch entire metabolic pathways.

Organisms are either relative generalists or specialists with respect to dietary range. For the generalist macroorganism with diverse prey items or the generalist microbe able to use many substrates, 'food' is potentially easier to find, substitutable resources can be alternated, search times are potentially shorter, and starvation is less likely. When food sources exert evolutionary pressure by their structural, behavioral, or physiological complexity, a specialized response by the consumer, such as restriction in dietary breadth, is a common result. This may be seen, for example, among individual bumblebees specializing as foragers on a particular flower type, by the coevolution of parasites with their hosts, and by specialist microbial strains that will not grow in the absence of a specific energy source. Broadly speaking, specializations offer advantages (optimization at doing certain things) but also impose disadvantages (fewer options). Evolution tends to move organisms towards increasing specialization, thereby narrowing options and limiting what they can do. Some of the best examples of specialization are found in coevolutionary relationships, where iterations involving the change in one partner (such as evolution of host resistance to parasites) are matched by corresponding changes in the other (evolution of virulence factors).

A corollary to optimal foraging theory is that generalists are predicted to be less efficient in finding and/or exploiting any particular resource than specialists on that resource. This has generally been accepted intuitively as reflected in the adage 'the jack-of-all-trades is master of none' and it has been documented with theoretical models going back to the early work of Robert MacArthur, among others. Nevertheless, trade-offs are not necessarily inherent and in some cases traits promoting performance in one environment do so also in others. The supporting evidence from field studies for trade-offs is mixed and subject to caveats. For example, there is difficulty in isolating the variable of interest (feeding range) with experimental systems of macroorganisms and, for microorganisms, in designing critical experiments (such as in determining how long it will take for trade-offs to manifest themselves), and in extrapolating the results of competition experiments between microbial strains judged to be generalists versus specialists under laboratory conditions to nature. A broader limitation is that organisms perform multiple tasks simultaneously and overall fitness is some aggregate function of which foraging is only one component.

Suggested Additional Reading

Alberts, B. et al. 2015. Molecular Biology of the Cell, 6th Ed. Garland Science, N.Y. *Excellent overview of cellular biochemistry, structure, and genetics, including comparisons of prokaryotes and eukaryotes.*

Caron, D.A. et al. 2017. Probing the evolution, ecology and physiology of marine protists using transcriptomics. Nature Rev. Microbiol. 15: 6–20. *An excellent synthesis of marine protists and their role in the ecology of the oceans, including mixotrophy.*

Gross, T. and H. Sayana (Eds.). 2009. Adaptive Networks: Theory, Models and Applications. Springer, NY. *Attributes and commonalities among abiotic and biotic networks and branching systems.*

Monk, J., J. Nogales, and B.O. Palsson. 2014. Optimizing genome-scale network reconstructions. Nature Biotechnol. 32: 447–452. *A comprehensive status report on knowledge of the projected metabolism for 78 species across the tree of life based largely on genome annotation.*

Oceanography, vol. 20 no.2, June 2007. Special Issue: A Sea of Microbes. *A good compendium of papers on the centrality of microorganisms in marine ecology. See also the compiled papers in "Microbial Carbon Pump in the Ocean" (eds. N. Jiao, F. Azam, S. Sanders) Supplement to Science, 13 May 2011; and M.A. Moran (2015), The global ocean microbiome. Science 350, doi:10.1126/science.aac8455*

Science, vol. 348, no.6237, pp.833–940, 22 May 2015. *Special issue on oceanic plankton, including abiotic as well as biotic factors.*

Size

© Springer Science+Business Media LLC 2017
J.H. Andrews, *Comparative Ecology of Microorganisms and Macroorganisms*,
DOI 10.1007/978-1-4939-6897-8_4

What is a microorganism? There is no simple answer to this question. The word 'microorganism' is not the name of a group of related organisms, as are the words 'plants' or 'invertebrates' or 'frogs'. The use of the word does, however, indicate that there is something special about small organisms; we use no special word to denote large animals or medium-sized ones.

W. R. Sistrom, 1969, p. 1

Not just any organism is possible.

F. Jacob, 1982, p. 21

4.1   Introduction

The most striking feature about an assemblage of different organisms is the distinction among species in size and architecture. Excluding entities such as viruses that reproduce but are generally regarded as nonliving, the size range spans at least 21 orders of magnitude from wall-less bacteria known as mycoplasmas at about 10^{-13} g to blue whales, which exceed 10^8 g. The blue whale, incidentally, is the largest animal ever known. It is about twice as big as the largest dinosaur *Brachiosaurus*, which probably weighed about 85 tons (ca. 8.5×10^7 g), and is equivalent in mass to about 40 or 50 elephants at approximately 3 tons each (2–3×10^7 g). The range as conventionally depicted (e.g., McMahon and Bonner 1983) also is noteworthy because it does not include plants (or massive fungi discussed in ►Chap. 5), the largest of which far outstrip even whales. For instance, the giant sequoias (*Sequoiadendron giganteum*), although shorter than the coastal redwoods (discussed in ►Chap. 7), weigh much more because of their larger girth and have been estimated at about 2×10^9 g (this includes only above-ground biomass, not the roots). And, a uniform stand of a clonal tree, aspen, estimated to comprise 47,000 stems over an area of 43 ha, is said to weigh perhaps 6×10^9 g (Grant et al. 1992). At the other end of the scale are the tiny prokaryotes; remarkably, even among them the size range is over many orders of magnitude from the nanobacteria to species that are macroscopically visible (Schulz et al. 1999; Schulz and Jorgensen 2001; Zinder and Dworkin 2013) (◻Fig. 4.1).

Two further observations on size are worth highlighting. First, it is common to refer to sizes of macroorganisms as the adult form, which is the life stage we mentally visualize as being the organism. With respect to size, the specific stage of the life cycle matters relatively little for a bacterium but a lot—more than an order of magnitude—for, say, a juvenile versus an adult elephant, and several orders if the mature form is compared with the zygote or seed or fetus. Bonner (1965) has compared the size of organisms at various stages of their life cycles. For all sexually reproducing species there is a point in the life cycle where the individual consists of a single cell, the zygote, and there are as well all the intervening sizes and shapes to maturity. The size counterpart in many asexually reproducing species is the unicellular spore. This aspect will be developed in ►Chap. 5.

Second, note that the scale referred to above is not readily applicable to modular organisms, such as bracken fern covering a hillside. Such clonal creatures, as genetic individuals, are indefinite in size. We will pursue this topic in ►Chap. 5, but it is worth commenting here that Oinonen (1967) observed clones of bracken fern (*Pteridium aquilinum*) extending almost 500 m across that were estimated at 700 years old, and he suggested that others may reach 1400 years. Clones of sea anemone may approximate several hundred meters in area

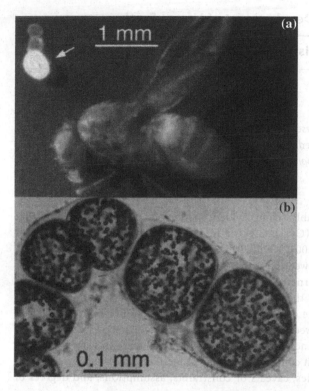

□ **Fig. 4.1** The giant sulfur bacterium *Thiomargarita namibiensis*, some cells of which can have diameters as large as 750 nm. (*top*) The *arrow points* to a single cell, appearing white because of sulfur inclusions. A fruit fly, 3-mm long, appears nearby to illustrate relative scale. (*bottom*) A chain of *Thiomargarita* cells as viewed by light microscopy. From Schulz et al. (1999). Reproduced from *Science* by permission of the American Association for the Advancement of Science ©1999; permission conveyed through the Copyright Clearance Center

and "probably persist for many decades" (Sebens 1983, p. 441). While as a result of environmental vicissitude clonal organisms may not come to reach the biomass of a blue whale, they can theoretically weigh more and they do potentially occupy very large areas. Clonal plants and animals escape many of the problems of large size, such as having to support a massive bulk. Such organisms may grow prostrate rather than in upright fashion, they tend to get large simply by the iteration of countless small parts, and often spread themselves about by fragmentation (►Chap. 5).

Here we examine from several angles how size affects organisms. First, since all forms of life were initially microorganisms, why, when, and how did macroorganisms evolve? To what extent do very small and very large organisms 'see' the world differently? And how small or how large can an organism even be and what sets the limits? While clearly there has been an increase in size overall in the evolutionary history of Earth, is this trend—and its correlate, increasing complexity—purely due to passive drift and the increase in variance in size of any taxon, or active and directed? Finally, to what extent are life history attributes such as dispersal and reproduction influenced by the size of a propagule or organism?

4.2   Changes in Size and Development of Life on Earth

4.2.1   Origins, Geobiochemistry, and Phylogenetic Benchmarks

》 *Life is older than organisms.*

G. Ehrensvärd (1962)

》 *It is not difficult to argue that research on the beginning of life represents one of the last bastions of classical science, defined by the significance of its central goal, its breadth of scope, and a ratio of hypothesis to fact approaching infinity.*

D. Deamer (1997)

There is complementary, reasonably strong evidence from geology and astronomy that Earth originated about 4.5 Gya (Gya = Giga or billion years ago) from a collision of planetesimals (Nisbet and Sleep 2001; Knoll 2015). Following a period of cooling, a crust was formed and temperatures allowed the existence of liquid water and other geochemical conditions conducive for the formation of the organic precursors of life. Life probably emerged sometime between the end of the last accretion impacts from meteorites (~4 Gya) and the formation of the earliest chemical fossils, which date to about 3.8 Gya (Barton et al. 2007). The broad sweep of early evolution of life on Earth is summarized in ◘Fig. 4.2. It should be kept in mind that all such reconstructions are to varying degrees contentious and whether based on combined or separate fossil and molecular approaches are subject to various uncertainties, caveats or constraints, assumptions, and degrees of inference.

A cornerstone of the current scientific paradigm for the early development of life is the so-called **'Universal Ancestor' (UA)**, which is a phylogenetic term for origin of the 'tree of life' and the nebulous entity(ies) from which all known life presumptively evolved (Woese 1998a). A common evolutionary origin is the most direct (parsimonious) inference from the fact that living creatures exhibit universal homologies, most notably a standard biochemical imprint of informational and building block molecules and cellular attributes summarized in ▶Chap. 3. Arguably, the UA was not a singular entity as visualized by Darwin, but may have been a fluid community of diverse, freely intercommunicating primordial cells variously termed precells, protocells, or progenotes. Although the word **'progenote'** has come to represent different things, as originally articulated by Woese in several papers (e.g., 1987, 1998a, b) it was, as distinct from "genote", a cellular construct that had an incomplete linkage between its genotype and phenotype and other features such as a short-stranded RNA as the information molecule; few disjointed genes; no cell wall; simple metabolism; and free cross-feeding with the environment and other protocells. Interestingly, if correct, this interpretation implies that contemporary horizontal gene transfer in prokaryotes and eukaryotes (▶Chap. 2) may be a vestige of the profligate rudimentary horizontal transfer of the primordial world.

There is no unambiguous fossil record for the cryptic progenote and placement of the era is by extrapolation. The lower boundary is set from the estimated age of ancient prebiotic assemblages obtained by geologic biosignatures in rock dated radiometrically; the upper boundary is marked by the first occurrence of putative stromatolites indicative of microbial

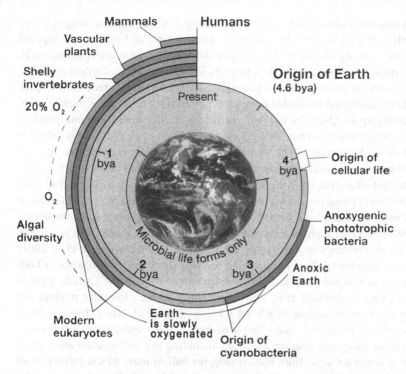

☐ Fig. 4.2 Overview of the history of life from the origin of Earth 4.5–4.6 billion years ago (bya or Gya) to the present. Cellular life probably arose ca. 3.8–4.0 bya and for half or more of the history of life on Earth was exclusively microbial. Very early life forms were anoxygenic chemotrophs and phototrophs followed by oxygenic (oxygen-generating), phototrophic cyanobacteria, which set the stage for evolution of increasingly complex, multicellular eukaryotes. For synopsis, see Hedges (2002). Reproduced from: Madigan, Michael T.; Martinko, John M.; Stahl, David A.; Clark, David P. *Brock Biology of Microorganisms, 13th Ed.* ©2012. Printed and electronically reproduced by permission of Pearson Education, Inc. New York, New York. Image of Earth from space produced by M. Jentoft-Nilsen, F. Hasler, D. Chesters (NASA/Goddard) and T. Nielsen (Univ. Hawaii)

(mainly cyanobacterial[1]) activity, some arguably as old as ~3.4–3.7 Gya (Tice and Lowe 2004; Schopf et al. 2007; Nutman et al. 2016). Nisbet and Sleep (2001, p. 1086) define **stromatolites** as *"organosedimentary structures produced by microbial trapping, binding, and precipitation, generally but not always photosynthetic"*. The reliability of many such early traces is vigorously contested (Brasier et al. 2006). Indeed, as alluded to above, all models for the origin of life are controversial, but in most of them the UA preceded the first simple, unicellular prokaryotes. This means that it would have been older than ~3.5 Gya. As noted at the outset, the earliest chemical fossils date to about 3.8 Gya in the late Archaean Eon (3.8–2.5 Gya) with ostensible biosignatures and microfossils variously attributed to hypothesized anoxygenic phototrophs or iron/S oxidizers by the mid-Archaean.

---

1   Cyanobacteria or blue-green algae comprise a large, ecologically and morphologically diverse group of oxygenic, phototrophic prokaryotes containing chlorophyll *a* and phycobilins, from which chloroplasts evolved in plants. They were primarily responsible for the oxygenation of early Earth beginning in the mid-late Archaean Eon (Schirrmeister et al. 2011; Madigan et al. 2015).

Geochemical markers indicative of oxidized rock imply that atmospheric oxygen levels, attributable largely to the photosynthetic activity of cyanobacteria and sufficient to support aerobic respiration, were significant by 2.8–2.2 Gya (Nisbet and Sleep 2001; Bekker et al. 2004). In other words, for about the first billion years Earth was anoxic. Oxygen availability was a critical event for several reasons, in major part because the absence of complex eukaryotes has been attributed to limited oxygen (King 2004). Presumptive multicellular, colonial macrofossils up to 12-cm in size and ascribed as living aerobically and possibly in mat-like arrays have been reported from 2.1 Gya (El Albani et al. 2010). This report is debatable and it is unclear whether the structures, if representing macroscopic life, are prokaryotic or eukaryotic. Unicellular eukaryotes (protists) arguably arose more or less concurrently with mitochondria, likely by ~1.5 Gya and possibly much earlier (Dyall et al. 2004). Pace (2009) infers based on molecular phylogenies that the eukaryote *nuclear* line of descent is as old as the archaeal, i.e., that the two forms have been present since the beginning of recognized life. What such an incipient eukaryote may have looked like is unclear. Fossils interpreted as presumptive eukaryotes are dated at 1.45 Gya (Javaux et al. 2001) and an authoritative benchmark for a multicellular red alga (implying the presence of both mitochondria and plastids) was recovered from sedimentary rock in the Canadian arctic dated at 0.75–1.25 Gya (Butterfield et al. 1990; Butterfield 2009). Molecular phylogenies now suggest that the green algae date to at least 1.5 Gya (Yoon et al. 2004). Thus the fossil and molecular evidence point to a major burst of multicellular eukaryotes dated about 1 Gya (King 2004; Grosberg and Strathmann 2007), including the multicellular ancestor of modern animals at about 0.6 Gya. **Thus, remarkably, for half or more of the history of all life on Earth the biota was exclusively microorganisms, and overwhelmingly if not exclusively, prokaryotes.**

How did the earliest life forms provide for themselves? Among the innumerable possibilities, there are three classical hypotheses (not mutually exclusive) for their functional or trophic nature. Because of the complexity of chlorophyll or similar pigments as well as of the associated energy transduction pathways, all three models visualize the incipient forms as being anaerobic chemotrophs rather than phototrophs. Beyond that, they are postulated as being either: (i) organoheterotrophs ("prebiotic or organic soup theory"; Bada 2004); (ii) lithoautotroph producers, possibly hyperthermophiles ("metabolist theory"; Wächtershäuser 2006); or (iii) lithoheterotroph organic conservers (strictly neither producers nor conservers; Ferry and House 2006; for terminology see ▶Chap. 3). Whatever their exact nature, all would have required then, as now, chemical substrates both for assimilation into monomers and polymers of cell biomass, as well as for dissimilation for energy generation. This, in turn, depends on linked redox reactions, i.e., nonequilibrium electron transfers from donors to acceptors. It is logical that because of the complexity of needed incipient biogeochemical pathways and the expense of autotrophic organosynthesis, the earliest cells probably acquired their building blocks from geochemical or extraterrestrial sources. The probable route for energy generation was via simple inorganic electron donors such as cyanide, CO, H_2, and FeS arising from cold seeps or thermal vents (for examples, see Amend and Shock 2001; Martin et al. 2008; Knoll 2015). Regardless of the specific scenario, it is evident **that emerging catalytic (enzymatic) control of chemical disequilibrium thermodynamic forces was the bridge between chemical and biological evolution. This critical step is probably the most fundamental common denominator in the origin of life.**

4.2.2 From Unicell to Multicell; Microorganism to Macroorganism

The first prokaryotes emerging from the progenote presumptively were unicellular (Kaiser 2001) though they may have existed in more-or-less organized colonies (Koonin and Martin 2005). Their nature is obscure because the very early fossil records are nonexistent or largely obliterated for various reasons (Brasier et al. 2006; Schopf et al. 2007). As noted above, to date the first relatively convincing physical evidence of microbial life is from stromatolites suggestive of filamentous and mat-forming cyanobacteria and other forms, followed by evidence for incipient cellular differentiation and diversification about 2 Gya (see comments on Shape below; Nisbet and Sleep 2001; Tomitani et al. 2006; Schirrmeister et al. 2011). So, progression to the ultimately intricate forms of multicellularity really has modest roots among the unicellular and filamentous *Bacteria* and *Archaea* (Spang et al. 2015) and, subsequently, the unicellular protists.

Was the primordial shape spherical and if so, why? Although extant prokaryotes are preponderantly unicells, multicellular and in many cases differentiated forms such as the filamentous cyanobacteria, myxobacteria, and actinomycetes have evolved independently multiple times within and among the bacterial clades (Brun and Janakiraman 2000; Flärdh and Buttner 2009; Flores and Herrero 2010; Claessen et al. 2014). The longstanding dogma is that bacterial cells are coccoid or rod-shaped and that the primordial prokaryote was spheroidal. Of all possible designs, a true sphere encloses the greatest volume with the least surface area (disadvantageous for transfer processes as we shall see later, but advantageous structurally to restrain turgor pressure). Hence, a spherical shell is robust, spacious, and economical. It seems a logical starting point for elaborations on the architecture of organisms, including the cylindrical form and embellishments thereof (Young 2006; see also Dusenbery 1998), also common among microbes.[2]

Despite their common simplicity of form, however, the extant species of prokaryotes actually span the realm of wondrous cell shapes; inference from molecular phylogenetics suggests the earliest cells were probably rods or filaments and the cocci are a derived morphotype that arose multiple times as an end-state in various bacterial clades (Siefert and Fox 1998; Young 2006). (Selective forces and bacterial shape are discussed at length by Young 2006.) Even the 'conventional', unicellular bacteria can modulate growth form advantageously and operate effectively as multicellular organisms in many ways (Shapiro 1998; van Gestel et al. 2015).

---

2   With some exceptions, small organisms and the propagules of most organisms assume simple, often spherical or sub-spherical, shapes. Such forms are exceedingly rare in the external morphology of macroorganisms. One reason that a curved (oblate) shape is a good design is that the container, whether a cell wall or the steel skin of storage tank, must withstand tensile but not bending stresses. Furthermore, because the covering is under the same tension per unit length throughout, there is no region more likely to rupture than any other (Chap. 3 in Thompson 1961; Chap. 6 in McMahon and Bonner 1983).

   Not far removed from the sphere, an example of a shape universal in both the abiotic and biotic worlds is cubic symmetry. A commonly replicated form is the icosahedron. Icosahedra are regular polyhedra with 20 faces, 12 vertices, and 30 edges. Built upon the fundamental unit of triangles (the only two-dimensional structures that do not deform under pressure), the icosahedron is robust, encloses considerable volume, and can be iterated almost without limit. There are no known icosahedral bacteria, but the shape is found in objects as small as the outer shell of many viruses, ranging up to the complex architectures of plants and animals. As early as the 1960s, Caspar and Klug (1962), in what was destined to be a landmark paper, described the role of icosahedra in viral form; Allen and Hoekstra (1992, pp. 174–188; see also Vogel 1988; Stewart 1998) recount how they are a fundamental building block in the assembly of macroorganisms.

Multicellular embellishment of shape The path to true multicellularity could have arisen by aggregation of cells embedded within mucilage (primordial biofilm); by incomplete cell fission producing chains; or by formation of syncytial filaments, i.e., from an initial condition of a common multinucleated cytoplasm that gave rise subsequently to branching filaments with cross-walls (Claessen et al. 2014). Examples are discussed later in the section Ways to Become Large. It is largely subjective whether a 'multicellular entity' such as some of the early or even extant colonial forms really operated as a multicellular individual, as opposed to being little more than a biofilm or an aggregated, more-or-less uncoordinated cell mass of indeterminate shape (for the developmental stages of a contemporary biofilm, see Vlamakis et al. 2013). The trend to multicellularity (an emergent individual with higher order patterns) was associated with many related and perhaps prerequisite biological changes facilitating coordination and mutual dependency. These changes include such attributes as cell-to-cell adhesion, intercellular connections, differentiation, and communication (and progressively more complex, integrated functions including developmental signaling, complex gene regulatory networks, and programmed cell death in some lineages) (Grosberg and Strathmann 2007; Rokas 2008a, b; Butterfield 2009; Srivastava et al. 2010; Abedin and King 2010). These important changes evidently were staged against a backdrop of multiple forces and converging conditions: biogeochemical, natural selection, genomic, and historical contingency (Carroll 2001; King 2004). Increase in complexity and interdependence is nicely illustrated by the volvocine algae discussed later.

General phylogenetic trends The major multicellular lineages evidently all arose and developed separately from *different* unicellular progenitors (Ruiz-Trillo et al. 2007; Abedin and King 2010): For example, land plants (embryophytes) probably originated from a unicellular, flagellated eukaryotic photoautotroph (likely from similar but genetically distinct flagellates that evolved at different times) and subsequently through a multicellular clade of charophyte green algae to the bryophytes (liverworts, mosses, hornworts) (Niklas and Kutschera 2009; Pires and Dolan 2012). The last common ancestor of animals, the so-called Urmetazoan, was multicellular and evolved presumptively from a colonial flagellate, which in turn evolved from a unicellular flagellate, likely a choanoflagellate, the closest known relatives to animals (King 2004; King et al. 2008; Richter and King 2013). Although the origins of the Kingdom *Fungi* remain more contentious, molecular phylogenies now convincingly support the interpretation that the fungi and animals shared a unicellular opisthokont ancestor (the fungi likely arising from a nucleariid-like amoeboid protist; Steenkamp et al. 2006) and, furthermore, it appears that the incipient animal and fungal lineages subsequently diverged with each having its own unicellular origin (Ruiz-Trillo et al. 2007; Knoll 2011).

Overall, eukaryotic multicellularity is estimated to have arisen *independently* at least 25 times—sometimes through an intermediary colonial condition (King 2004)—and with occasional reversals (Buss 1987; Grosberg and Strathmann 2007; Niklas 2014). That origins are 'independent' has to be inferred, usually from molecular phylogenies supplemented in some cases by evidence from the pattern and placement of cells. Among extant organisms, the multicellular condition appears in at least 16 independent eukaryotic lineages (King 2004; Rokas 2008; Knoll 2011) (◘Fig. 4.3). A few of these eukaryotes, such as the dimorphic species of fungi, can oscillate between a unicellular and multicellular state dependent on environmental conditions. Others, such as the Myxomycota (including the dictyostelid and acrasid cellular slime molds or social amoebas), as well as the true (plasmodial) slime molds (myxomycetes), have a developmental program featuring both a unicellular and multicellular, differentiated phase. Clues to the origins of multicellularity and incipient differentiation probably reside in

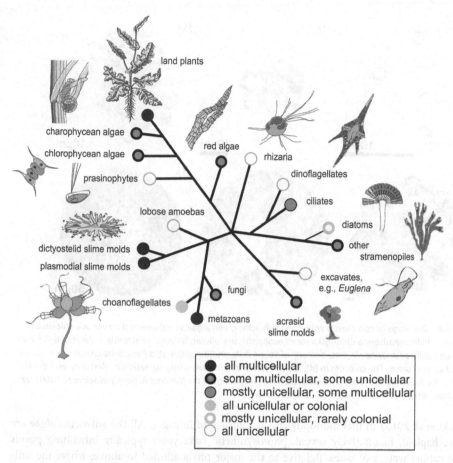

☐ **Fig. 4.3** Diverse representation of multicellularity among the major eukaryote clades shown on a simplified, unrooted phylogenetic tree. Some lineages are entirely unicellular or multicellular but most are mixed. For discussion of multicellularity among prokaryotes, see text and Claessen et al. (2014); for discussion of animals, see text and King (2004). From Niklas (2014); figure reproduced from the *American Journal of Botany* by permission of Karl Niklas and The Botanical Society of America © 2014

some of these borderline, metastable taxa. 'Metastable' is used here in the sense that a cell aggregate can differentiate to somatic stalk cells and totipotent spores in a spore-bearing structure (sorocarp) but such states may revert to a multicellular, undifferentiated form with few if any somatic cells (Buss 1982, 1987). (See Tarnita et al. (2015) for some of the genetic and ecological consequences of this extraordinary life cycle.) The remarkable fact is that of the many explorations of the multicellular lifestyle, in only three groups—plants, fungi, and animals—does cellular differentiation occur among many species. Why this major transition to a differentiated entity was closed to all but a few taxa remains a major mystery (Buss 1987, pp. 69–77).

Multicellular origins and the volvocine algal paradigm The green alga *Volvox* and its relatives (three families and about 50 species) provide a fascinating case study in the transition from unicellularity to multicellularity and extensive differentiation, including many examples of inferred multiple origins and reversals (Herron and Michod 2008; Herron et al. 2009;

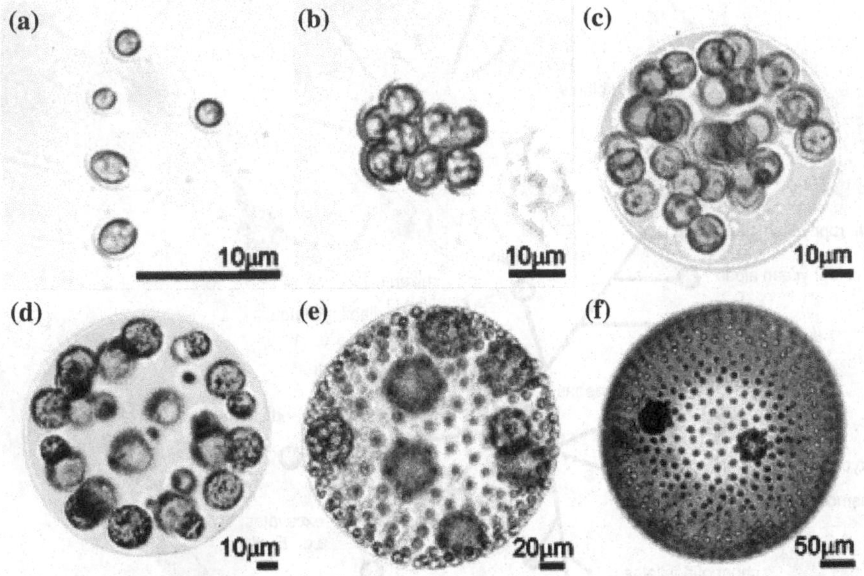

Fig. 4.4   The range of complexity within the volvocine green algae as represented by various unicellular and multicellular members: **a** *Chlamydomonas reinhardtii*, unicellular; **b** *Gonium pectorale*, a sheet of 8–32 undifferentiated cells; **c** *Eudorina elegans*, a colony of 16–64 cells, undifferentiated; **d** *Pleodorina californica*; **e** *Volvox carteri*; **f** *Volvox aureus*. The colonies in (d), (e), and (f) are differentiated into somatic (smaller) and reproductive (larger) cells. From Michod (2007); reproduced from *Proceedings of the National Academy of Sciences* ©2007 by permission of the National Academy of Sciences, USA

Arakaki et al. 2013) in this monophyletic assemblage of lineages. All the volvocine algae are motile, haploid, facultatively sexual, photosynthetic eukaryotes typically inhabiting ponds or the calmer waters of lakes. Relative to the major phyla alluded to above, where the unicellular to multicellular transition is much more ancient (on the order of 1 Gya), these algae evolved relatively recently (~200–300 million years ago = Mya). The forms spanned single-celled ancestors probably like present-day *Chlamydomonas* or *Vitreochlamys* to multicellular species varying from a few up to ~50,000 cells in the relatively complex, differentiated *Volvox* (Herron et al. 2009) (**Fig. 4.4**). The colonial forms range from relatively simple, totipotent, undifferentiated clumps of cells that remain attached after division (such as *Gonium* or *Eudorina*), to small colonies of various shapes (*Pandorina*), to a large sphere composed of dozens of cells with incipient differentiation in cell size and function (*Eudorina, Pleodorina*), culminating in thousands of biflagellate, peripheral somatic cells surrounding a few internal germ cells embedded in transparent, glycoprotein-rich extracellular matrix (*Volvox* noted above) (Kirk 2005). However, the intermediate gradations in organization are best interpreted *not* as a linear progression of transitional states but as alternative stable states (Larson et al. 1992).

Along with increase in cell number in the volvocine lineage have come increased integration, intercellular communication, germ/soma specialization (division of labor), complexity, and individuality, as previously individual cells or small groups of cells coalesced to become higher order individuals (see comments in ▶Chap. 1 and Michod 2007; Herron and Michod 2008). Genomic comparative analysis of the unicellular *Chlamydomonas reinhardtii* and its multicellular relative *Volvox carteri* shows that, with few exceptions, the increase in developmental complexity evidently centered on expansion of lineage-specific proteins rather than extensive protein-coding innovation (for analogous comparison with animals, see King et al. 2008). In other words, it did not involve major change in the original protein-coding repertoire, suggesting that ancestral genes were co-opted into new processes (Prochnik et al. 2010).

The evolution of germ and soma in the volvocaceans illustrates nicely the general principle of increasing complexity associated with multicellularity and the trade-offs inherent in cell specialization and associated division of labor. Because the algal cells are denser than water they need motility to maintain an optimal position in the photic zone; extensive day/night migrations occur, presumably in response to nutrient fluctuations and too much or too little light (Koufopanou 1994). For the larger colonies, organized flagellar beating also provides stirring of the boundary layer, which facilitates nutrient transfer at rates higher than diffusion alone (Solari et al. 2006a, b). However, there is a constraint between cell division and locomotion imposed by the unique structure of flagella. These organelles are essentially microtubular assemblages attached to the cell through their basal bodies; they also are connected to the nuclear region where a 'microtubule organizing center' (centriole or centrosome in some terminology) plays a key role in cell division by acting as the mitotic spindle. The centers cannot function in both capacities simultaneously. (A similar division and ciliation constraint applies in metazoans; Buss 1987, pp. 35–44.) Flagella can continue to beat without their basal bodies for only about five rounds of cell division (32 cells). Thus, to remain suspended, the larger colonies (implying multiple rounds of cell division) would sink before completing development. An evolutionary solution to this problem was a division of labor whereby such colonies consist of peripheral, sterile, and ultimately mortal somatic tissue with functional flagella to maintain buoyancy and propulsion, supporting non-flagellated, reproductive, immortal germ cells internal to the somatic layer (in *Volvox*).

Somatic tissue, i.e., differentiation, originates in colonies of 32–128 cells (Koufopanou 1994) and as colony size increases in the Volvocaceae the ratio of somatic to germ cells increases. Detailed hydrodynamic analyses (Solari et al. 2006a, b) following up the work of Koufopanou (1994) show that the relatively increased soma is needed to maintain buoyancy/motility and internalization of germ cells decreases colony drag. A greater allocation of biomass to germ tissue increases colony fecundity but also increases gravitational force and decreases swimming speed. Thus, there are trade-offs among reproduction, motility, and size. Despite these trade-offs, multicellularity along with differentiation arose several times in *Volvox*, attesting to the advantages of the multicellular and differentiated condition.

Natural selection and the multicellular condition The budding yeast *Saccharomyces cerevisiae*, well known in industry as well as in genetics and molecular biology, is preponderantly unicellular (Brückner and Mösch 2012). Using a population of initially unicellular cells, Ratcliff et al. (2012) reenacted in contemporary time the evolutionary transition to multicellularity over geological time. They show in principle that at least the incipient stages can occur relatively quickly under appropriate selective conditions. Having arisen in the prehistoric microbiological world, why might multicellularity have been maintained by natural selection? The main benefits attributed to increased size involve, broadly speaking, more effective resource acquisition or avoidance of size-selective predation (phagotrophy). Additional advantages probably include (i) provision of storage reserves; (ii) provision of an internal environment protected by outer layer(s) of cells; (iii) allowance of novel metabolic innovations; (iv) enhancement of motility (Grosberg and Strathmann 2007). Bell and Koufopanou (1991) approach the question by defining the benefits of small size and then attributing the selection for large size to failure of the assumptions for benefit of smallness. Their most simple theoretical case involves a life cycle of rapid growth from an initial to final size followed by fission. In a uniform, invariant environment with no interactions among individuals, fitness would approximate the growth rate r, with r approaching r_{max} in optimal conditions, being greatest for small organisms and decreasing with size (see comments on r-selection, later). Effectively this describes bacterial exponential growth dynamics in continuous culture. In more realistic cases where the environment varies in time and space and individual interactions occur, larger forms typically have several advantages.

The multicellular gliding bacterium *Myxococcus* is a nice example of how a consortium of cells can operate advantageously. *Myxococcus* secretes various enzymes that degrade other bacteria on which it feeds in its soil habitat. Likely the organism functions better as a collective than it would as individual foraging cells. In fact, it is social in its eating habits: as Kaiser (1986, p. 541) says, "*Myxococcus* feeds as a pack of microbial wolves, with each cell benefitting from the enzymes secreted by its mates". Each cyst is a package of cells, ready upon germination for activity as a cooperative group, and the fruiting body is of sufficient size (~ 0.2 mm; Kaiser 2001) that it is readily transported by small animals in the soil. Such groups of cells and eventually multicellular assemblages are also more resistant to abiotic disturbance (erosion from surfaces, desiccation, UV injury, etc.) as is apparent from the biology of bacterial biofilms (Teschler et al. 2015).

An important consequence of the transition from unicellularity to differentiated multicellularity was the transfer in fitness from lower level units (cells or groups of cells) to the higher level individual (Buss 1987; Michod 2006, 2007; see also comments in ►Chap. 1). This arose inevitably as a compromise associated with cell specialization wherein formerly totipotent, independent cells gave up their autonomy to collaborate, thereby becoming subordinate and dependent on the group for survival and reproduction. This had some very important evolutionary consequences. Some cells acquired only a somatic role and could no longer form a new individual. Furthermore, the division of labor allowed fitness of the integrated group to be augmented over that of the average fitness of its component units.

Moreover, the origin of cooperation was followed by cheating, wherein a mutant cell lineage proliferates at the expense of the whole individual (Hammerschmidt et al. 2014). This situation arises by mutation in the original clonal lineage; an analogous threat arises when different genotypes fuse and one overtakes the other (Buss 1982, 1987; Rainey and De Monte 2014). Such genetic conflicts would tend to counteract the benefits of size increase (Grosberg and Strathmann 2007). That there are multiple defenses against such defections or invasion by other genotypes and that they are phylogenetically widespread, even into the most primitive of taxa, testifies to the threat that genomic conflict poses to the evolutionary individual. These defenses include: passage through a single-cell stage typical of life histories, germ-line sequestration in many organisms, and various self/non-self recognition systems. Even in the earliest eukaryotes such countermeasures exist: In the cellular slime molds noted earlier, 'the individual' spends much of its life in disaggregated form as independently foraging cells on the forest floor and the individual reunites as a consolidated entity (and hence potentially vulnerable) for a relatively brief phase associated with reproduction; whereas in *Volvox*, defense takes the form of early differentiation of germ cells so that invasion by variants is limited to the brief phase of the first few cell generations early in morphogenesis (not unlike the embryogenesis pattern in many animals (see Buss 1987 and Chap. 5).

Ways to become large ◾Fig. 4.5 shows conceptually in simplistic terms the options for increase in size, with some examples from extant organisms. Up to a point, presumably set by physicochemical factors related to surface:volume exchange processes (section, Seeing the World, below), unicells can simply increase their volume. Indeed, as alluded to at the outset,

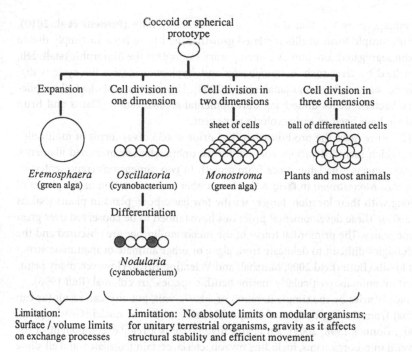

Fig. 4.5 Four simplified, hypothetical examples of how increase in size, multicellularity, and differentiation could have evolved. Spherical geometry is shown for simplicity; a very early shape also was filamentous. In addition to incipient multicellularity occurring from cohesion of the products of cell division, it could have resulted from aggregation of compatible cells. Representative genera are extant forms

a few species of prokaryotes are relatively immense, but generally unicellular eukaryotes are larger. Multicellularity confers the advantages of large size achieved by the building block approach. This method resolves metabolic concerns related to cell surface area:volume scaling by delimiting the size of the individual blocks, and allows for regularized genetic oversight of the developing soma by mitosis (Chap. 4 in Bonner 1988). A developmental constraint here is the nature of the starting material: Walled forms of green algae developed extensive multicellularity, whereas naked forms did not (Bell and Koufopanou 1991; Koufopanou 1994). A simple extension of the unicellular plan was for cells to divide but rather than separating be retained in one dimension (see **Fig. 4.5). This results in unbranched filamentous growth. McShea (2001) describes insightfully for hierarchical organisms how clones and colonial organisms may have assembled and a possible geological timeframe for the events.

The earliest multicellular eukaryotes were on the scale of millimeters and also took the form of linear or branched filaments (Carroll 2001). Presumably undifferentiated multicellular clusters (Pfeiffer and Bonhoeffer 2003) or filaments occurred first, followed by differentiated cells, such as heterocysts (specialized for nitrogen fixation) or resting spores known as akinetes in cyanobacteria (Rossetti et al. 2010; Flores and Herrero 2010; Schirrmeister et al. 2011). Because nitrogen fixation is an anaerobic process destroyed by even trace amounts of oxygen, unicellular cyanobacteria such as *Gloeothece* must alternate photosynthesis by day with nitrogen fixation at night. However, their differentiated, multicellular relatives such as *Nostoc* and *Anabaena* can simultaneously conduct photosynthesis in vegetative cells and fix nitrogen in the heterocysts. There is some evidence that the multicellular species have a competitive advantage provided that day length is sufficiently long (Rossetti et al. 2010). A morphologically simple form of differentiated growth would have been to simply divide into two cells with segregated function. A contemporary example is the dimorphic (stalk cell, swarmer cell) stalked bacteria such as *Caulobacter* and *Hyphomonas* (this life cycle is discussed in ▶Chap. 6; see also Brun and Janakiraman 2000). However, morphological simplicity of the stalked bacteria belies underlying developmental sophistication (Curtis and Brun 2010) and could not have been an early evolutionary event.

Butterfield (2009) reviews the pre-Edicarian (i.e., prior to 635 Mya) forms of multicellular organization, which included various colonial forms, unbranched or branched filaments, and monostromatic sheets. The latter arose from growth in two dimensions (represented by the extant green alga *Monostroma* in ◘Fig. 4.5). The number and orientation of the planes of cell division, along with their location, largely set the few basic body plans in plants (Niklas and Kutschera 2009). These developmental processes have tended to be conserved over great evolutionary time scales. The primordial traces of the metazoan lineage are obscured and the incipient morphologies difficult to delineate from algae or other animate or inanimate structures in ancient fossils (Butterfield 2009; Marshall and Valentine 2010). However, many primitive as well as extant animals, particularly marine benthic species, are colonial (Raff 1996).

Colonies typically arise by the simple iteration of a basic unit, but also can result (as can other body forms) from the aggregation of compatible cells, tissues, or nuclei (Grosberg and Strathmann 2007; Bonner 1998, 2009). An aggregative phase is a regular component of the life cycles of several microorganisms, including myxobacteria, certain flagellates, and all slime molds (but especially the cellular slime molds), noted above. An extension of this aggregative style occurs in fusion chimeras, where different clones (usually of sessile organisms such as benthic invertebrates or terrestrial fungi) meet fortuitously and, if genetically compatible, merge (Buss 1982; Grosberg 1988). Thus, in conceptual and developmental terms, multicellularity can arise in two ways: either by clonal processes (repeated division of an original cell or zygote) or by aggregative processes. Bonner (1998) argues that the inception of multicellularity in aquatic lineages occurred when the products of cell division failed to separate, whereas in most terrestrial lineages multicellularity originated by the aggregation of cells or of nuclei in a multinucleate syncytium.

The eukaryotic cell as a pivotal development Evolution of the eukaryotic cell was a major event in the development of macroorganisms. The details of how this novel cell type arose remain unclear and there are no known intermediate types in the fossil record. Although the differences between prokaryotic and eukaryotic cells are many, the most significant overall advance was probably the compartmentation of the eukaryotic cytoplasm that allowed localization of cellular activities, together with an oxidative metabolism (in common with some prokaryotes) that improved energy extraction beyond that obtainable by fermentation. These activities were fostered by acquisition of endosymbionts capable of photosynthesis and respiration. Eukaryotic cells are on average 1000-fold larger in volume than prokaryotic cells. Even a unicellular eukaryote carries more DNA and in general more structural and genetic information than does a prokaryote. A stunning array of morphological forms is apparent even among the protozoa and single-celled algae.

The eukaryote design must have been a good building block for multicellular architecture, for it seems more than coincidental that the cells of all the more complex organisms are eukaryotic. Apart from establishing a lineage (the unicellular protists) that could obtain food by engulfing it, the primitive eukaryotic cell type gave rise to three fundamentally diverse

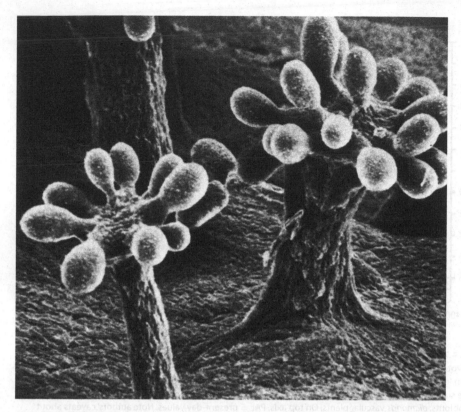

⬛ Fig. 4.6 Differentiation and elegant forms of rudimentary multicellularity among the bacteria: Mature fruiting bodies of the myxobacterium *Chrondromyces crocatus* showing elevated multiple sporangia (cysts), each of which contains resting cells embedded in slime. Photo courtesy of Patricia Grilione, San Jose State University

body types: (i) multicellular, soft-celled, heterotrophic forms that digested food internally (primitive invertebrates); (ii) multicellular, hard-walled forms that digested food externally (the fungi); and to (iii) multicellular forms that, by symbiosis with bacteria, came to manufacture their own food (green algae and the higher plants). Prokaryotes approach this multicellular, differentiated condition in the complex structures of some of the contemporary fruiting myxobacteria such as *Chondromyces* (⬛Fig. 4.6; close relative of *Myxococcus* noted earlier). Nevertheless, the multicellular route for bacteria was little more than a detour and ultimately an evolutionary dead end. The eukaryotic avenue led to architecturally elegant forms of cell differentiation and the radiation of life. One of the earliest and most important developments was probably a boundary layer (the epithelium) separating the interior of an organism from the exterior and, in the case of plants and fungi, a complex and rigid but dynamic cell wall (Domozych and Domozych 2014). The culmination in cell specialization is evident in the exquisite architecture and division of labor among the some 200 cell types of a vertebrate. The driving force behind this series of innovative events was that first the larger cell, then the multicellular macroorganism, by virtue of their size, were able to exploit some circumstances (environments, resources) better than or at least differently from the microbe. The new forms, through growth and activity, shaped their surroundings and in turn evolved in ways that their progenitors could not.

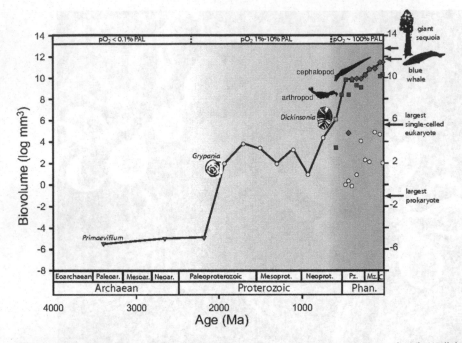

☐ Fig. 4.7   Size plot of the largest fossils over geological time, with representative examples of unicellular eukaryotes, animals, and vascular plants. Sharp increases in size correspond approximately to increases in oxygen levels (note top axis). Symbols represent: *triangles* prokaryotes; *circles* protists; *squares* animals, except for the square labeled *Dickinsonia*, which represents pre-Cambrian, enigmatic, multicellular eukaryotes called vendobionts; *diamonds* vascular plants. On top axis, PAL = present-day values. Note authors' caveats about organisms included and what is meant by size of the 'individual' [size here excludes colonial organisms; for unicellular organisms, it is cell size, not clone size]. From Payne et al. (2009); reproduced from *Proceedings of the National Academy of Sciences* by permission of the National Academy of Sciences, USA

Cope's rule and the evolution of larger organisms There has been an increase in size during evolution in the sense that among biota as a whole both the mean and upper size limits have increased through geological time (Carroll 2001; for synopsis see Chap. 2 in Bonner 1988). For example, Payne et al. (2009) compared the largest known fossil organisms from the origin of life in the Archaean (4000–2500 Mya) through the Phanerozoic (542–0 Mya) eons and found that maximum size as represented by their data set (cf. McShea 2001) increased 16 orders of magnitude over that approximate 3.5 billion years (☐Fig. 4.7). Most of this increase in size occurred in two episodes; not surprisingly, the first was associated with the advent of the eukaryotic cell and the second with multicellularity.

The above generality does not mean that the paleontological record is without periods or lineages of size decrease, and the rate of change of size tends to vary inversely with the duration over which it is studied (Gingerich 1983). For instance, organisms such as the birds, amphibians, and rotifers are smaller than their ancestors, and at least some of the extant unicellular fungi apparently are derived secondarily from multicellular ancestors. The case for birds is particularly interesting. Extinction was not the fate of all the dinosaurs because extant birds share a common ancestor with them. This ancestral creature appears to have been perhaps 1-m or more long, some 20–30 kg in mass, and a bipedal carnivore of the early Triassic Period (ca. 250 Mya). While the other dinosaur clades were marked by size increase and in some cases to titanic proportions, the only sustained trend to size decrease was in the evolution of the avian lineage (Sereno 1999). Likewise, the modern club mosses (lycopods) are a few centimeters high compared with their ancestors, many meters in height, which proliferated in the luxuriant Carboniferous coal forests of 300 million years ago.

The tendency for most animal lineages to evolve towards large body size is epitomized by Cope's Rule, in honor of the paleontologist Edward Drinker Cope who did not enunciate it as such but the principle emerged from his work during the late 1800's (Stanley 1973; McKinney 1990). An important caveat is that Cope's studies were focused on mammals over the Cenozoic (65 Mya–present) and he ascribed the mechanism specifically to an active, directional replacement of older (smaller) forms by larger ones under the influence of natural selection. Thus, the minimum size also increases and the explanation implies that larger organisms would be fitter. Lineages can also increase in body size by a passive trend, however, without unidirectional selection (see later comments on Gould's [1996] argument for random "diffusion off the left wall"). In this case, larger descendants arise but the lower size category characteristic of the ancestors still remains, i.e., trait variance increases (Gould 1996; Carroll 2001).

So, while lineages generally increase in size and complexity through geological time, this is not invariable, and there is considerable debate over whether an active or passive mechanism is primarily operative in a given instance and whether there is also a wall on the right (see Knoll and Bambach 2000). Jablonski (1996) argued that for species of various molluscs in the late Cretaceous that there was an *active* trend for *both size increase and size decrease.* In a major critique of Cope's Rule, Stanley (1973) began with the observation that if net size increase has prevailed in evolution, a prediction, which is borne out as a generality, would be that most higher taxa should have arisen at small (relatively unspecialized) body size relative to the optimum for the group. Larger (more specialized) taxa subsequently emerge, typically during speciation events, and the descendants fill new adaptive zones. Thus, a size increase or size decrease actually may occur, depending on whether the mean size of the original species is smaller or larger than the optimum for the occupied niche. Major adaptive breakthroughs would tend to be restricted to small body sizes because large species typically are locked into a morphologically specialized body plan dictated by space:volume scaling constraints (following section), which is inherently unsuited for descendent taxa specialized in different ways (Stanley 1973).

4.3 On Seeing the World as an Elephant or a Mycoplasma

4.3.1 Lower Limits and Upper Limits on Cell and Organism Size

Physical and chemical laws ultimately set the lower and upper constraints on life. At the low end, the metabolizing, self-reproducing, unicellular prokaryote has to be large enough to accommodate operational genetic and metabolic equipment, including the genome, catalytic enzymes, and ribosomes, etc., as well as for molecular traffic. Polymerases need to have access to the genome. Ribosomes must have space to produce proteins. In reality, for free-living cells in fluctuating environments, this minimum increases somewhat because of additional machinery to contend with catastrophes (Koch 1996). Prokaryotes vary in diameter from 0.2 μm (certain mycoplasmas) to more than 700 μm (*Thiomargarita*, a sulfur chemolithotroph; shown earlier in $\blacksquare$Fig. 4.1). The typical bacterium in laboratory culture, for example

Escherichia coli, is about $1 \times 2 \mu m$, but under natural conditions where nutrient stress is prevalent, cell sizes frequently are smaller. The average bacterial size in soils and lake water is $0.1 \mu m^3$ and they are even smaller in oligotrophic environments such as the open oceans (Schulz and Jorgensen 2001), close to the estimated minimum theoretically possible of a sphere of about 300 nm diameter (Young 2006). Transport of substrate to the cells is by diffusion and by advection (movement of liquid or of the cell by swimming). The implications of small scale and the non-bounded compartmentation of prokaryotic cells to diffusion and other aspects of cell physiology are fascinating (see, e.g., Koch 1996; Schulz and Jorgensen 2001; Young 2006). Depending on external substrate concentration, bacteria may be uptake-limited or diffusion-limited. Generally speaking, motile cells must be larger than $10 \mu m$ before they can gain more substrate by swimming (for upper limits and cell scaling issues, see below). Due to typically rapid diffusion at scales of about $1 \mu m$, bacteria are surrounded by a substrate-depleted microenvironment from which they cannot escape: a 'depletion halo' follows them to some extent as fast as they can swim. Movement serves only to get them into somewhat more favorable nutrient or redox patches (Schulz and Jorgensen 2001; Barbara and Mitchell 2003).

At the other extreme, why are there not any really large cells? Almost all have volumes in the range of 1–$1000 \mu m^3$. Apparent exceptions to this size spectrum generally can be explained by unusual internal structure or specialized function. For example, large water-containing vacuoles may occupy much of the volume of certain plant cells and the cytoplasm in the huge cells of the prokaryote *Thiomargarita* occupies only a very thin peripheral layer. The extraordinary size of *Thiomargarita* may have evolved to store nitrate, which serves the bacterium as an electron acceptor and is only sporadically available in its habitat (Schulz and Jorgensen 2001). Various plant and animal cells are inert at maturity, serving for transport or structural support, or they may contain inclusions or food as in the case of the ostrich egg, which reaches $10^{15} \mu m^3$.

The upper limit to a unicellular body plan appears to be set mainly by the declining surface area-to-volume relationship with increasing size expressed as the '2/3 power law', also known as the **Principle (or Law) of Similitude** (Thompson 1961), where $S \propto V^{0.67}$ (Niklas 2000; see later comments on scaling). This relationship is illustrated perhaps most graphically by Vogel (2003; see his Chap. 3) who says that the main difference between a bacterium and a whale is that the bacterium has 10^8 times as much surface area, relative to its volume, as does the whale, with the amusing quip that the former has a lot outside with not much inside and the latter a lot inside and little outside. The S/V relationship inevitably takes a toll on exchange processes of nutrients and waste products. The operative diffusion rate limit is expressed quantitatively in Fick's Law (Koch 1971, 1990, 1996), and is relevant to both microorganisms and macroorganisms, though the latter augment diffusion in various ways by transport mechanisms, tissue systems, and cell design (see scaling issues, below). Transportation of materials up to or within the cell, specific metabolic rate, and removal of wastes, all decrease with increasing cell size. For example, Koch (1996) shows mathematically for the single, spherical prokaryotic cell that, subject to certain simplifying assumptions, the efficiency (E) of clearance (measured by the cell volumes of medium cleared per second) is expressed as $E = 3D/r^2$, where D is the diffusion constant of an exogenous small molecule and r is the radius of the cell. So, halving the size of a cell quadruples its efficiency.

In terms of evolutionary design, the foregoing implies that the size of unicellular microbes cannot increase very much from being truly 'micro' without incurring a severe penalty in diminished diffusion rate sufficient to supply nutrients fast enough to sustain rapid (competitive) growth. Niklas (2000) argues that maximal body (cell) size in the unicellular algae is set by size-dependent variations in surface area, intracellular metabolites and metabolic machinery, and reproductive rates. Such limits likely pertain more broadly to all unicellular organisms, though cell shape can vary, as discussed later. An alternative avenue open to natural selection to increase size would then have been through independent experiments on aggregating moderately sized cells into a multicellular body plan, as discussed above.

With respect to constraints on multicellular *organism* size as opposed to *cell* size, the upper bound seems less rigorously demarcated than is the lower limit. While there are advantages to being larger (discussed later), mechanical and physiological problems ensue if an animal or plant becomes too big. Huxley famously said long ago (1958, p. 24) *"It is impossible to construct an efficient terrestrial animal much larger than an elephant"*. However, the size of what can be constructed really hinges on growth *form*, that is, whether the organism is of unitary or modular design (▶Chap. 5). Modular life forms such as plants are able to add components indefinitely. In effect they use geometry to defeat some of the constraints of gravity. Organizing a soma in this fashion allows an organism to increase biomass up to a point without transgressing critical morphological limits (Hughes and Cancino 1985; see also ▶Chap. 5). Nevertheless, there remain ultimate limits: very tall trees, for instance, eventually encounter hydraulics problems (Koch et al. 2004) and will buckle under their own weight if they exceed certain critical heights (see following section on scaling relationships and Niklas 1994a, pp. 164–186).

4.3.2 The Organism as Geometrician

Before going on to contrast specific organisms with respect to size, we consider briefly the role of shape and geometry as well as certain simple mathematical relationships in such comparisons. *Geometrically similar* bodies, i.e., those in which all corresponding linear dimensions are related in the same constant proportions, are said to be *isometric* (Schmidt-Nielsen 1984). This would be the case if two cubes of different sizes were compared, or two triangles that had corresponding angles equal and corresponding sides in a constant ratio. It is a mathematical inevitability (the Principle of Similitude noted earlier) that in isometric bodies area-related attributes increase with the square of linear dimensions, whereas volume-related features increase as the cube (Thompson 1961). In large taxa, critical physiological processes and associated structures, such as those involved in musculature, respiration, and physical support, typically show adaptations (examples below) to partially offset such geometric realities, but these compensations are ultimately limited by size. This relationship also underpins the generality that larger taxa tend to be more specialized than smaller taxa.

A particular area:volume problem concerns efficient exchange of metabolites such as nutrients and various gases and waste products. Haldane (1956, p. 954) said that *"comparative anatomy is largely the story of the struggle to increase surface in proportion to volume"*. How is this achieved? The three evolutionary structural 'solutions' (Gould 1966) bearing wholly or in part on this problem have been: (i) a differential increase in surfaces among the more advanced animals (e.g., fish gills; lung alveoli; intestinal villi); (ii) a change in shape by flattening or attenuation without structural elaboration (e.g., the tapeworms); (iii) the incorporation of inactive organic matter within the soma (e.g., jelly in the case of Coelenterates; wood fibers in plants). To Gould's list of solutions could be added the efficient partitioning of surface

■ **Table 4.1**   The law of similitude: the mathematics dictating declining surface area to volume for five cubes of increasing size

Length of one face	1	2	3	4	5
Surface of cube (area of 1 face × 6)	6	24	54	96	150
Volume of cube (length of 1 face^3)	1	8	27	64	125
Area/volume	6	3	2	1.5	1.2

area:volume for the capture of resources—organisms as diverse as the fungi, sponges, plants, and corals all show essentially this same feature (►Chap. 5).

Macroorganisms are rarely isometric, though they may appear superficially similar in appearance. The term **allometric or non-isometric scaling** denotes the regular changes in certain proportions or traits as a function of changes in size (Chap. 2 in Schmidt-Nielsen 1984; see also West et al. 1997; Brown and West 2000). As most parents understand intuitively, when humans grow, their body proportions change—babies, for instance, have much larger heads relative to their size than do adults.

The general scaling relationship generally takes the form

$$Y = aW^b \qquad \log Y = \log a + b \log W,$$

where Y is the biological variable of interest such as metabolic rate; W is a measure of body size such as mass; a is the normalization constant derived empirically and typical of the kind of organism; b is the scaling constant, also derived empirically. The equation in logarithmic form is a straight line with slope b and intercept $\log a$.

The exponent b defines the general nature of the relationship as follows (Lindstedt and Swain 1988; Brown et al. 2000): **When b is at or very near 1.0**, Y varies as a fixed percent of body mass and would plot as a straight line with a slope of 1 on either linear or log axes. Indeed, such situations would be the sole instance where changes are strictly proportionate, i.e., isometric. Examples include mammalian heart size, and lung, gut, and blood volume, all of which increase closely in proportion to body size. If b is other than unity, the relationship is allometric and plots as a curve on linear axes. **When $b > 1.0$**, the increase in Y becomes relatively greater as size increases. The skeleton in a shrew is about 4% of the animal's body mass but accounts for 25% of the body mass of an elephant (Lindstedt and Swain 1988). How thick does the trunk of a tall tree (or a flagpole, for that matter) need to be to prevent it from buckling under its own weight or collapsing in an ice storm? The general mathematical result is that the diameter for taller trees should increase as the 3/2 power of the height, i.e., larger trees are disproportionately thick and this relationship holds throughout the life cycle (see later comments about the stems of small organisms).

When b is between 0 and 1, Y increases but only fractionally for each unit of increase in body mass. The famous 2/3 power law, $S \propto V^{0.67}$, was introduced earlier where we introduced the Law of Similitude. This can be visualized by comparing cubes of different sizes (■Table 4.1 and ■Fig. 4.8). **Many if not most biological scaling relationships are based actually on quarter powers of body mass** (Delong et al. 2010). For instance, metabolic

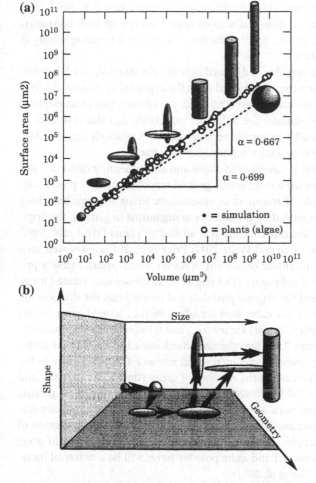

□ Fig. 4.8 Implications of the Law of Similitude. **a** Relationship between cell surface S and volume V for 57 species of unicellular algae (*the straight line plot* with slope 0.669). This is compared with a plot for spheres of different sizes (*dashed line*, slope 0.667) and with a computer-generated series of geometries (*spherical to cylindrical*) that maximize S to V as size increases. **b** Within this data set, the smallest unicells have spheroidal geometries and the largest are cylindrical. Within each category of geometry larger species tend to be spheroidal (*flatter*) than the smaller or more slender cylinders. From Niklas (2000); reproduced from *Annals of Botany* by permission of Karl Niklas and Oxford University Press, ©2000

rate α $M^{3/4}$ may pertain to poikilothermic as well as homeothermic animals and, as originally proposed, to plants and microorganisms (West et al. 1997; Brown et al. 2000; however, see Reich et al. 2006; Price et al. 2012, and caveats below). This means that for each 10-fold increase in size, metabolic rate increases, but by only 7.5-fold. In other words, weight-specific metabolism is higher in smaller animals. Indeed the approximate ¾ power metabolic scaling relationship was once thought to be universal and became elevated to the status of a law, as 'Kleiber's Law', in honor of Max Kleiber who did seminal research on metabolic scaling in the

1930s and 1940s. Based on experiments with a broad survey of plants, however, Reich et al. (2006) conclude that whole-plant respiration scales about isometrically with total body mass (i.e., with a scaling component of 1, not ¾) and that the mechanisms underpinning scaling in plants are distinct from those in animals.

When *b* is negative, the slope is negative and accordingly the absolute value of Y is maximized in the smallest creatures. For various biological rates the exponent is close to -¼. This means that while the heart of a shrew can exceed 1000 beats per minute, that of an elephant beats at only about 25–30 times per minute. The difference reflects the fact that resistance to blood flow is relatively higher in smaller mammals because they have relatively larger surface areas (supplied by capillaries where resistance is highest) for their size.

The argument of West et al. (1997), summarized above and subsequently extended (West et al. 2000; Brown et al. 2000), proposed a mechanistic general model for quarter-power scaling based on the common principle of transport of essential materials through branching fractal-like networks. Though extrapolated by the authors to organisms in general, the original model emphasized metazoans and subsequently was applied to plants (West et al. 1997; but see criticism and alternative by Price and Weitz 2012; Price et al. 2012). How well such a network model might apply to unicellular organisms, however, seems questionable a priori. Subsequent work by Brown and colleagues (DeLong et al. 2010; see also Huete-Ortega et al. 2012 for phytoplankton) revised the original postulate and showed that the slopes of log plots of active metabolic rate versus body mass were actually 1.96 (i.e., termed super-linear) for heterotrophic prokaryotes, 1.06 (i.e., linear) for protists, and 0.79 (sublinear; approximating ¾ power function) for metazoans. They hypothesized mechanisms to account for metabolic and body size scaling in each group: Rapid increase in rate as a function of size in the prokaryotes was attributed to an increase in the number of genes (thus more enzymes and biochemical networks and hence metabolic power); the approximate linear scaling in protists was arguably due to a linear increase with size in respiratory complexes in volume of mitochondria; sublinear scaling of metazoans reflected constraints imposed by differentiation of vascular and skeletal systems to supply increasingly large and complicated bodies. In overview, the consensus is that there is not yet and quite possibly never will be a universal metabolically based theory of ecology (Price et al. 2012).

In addition to the decreasing area-to-volume ratios as a consequence of increasing size, gravity constrains terrestrial bodies like flag poles and large organisms (Thompson 1961; Gould 1966). The oceans provide neutral buoyancy for whales. The giant sequoia tree has hard cell walls and massive arrays of supporting fibers and hence does not collapse under its own weight. Almost all animals have some sort of skeleton (hydrostatic, earthworms; exoskeleton, arthropods; endoskeleton, vertebrates). As alluded to earlier with respect to similitude (diameters increasing as the 3/2 power of height; or, equivalently, Height α Diameter$^{2/3}$), bones in animals and stems in plants must become disproportionately thick with increasing mass or height, if additional mass is to be accommodated (Chap. 5 in McMahon and Bonner 1983; Niklas and Spatz 2004). The wings of birds are designed to provide lift and to be strong yet light in weight. Because weight varies as volume while strength of a device to support it varies as cross-sectional area, engineering considerations mean that to remain strong the arthropod exoskeleton must be thicker in larger animals. But beyond a certain point the shell would be cumbersome and would decrease survival. It is probably because of this structural constraint that insects were not able to evolve into forms able to occupy the niche now filled by passerine birds.

Microorganisms, in contrast, largely escape the scaling problems related to gravity, large size, and strength. Despite considerable variation in the size of fruiting bodies of the cellular slime molds, gravity has little effect on the relative diameter of their supporting stalks. Weight in this case has probably been negligible as an evolutionary consideration (Bonner 1982a), the stalks being geometrically similar, unlike the stalk of the relatively much larger mushroom or of a large tree discussed above, which become disproportionately thicker in the taller forms. Hence, Bonner and Horn (1982) have said that small organisms tend to have **geometric similarity (isometry)**; large organisms have **elastic similarity**. The former organisms scale essentially as the function of one axis, that is, the diameter is directly proportional to the length ($d \propto l$) and a plot would have a slope of 1; the latter vary as two axes such that often but not invariably $d \propto l^{3/2}$ (McMahon 1973; McMahon and Bonner 1983; Niklas and Spatz 2004). How this 3/2 relationship arose during evolution of large organisms is explored elsewhere (Chap. 4 in Bonner 1988) and is one example of how selection for size in macroorganisms also affects shape (for general comments on engineering as it relates to organisms, see Wainwright et al. 1976).

By definition, a microorganism is small and its environment will obviously be small in absolute terms relative to that of a macroorganism. (Likewise, the impact of smaller organisms on their environment generally will be less than larger organisms.) This is clear for unicellular microbes such as bacteria and yeasts, as well as the unicellular protists and some small invertebrates such as the nematodes and rotifers. It is less straightforward for filamentous microbes, which technically are microscopic only by virtue of their narrow cross-section. If the component strands of the mycelial network of a fungus aggregate, however, the organism then becomes macroscopic, as happens when fruiting structures (mushrooms), mycelial sheets, or root-like rhizomorphs develop. Mushrooms form on a wet lawn in a matter of hours, moving the fungus through size and related changes in shape and gravity effects very quickly. So, size changes can occur abruptly. Nutrient signals appear to be the main trigger for the cellular slime mold to change from a disaggregated state of solitary, grazing amoebae to the social organism comprised of some 100,000 organized, aggregated cells (Kessin 2001).

Microbes are governed by the forces of diffusion, surface tension, viscosity, and Brownian movement (see Movement in a Fluid, below, and McMahon and Bonner 1983; Young 2006). Theirs is a world of molecular phenomena not noticed by macroorganisms any more than bacteria experience gravity. Even exclusively terrestrial microbes, including many protists and nematodes, are usually associated with liquid in some form. This may be mucilage or other secretions of their own making; soil capillary water; the interiors of plants and animals; or boundary layer films of various origins. Free water is required almost invariably for such activities as mobility, growth, and reproduction. Propagules are frequently released into or must escape through a liquid film. Surface tension can even be a major factor for the smaller macroorganisms. Haldane (1956) observed that whereas a human emerging from a bath carries only a thin film of water weighing approximately one pound, a wet mouse has to carry about its own weight of water and a wet fly is in very serious trouble... *"an insect going for a drink is in as great a danger as a man leaning out over a precipice in search of food"* (p. 953).

Size and some metabolic implications Peters (1983, see especially his ►Chap. 3) reviewed the literature and discussed at length the metabolic consequences associated with a particular size. Only a few of his many interesting points can be summarized here. A note on methodological differences should be made first. For higher animals, the power required to just maintain life is estimated from some minimum, so-called *basal* metabolic rate; this amount is supplemented by estimates of additional power required for each type of activity. For other taxa, such as the microbes, the term *standard* metabolic rate replaces basal rate, implying that the data are obtained under standard but not necessarily minimal conditions.

Peters has produced useful comparisons for unicellular organisms, poikilotherms, and homeotherms. The first of these is that the relationship between metabolism and body mass is similar for the three groups. Consider again the general equation, $Y = aW^b$. Because the value of a above is highest for homeotherms, declining in turn for poikilotherms, and then for unicells, metabolic rate for a hypothetical 1-kg organism in each group declines comparably. As Peters remarked, one ecological implication is that the relative demands of the three types of organisms on their bodies and on the environment must decline in similar fashion. Consequently, homeotherms need high resource levels and have to be relatively efficient in resource utilization. Heat loss or heat gain is a size-related consideration for homeothermic macroorganisms. A human, for example, consumes about 1/50 of its weight per day, a mouse 1/2 its weight. The design of a warm-blooded animal much smaller than a mouse becomes impossible because the organism could not maintain a constant body temperature (Chap. 2 in Thompson 1961). Peters (1983, p. 33) calculated that small mammals degrade about 10 times as much chemical energy per unit time as would an equivalent mass of large mammals. In contrast, the constraint at the upper extreme pertains to heat dissipation because the larger the animal the greater its heat production relative to heat loss. This is manifested by lethargy and various heat exchange devices or behaviors of various kinds to promote cooling. Heat production for the mouse and heat dissipation for the elephant is both a consequence of the fact that heat generating capacity varies as the cube of the linear dimension while loss depends on surface area, which varies as the square.

If power production is expressed instead in specific terms (watts kg^{-1}), the rate of energy consumption decreases with body size. This means that within each of the three groups the maintenance cost for large organisms is less than that for the same amount of smaller organisms. Another way to look at the same issue is to ask what the maximum amount of biomass is that could be supported per unit of energy supply (kg watt^{-1}). Thus, it turns out that the same amount of energy could support about 30 times more poikilotherms (and still more unicells) than homeotherms. Within each class, a greater biomass of the larger than the smaller organisms could be sustained per unit of energy flow.

Crude estimates can be made of turnover time, i.e., the time needed to metabolize an amount of energy equivalent to the energetic content of tissues (Peters 1983, pp. 33–37). Again, among organisms hypothetically of equivalent weight, turnover times are shortest for homeotherms, and progressively longer for poikilotherms and unicells in which energy is mobilized more slowly. Within each metabolic group, energetic reserves of the larger organisms last longer than those of the smaller.

4.3.3 Movement in a Fluid

Planktonic microorganisms such as bacteria, diatoms, dinoflagellates, and many of the green algae live in a world dominated by fluid mechanics. Likewise, intestinal and soil- or sediment-inhabiting bacteria live in viscous environments quite unlike those typically inhabited by macroorganisms. One criterion of life in such habitats is the Reynolds number

(Re), a dimensionless or relative velocity related to movement in a fluid that expresses the ratio of inertial forces (i.e., forces required to accelerate masses; numerator) to viscous forces (forces required to cause shear; denominator) as follows:

$$Re = \frac{\rho l v}{\mu}$$

where

Re   Reynolds number
ρ   fluid density, g/cm
l   characteristic organism length, cm
v   characteristic organism speed, cm/s
μ   fluid viscosity, g/cm s

McMahon and Bonner (1983, see their Chaps. 5, 6; Berg 2004; Stocker and Seymour 2012) discuss some interesting implications of Reynolds numbers, particularly as they apply to the propulsion of microorganisms. For the movement of small organisms (i.e., with short lengths and speed), viscous forces (density and frictional components) dominate, and the Re is low. Conversely, for large organisms, inertial forces dominate and Re is high. The blue whale, because of its huge size and relatively fast speed, swims at a Re of about 10^8, the porpoise of 10^5, while that for a moving bacterium is comparatively miniscule at about 10^{-6}. To small organisms water appears very viscous, perhaps analogous to how quicksand would appear to humans; motility is further constrained where increased viscosity, surface effects, or porosity impair swimming (Fenchel 2008). Interestingly, some microorganisms (e.g., some ciliates, copepods) have mechanisms to momentarily escape a low-Reynolds number world by jumping to enhance nutrient uptake or to attack prey. Guasto et al. (2012) discuss these adaptations and the many ramifications of life in fluid environments.

For planktonic microorganisms, morphology and flagellar propulsion systems appear to be optimized for efficient movement. Inertial forces of water, used to advantage by fish for movement, are of little consequence at a low Re. A microscopic whale would get nowhere by its mode of propulsion. Conversely, a bacterium or protozoan could not swim like a whale. Their motion is dominated by the viscosity component which also means that, unlike fish, they have essentially no glide distance once their propulsive engines have stopped. It does not coast. Furthermore, at a low Re, swimming is completely reversible. In theory, if a microorganism moved ahead and then back by exactly the same number of propulsive movements, not only would it return to the identical spot but all the displaced water molecules would also return to their original places. Finally and most fascinating of all are the implications for the design of the propulsive equipment. The problem, because of the reversibility of movement at low Re, is that if microbes had to move by solid 'oars' they would go nowhere (forward on the thrust, followed by an equal distance back on the return stroke). The novel solution is provided by the form of the propulsive equipment: flagella in bacteria, which are anchored at their base and rotate like propellers, and flexible oars in the form of related but structurally distinct cilia for larger microorganisms such as the protozoa (Son et al. 2015). The cilium is held straight out on the power stroke but collapses parallel to the body for the return, which reduces drag. (It is noteworthy as an aside that although ciliated organisms range in size over two orders of magnitude, in general, cilia length and frequency of beating remain approximately constant. This means that most swim at about the same speed of about 1 mm/s or approximately 10-fold that of a flagellated bacterium.)

To summarize, physical constraints imposed on an organism vary with its size. The elephant with a huge bulk supported by a massive skeletal system experiences a world dominated by gravity, area:volume scaling, and thermodynamics. The tiny mycoplasma, without even a cell wall for support, knows nothing about gravity or the complexities of multicellular homeostasis, and exists partially suspended in a world governed by fluid dynamics and diffusion phenomena. Such physical determinants are probably as important in evolution as they are in civil engineering (Lindstedt and Swain 1988). These constraints influence shape—shapes possible for one organism are not options for another—means of locomotion, speed, and many other related features that are discussed further below. There is probably an optimal size associated with each type of activity (Haldane 1956; Pirie 1973).

4.4   Some Correlates of Size

4.4.1   Complexity

To function efficiently, a large organization has many parts and depends on a division of labor or specialization among its component units. This is intuitively true and is evident whether we compare a water flea with a rhinoceros, or a prokaryote with a eukaryote. 'Complexity' of organisms is a nebulous, controversial term probably most tangibly based on the number of cell *types* (which usually have different, specialized functions; Chap. 5 in Bonner 1988; Carroll 2001). Across the broad sweep of biotic diversity, as well as within specific lineages, life has moved toward increasing complexity: There is a direct correlation between number of cell types and size (◘Fig. 4.9 and ◘Table 4.2), although for either a given size or number of cell types there is considerable variation in the other parameter. For example, the maximum number of cell types in prokaryotes and protists is about four (prokaryotes are typically of one or two cell types but can reach as many as four in some cyanobacteria; Flores and Herrero 2010) and rises to about 30 in plants, 50 in nematodes and fruit flies, and over 100 in vertebrates.

Gene number (also shown in ◘Table 4.2) is sometimes used as a measure of complexity, but is subject to variation from gene duplications and losses, can be difficult to determine accurately, and has several limitations, so is not particularly informative (Carroll 2001). While it is broadly true that eukaryotes have more genes than prokaryotes, gene number varies roughly 10-fold even among prokaryotes, and among eukaryotes the organisms with fewer cell types may have appreciably more genes. Nevertheless, the evolutionary transition to increased size and complexity runs in tandem with phylogenetic increase in nuclear DNA content. For instance, as is well known, the amount of haploid DNA in mammalian cells is, to an order of magnitude, about 10^3-fold greater than that of bacteria (summarized in Watson et al. 2008, p. 139). These values are approximate and the correspondence between DNA content and complexity is not perfect—some amphibians have 25 times the amount of DNA per cell than do mammals. The pattern must be more than coincidental and suggests that, in general, gene amplification played a major role in phylogenetic change (►Chap. 2 and Stebbins 1968; Griffiths et al. 2015). By virtue of providing an extra

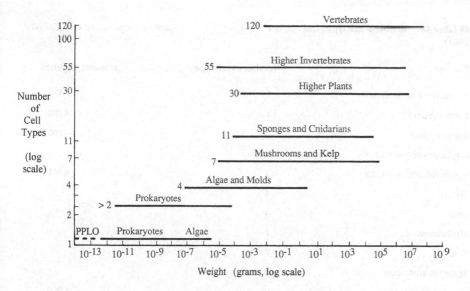

◘ **Fig. 4.9**   Number of cell types (Y axis) as a function of increasing size (weight). PPLO—pleuropneumonia-like organisms are now called mycoplasmas. 'Algae' (*at bottom*) refers to small cyanobacteria and the green alga *Protococcus*. The prokaryotes appearing as more than two cell types (second horizontal line from the bottom) include the large cyanobacteria, the spore-formers, and other multicellular bacteria. From Bonner (1988); reproduced from *The Evolution of Complexity by Means of Natural Selection* by John Tyler Bonner, by permission of Princeton University Press, ©1988

gene copy for evolutionary 'experiments', duplications foster flexibility and allow novelty to arise in formerly tightly linked traits and pathways (Vermeij 1999). Within phyla, however, there is no correlation between DNA content and complexity. Stebbins (1968) attributed this apparent anomaly to the possibility that the origin of new phyla and the associated requirement for new cells and organs necessitate an increase in different enzyme systems. In contrast, evolution within a phylum may be more a matter of integration of function or alteration of conformation, processes that could be accomplished by mutation and recombination of existing genes.

Probably the most important event spurring evolution of greater complexity arguably was the transition from unicellularity to multicellularity, discussed earlier. Unicells can differentiate only in time, not in space, i.e., they can, at a given time, only be one thing or another, unlike even a two-celled entity where the two members can be structurally or functionally different from one another simultaneously. Once multicellularity arose, it was comparatively easy to embellish rudimentary division of labor and specialization into increasingly sophisticated forms. As these entities and partnerships evolved, new interactions and mutualisms at multiple levels became possible in ways that were not available to their precursors (Vermeij 1999). Another major event advancing complexity was the transition from an aquatic to a terrestrial existence. This led to the embellishment of animal body plans already established in the oceans and to a major radiation of plant life leading to, for example, massive arboreal forms and seed plants. For a fascinating discussion of these points, see Raff (1996 for animals) and Corner (1964) or Graham (1993 for plants).

4

□ Table 4.2 Number of cell types and number of genes in various organisms (abbreviated from Carroll 2001)

Species	Number of cell types	Number of genes
Mycoplasma genitalium	1	470
Escherichia coli	1	4288
Bacillus subtilis	2	~4100
Caulobacter crescentus	2	
Saccharomyces cerevisiae	3	6241
Volvox	4	
Ulva	4	
Mushrooms	7	
Kelp	7	
Sponges, cnidarians	~11	
Arabidopsis thaliana	~30	~24,000
Caenorhabditis elegans	~50	18,424
Human	~120	~80,000–100,000[a]

[a]This is much higher than the current estimate of about 21,000 protein-coding genes (Griffiths et al. 2015)

As the number of distinct parts increases so does the number of different interactions among them. Natural selection can act on size, shape, or complexity, yet a significant change in one influences the others (Bonner 1988, p. 226). Bonner notes that selection acting on complexity is probably more important than that acting on size because increase in number of cell types opens the way for a large increase in size. In contrast, while size alone is somewhat plastic, its upward movement in the absence of an associated increase in complexity is limited by losses in efficiency. One of the implications of the size-complexity issue is that a large organism is locked into an intricate developmental pattern, each step of which is influenced and in many cases predetermined by what has gone before (recall comments in ►Chap. 1 about ontogenetic constraints). Microorganisms, being small and less complex in number of cell types, escape this complicated ontogeny.

A body plan built on specialized, interdependent subunits offers the benefits of high efficiency at particular tasks (e.g., sight, taste, translocation, support, and defense) at the cost of vulnerability, often death, if an intricate system fails. Such failure occurs, for instance, when tetanus exotoxin binds specifically to one of the lipids of human nerve synapses in the central nervous system; when T-4 lymphocyte helper cells are killed by the AIDS virus; or when propagules of the fungus causing Dutch elm disease block water-conducting xylem vessels of the host. Organs can fail also for hereditary and environmental reasons. By having fewer and less complicated parts, microbes lack the advantages but avoid the shortcomings of life based on a complex blueprint. The colonial form of existence for certain organisms is perhaps a compromise between the two alternatives.

■ Table 4.3   Major evolutionary steps in the history of life

'Major Transitions' *(Maynard Smith and Száthmary 1995, p. 6)*

1. Replicating molecules to populations of molecules in compartments

2. Independent replicators to chromosomes

3. RNA as gene and enzyme to DNA + protein (genetic code)

4. Prokaryotes to eukaryotes

5. Asexual clones to sexual populations

6. Protists to animals, plants, fungi (cell differentiation)

7. Solitary individuals to colonies (non-reproductive castes)

8. Primate societies to human societies (language)

'Megatrajectories' *(Knoll and Bambach 2000)*

1. Increase in efficiency of life processes from inception to last common ancestor

2. Metabolic diversification of prokaryotes

3. Evolution of the eukaryote cell

4. Multicellularity

5. Invasion of the land

6. Intelligence and technology

Diffusion off the left wall In his book *Full House*, Steven Gould (1996) argues that the evolutionary record of increasing size and complexity is best explained not in terms of directional drive but as resulting from unpredictability, contingency, and accumulating variance. Thus, what could be called the phylogenetic 'march of life' might be viewed better as a 'random walk of life'; it is visualized by Gould as emerging from an incipient form of minimal size and complexity, i.e., the primordial prokaryote, and diffusing away from that boundary. His metaphor is the 'drunkard's walk' away from a bar (left wall): the person will sooner or later end up in the gutter simply due to random movement. All that is needed is an impenetrable limitation on the left (represented in biology by irreducible simplicity) and time. There will be inevitable progression overall to the right, though this is purely from randomness conveying the appearance of directionality. Gould's interpretation has received mixed support and has been variously modified, for example, by imposition of a 'penetrable' wall on the right (Knoll and Bambach 2000). The right wall would represent evolutionary limitations imposed by physiology or architecture that halt further diversification for an extended period. Examples of such right walls could include the evolution of hyperthermophily in certain bacteria, or the evolution of the eukaryotic cell, or multicellularity. Indeed, any of the evolutionary 'major transitions' in life postulated by Maynard Smith and Szathmáry (1995), or the 'megatrajectories' proposed by Knoll and Bambach (2000) (■Table 4.3), would qualify as right walls. While, overall, biotic diversification within each of the 'plateaus' following an evolutionary breakthrough accords with an interpretation of increasing variance, the major transitions themselves in effect reset the evolutionary clock and impose a pattern of directionality within clades (Knoll and Bambach 2000; see also Vermeij 1999).

4.4.2 Chronological Versus Physiological (Metabolic) Time

Increase in size can be viewed as an integral of steps through the life cycle; it follows that more steps, hence more absolute or chronological time, are required to produce a larger than a smaller organism. This in turn means that the generation time (■Fig. 4.10) or the time to reach sexual maturity is usually longer, as must be life span in general terms, for larger organisms. Generation times range from on the order of a few minutes for bacteria growing under favorable conditions, to a few hours for protozoa, a few days for the house fly, about 20 years for humans, and about 60 years (albeit with high variation) for the giant sequoia. This relationship has obvious implications for potential population density (next section) and for the amount of absolute time and number of environmental events that a generation of any given size-class of organisms experiences.

There is also another time scale, one that has a physiological basis and is size-dependent. As early as 1950, Hill speculated that all physiological events in an animal might be set by a clock that runs according to body size. Thus, small animals have a fast pace of life but do not live long. On balance, both the shrew and the elephant experience about the same number of physiological events or actions per life span (Chap. 6; see also Chap. 12 in Schmidt-Nielsen 1984). Almost all biological times in birds and mammals (e.g., muscle contraction, blood circulation, respiratory cycle, cardiac cycle) vary with the same body-mass exponent

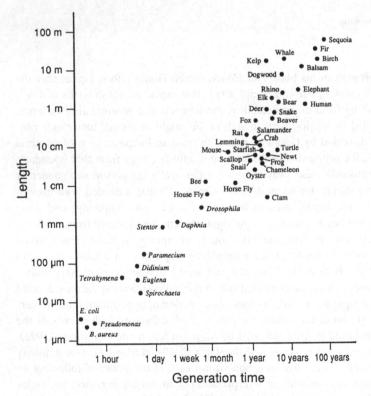

■ **Fig. 4.10** The implications of size to reproductive potential: Lengths of micro- and macroorganisms versus their generation times. Redrawn from Bonner (1965); reproduced from *Size and Cycle: An Essay on the Structure of Biology* by John Tyler Bonner, by permission of Princeton University Press, ©1965 (mean = 0.24; Lindstedt and Calder 1981; Brown et al. 2000), regardless of the tissues, organs, or systems involved. The relationship also holds for a range of life history or ecological times, for example, life span, time to reproductive maturity, time for population doubling. It follows that the *ratios* of any two biological times (e.g., respiration cycle:heart cycle, or time to reproductive maturity:blood circulation time) versus body size plot as a horizontal line (Lindstedt and Swain 1988). In other words, just as the number of minutes per hour is constant in absolute time, so are biological activities per unit of physiological time: All mammals have 4–5 heartbeats *per breath*, and use about the same number of calories per unit weight *per lifetime*. Because physiological time accounts for body mass, hence how each organism sees the world, in a way that absolute time cannot, it may be the most appropriate criterion for interspecific comparisons at least among warm-blooded animals (Lindstedt and Swain 1988; Lindstedt and Calder 1981). However, physiological time would appear to have little value in comparisons involving cold-blooded animals such as frogs, fish, and invertebrates. Their energy expended per unit time or maintenance metabolic rate is more variable and highly environmentally dependent.

4.4.3 Density Relationships: Body Size and Numbers of Species and of Individuals

The so-called species abundance relationship (the number of species in different abundance classes), first explored in depth by Preston (1948), and the species body size distribution (the number of species of a particular body size in a community), popularized by Hutchinson and MacArthur (1959), are among the most prominent of general patterns in ecology. Here we focus on size relationships.

Generally speaking, there are many more species in small-to-intermediate size classes of organisms than in the larger body size classes. Even on a log scale the frequency distribution of species body sizes (species number versus log characteristic length or body mass) is typically right-skewed (Kozlowski and Gawelczyk 2002). This generality is subject to several qualifications, among them that: (i) most of the evidence is for terrestrial, aclonal life forms; (ii) size distributions can also be symmetric or, alternatively, left-skewed; (iii) innumerable very small animals, protists, and microbes in particular remain unsampled and unidentified, implying that at least for very small organisms the plot may be artefactual; and relatedly that (iv) conventional taxonomic criteria for the smallest of organisms (prokaryotes, see below) are highly controversial and may be without real biological meaning. Within the length range of about 10 to 10^4 mm, the number of terrestrial animal species (S) varies roughly with characteristic length (L) by the relationship $S \propto L^{-2}$ (May 1978). This means empirically as a general trend that for each 10-fold reduction in length a 100-fold increase in species would be expected. However, it was argued that the relationship does not appear to hold for organisms whose body length is less than about 10 mm (species number decreases). Most evidence accumulated over the past ca. 60 years has been for terrestrial animals (Hutchinson and MacArthur 1959; Van Valen 1973; May 1978; Peters 1983; Gaston and Blackburn 2000) and plants (Aarssen et al. 2006; Dombroskie and Aarssen 2010). Peters (1983, p. 179) subsequently expressed the relationship as a function of body mass, $S \propto W^{-0.67}$.

Bonner (1988) extended the size-frequency generality to all life forms, from bacteria to the largest macroorganisms. As shown (◘Fig. 4.11), his plot also depicts a reduced number of species at both ends. Based on the distinctive attributes of microorganisms, Fenchel

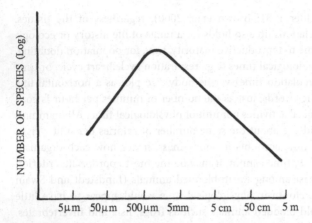

Fig. 4.11 Postulated relationship between the number of species of all organisms as a function of the typical length of the constituent individuals (log/log scale). See text for caveats. From Bonner (1988); reproduced from *The Evolution of Complexity by Means of Natural Selection* by John Tyler Bonner, by permission of Princeton University Press, ©1988

(1993), among others (e.g., Finlay and Esteban 2007), explain species trends among the protists and prokaryotes, and distinguish between global and local (within community) dynamics. For example, Fenchel emphasizes the cosmopolitan distribution of microorganisms (high migration rates), their lower rates of global and local extinction, and supposed lower rates of allopatric speciation compared to larger organisms. Thus, because of very large population sizes and dispersal ability, together with prolonged quiescence capability of microbes, the dynamics of species extinction are likely quite different from macroorganisms. Indeed, it is even questionable whether the terms allopatry and sympatry as used by macroecologists have meaning in a microbial context.

Fenchel's (1993) analysis tends to discount poor taxonomic resolution as an explanation for reduced species number at the low (bacterial) end of the species size distribution. However, it is worth noting that because of the atypical sexuality of prokaryotes (reviewed in ▶Chap. 2), the standard 'biological species concept' familiar to plant and animal ecologists does not apply. Bacteriologists use instead a '*phylogenetic* species concept' whereby a species is defined operationally as a group of strains that cluster closely and distinctively from other clusters. This somewhat arbitrary arrangement is based on physiological and, increasingly, on solely molecular criteria, e.g., sequence similarity in alignments of DNA of a phylogenetically relevant gene or multigenes. This has given rise to the term 'operational taxonomic unit', arbitrarily defined, as a euphemism for 'species'. The current standard for grouping strains together as a species is 70% or greater genomic DNA–DNA hybridization, together with 97% or greater identity in 16S rRNA gene sequence (Madigan et al. 2015). A move to molecular criteria implies increasingly refined specificity in delineating new species. Will advancing taxonomic techniques and discoveries from relatively unexplored microbial habitats such as the seabed, deserts, caves, and the lithosphere alter the species size distribution for all species? In terms of total number of *microbial* species on Earth, a recent, extreme estimate is, remarkably, ~1 trillion (10^{12}) (Locey and Lennon 2016a). (They include in their tally of "microbial species" bacterial, archaeal, and "microscopic" fungi; they make their projection based on scaling laws and extrapolations, among other key assumptions.) Such extrapolations are at best risky and may be invalid. Their methods have been challenged (Willis 2016) and the authors have attempted to rebut the criticism (Locey and Lennon 2016b).

The interesting biological questions are: why should there generally be more species of smaller than larger organisms? And, why does the frequency distribution appear not to hold for the smallest animal species, either overall or, for that matter, within a taxon? Multiple hypotheses have been put forward and most if not all are controversial. These began with the postulate of Hutchinson and MacArthur (1959) that smaller organisms could more finely divide a habitat than larger ones; with more such niches available, there would be more species of smaller than larger organisms. Though extrapolated to animals in general, their theoretical paper was focused on mammals. Moreover, the view of habitat complexity has expanded and become more tractable since then (e.g., Morse et al. 1985). Nevertheless, how organisms of different sizes may see their habitat remains hotly contested (see ▶Chap. 7). Also, a key prediction of the Hutchinson and MacArthur theoretical model, that the smallest categories of organisms either overall or within a taxon would be the most speciose, has not been realized. Rather, analysis of relationships within related assemblages (e.g., families and orders) suggests that it is not the smallest organisms that are the most species-diverse, but those averaging 38% larger than the smallest (Dial and Marzluff 1988). In other words, the small-to-medium sized taxa may be the most numerous. They analyzed trends in 46 assemblages but excluded microorganisms. The various other explanations emphasize, for example, community dynamics (May 1978, 1988), energetic relations and fitness (Brown et al. 1993, 2004), inter- and intraspecific size optimization trade-offs (Gaston and Blackburn 2000), and the purely mathematical role of population size and size-dependent extinction and speciation processes (Stanley 1973); for general summaries see McKinney (1990), Gaston and Blackburn (2000), Kozlowski and Gawelczyk (2002), Purvis et al. (2003), Dombroskie and Aarssen 2010). Gaston et al. (1993) compare the distribution of automobile 'species' of different sizes with that of beetle species varying in size. They find that despite quantitative differences, the trends are the same qualitatively in the biotic and abiotic systems. This can be explained in various ways but may mean that these coincident patterns result simply from macroscale properties common to large systems in general, whether ecological, economic, or geological (see Blonder et al. 2014). Similar inanimate/animate comparisons are taken up again in ▶Chap. 8.

In terms of number of individuals, population density tends to vary inversely with organism size. Relatively speaking, based on the wet weight of single cells, Brock (1966, p. 112) estimated that a given amount of nutrient can support 10^9 small bacteria, 10^7 yeasts, 10^5 amoebae, or 10^3 paramecia. A more contemporary estimate from actual populations in nature (a freshwater pond) shows a relative order of magnitude ranking of 10^{18} bacteria, 10^{16} protists, and 10^{11} small animals (reviewed in Finlay and Esteban 2007). Among macroorganisms there is also good evidence that the population density of aclonal animals decreases with increasing body size (Chap. 10 in Peters 1983). Damuth (1981; see also Carbone and Gittleman 2002; Brown et al. 2004) expressed the relationship for mammals as follows:

$$\text{Population density} \propto (\text{body mass})^{-0.75}$$

4.5   Some Ecological Consequences of Size

4.5.1   Benefits and Costs

Size effects can be considered at multiple levels—among individuals, populations, or taxa. At the individual level, large size can increase competitive ability, facilitate success as a predator, and deter predation. Harper (Chap. 22, 1977), for example, notes that for plants (unlike most animals) the *oldest* individuals may be the largest, have the greatest reproductive output, and also control the recruitment of seedlings. Among mobile taxa and within limits, speed *tends* to increase with size regardless whether the form of locomotion is by running, swimming, or flying (Bonner 1965; McMahon and Bonner 1983; Peters 1983). Interestingly, dolphins move at the same speed (1030 cm/s) as blue whales (Bonner 1965, his Table 2, pp. 185–187). Among land animals there is an eventual trade-off between size and speed because of increasing weight with volume. If swimming speed is expressed in body lengths per unit time (i.e., relative to the amount of new environment sampled by the organism), however, rather than in absolute terms, a motile bacterium explores at the same rate as, say, a dolphin. Large species consume more energy per unit time or distance, but specific costs (i.e., per unit weight) on either basis decrease.

Because generation time increases with size, the individual in a larger bodied species may be more prone to die before reaching sexual maturity and hence leave no descendants. To function, the macroorganism depends on complex developmental programs and integrity of specialized, interrelated cell types, the failure of any one of which could have lethal consequences. Adaptation to change tends to be slower in larger animal species as is repopulation, dispersal ability in general, and the ability to colonize new habitats. Peters (see his Chap. 8, 1983) examined colonization by assuming that a catastrophe depressed populations of micro- and macroorganisms to an arbitrarily low (1 gm/km^2) density (◖Fig. 4.12). He asked how long it would take for equivalent recovery of biomass to 100 kg/ km^2, assuming each species increased at its maximum intrinsic growth rate (r_{max}). The time ranged from an order of one day for the bacteria to one century for the large vertebrates. The tendency for body size to increase during evolution is accompanied by higher extinction rates among larger taxa (Chap. 9 in Stanley 1979; for caveats and details see Pimm et al. 1988). The mechanisms of extinction are complex and among the most controversial areas of biology. Ultimately extinction is due in varying degrees to intrinsic and extrinsic forces—as Raup (1991) has put it, to bad genes or bad luck. Therefore it is not surprising that understanding why extinction rates should generally be higher for larger species is complicated. But part of the answer likely involves their relative geographic insularity and vulnerability to environmental fluctuations that drive populations down to levels where they can go extinct from demographic stochasticity (Fenchel 1993; see 'The Gambler's Ruin' problem, Chap. 3 in Raup 1991). In contrast, species of smaller organisms in general and microorganisms in particular are buffered from such extinction events probably because of their relatively much larger population sizes, capacity for prolonged dormancy, and cosmopolitan distribution. (An interesting postscript is that essentially nothing is known about species extinction rates under natural conditions for microorganisms, especially bacteria.)

Peters' (1983) recovery time example above also means that although microorganisms respond faster, the duration for which they are able to grow potentially at a maximum rate is orders of magnitude lower than that for the largest vertebrates. The short generation

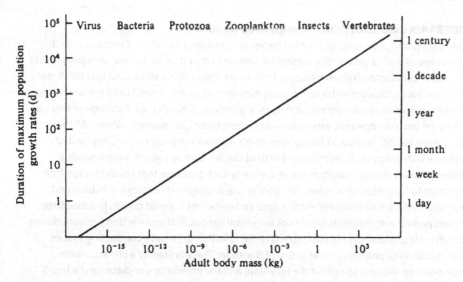

Fig. 4.12 Estimated relative influence of body size on potential duration of maximum (exponential) population growth rates (r_{max}) and hence of relative colonizing potential of micro- and macroorganisms. For bacteria, it takes on the order of days and for insects on the order of years (right axis) to reestablish a hypothetical population biomass of 100 kg/km² (average density rounded to the nearest order of magnitude of many animal populations) starting from a negligible starting density of 1 g/km². From Peters (1983); reproduced from *The Ecological Implications of Body Size* by R.H. Peters, by permission of Cambridge University Press, ©1983

times for microbes that lead rapidly to large population sizes also imply the relatively early onset of crowding effects. The implications for temporal scale in biology are far-reaching, as Peters notes (see also Allen 1977). One consequence is that the apparent relative stability of the larger macroorganisms may be simply because they cannot fluctuate appreciably during the time course chosen for their observation. A wildlife ecologist studying moose, lions, or elephants may make observations at monthly or yearly intervals for decades. At the other extreme, observations on bacteria or phytoplankton kinetics typically occur over hours, possibly weeks, perhaps for a month or a growing season, depending on the experiment. Choice of the appropriate interval is critical, particularly when interpreting microbial succession on natural substrata. Finally, concepts such as stability or community equilibrium may well be an illusion: they depend on the frame of reference, the scope of the community under scrutiny, the length of life of the organisms concerned, the frequency of the observations, and the duration of the study. It may be that our concept of time is only relevant when the parameters by which we measure it (sun/moon; night/day; seasons) are also responded to by the organism of interest (see ►Chap. 7).

Some organisms can adjust their size as circumstances dictate. Pufferfish deter predators in part by inflating themselves. Analogously, but in terms of size reduction, vascular wilt fungi can produce a small form of spore that facilitates movement within the conducting vessels of their living hosts, and thereby colonization. Also, because of space limitations, such spores can be smaller (microspores), or in some cases yeast-like cells produced in situ by budding, rather than on conventional stalks or conidiophores (see **Sidebar**, below). Interestingly, many systemic fungal pathogens of humans and animals are dimorphic (Nemecek et al. 2006), alternating between a single-celled, yeast-like, parasitic state, and a saprophytic mycelial form.

4

SIDEBAR: A Case Study: When it Helps to Be Small

The plant pathogenic fungi (primarily species of *Fusarium, Verticillium, Ceratocystis,* and *Ophiostoma*) that cause vascular wilting are unique in that they set up housekeeping within the water-conducting (xylem) tissues of their hosts. They remain there (and specifically for the most part within the water-conducting elements or xylem vessels) until the late stages of disease development whereupon they may grow into other tissues. Although details of life cycles and pathogenesis vary with each system, there is a common pattern. All these fungi must be able to grow at low oxygen levels within a relatively narrow, segmented pipeline with rough walls, interspersed with obstacles, through which is continuously flowing a dilute nutrient solution. There are many such pipelines that conduct water from the roots to the uppermost leaves. The 'goal' of the pathogen is to seize the habitat and reproduce. In such an environment the time and resources it would take to produce large spores packed with nutrients would not be advantageous. Rather, selection pressure during the lifecycle phase when the pathogen is residing within the plant's conducting tissues is on numbers of propagules and speed of transport: There is literally a life-and-death race between the host to delimit the infection, and the invader to out-distance the host's response and seize territory. How can the pathogen win the race?

It seems more than coincidental that the wilt-inducing fungi are dimorphic (Puhalla and Bell 1981; see also Nemecek et al. 2006), that is, capable of growing in two distinct forms closely aligned with the parasitic and saprophytic phases of the life cycle. The parasitic phase is typified by single-cell, yeast-like propagules (as bud cells or as microspores borne from simplified stalks or conidiophores) and hyphae, whereas in the saprophytic phase these fungi may produce extensive hyphae and conventional spores, some of which are modified for dormancy or dispersal. The obvious adaptive value of the yeast or micro-conidial phase in xylem vessels is that fungal biomass is channeled directly into many small cells that move passively upward in the xylem fluid to systemically colonize the host. Numbers may reach about 7000 conidia per mL of xylem fluid (for *Verticillium albo-atrum* in the hop plant; Sewell and Wilson 1964).

A given length of pipe (technically, vessel) rarely extends the full length of the plant and water must pass through a thin cell wall (pit membrane) as it moves laterally from one vessel to the overlapping portion of the next. These walls block passage of spores. Vessels, in turn, are composed of many segments of pipe (vessel elements) stacked one upon the other. The end walls (perforation plates) where the segments join are sufficiently open that water flow is not impeded appreciably but, depending on species of plant, rather than containing one large opening may consist of many smaller ones that entrain fungal propagules. The common strategy of the wilt fungi is to release into the sap stream abundant spores that travel for a distance before becoming obstructed; they then germinate quickly to produce hyphae that grow through the obstruction, produce another round of yeast-like cells or microconidia, and the sequence is repeated. Thus ensues a pitched battle between the host—which attempts to restrict the invasion by various mechanical and biochemical mechanisms—and the pathogen, which must advance rapidly if it is to escape the counterattack. It does so in successive waves of propagule release followed by growth through obstacles, much like an infantry battalion advances to secure successive beachheads. This battle is illustrated in ◨ Fig. 4.13.

A related and interesting side note: wilt diseases are classically a problem of Angiosperms and not Gymnosperms because of the anatomical difference in their water-conducting elements (Dimond 1970). For additional information, see Esau (1965), Mace et al. (1981), Pegg (1985), and Evert (2006).

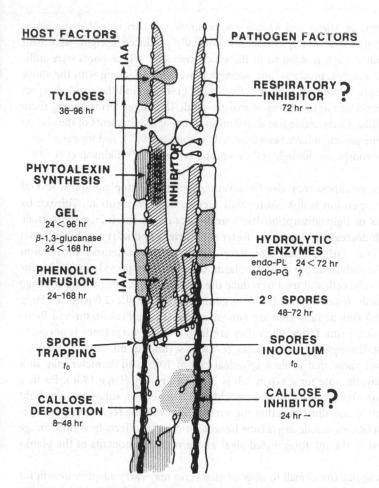

HOST FACTORS

IAA

TYLOSES
36–96 hr

PHYTOALEXIN
SYNTHESIS

GEL
24 < 96 hr
β-1,3-glucanase
24 < 168 hr

PHENOLIC
INFUSION
24–168 hr

IAA

SPORE
TRAPPING
t_0

CALLOSE
DEPOSITION
8–48 hr

INHIBITOR

PATHOGEN FACTORS

RESPIRATORY
INHIBITOR **?**
72 hr →

HYDROLYTIC
ENZYMES
endo-PL   24 < 72 hr
endo-PG   **?**

2° SPORES
48–72 hr

SPORES
INOCULUM
t_0

CALLOSE
INHIBITOR **?**
24 hr →

◻ Fig. 4.13 Model showing in longitudinal view how the plant host and vascular wilt pathogen confront each other in a portion of the water-conducting (xylem) tissue. Events shown occur over the initial 7-day time period after infection; a race ensues between the pathogen, which produces factors to promote systemic colonization; and the host, which attempts to physically and chemically contain the pathogen. The large central 'pipeline' is a portion of a vessel showing many small pathogen spores, some of which are entrained at a vessel element end wall (perforation plate), subsequent invasion of the adjacent vessel element by spore germination, followed by secondary rounds of sporulation. Spores are small and easily moved upwards in the transpiration stream though impeded by obstacles produced by the host. The difference between a resistant and susceptible plant depends critically on speed or response of the host. Tyloses = balloon-like invaginations of parenchyma cells adjoining the vessel elements into the lumen of the vessels; IAA = plant hormone indoleacetic acid; phytoalexin = specific type of inhibitory chemicals elucidated by host to inhibit pathogen. Reproduced from Beckman, C.H. 1987. *The Nature of Wilt Diseases of Plants*, by permission of the American Phytopathological Society, St. Paul, MN, ©1987

As early as the 1930s Kubiena (1932, 1938) made fascinating observations pertaining to the size of fungi in soil. Some organisms were limited in occurrence to large spaces between soil particles because they were too big to develop in the smaller pores. The fruiting bodies of *Cunninghamella* occurred only in spaces of diameter exceeding 600 μm. Conidiophores of *Botrytis* and sporangia of *Mucor* and *Rhizopus* were similarly restricted. The sporangio-phores (spore-bearing stalks) of *Rhizopus nodosus*, normally produced upright, were bent and occasionally spirally curled, molded to fit the space available. Where pores were suffi-ciently large, soil insects, mites, protozoa, and springtails were found, along with the above molds and a larger fungus, *Humicola*. Likewise, Swedmark (1964) found the smallest repre-sentatives of several phyla in the interstices of marine sand. The larger forms among them were frequently threadlike. Comparable size distribution limits of the filaments of the gliding sulfur bacteria in marine sediment have been reported (Jorgensen 1977), and for percolation of bacteria and fungal zoospores through soil or aquatic sediments (Wilkinson et al. 1981; Fenchel 2008).

In aquatic systems, smallness may also be advantageous for phytoplankton in several ways: by promoting suspension in the photic zone (see earlier comments on *Volvox*); by increasing the process of light absorption itself; and, because of the increasing surface area:volume ratio with decreasing volume, by fostering nutrient uptake (Fogg 1986; Raven 1986b, 1998; Moore et al. 2013). Small size, irregular shape, and extensive vacuolation increase the area of plasmalemma per unit cytoplasm; these changes provide more sites for nutrient transport into the cell, and tend to reduce the sinking rate. Cells in the size range of picoplankton (broadly speaking, planktonic organisms between 0.2–2.0 μm; see Fogg for caveats and details) sink at rates that are almost imperceptible (Takahashi and Bien-fang 1983). Phytoplankton sink faster when they are larger, and if turbulence is adequate they will cycle rapidly through the whole water column, scrubbing nutrients as they go. Theoretical calculations show that while a spherical cell of 10 μm in diameter may sink at about 25 cm per day, the rate for a 1-μm cell is 2.5 mm per day (Fogg 1986). Further reduction in size below 1 μm appears to have no additional benefit on sinking rate or light absorption. This example also illustrates that the surface area:volume relationship is espe-cially important for relatively sessile organisms because it directly affects how they forage (►Chap. 5), in contrast to the multiflagellated algal and protist components of the plank-ton.

Although changes in size from small to large or vice versa may carry adaptive benefit for the organism, it is worth noting in passing Gould's (1966) observation that this does not nec-essarily apply to specific structures, which must be above a certain minimum size to function at all. There are examples both among the micro- and macroorganisms. To insure effective spore dispersal, mushrooms must be sufficiently high above the ground. The stalks (stipes) of the larger fruiting bodies need not exceed this length and hence on a biomass basis are pro-portionately smaller than those of the small species (Ingold 1946). Among animals, the size of rods and cones of the eye does not vary with organism size, but is evidently set by optical properties (Haldane 1956; Thompson 1961, pp. 34–35).

4.6   Size and Life History Theory

4.6.1   Size of Dispersal Units

Salisbury (1942) developed an influential proposition about seed size in flowering plants and nutritional resources that was extended by Garrett (1973) to plant pathogenic fungi. The idea in Garrett's words (p. 3) is that *"the average size of a reproductive propagule is determined by the nutritional needs for establishment of a new young individual of the species in its typical habitat"*. Salisbury's thesis was that seed size is determined by the length of time during which a seedling must be self-supporting before it can supply its own needs by photosynthesis. Salisbury's generalization was based on data on seed and fruit production by 240 species of British flowering plants. He found that the spectrum in seed size could be correlated with habitat type: Species with the smallest seeds were characteristic of open habitats (early successional); at the other extreme were the heaviest seeds from shade-adapted woodland flora (later successional). By evolutionary adjustment of seed size to habitat type, it was argued, any plant species could efficiently allocate resources, providing the appropriate level of reserve in each instance. This story is analogous to Lack's (1947) and Cody's (1966) famous work on allocation of resources and clutch size in birds. Though Salisbury's logic is appealing (perhaps deceptively so), subsequent studies on the evolution of seed size have shown the issue to be considerably more complicated and related to many attributes, among them plant size and longevity, and not only survival through the juvenile phase but through reproduction (e.g., Chap. 21 in Harper 1977; Moles and Westoby 2006; Rees and Venable 2007). An important caveat must also be added in how this logic is worded and interpreted. Ultimately at issue here is what compromise strategy maximizes fitness. Reproductive capacity does not evolve to match the hazards of the environment because as Harper says (1977, p. 648) ... *"it is impossible for an organism to evolve to a state in which it leaves enough offspring to replace itself from a condition in which it did not leave enough"*. Rather, the interpretation (and obviously that of Garrett, below) needs to be made from the standpoint of natural selection favoring those individuals that contribute more descendants than their competitors to subsequent generations (see comments ►Chap. 1).

Garrett (1973) proposed that the fungi *Botrytis* and *Fusarium* illustrated the principle that the level of endogenous reserves in fungal spores appears to be adjusted to take advantage of supplementary exogenous nutrients supplied in plant exudates. *B. fabae* produces spores about nine times the size of (and with proportionately more nutrient reserves than) those of *B. cinerea*. The former pathogen is able to overcome the resistance of healthy, vigorous leaves. In contrast, *B. cinerea* does not preferentially infect healthy leaves, except under extreme conditions of inoculum pressure or in the presence of additional nutrients (e.g., as may be leached from pollen deposited on leaf surfaces). This species is well known to be an opportunistic parasite of senescing, wounded, or weakened plants. Its typical infection courts are debilitated leaves, floral organs, overripe fruit—all of which are predisposed to infection by virtue of nutrient leakage or reduced resistance or both (Ngugi and Scherm 2006). So Haldane's (1956) maxim "on being the right size" for *Botrytis* may mean one of two strategies: The organism could produce numerous small propagules of limited infectivity that rely on nutrients from a conducive host structure or condition to compensate for their small nutrient reserve (analogous to Salisbury's theme of smaller but more abundant seeds characteristic of plants from open habitats). Alternatively, fewer but large, well-supplied spores could be produced, capable of overcoming highly resistant host organs. The benefit would be a propagule of higher infectivity, less subject to certain external conditions; the cost is reduced output. Both strategies are evidently successful, but it is interesting that *B. cinerea* is the more abundant and widespread of the two species (Garrett 1973), suggesting that it may be ecologically more successful. Among *Botrytis* species, *B. cinerea* is considered to be a polyphagous generalist attacking over 200 eudicot species, whereas *B. fabae* is a specialist restricted to certain hosts in the Fabaceae (pea family) (Elad et al. 2004; Staats et al. 2005), although human influences on host manipulation complicate the interpretation somewhat.

Salisbury (1942) extended his seed size hypothesis to encompass vegetative propagules such as rhizomes and stolons, insofar as these were another means of nutritional support from the parent plant. Analogously, Garrett (1973) recognized the striking visual and functional parallel presented by certain structures of root-infecting fungi such as species of *Fomes, Armillaria*, and *Phymatotrichum*. Mycelial strands and rhizomorphs differ in detail but all are basically subterranean, macroscopic (often several mm in diameter), multistranded cables of hyphae that may extend for dozens of meters. In *Armillaria*, clones extending at least 450 m have been documented (Anderson et al. 1979) and the maximum size of a single clonal population can be immense (Smith et al. 1992). All such strands or rhizomorphs function in translocating nutrients from a food base to the growing apex, ultimately to the point of infection. Why does an essentially microscopic organism consisting of fine mycelial threads allocate biomass to produce such a massive, elaborate structure? In part, these are effective organs for resource exploration (Boddy et al. 2009). Furthermore, unlike the roots of herbaceous plants that are relatively vulnerable to infection by single spores, those of undamaged, woody hosts resist invasion by physical and chemical defenses. By aggregating multiple hyphae as a rhizomorph, the pathogen can breach the defenses of a mature tree that would be impenetrable by a single hypha. Rhizomorphs also offer the fungus some insularity from adverse environments or nutrient leakage while the organism is in an exploratory mode.

Harper (1977, pp. 672–673) makes some most interesting points relating to plants with very small seeds that complement Salisbury's story. This pertains to certain symbiotic plants, such as the saprophytic *Monotropa* and the parasitic *Orobanche*. Unlike some of the larger seeded symbionts (dodder and mistletoes, which must either grow to locate a host or penetrate its bark, respectively), these small-seeded plants germinate only when triggered to do so by a host chemical. They are thus assured of an external food source immediately upon germination. In evolutionary terms, this removed the need for an internal food reserve and natural selection acted to increase seed number rather than size. The species have acted opportunistically, seizing on *"an alternative mode of embryo nutrition to reduce seed size to tiny dried bags of DNA and expand their reproductive capacity to a new limit"* (p. 673). This is in contrast to the conventional interpretation that they "need" to produce more seeds to "find" a host—an explanation which, as Harper points out, is wrong, for reasons noted above. In this context it is also important to recall that **number of progeny** (i.e., seeds in the present context) **is not the same thing as number of descendants**, and that the currency of natural selection is the latter (▶Chap. 1).

4.6.2 Microorganisms are Not Necessarily *r*-Strategists

Salisbury's and Garrett's hypotheses prompt the more general question of how body size may influence life history traits such as reproductive rate and competitive ability. To what extent do environments influence these traits and what, if any, are the trade-offs? Do microorganisms behave fundamentally differently from macroorganisms? The original and still most general concept that deals with this issue, albeit it indirectly, is *r*- and *K*-selection (MacArthur

▢ **Table 4.4** Estimated maximal instantaneous rates (r_{max} per capita per day) of increase realizable for various organisms (abbreviated from Pianka 2000, p. 151)[a]		
Organism	r_{max}	**Generation time (T) (days)**
Escherichia coli (bacterium)	~60	0.014
Paramecium aurelia (protist)	1.24	0.33–0.50
Tribolium confusum (insect)	0.120	~80
Ptinus sexpunctatus (insect)	0.006	215
Rattus norwegicus (mammal)	0.015	150
Canis domesticus (mammal)	0.009	~1000
Homo sapiens (mammal)	0.0003	~7000

[a]The components of r_{max} are the instantaneous birth and death rates per capita under optimal environmental conditions in a closed population (no immigration or emigration). The term μ_{MAX} is usually used in place of r_{max} in bacterial semantics. For details, see Pianka (2000 pp. 150–153) or a more lengthy discussion in Pianka's 2nd ed. (1978, pp. 110–118); see Andrews and Harris (1986) for microbial ecology implications. Note that r_{max} varies by several orders of magnitude among organisms

and Wilson 1967; Pianka 1970, 2000; Boyce 1984; Andrews and Harris 1986). MacArthur and Wilson (1963, Chap. 7 in 1967) in articulating their elegant theory of island biogeography considered that at the extremes there were two categories of colonists associated with two different selection regimes. Where islands are new (untenanted), relatively free of competitors, and with relatively abundant resources, the selection pressure should favor those immigrants that reproduce quickly, i.e., what became known as **r-strategists**. Over time, as the island becomes increasingly crowded, richer in species, and resource-limiting, selection should favor genotypes that are more competitive or **K-strategists**.

Consideration of *r*- and *K*-selection and the related terminology require first a brief review of population growth. In **exponential** (unlimited) growth of a closed population (where immigration and emigration can be excluded), the instantaneous change is usually written as $dN/dt = (b - d)N$, where N is number of individuals and b and d are the average birth and death rates per individual per unit time. Where growth is limited, the **logistic** equation replaces the exponential and is often written as $dN/dt = rN(K - N)/K$ where r is maximal and fixed and taken to be the difference between instantaneous birth and death rates as population densities approach zero. Thus, *r is usually defined as the intrinsic rate of increase* in numbers per unit time per individual and in effect represents the r_{max} for a specific organism, i.e., the maximum instantaneous rate of increase possible under optimal conditions where r is unconstrained. The r_{max} for various species is shown in ▢Table 4.4 and discussed by Pianka (1970). At the other extreme, when $b = d$, the population maintains a stable size and this situation is referred to as *the carrying capacity or K*. This is the dynamic upper boundary of the population; if $N > K$ the population will decrease; if $N < K$, the population will increase. Note that in practice as birth and death rates change the *actual* instantaneous rate of change per individual, r_a, is a variable and will change as a function of r_{max} (the r in the logistic), N, and K.

Thus, the logistic expression describes overall population growth and regulation in a limiting environment. Such growth curves are well known and appear in some form in every introductory ecology and microbiology book as linear (log-transformed) or sigmoid (arithmetic) plots of change in population density versus time. The simplifying assumptions of the logistic expression have long been acknowledged. However, though inherent to the expression, the point is rarely made that the equation is purely descriptive of events, not explanatory. The following discussion summarizes some implications for macroorganism as well as microbial ecology (for details see Andrews and Harris 1986). In particular, the implications of 'crowding' should be recognized because, especially in plant and animal ecology, they are often interpreted narrowly to involve only food density, i.e., competition for resources. But crowding implies all factors that change in density-dependent fashion, including parasitism, predation, and—especially in microbiology—production of inhibitory metabolites.

Although both r and K are subject to evolutionary adjustment, the ecological dogma is that a high r occurs at the expense of a low K. In other words, there are inherent trade-offs: an individual cannot maximize both parameters (Roughgarden 1971). From well-controlled experiments with bacterial populations, albeit in non-structured or homogeneous environments (below), there is evidence both for and against the trade-off hypothesis under conditions of resource abundance and scarcity (Velicer and Lenski 1999). This is not surprising in part because whether there is a trade-off is likely trait-dependent. In the formative stages of r/K theory, MacArthur (1972, pp. 226–230), using populations of two alleles, illustrated situations in which r- and K-selection coincided and situations where they did not. He also noted situations where populations could alternate between r- and K-selected environments. Furthermore, a major complication in microbial systems is that in non-structured environments, which are the standard condition in laboratory tests and include suspension cultures and chemostats, a rapidly growing strain will quickly displace a slower competitor before any trade-offs such as between growth rate and growth yield (cells per unit substrate) can be manifested. An emulsion-based culture system, which imposes a form of structure, may provide for more realistic experimentation, and has led to such trade-offs becoming apparent (Bachmann et al. 2013) (Recall the earlier discussion of growth [ATP] rate versus growth yield in ▶Chap. 3. These attributes are relevant here because one attribute of high r is the ability to generate ATP rapidly, if wastefully; a key attribute of a K organism is the ability to harvest resources efficiently.)

The central question is what life history traits are associated with purportedly high r- or K-selecting environments. r-selected individuals are predicted to be smaller in size, to mature earlier, and to have more and smaller progeny. In contrast, K-selected individuals should be larger, show delayed reproduction, allocate resources more to growth (size) for competitive advantage, and hence have fewer but larger offspring. Pianka (1970, 2000) recognized several correlates (◻Table 4.5; however, see criticism by Boyce 1984). On balance it is clear from the table that small organisms, microbes in particular, would be shifted toward the r-end of an r-K spectrum. It is important to recognize, however, that microorganisms themselves appear to be relatively r- or K-selected based on their relevant life history traits (◻Table 4.6; see also Swift 1976; Andrews and Rouse 1982; Andrews and Harris 1986; Andrews 1991).

◼ **Table 4.5** Some correlates of *r*- and *K*-selection (after Pianka 1970)

Attribute	*r*-selection	*K*-selection
Climate	Variable and/or unpredictable: uncertain	Fairly constant and/or predictable: more certain
Mortality	Often catastrophic; nondirected; density-independent	More directed; density-dependent
Population	Variable, nonequilibrium, usually well below carrying capacity; unsaturated communities or portions thereof; recolonization annually	Fairly constant, equilibrium; at or near carrying capacity; saturated communities; no recolonization
Competition	Variable, often lax	Usually keen
Selection favors	Rapid development; high r_{max}; early reproduction; small body size	Slower development; greater competitive ability; lower resource thresholds; delayed reproduction; larger bodies
Length of life	Short; usually less than 1 year	Longer; usually more than 1 year
Leads to	Productivity	Efficiency

While *r*- and *K*-comparisons across widely different taxa can be informative in a general sense, several limitations need to be kept in mind. First, the concept of reproductive value (►Chap. 6), and relatedly age-specific models that are familiar to macro-ecologists, are foreign to microbiologists. This is a measure of contemporary reproductive output and residual reproductive value; for macroorganisms it entails life table statistics based on the likelihood of survival and reproduction for specific age classes (juveniles versus adults). A particular life history exhibited by an organism should theoretically be one that has the greatest overall reproductive value. Although the *r/K* model assumes that environmental fluctuations affect all age classes equally, competing models (e.g., bet-hedging) do not. The issue is moot in microbial ecology because of the short generation times and the plasticity and totipotency of microorganisms, which preclude any analog to adolescence or pre- and post-reproductive classes used in plant and animal biology.

Second is the distinction between modular and unitary organisms (►Chap. 5). By growing in modular fashion, the genetic individual or genet can increase indefinitely in size by adding modules (as in flowers, branches, or leaves of a tree; hyphal tips of a fungus). Population growth for these organisms may thus be exponential and, unlike the case for unitary organisms, is not necessarily delayed by postponing reproduction. The predictions from *r/K* theory for modular as opposed to unitary organisms may be quite different (Sackville Hamilton et al. 1987) as discussed in ►Chap. 5.

Third, comparisons are best made among species on the same trophic level, where possible, because whether resources or some other mechanism typically acts to limit organisms may depend on their position in the trophic web.

Finally, when organisms are compared, body size should be considered to determine whether natural selection influences reproductive output independently of size. In other words, does selection act primarily on size (Stearns 1983) with a corresponding rate of

4

□ Table 4.6 Some life history features of *r*- or *K*-selected microorganisms (Andrews and Harris 1986)

Trait	Organism	
	r-strategist	*K*-strategist
Longevity of growth phase	Short	Long
Rate of growth under uncrowded conditions	High	Low
Relative food allocation during transition from uncrowded to crowded conditions	Shift from growth and maintenance to reproduction (spores)	Growth and maintenance
Population density dynamics under crowded conditions	High population density of resting biomass; high initial density compensates for death loss	High equilibrium population density of highly competitive, efficient, growing biomass; growth replacement compensates for death loss
Response to enrichment	Fast growth after variable lag	Slow growth after variable lag
Mortality	Often catastrophic; density-independent	Variable
Migratory tendency	High	Variable

increase (r_{max}) following as a consequence? For engineering reasons noted earlier related to complexity, it takes less time to construct a small than a large organism. Small organisms mature and breed earlier than do large organisms, hence the correlation between smaller body size and a higher r_{max}. Or, alternatively, is selection mainly for a particular r_{max}, which then dictates a given size because of the relationship between the two parameters (Ross 1988)? Ross examined reproductive patterns in 58 primate species and found generally that after size differences are factored out, the species inhabiting unpredictable, r-environments had a high r_{max}. Selection has evidently acted directly to increase r_{max} of these species rather than indirectly by decreasing body weight. There was no evidence, however, that either the raw or relative r_{max} values were as predicted for species in predictable environments. This anomaly remains to be reconciled but may be explained by imprecise classification schemes for the respective environments (Ross 1988). Such formal comparisons have not been made among the microbes. However, because microorganisms of similar size behave very differently in different environments (Andrews and Harris 1986) and also have very different intrinsic growth rates (Brock 1966, p. 95), it seems that selection has been primarily for an r- or K-type strategy rather than indirectly as an unavoidable consequence of size.

The paragraphs above consolidate to this: r- and K-selection can be informative in comparing taxa broadly with respect to life histories along a species continuum, as originally done by Pianka (1970). Such applications are roughly analogous to placing species on a plot of r_{max} versus generation time as done also by Pianka (1970). The r/K theory does not explain or account for all life history traits and there are organisms that appear not to fit the scheme (see concluding remarks). It has more conceptual value when comparisons are made in a narrower context, such as with respect to strategies among various microbial species

That the logistic equation underpinning r- and K-selection has no mechanistic basis is an obvious limitation. The complexity of most plant and animal systems means that it is difficult if not impossible in population dynamics to delve deeper than gross phenotypic expressions (birth, death, growth). Microbial systems, particularly those with bacteria, can provide for a causative interpretation of the effect of crowding. Subject to the appropriate design noted earlier (e.g., Bachmann et al. 2013), experiments can be conducted under controlled as well as uncontrolled conditions with organisms that are relatively well characterized in terms of their genetics, growth rates and efficiencies, nutritional requirements, and metabolic pathways.

In overview, like many theories in ecology and other sciences, the r/K scheme had a period of ascendancy followed by decline in the wake of critical reassessment; after various refinements, stability occurred at some lower level of influence, with both adherents and skeptics. Disenchantment with the theory (see, e.g., Wilbur et al. 1974; Boyce 1984; Stearns 1977, 1992) has probably stemmed mainly from attempts to overinterpret it. Responses to crowding are one of many that ultimately shape life history traits. Charlesworth (1994, pp. 265–266) has pointed out that the emphasis on logistic growth in formulation of the theory has neglected the consideration of age-structure in populations, thereby omitting some important demographic factors shaping selection. There has been ambiguity in the literature about what r/K is supposed to be describing: response to crowding in terms of resource acquisition only? to density effects broadly? to particular environments broadly?

It is important to recognize that the organism and selective regime attributes identified by Pianka (1970) and others are **correlates** (as Pianka correctly stated). They do not imply causation. While such caveats were implicit in many of the renditions of r- and K-theory, they were rarely made explicit. As noted here, the theory has not gone beyond phenotypic attributes to address a mechanistic basis for their occurrence. Nevertheless, as a conceptual vehicle for organizing knowledge, describing some general differences among organisms, and prompting tests, the r/K postulate has served us well (though not all ecologists would share this opinion)!

4.7   Summary

A striking feature of life is the immense difference, exceeding some 21 orders of magnitude by weight, in size among species from bacteria to whales and large trees. Even among the smallest of organisms, the prokaryotes, size range exceeds several orders of magnitude. Size of a particular individual also varies extensively as a function of its life cycle stage from inception as a single cell to death as an adult. These differences have far-reaching consequences on how an organism sees its world and on attributes such as shape and complexity.

Reasonably strong evidence exists that Earth originated about 4.5 billion years ago. Life probably arose between that time and the occurrence of the earliest chemical fossils (~3.8 Gya). A common evolutionary origin for all life is inferred from universal homologies, most notably a standard biochemical recipe. The incipient primordial entity and base of the 'tree of life' called the Universal Ancestor probably arose about 3.6–3.4 Gya and possibly existed as a community of primordial cells often referred to as progenotes.

The evolutionary history of life is replete with experimental transitions from unicellularity to multicellularity. The paleontological record suggests multicellular (filamentous) prokaryotes and cell differentiation by about 2 Gya and the first unicellular eukaryotes at about 1.5 Gya (possibly much earlier). The major multicellular phyla evidently all arose and developed from different unicellular progenitors. These transitions apparently occurred at least 25 times and in all three major domains of life, sometimes through a colonial intermediary and with occasional reversals to unicellularity. Details of the timetable for these events are vague and probably always will be controversial. It is nevertheless broadly accepted that, for about the first 2–3 billion years, life on Earth was microscopic and simple, preponderantly if not exclusively prokaryotic, with only basic forms of differentiation occurring, if at all. Conventional multicellularity was ultimately an evolutionary dead-end for prokaryotes; eukaryotes exploited growth form, in large part enabled by a fundamentally different cell structure as the building block, in a way that prokaryotes could not.

The contemporary volvocine algae (*Volvox* and its relatives, in all about 50 species), which evolved from unicellular ancestors about 200–300 Mya, provide an instructive case study for size increase and evolving complexity. The range of species shows marked differences in cell number, states of differentiation (e.g., of germ and soma tissue in some members), and emergence of individuality through division of labor and functional integration.

Size increase by the emergence of larger cells and the transition to multicellularity (either by clonal division or aggregation) would have provided for an alternative and arguably more effective method of resource acquisition, and avoidance of predation by size alone or through associated traits such as increased mobility. The evolution of multicellularity opened the way for cell specialization and enhanced intercellular cooperation, which in turn allowed for many attributes or functions such as metabolic innovation, storage of reserves, provision of an internal environment for the organism, and buffering from the external environment. Natural selection for multicellularity may be opposed by genetic variants (cheater or defector cell lines) that promote their own multiplication at the expense of the integrated individual. That such potential genetic conflicts are dangerous to the integrity of the multicellular individual are evidenced by the numerous and phylogenetically widespread defenses against aberrant cell lines. These include passage through a single-cell (and frequently haploid) developmental stage, germ-line sequestration, and various self/non-self-recognition systems.

Physical and chemical laws ultimately set the theoretical lower and upper size constraints on life. The low end, approached by some very small bacteria, is established by the minimum package size to contain cell machinery and associated molecular traffic; the upper limit on cell size is set mainly by the declining surface area-to-volume ratio as size increases, which affects exchange processes, among other biophysical relationships. The small size of prokaryotes (and microorganisms in general) means that their world is dominated by molecular phenomena such as diffusion, surface tension, viscosity, and Brownian movement. Macroorganisms, in contrast, are conditioned primarily by gravity and scaling relationships. Extremely large size can be attained in some circumstances where gravitational and other constraints can be mitigated, such as by clonal growth on land, which can diffuse bulk over a broad area, or by exploiting buoyancy afforded by life in a medium denser than air, the seas (whales).

Allometric scaling denotes regular changes in certain proportions or traits as a function of size according to the general relationship of $Y = aW^b$ where Y is the biological variable of interest, W is a measure of body size such as mass, a is the normalization constant typical of the kind of organism, and exponent b is the scaling constant. The general nature of the relationship is defined by b. A broadly based and biologically important example is the surface area law or the 2/3-power law (exponent b is 2/3) dictating the relative decline in surface area as volume increases (Principle of Similitude). This means that if the rate of a metabolic process associated with volume depends on surface area, then the rate of that process must also scale as $V^{2/3}$. Thus, organisms are inherently constrained in innumerable ways relative to surface area functions. They may compensate in certain key functions by scaling adjustments so that surface area and volume scale in direct proportion (exponent b is unity) as size increases. A related example is the 3/4 power law (exponent b is 3/4) depicting the relative decline in metabolic function with size increase.

Both mean and upper size limits among biota as a whole have increased over geological time. While specific animal *lineages* also generally show overall trends of size increase (Cope's Rule) this is not invariable, and the mechanism (active, directional replacement of smaller forms versus a passive trend marked by increasing variance in size) is controversial. With increase in size has come increase in complexity (e.g., number of cell *types*). The most important event spurring evolution of greater complexity arguably was the transition from unicellularity to multicellularity. This enabled organisms to increase in size, differentiate to segregate function by specialized cell type, and to develop increasingly sophisticated forms of intercellular communication and division of labor.

Population densities are higher for smaller than larger organisms. Generally speaking, habitats also contain many more species of smaller organisms than large ones. Why this should be the case is controversial but in part relates to the ability of smaller organisms to partition resources more effectively than larger organisms (finer dissection, more niches available) and in part because many larger species have a restricted distribution. This situation, combined with their lower rates of population increase and lower densities, means that larger species are more vulnerable to extinction.

Various models, hypotheses, and theories have been developed to interpret life history traits such as reproductive rate and longevity (competitiveness) as a function of body size. r- and K-selection, although not focusing on size specifically, addresses it indirectly. Using terms taken from the logistic equation, the concept emerged from MacArthur and Wilson's theory of island biogeography originally developed to contrast the nature of selection operating at the initial stages of island colonization (r-selection that would favor population growth rate) as opposed to later as the island became saturated with species (K-selection for competitive ability). r-selection favors a high maximal rate of increase, small size, early reproduction, and many small offspring, among other attributes. Correlates of K-selection are greater competitive ability, larger size, delayed reproduction, and fewer, larger progeny. Organisms adapted for success in either environment became known as r- or K-strategists on a relative basis along an r-K continuum. While microorganisms are in general the ultimate r-strategists, it is possible to recognize a spectrum even within this group. r- and K-theory does not provide a mechanistic explanation for the population behavior and traits identified as r- or K-related may have arisen for other reasons. Definitive experiments to test predictions of the theory either with microorganisms or macroorganisms have proven elusive. Nevertheless, r- and K-selection has provided a relatively simple, useful conceptual synopsis of many of the differences among organisms.

The vast size differences between the tiny microorganism and the comparatively giant macroorganism mean that they see different worlds shaped to varying degrees by different forces. Are the differences that size imposes so profound that they overwhelm meaningful ecological commonalities between the two groups? On the contrary, we saw in ▶Chap. 2 that their ecological genetics operates fundamentally from the same playbook though the tempo may vary, and in ▶Chap. 3 that there are close analogies in modes of energetics, carbon dynamics, and optimal foraging. And in the next chapter we shall see that there are also close approximations in growth form and population biology.

Suggested Additional Reading

Alegado, R.A. and N. King (Organizers). 2014. The Origin and Evolution of Eukaryotes. Cold Spring Harbor Perspect. Biol. Doi:10.1101/cshperspect.a016162. *Collected papers on the evolutionary transitions to multicellularity and complexity.*

Bonner, J.T. 1965. Size and Cycle: An Essay on the Structure of Biology. Princeton Univ. Press, Princeton, NJ. *A fascinating, eloquent account, timeless in its relevance, of how size affects all creatures with emphasis on how size of the organism changes during the course of the life cycle.*

Bonner, J.T. 1988. The Evolution of Complexity by Means of Natural Selection. Princeton Univ. Press, Princeton, NJ. *An excellent, stimulating synthesis on why there has been a progressive increase in size and complexity from bacteria to plants and animals.*

Carroll, S.B. 2001. Chance and necessity: the evolution of morphological complexity and diversity. Science 409: 1102–1109. *An insightful and authoritative synthesis on the evolution of life, including the evolutionary increase in size.*

Hedges, S.B. 2002. The origin and evolution of model organisms. Nature Rev. Genet.3: 838–849. *An interesting and informative synopsis of times of divergence of the prokaryotes, protists, plants, fungi, and animals, subject to the caveat that all such estimates are works in progress.*

Knoll, A.H., D.E. Canfield, and K.O. Konhauser (eds.). 2012. Fundamentals of Geobiology. Oxford Univ. Press, UK. *An excellent, well-illustrated synthesis by multiple authors on the early history of life on Earth.*

Thompson, D'A. W. 1961. On Growth and Form. (Abridged edition edited by J.T. Bonner of the original 1942 text.) Cambridge University Press, Cambridge, England. *This classic book remains a benchmark of excellence on the analysis of form.*

Vogel, S. 2003. Comparative Biomechanics: Life's Physical World. Princeton Univ. Press. *An interesting, readable, witty, and authoritative explanation pitched at the undergraduate level with innumerable fascinating examples.*

Growth and Growth Form

© Springer Science+Business Media LLC 2017
J.H. Andrews, *Comparative Ecology of Microorganisms and Macroorganisms*,
DOI 10.1007/978-1-4939-6897-8_5

Geometry is the most obvious framework upon which nature works to keep her scale in 'designing'.
She relates things to each other and to the whole, while meantime she gives to your eye most subtle,
mysterious and apparently spontaneous irregularities in effects.

Frank Lloyd Wright, 1953, p. 53.

5.1   Introduction

Growth form, that is, the shape and mode of construction of an organism, together with size sets fundamental opportunities as well as limits on the biology of living things. We consider here the case that organisms can be viewed as either basically unitary or modular in construction. This division pertains broadly to whether or not the life cycle consists essentially of repeated ontogenies at multiple levels. Being designed according to one plan or the other carries numerous implications, the most important of which are evolutionary consequences pertaining to fecundity and the transfer and expression of genetic variation. Further, it is argued that at least some microorganisms, notably the bacteria and fungi, are inherently modular in construction and share with macroorganisms the ecological properties that emerge from this design. Thus, the ecology and evolutionary biology of modular macroorganisms should be instructive to microbial ecologists. Conversely, experiments with the relevant microorganisms into the theoretical predictions of modularity may inform plant and animal ecologists.

How and when in geological time did this major morphological and developmental demarcation occur? As discussed in ▶Chap. 4, multicellularity arose multiple times, independently, in various prokaryotic and eukaryotic lineages. As we saw, the last common ancestor of plants and animals very likely was unicellular and the genetic tools for the development of multicellular forms are traceable, in part, to unicellular ancestors (King et al. 2008; Rokas 2008a, b). However, development at the molecular level proceeded independently in the plant and animal lineages (Meyerowitz 2002; Knoll 2011). But even more interesting in our present context is that, at a larger scale, plants and animals independently developed modular and, in some cases, unitary growth forms. Arguably, a similar bifurcation occurred among the microorganisms.

5.2   Unitary and Modular Organisms: An Overview

5.2.1   Design of the Unitary Organism

Unitary organisms, represented by most mobile animals, have a standard number of appendages fixed early in ontogeny and follow sequential life cycle phases predictably. These are the sort of organisms that are the favorite subjects at zoos and for televised nature programs. Their growth is non-iterative, i.e., not based on a repeated multicellular unit of construction, and is typically determinate, i.e., of strictly limited duration. They display generalized or systemic senescence, and are generally unbranched. **Reproductive value** (a measure of current fecundity and anticipated future survival and fecundity; see ▶Chap. 6) increases with age to some peak and then declines.

The most important evolutionary implication of a unitary design is that the genetic individual (genet of ▶Chaps. 1 and 2) and the physiological individual are the same entity. The unitary organism begins only at the start of the life cycle from a single cell, typically a zygote.

Harper (1977, pp. 515–516) illustrated the direct correspondence between the genetic and numerical individual by observing that a count of rabbits measures the number of genotypes and also gives an estimate of biomass. Likewise, within a factor of about 10, it is possible to estimate roughly the number of individuals if the biomass is known. Because form is determinate in unitary organisms (as is cell number within relatively strict limits), appendages are fixed in number: you could determine the number of rabbits by counting the legs in a population and dividing by four or the ears and dividing by two.

5.2.2 Design of the Modular Organism

In contrast to unitary organisms, the zygote in modular organisms produces a basic multicellular unit of construction (module) that is iterated indefinitely, giving progressively more modules, which in aggregate constitute the genetic individual. Plants and sessile benthic invertebrates are classic examples of modular organisms. A typical, easily visualized module[1] in the case of plants is the leaf together with its axillary bud. White (1979) captured its essence when he referred to "the plant as a metapopulation." The number of modules— whether they are leaves or roots on a tree, hyphae of a fungal thallus, bacterial cells in a colony, or polyps on a coral—is indeterminate and the organism has extreme phenotypic plasticity. It is for this reason that a simple count of plants, unless their size is known, provides so much less information than a count of unitary animals (Harper 1977, pp. 25–27). The major contrasting features between unitary and modular organisms are summarized in ◘Table 5.1 and several modular organisms are shown in ◘Fig. 5.1 (Harper et al. 1986; Hughes 2005). The inherently modular physical characteristics of fungi and bacteria are apparent when compared with the modular macroorganisms in this simplified figure. As will be developed in the following sections, **whether an organism is unitary and mobile, or modular and sessile, probably has as great or greater biological significance than its size (microorganism vs. macroorganism).**

Modules may remain attached and contribute substantially to the organism's architecture. This is the case for trees and corals, much of which often consist of accumulated dead modules. Alternatively, in the case of some organisms, modules either individually or in clumps can operate as physiological individuals, i.e., ramets, at a level of organization that they are or can be functional on their own when detached (strawberry or the creeping buttercup, *Ranunculus repens*), or naturally budded or sloughed off from the parent as part of the developmental program (lichens; corals; the floating aquatic plant *Lemna*). The ramets in aggregate constitute the genetic individual or genet; in other words, the genotype is fragmented (◘Fig. 5.2 and Harper and Bell 1979). Thus, modular organisms frequently grow **clonally**, that is, by the formation of individuals descending entirely from a common ancestor and thus of nearly identical genetic composition (though subject to genetic change over time; recall clonal discussion in ►Chap. 2 and see Jackson et al. 1985; Hughes 1989; Monro and Poore 2009b). In such cases the ramets are typically the 'countable units' in a pasture of white clover, *Trifolium repens*, or that arise from plating a bacterial suspension culture onto

---

1   The term 'module' as used by ecologists differs from the module of developmental biologists, which is usually taken to be a discrete subunit with respect to gene expression (see Raff 1996, pp. 326–334), though modularity has multiple connotations from the molecular through the organismal level of organization (Klingenberg 2008).

5

◘ **Table 5.1**   Major attributes of unitary and modular organisms[a] (Andrews 1994)

Attribute	Unitary organisms	Modular organisms
Branching	Generally non-branched	Generally branched
Mobility	Mobile; active	Nonmotile;[b] sessile
Germplasm	Segregated from soma	Not segregated
Development[c]	Typically preformistic	Typically somatic embryogenesis
Growth pattern	Non-iterative; determinate	Iterative; indeterminate
Internal age structure	Absent	Present
Reproductive value	Increases with age, then decreases; generalized senescence	Increases; senescence delayed or absent; directed at module
Role of environment in development	Relatively minor	Relatively major
Examples	Mobile animals generally, especially the vertebrates	Many of the sessile invertebrates such as hydroids; corals; colonial ascidians; also plants; fungi; bacteria

[a]These are generalizations. There are some exceptions (see text)
[b]Juvenile or dispersal phases passively or actively mobile. Many bacteria and protists motile or have motile stages
[c]Refers to degree to which embryonic cells are irreversibly determined. **Preformistic** = all cell lineages so determined in early ontogeny; **somatic embryogenesis** = organisms capable of regenerating a new individual from some cells at any life stage (cells totipotent or pluripotent). See Buss (1983)

an agar medium in a petri dish. The pasture would consist of multiple clones of white clover, each represented by many individual ramets, which are the 'plants' or 'plantlets' of white clover. Likewise, the petri dish would consist of a clone of the bacterium represented by dozens of ramets or bacterial colonies, each in principle having arisen from a single cell (see below and later discussion of bacteria and fungi). Perhaps, the most eloquently stated impression of a clone is that by Janzen (1977b) who viewed a dandelion population as one that "*contains a small number of highly divided EIs* [evolutionary individuals] *with very long lives and very low population growth rates and which exist through the harvest of a highly predictable resource (p. 586) … the EI dandelion is a very large tree with no investment in trunk, major branches, or perennial roots. It has a highly diffuse crown*" (p. 587) … [note that although the dandelion plant has flowers and forms seeds, many populations of dandelion are exclusively asexual, a product of apomixis; ►Chap. 2].

Before taking up the functional and evolutionary implications of modularity, a brief detour into terminology and definitions is necessary because of the very similar and in some cases overlapping attributes of modularity, clonality, and colonial growth habits. Although some ecologists use the terms clonal and modular interchangeably (Jackson and Coates 1986; West et al. 2011), modular growth is more appropriately regarded as being in some ways analogous to but distinct from clonal growth. Asexuality and clones were discussed in some detail in ►Chap. 2. Organisms that reproduce asexually are clonal and often, but not invariably, modular.

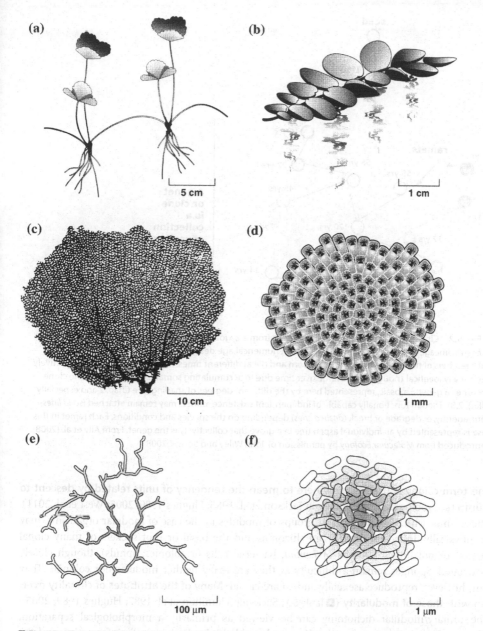

⬛ Fig. 5.1 Some modular microorganisms and macroorganisms. **a** A clonal terrestrial plant (strawberry) showing two ramets; compare with ⬛Fig. 5.2; **b** a clonal, floating aquatic plant (*Salvinia*); **c** a sea fan coral (*Gorgonia*); **d** a colonial bryozoan (*Membranipora*); **e** mycelium of a fungus; **f** microcolony of bacteria. Note the modular architecture of the fungus and bacterium

Conversely, plants are clearly modular but only some grow clonally. And, there are some clonal unitary creatures. Examples include mobile animals that reproduce parthenogenetically or apomictically, such as aphids, rotifers, lizards, and earthworms. Finally, clonal organisms may be non-colonial (solitary, often dispersed over large areas) or **colonial** in growth habit.

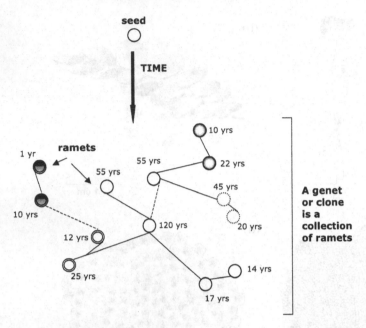

◘ Fig. 5.2 Clonal growth through time originating from a zygote (*seed*) whose genotype sets the basic ancestral lineage of the clone. Ramets (*circles* with numerical age designations) are functionally independent members of the genet or clone and are born and die at different times. Initially, the ramets are effectively genetically identical though they diverge over time due to accumulating somatic mutations and the clone becomes a genetic mosaic, represented here by the different degrees of shading (see Chap. 2 and especially ◘Fig. 2.8). Though functionally capable of independent existence, ramets may remain attached (*solid lines*) permanently, or degrade or break (*dashed lines*) depending on the species and conditions. Each ramet in this case is represented by an individual aspen tree in a grove that collectively is the genet. From Ally et al. (2008). Reproduced from *Molecular Ecology* by permission of John Wiley and Sons, ©2008

The term colony is used broadly here to mean the tendency of units related by descent to clump (see zoological terminology in Jackson et al. 1985; Hughes 1989, 2005; West et al. 2011). These units, which are modules or groups of modules in the case of modular organisms, may be physically linked together (as in rhizomes and the basal or root suckers of many clonal plants) or not linked (clumps of *Lemna*; bacterial cells or colonies; aphids), though closely associated. Sponges remain a difficulty as they are really neither modular nor colonial. They can, however, reproduce asexually and so are clonal. **Many of the attributes of clonality overlap with those of modularity** (◘Table 5.1; Sackville Hamilton et al. 1987; Hughes 1989, 2005). The unitary/modular dichotomy can be viewed as primarily a morphological separation, whereas demarcation based on a solitary/colonial habit is physiologically informative; and the aclonal/clonal distinction carries demographic implications (for historical review of terminology and life history insights, see Grosberg and Patterson 1989).

Implicit in the above remarks is that, generally speaking, what constitutes an individual for modular, clonal, or colonial organisms cannot be sharply defined. Predictably, this ambiguity has spawned considerable debate and sematic difficulties going back at least to the 1970s (Boardman et al. 1973; van Valen 1978; Larwood and Rosen 1979). While proceedings from major symposia in the 1980s on clonal organisms (Jackson et al. 1985) and modular organisms (Harper 1986; Harper et al. 1986), and subsequent debate (Hughes 2005; West et al. 2011) provide thought-provoking perspectives, the gulf between approaches and terminology of the botanists and zoologists is quite evident. The historical confusion exists primarily because of two levels of complication. First, for clonal macroorganisms, the physiologically functional individual does not correspond to the genetic individual. The ratio of genets:ramets is 1:many. The modular discussion above was focused on plants as an example, but applies equally to clonal invertebrates (see, e.g., Sanchez et al. 2004; Hughes 2005). Among corals, logical arguments could be made depending on a particular context for the individual to be represented by the polyp, by numerous polyps together forming a colony, by the zooid, or even by an entire reef (reviewed by Rosen 1986).

Second, the genetically mosaic nature of bacterial and many fungal individuals confounds strict boundaries. In these microbes, the genet itself is frequently discontinuous: The organism through its life cycle may be in a state of genetic flux, comprising both many ramets and frequently more than one genet as the original clonal lineage (especially for bacteria) becomes genetically diffuse due to somatic mutation and other more extreme genetic changes such as horizontal gene transmission (bacteria) or heterokaryosis (fungi; see below and ▶Chap. 2). Occasionally, genetic change occurs in more regular fashion as in a balanced dikaryosis in some fungi (▶Chap. 2) or an event in a complex life cycle (▶Chap. 6). These 'complications' need to be taken into account in considering their evolutionary biology and the extent to which they are inherently modular.

5.3 Fungi as Modular Organisms

Fungi characteristically grow as a network of branching tubes (hyphae), collectively called a mycelium (Andrews 1995; Bebber et al. 2007). Knowledge of fungal architecture and hyphal growth patterns has been obtained almost exclusively from laboratory culture of representative ascomycetes and basidiomycetes (and prokaryotic actinomycetes) on solid nutrient media (Harold 1990; Riquelme et al. 2011). From these relatively uniform and highly controlled conditions the following general pattern emerges. Hyphae initiated from a germinating spore grow radially outwards, branch, occasionally fuse, and show mutual avoidance reactions (Gow et al. 1999; Glass et al. 2004). As long as growth remains unrestricted, the ratio between total hyphal length and number of branches eventually becomes constant for any particular strain, a value referred to as the **hyphal growth unit** (Trinci 1984; Trinci and Cutter 1986). (This phenomenon, incidentally, is characteristic of branching systems in general and is not unique to fungi.) Thus, under ideal conditions, the mycelium can be regarded as resulting from duplication of this hypothetical growth unit comprising a hyphal tip and an associated growing mass of constant size. Conceptually, this is a convenient vegetative module, as is the leaf and its axillary bud a unit of plants. Most fungi (and actinomycetes) go on to produce other forms of vegetative (asexual spores) and sexual modules, just as plants develop vegetative modules that are later alternated with or replaced by sexual modules such as flowers, stamens, and carpels. Superficially similar branching patterns are evident among other sessile organisms such as the stoloniferous colonies of the marine hydroid *Hydractinia* (Buss 1986). Details of the fungal architecture and the case that fungi are modular organisms are discussed at length in Andrews (1994, 1995).

While in its diffuse feeding mode the branching mycelium is conspicuously modular in construction, it can differentiate to function in two other roles, namely survival and reproduction/dispersal. In both cases, hyphae that were formerly kept separate by some

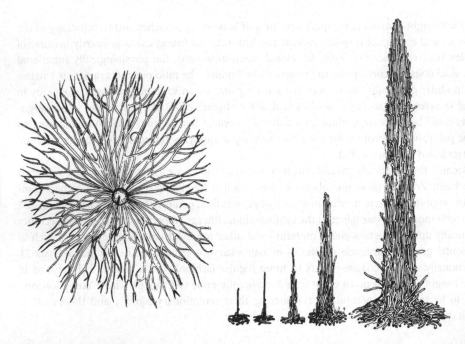

■ **Fig. 5.3** Fungi as modular organisms. *Left*, *Achlya*, a so-called water-mold (technically a member of the Phylum Oomycota, studied by mycologists but not phylogenetically a 'true' fungus) growing in diffuse fashion, allocating biomass primarily to foraging hyphae (from Bonner 1974, p. 97; redrawn from J.R. Raper). *Right*, *Pterula gracilis*, a basidiomycete, growing as an aggregate hyphal unit in the process of forming a fruiting body, allocating biomass primarily to reproduction. From Bonner (1974, p. 96; redrawn from E.J.H. Corner). Reprinted by permission from *On Development: The Biology of Form* by John Tyler Bonner, Cambridge, Mass.: Harvard University Press, copyright ©1974 by the President and Fellows of Harvard College

repulsive mechanism come together under control to produce aggregated structures (■Fig. 5.3). These forms often have architectures unlike anything else on earth. The frequently complex and massive fruiting bodies (basidiocarps) of the basidiomycetes represent the extreme in such morphogenesis. In such cases the mycelium performs a structural and transport function; here it plays no role in nutrient uptake, serving only to provide an exposed surface for spore dissemination. In some (including the common mushrooms), the basidiocarps (fruiting bodies) are fleshy, transient, and rely mainly on turgor pressure for support. In contrast, those of the bracket fungi are large, and may be crust-like or woody and perennial (*Phellinus*, *Ganoderma*). Specimens of the well-known 'artist's conk' (*Ganoderma applanatum*) are recorded up to 1 m across and approximately 10 kg. In this sense they are analogous to trees in accumulating structural dead biomass. The important generality with respect to architecture is that the fungi seem to exploit two major patterns: a branched, foraging mode, and one that is organized and consolidated for reproductive or other purposes. That substantial biomass can be diverted away from resource acquisition to survival or reproductive structures attests to the importance of dormancy and relatively long-range dispersal of new genets, as opposed to localized growth by mycelial extension, in the life histories of these fungi (see also ▶Chaps. 6 and 7 and Andrews 1992).

5.4   Bacteria as Modular Organisms

Bacteria range in complexity from unicellular rods and cocci (the typical 'textbook' bacteria) of more-or-less spherical and occasionally pleomorphic shape, to multicellular, branched (the actinomycetes), or unbranched (e.g., *Leucothrix*) filaments (Young 2006; Zinder and Dworkin 2013). Some filamentous forms may show gliding motility or are enclosed by a sheath. The myxobacteria (►Chaps. 4 and 7; Dworkin 1985) are typically unicellular rods that aggregate and produce ornate fruiting bodies, usually containing specialized resting cells. Despite occasional complex morphology, developmental cycles, or aggregation patterns, all bacterial types can be resolved essentially into growth patterns based on either the unicell or a filamentous unit.

Bacteria may exist in nature as single cells (as in bloodstream infections, septicemia), but apparently this is unusual (Shapiro 1985, 1998). Microcolonies (aggregations visible by light microscopy) form when cell division is not followed by separation of the daughters and dispersal. As early as 1949, Winogradsky was examining such cellular consortia on soil particles, which he termed "families". They have since been observed in many other habitats including the surfaces of plants and animals; the mucosal walls of the colon; dental plaque; and as suspended particulate matter in lakes and oceans. The form of a microcolony is affected by the way a cell divides and the nature of its surface. Macroscopically visible colonies of the sort typifying growth in a petri dish are relatively uncommon in nature (Carlile 1980; Pfennig 1989). Conspicuous exceptions are the massive growths in sulfur hot springs, or occasionally in stagnant water (e.g., Pfennig 1989). Where moisture and free water are present at surfaces, structurally complex and typically multispecies biofilms form (Hall-Stoodley et al. 2004; Raaijmakers and Mazzola 2012; Drescher et al. 2016). Indeed, many such multicellular aggregations of bacteria are not grossly dissimilar visually from colonies of fungi, lichens, and bryozoans (Andrews 1998).

Because bacteria tend to form clumps, for the most part they function ecologically as multicells. The disadvantage of this state of affairs for the bacterial cell is largely one of intraspecific competitions for space and nutrients; however, the advantages may include enhanced return from concerted activity (e.g., migration in the case of gliding bacteria or collective enzymatic degradation), or the resistance of pathogenic microbes to host defense mechanisms (phagocytosis; antibodies) and of free-living species to abiotic factors (desiccation; UV light; mechanical erosion). As cell division continues, the clonal aggregation will fragment periodically under the influence of various erosive and dispersal mechanisms. **The bacterial equivalent of the genet is thereby dispersed and this asexual process is directly analogous to the shedding of plantlets by various floating aquatic plants** such as *Lemna*, *Azolla*, and *Salvinia*, noted earlier. Some bacterial clones are distributed across continents or even globally and may be genetically intact as lineages over hundreds or thousands of years (Spratt and Maiden 1999; Tibayrenc and Ayala 2012).

Given the above levels of organization of a bacterial genet, conceptually it is possible to extend the modular framework to bacteria as organisms. The basic module of a colony of unicells is thus the individual bacterial cell, with successively higher units being microcolonies and macrocolonies and clonal aggregations. *The bacterial cell qualifies as the module, whereas the individual cell of unitary organisms does not, because it is iterated indefinitely.* Although cell number in unitary organisms changes (e.g., in response to disease; by regular turnover as in replacement of blood and epidermal cells), the number is defined within fairly tight limits by intrinsic genetic and developmental factors. By convention, the term module in eukaryotes is reserved for multicellular structures; while its use in the bacterial context is thus strictly a departure, the relationship of a single cell to the bacterial clone with that, say, of a multicellular *Salvinia* plantlet to the *Salvinia* clone is analogous.

Why should one bother including microorganisms in the modular paradigm? The most important reason is that to understand their biology in the real world, we need to consider these organisms in their entirety—as holistic, evolving genetic entities (albeit typically fragmented), rather than from the customary reductionist focus on the individual cell or local population of cells in a test tube or petri dish. It is the clone that changes, disperses, forms mutualistic or parasitic relationships, causes global pandemics, or becomes locally extinct. **Attributes and evolutionary possibilities of modular macroorganisms are also those of these microorganisms and they are distinct from those of unitary organisms.**

5.5   Life Histories of Modular Versus Unitary Organisms

5.5.1   Growth of the Individual and Size of Populations

For modular organisms there are two levels of population structure in a community (Chap. 1 in Harper 1977; Harper and Bell 1979): The genets, formally equivalent to the original zygotes (called N in animal population dynamics) and the quantity of modules (called η in plant population dynamics), a variable number of which occurs per genet. In theory both N and $N \cdot \eta$ (i.e., the number of modules in a given population) can be quantified. For aclonal modular organisms such as annual plants and most tree species, N can be counted easily. The difficulty occurs with clonally spreading perennials or among microorganisms and invertebrates where intermingling, sporulating, budding, or fragmenting clones necessitate either a logistically feasible marker or a mapping system to separate genets from ramets (discussed later under Longevity). Individuals tallied where possible as genets are meaningful to the geneticist or evolutionist interested in the genetic variation of a population. The production agriculturalist or population biologist or microbiologist is more interested in the number of ramets, that is, functional individuals, which gives a rough idea of biomass. The important point is that the basic demographic equation

$$N_{t+1} = N_t + \text{births} - \text{deaths} + \text{immigrants} - \text{emigrants}$$

applies at both the genet and module levels of the population.

The most obvious difference, then, between a modular organism and a unitary organism with respect to growth is that growth of the former is a *population* event: There is an increase in number of modules, whether these are multicellular units of construction (e.g., leaves) or units of clonal growth capable of separate existence (i.e., ramets such as fronds of bracken fern or bacterial cells; see again ◻Fig. 5.2). Growth (enlargement) of the *module* is formally equivalent to growth (enlargement) of the entire soma in unitary organisms (a single deer, a single bear, etc.), and this is distinct from population growth, which is the aggregate expression of each unit of construction. Hence, while each cell of the bacterial or yeast clone increases in mass more-or-less arithmetically prior to fission or budding, the population of such cells increases logarithmically, at least initially. Similarly, for the fungi, increase in germ tube length is initially exponential, then linear; however, because new branches are formed continuously the germling as a whole continues to expand exponentially, both in terms of total length of mycelium and number of branches. As summarized by Harper and Bell (1979, p. 31) *"the parts of modular organisms have their own birth and death rates; a genet has its own internal population dynamics"*.

Where modular organisms are also clonal, growth of the genet can be especially rapid. Thus, under favorable conditions, the number of ramets may increase (more poplar trees), or the number of modules per ramet (more leaves per poplar stem), or both. Conversely, 'degrowth' or shrinking can occur at either or both levels (Hughes and Jackson 1980; Hughes and Cancino 1985; Sebens 1987). For colonial clones, such as bryozoans and bacteria, there may be larger colonies, a larger number of colonies, or both. The prodigious, in some cases exponential, multiplication rates of bacteria are the extreme example.

Some implications of the population biology of modular organisms are nicely illustrated by the growth of *Lolium* grass (Harper and Bell 1979). The seed sown represents the new genets (N) and the young plants expand by producing tillers (branches from basal nodes) that represent new modules (η). The number of original zygotes will decline due to density-dependent controls, while the number of tillers will increase initially and then decline. In other words, the early effects of density will be reflected in the death rate of genets and the birth rate of modules. When this grass sward flowers and produces seed, the progeny will not reflect the balance of genes in the original population because some of the original genets will have died. There will also have been differential growth and reproduction of the tillers, hence flowers and seeds, on the surviving genets (Harper and Bell 1979). This thinking can be extended to modular organisms in general, including microorganisms. For example, the number of clones of a bacterial or fungal species in an area would be represented by N, while η would represent the number of cells (or, alternatively, colonies) per clone. As the population grows the genetic structure will change as some clones outcompete and displace others, while yet others continue to arrive (Reeves 1992; Andrews 1998). This, of course, is a well-known phenomenon in plant pathogen and medical epidemiology (recall the worldwide ebb and flow of pathogenic bacterial clones carrying antibiotic resistance genes described in Sidebar in ▶Chap. 2; e.g., Hawkey and Jones 2009).

5.5.2 Phenotypic Plasticity and Somatic Polymorphism

Phenotypic plasticity is treated in broader terms in ▶Chap. 7 but brief comment here should be made with respect to the role that it plays specifically in the biology of modular organisms. As noted previously, modular organisms, especially those that are clonal, tend to be highly plastic in size, shape, and fecundity because the number of modules can change readily by birth and death processes. Harper and Bell (1979) state in part (p. 31) *"the placement of modular units determines the form of the organism. Form is a consequence of the dynamics."* This is particularly evident in the case of plants where the iterative process determines the general size, shape, arrangement of leaves, branch angles, etc. The modular growth and death process is analogous to a child playing with Lego blocks, adding pieces in some places and removing them in others.

Sessile modular organisms tend to grow either predominantly vertically or horizontally. For many plants and in particular trees, the competitive race may be to stack modules vertically and to retain the genet intact thereby shading one's competing neighbors. Alternatively, for clonal plants such as clover or bracken fern, it may be to capture light resources by lateral expansion and perhaps in so doing avoid some of the mechanical limitations of vertical expansion (recall scaling in ▶Chap. 4; also Watkinson and White 1985). A compromise between the two would be a clonal tree like quaking aspen (*Populus tremuloides*), which has an erect form but spreads laterally by root suckers (Mitton and Grant 1996). Other modular organisms may have a dissociated genet as in the case of floating aquatic plants such as *Lemna* noted earlier. In rhizomatous or stoloniferous plants new root systems typically are produced at nodes. The intervening segments may rot, disintegrate, or break under the influence of trampling hooves, so that the original zygote becomes represented by physiologically independent individuals. Harper worded these processes imaginatively as follows: "*A clone of Lemna or Pistia expresses itself as a genetic individual by continually falling to pieces*" (1978, p. 28) and "*The ability of some plant species to form fragmented phenotypes of a single genotype is just one of the variety of successful ways of playing the game of being a plant* (1977, p. 27) ... *a zygote of* Hydra *does just the same, and, at the end of a season of growth, many daughter zygotes may be the descendants of a single parental zygote but formed from a fragmented phenotype of independently living polyps*" (Harper 1978, p. 28). So, ultimately the number of modules produced—whether they remain attached or become separate—will affect the number of progeny produced, in turn the number of descendants, and hence fitness. In some circumstances, evidently fitness is increased by retention of the intact genet; in others, by a dissociated genet (Harper 1981a). Whether modules remain intact and integrated, or become dissociated and function independently, is an interesting developmental question that has received relatively little attention (however, see later comments in later section, *When to Divide?*).

Iteration means that different parts of the same genet are affected by different environments and consequently are subjected to different selection pressures (Harper 1985; Harper et al. 1986). One segment of the genet may be expanding while the other is contracting and progeny typically are produced from those portions of the genet that are most successful. This situation is not possible for unitary organisms. The portions might be segments that have adequate nutrition or avoid being eaten by predators or, in the case of a pathogenic fungus, those clonal spores that contact a suitable host. Thus, for modular organisms, the testing of a particular gene combination is achieved by growth, i.e., by a particular architecture or geometry, whereas for unitary organisms it is frequently by mobility within the population. Phrased differently, growth amounts to movement for modular organisms.

Somatic polymorphism (a type of phenotypic plasticity; ▶Chap. 7) has little real meaning in unitary organisms. It does occur modular organisms to the extent that a given genotype presents different phenotypes adapted to different conditions. Examples include the aerial versus submerged leaves of the same aquatic plant, or differences in leaf morphology between wetter and drier seasons (large vs. small leaf sets, respectively) of some desert plants. The form and arrangement of leaves may be quite different on juvenile and mature branches of *Eucalyptus*. There are at least two important consequences of somatic polymorphism: The first is on the variable number of modules allocated to a particular reproductive or vegetative role at a given time. The second is that the phenotypic plasticity that results from this allocation pattern also influences the behavior of associated organisms. Insofar as parasites and herbivores are concerned, flowers are quite different from leaves or carpels or roots. Rather than the entire genet being directly affected (as is the case with unitary organisms), the potentially damaging activity is concentrated on the module, which may be destroyed and replaced, potentially without great consequence to the organism. Here is yet another example of the issue of trade-offs—how natural selection balances among the available polymorphisms to maximize fitness of the organism given competing demands and with any option entailing costs in terms of resource allocation.

5.5.3 Longevity, 'Potential Immortality', and Somatic Mutation

Age-specific schedules, important in life tables for the population biology of unitary animals, cannot be applied directly to genets of modular organisms (for unitary/modular comparisons, see Chap. 4 in Begon et al. 1996). This is because the inherent variability in modular organisms means that age is not a strong predictor of probability of death, reproduction, or growth. However, modules have an age structure with associated physiological characteristics and life tables for various modules, such as leaves or lateral branches, have been constructed (Reich et al. 2004). Particularly in clonal modular organisms, the number of reproductive units tends to increase exponentially with age and this exponential increase can continue indefinitely (Harper and Bell 1979; Watkinson and White 1985; Harper et al. 1986). As long as the birth rate of ramets exceeds their death rate, however, the genet not only persists but expands. This state of affairs has been called "potential immortality" and is taken up under Senescence in ►Chap. 6. If true, Hamilton's (1966) postulate that senescence is an inevitable consequence of natural selection does not apply.

Worded differently with somewhat different emphases, this means that asexual multiplication not only increases the size of the genet, but spreads the risk of death of the entire genet (see later section, When Should a Clonal, Modular Organism Divide?). Another, perhaps unexpected, consequence of exponential growth is that where genetic individuals continually increase in fecundity, the frequency distribution of fecundities of genets tends to become log-normal, i.e., very few individuals contribute the great majority of the zygotes for the next generation (Harper 1977; Harper and Bell 1979).

Numerous direct (e.g., counting growth rings in trees) and indirect (e.g., measuring the area or diameter of clones) methods exist to determine the age of modular organisms (de Witte and Stöcklin 2010), although most involve degrees of inference and assumptions and are subject to various sources of error. For example, determining the age of a clone based on its size assumes several things: that age and size are closely correlated; that the full extent of the clone can be accurately identified as the same genetic entity; and that the clone expands as a perfect circle with a linearly expanding radius (Ally et al. 2008). Molecular techniques, such as the accumulated divergence in microsatellites[2] due to somatic mutations, are increasingly being used as 'ontogenetic clocks' to estimate the extent and age of clones. Mock et al. (2008) studied clones of trembling aspen (*Populus tremuloides*) in the western USA, while Ally et al. (2008) used 14 microsatellite loci in studies of the same species in southwestern Canada. Their results with respect to clone age/size are discussed later. While it is not simple to distinguish mutational changes from allelic variation, somatic mutation can be inferred presumptively when an individual ramet differs by only one allele at one locus but is otherwise identical to the preponderant clonal genotype. These along with test procedures and caveats are discussed well both by Ally et al. (2008) and Mock et al. (2008); for broader considerations, see Arnaud-Haond et al. (2007).

Large and extremely long-lived genets have been recorded, subject to the sources of error noted above, particularly among clonal plants (❍Table 5.2): an immense clonal patch of

---

2 Microsatellites are simple sequence tandem repeats, typically of 1–6 bp DNA. Dinucleotide repeats, which are often the focus of interest, are almost always in noncoding genes and, as such, the microsatellites are more likely to evolve neutrally than if changes in protein-coding genes are used. Because microsatellite loci usually have relatively high mutation rates, they have greater resolution for age determinations (Rahman et al. 2000; O'Connell and Ritland 2004; Ally et al. 2008).

◘ Table 5.2   Estimated size and longevity[a] of some clonal plants (abbreviated from de Witte and Stöcklin 2010 and references therein)

Plant	Size of genet	Age (oldest genet, yr.)
Populus tremuloides (tree)	43 ha	12,000
Larrea tridentata (shrub)	>7.8 m (radius)	11,700
Lomatia tasmanica (shrub)	1.2 km (linear extent)	43,600
Holcus mollus (grass)	1.6 km (fragmented expanse)	–
Pteridium aquilinum (bracken)	138,408 m²	1,400

[a]Age estimates are frequently extrapolations from size and various other caveats apply. See text and Ally et al. (2008)

quaking aspen (*Populus tremuloides*) has been estimated at 10,000+ years and to weigh on the order of 6×10^6 kg (Kemperman and Barnes 1976; Grant et al. 1992; Ally et al. 2008; however, see Mock et al. 2008); creosote bush (*Larrea tridentata*) in the Mojave Desert of California at approximately 11,000 years (Vasek 1980); and the rare Tasmanian shrub *Lomatia tasmanica* at 43,600 + years (Lynch et al. 1998). Watkinson and White (1985) suggest that among clonal plants the main causes of death of old genets include fire, disease, and competition. Dormancy, coupled with various forms of asexual multiplication and dispersal, would all act to reduce the risk of death to entire genets. Fungal clones can also be sizeable and long-lived. Dickman and Cook (1989) found several genets of the wood-rotting fungus *Phellinus* [=*Poria*] *weirii* estimated to be older than 1,000 years in mountain hemlock (*Tsuga mertensiana*) forests of Oregon. The fungus spreads outward from a focus of infection primarily by root-to-root contact. Sibling ramets of the various fungal genets survive forest fires that periodically destroy large areas of the stands. The facultative tree-root pathogen *Armillaria bulbosa*, which radiates through soil by cord-like rhizomorphs, is ascribed as being among the largest (occupying at least 15 ha and weighing in excess of 10,000 kg) and oldest (at least 1,500 years) organisms (Smith et al. 1992).

The issue of somatic genetic variation was introduced earlier (►Chap. 2). Some evolutionary biologists have downplayed its importance either by arguing that all somatic cells, including variants, ultimately arise from a proximal common ancestral source (the zygote) and therefore are not very different (pp. 244–247 in Maynard Smith and Szathmary 1995), or by questioning whether such variants are evolutionary meaningful (appreciably heritable) (Harper 1988). As noted in ►Chap. 2, the *evolutionary* importance of somatic variation depends on the ontogenetic program of the individual and ranges from being likely relatively negligible (organisms with preformistic development) to potentially significant (cases of somatic embryogenesis). If the mutated cells are destined to remain somatic (not transmitted to offspring), as is typical in the former case, then the changes of course are effectively evolutionary dead-ends. However, when the mutation confers a difference in the ability to survive and compete among other cell lineages during development, and become subject to either cell lineage (germ line) selection or gametic selection, it is potentially important (Otto and Hastings 1998). Cell lineage selection may lead to offspring through gametes or via various asexual processes such as fragmentation or sporulation. Gametic selection alludes to competition among gametes for fertilization and is probably significant in plants, which have a gametophyte generation (►Chap. 6). (Gametes in higher animals, in contrast, are typically a transient phase and express the diploid genotype of their progenitor cells; thus they do not vary as extensively phenotypically and are not subject to as extreme selection pressure; Otto and Hastings 1998.)

Most individuals are genetic mosaics because most are composed of great numbers of cells and genes, which are subject to mutations usually caused by damage over time or by replication errors (Gill et al. 1995; Otto and Hastings 1998). Examples of the kinds of mutation were discussed in ►Chap. 2, but broadly speaking they include any change to the genome of a cell that is transmitted to the progeny of that cell. Rates in general per gene per individual generation have been given as 10^{-7} to 10^{-4}, or as 10^{-5} to 10^{-4} per individual generation for mitotic recombination in plants (reviewed by Otto and Orive 1995; Otto and Hastings 1998; an 'individual generation' is generally taken to mean from a parent single-cell stage to an equivalent offspring single cell, i.e., from zygote to zygote). The impact of a mutation within a mosaic individual depends on its nature and the degree to which it accumulates, i.e., the number of cell divisions per individual generation. These vary with the organism (e.g., 50 such divisions occur in maize from zygote to zygote; reviewed by Otto and Hastings 1998). If the mutation is beneficial, germline selection can substantially increase its occurrence among the individual's reproductive cells, and hence its chance of ultimately being fixed in the population. At the level of the clone, variation can influence fitness (Monro and Poore 2004, 2009a) and potentially be of adaptive value (Folse and Roughgarden 2011).

Whitham and Slobodchikoff (1981) apparently were the first to postulate that as a result of somatic mutation trees are genetic mosaics. Significantly, they argued further that the somatic heterogeneity presented to insect pests may enable plants to counteract, on a relatively short-time scale not otherwise possible, the adaptations of their herbivores to overcome usual systemic resistance conferred by routine meiotic recombination. What came to be known as the 'genetic mosaicism hypothesis' was later extended to clonal animals (Gill et al. 1995). Somatic mutation is especially germane to clonal plants because such organisms: (i) are frequently very large (many somatic cells); (ii) are frequently long-lived, (mutations can potentially accumulate; however, see below); and (iii) do not have a segregated germ line, so mutations arising somatically can be transmitted to offspring, both sexually and asexually. The tissues of importance with respect to somatic mutation in plants are the meristems because they are, as the name implies, effectively the reservoir for the stem cells of plants[3] (Weigel and Jurgens 2002; Heidstra and Sabatini 2014).

The specific type of meristem associated with extension of the plant axis, and also typically the production of ramets, is the apical meristem of shoots and roots. The organization of such tissues and implications for the spread of somatic mutants will be briefly described in passing as an example of the difficulty in making explicit predictions of the consequences of

---

3   Plant and animal stem cells reside in 'stem cell niches' and respond to signals regulating the balance between renewal and generation of daughters that become differentiated into various tissues. In plants such niches are located within the meristems (Heidstra and Sabatini 2014). Plant stem cells have been largely overlooked, other than by botanists, in recent years by all notoriety surrounding stem cells of humans! The term 'stem' is perhaps best avoided in plants because of ambiguous usage; see Evert (2006, p. 143).

somatic mutation in large, potentially long-lived organisms. In multiple papers, Klekowski and colleagues (e.g., Klekowski et al. 1985; Klekowski 2003) have shown how organization of the plant apical meristem influences the mutational process, with **diplontic selection**[4] occurring among cell lineages in the meristem, and deleterious mutations accumulating within the individual genet potentially leading to 'mutational meltdown' (described below).

The most primitive group of vascular plants, the pteridophytes (ferns, horsetails, and lycopods), have a single, permanent apical cell. When this cell divides into two daughter cells, one remains as the meristem initial and the other initiates the lineage that differentiates ultimately into tissues of the plant body and, in some cases, ramets. Thus, if the initial mutates, the mutation (if nonlethal) will be passed on to the daughters and their clonal derivatives. Klekowski (2003) views the displacement of a mutant apical initial by adjoining wild-type cells as unlikely. This means that diplontic selection is not strong and the clone is prone to meltdown. Limited evidence, especially for ferns, supports this view (Klekowski 2003).

The second group of vascular plants comprises the gymnosperms (e.g., cycads, ginko, gnetophytes, conifers). Though the structural pattern varies among the groups of gymnosperms, the unifying meristematic feature is one involving multiple cells and of cytological zonation based on planes of cell division and activity, i.e., lack of a permanent apical initial. This pattern implies that diplontic selection is more likely to occur than in the pteridophytes and hence the impact of mutation to be less overall.

The third and most complex cellular arrangement applies mainly to the angiosperms (flowering plants), which typically have two clonally distinct meristematic tissue zones in the apical meristem: the tunica, consisting of one or more peripheral layers of cells; and the corpus, a mass of cells overlain by the tunica. The tunica is involved primarily in surface expansion, whereas the corpus contributes volume to the growing shoot (Chap. 5 in Esau 1965; Chap. 6 in Evert 2006). Because the pattern of cell division differs in the two areas, somewhat isolated subpopulations of meristematic cells develop. Although diplontic selection is maximized within the subpopulations, a mutant that arises in one layer can be sustained by the wild-type cells in a different layer nearby, so stable chimeras may form and persist (Klekowski 2003).

So, to what extent are somatic mutations accumulating in clonal populations and what are the consequences? Lynch et al. (1993) describe the phenomenon of **mutational meltdown** attributed to genetic drift in small populations where natural selection cannot purge slightly deleterious mutations, which accumulate and reduce fitness. This has been documented in small populations (Zeyl et al. 2001) of yeast asexually propagated under laboratory conditions. The consequence of mutations to bacterial clones was discussed in ▶Chap. 2. In clonal plants, random drift may overwhelm diplontic selection between mutated and wild-type cells in an apical meristem and eventually lead to the fixation of a somatic mutation in some or all of the ramet shoot apices of a particular genet. This is clear from theory summarized above and is also evident from natural observations and controlled experimentation (e.g., Gill et al. 1995; Klekowski et al. 1996). Plant anatomy and ontogeny evidently exert considerable influence on the likelihood of this outcome.

---

4   'Diplontic' in this context alludes to competition among cell lineages within the (diploid) plant axis. In organisms that are exclusively diplonts (typically multicellular animals), the haploid products of meiosis behave directly as gametes. Plants are typically 'diplohaplonts', alternating between haploid and diploid individuals. See ▶Chap. 6.

The genetic study of trembling aspen alluded to earlier included analysis of the immense (>40 ha) 'Pando' clone in central Utah (Mock et al. 2008; DeWoody et al. 2008). Remarkably, in the 256 ramet samples from this genet, only six variants of the predominant multi-locus genotype were detected and all were single-step mutations. This suggests remarkable homogeneity and possibly relatively young age notwithstanding earlier reports of great longevity of aspen and other clonal plants (◘Table 5.2). The boundaries of the clone detected genetically were, remarkably, almost identical to those detected earlier by morphological methods (Kemperman and Barnes 1976). An additional 40 genotypes were distributed along the periphery of 'Pando'. Unlike the single-step variants, these differed at 4–7 alleles, and it is unclear whether these are of somatic or sexual origin (DeWoody et al. 2008). Peripheral patches of genetically distinct clones were also detected in the Canadian study (Ally et al. 2008). The Canadian study also documented reduced male fertility (decline in average number of viable pollen grains per catkin per ramet) with clone age. This was attributed as probably being due to somatic mutations *that accumulate over time affecting sexual fitness only, with little to no effect on clonal growth* (Ally et al. 2010). The results are reviewed and the interesting implications discussed in ►Chap. 6 under Senescence.

5.5.4 Occupation of Space and Capture of Resources

As noted at the outset, modular organisms in general have a fundamentally iterative, branched structure. Most occupy relatively fixed positions and by growing capture space and thereby other resources. Branching networks can be characterized based on certain general properties (Gross and Sayana 2009) and in the living world have been studied extensively among both plants and animals (Stevens 1974; Harper 1985; Waller and Steingraeber 1985; Sanchez et al. 2004). One attribute, for example, is the degree to which the modular subunits are consolidated and the extent of interaction among subunits of the same and different individuals. Of particular interest is how the organization of subunits affects the efficiency of nutrient and energy capture *by the overall genet*. For plants, resources may take the form of light entering the canopy of a tree, or mineral nutrients in the rhizosphere; for a bryozoan, they are food particles in the surrounding water column; for a fungus, the resource may be a fallen tree.

For the modular sessile organism, modules act to capture resources, which sooner or later tend to become locally exhausted creating a **resource depletion zone** (RDZ). By overtopping rivals, the tree canopy of a superior competitor creates in effect a RDZ of light in the understory. Roots forage for and are plastic enough to respond to nutrient patches (Ruffel et al. 2011; Chen et al. 2016) in a manner analogous to the aerial portions of plants responding to light or filamentous fungal hyphae responding to resources in the leaf litter or soil. In the case of the rhizosphere (◘Fig. 5.4), the size of a RDZ, and if such a zone develops at all, depends largely on whether a particular solute is absorbed relatively faster than water (for dynamics of nutrient acquisition by roots, see Chapman et al. 2012). If it is, then the concentration at the root (module) surface will be lowered. In contrast, if water is taken up faster, the solute may accumulate at the root surface and diffuse away. (It should also be remembered that root zones can actually be areas of enrichment for certain nutrients, such as amino acids and carbohydrates, which are exuded by the host and are thereby 'hotspots' for microbial activity.) Thus, the onset and dimensions of RDZs depend, among other things, on how rapidly resources are used by the module and resupplied by the medium.

(a)

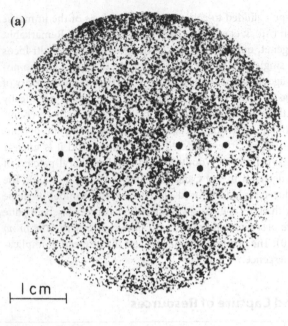

(b)

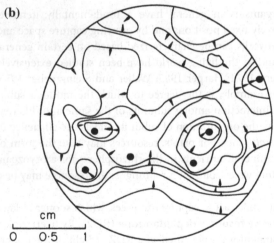

▣ **Fig. 5.4** (*Top*) Resource depletion zones appearing as halos around onion roots (*black dots*) in cross section in an autoradiograph of soil labeled with $^{35}SO_4$. (*bottom*) The concentration profile of $^{35}SO_4$ shown as contours of equivalent concentration based on densitometer tracings. Both figures from Sanders (1971); courtesy of F.E.T. Sanders, University of Leeds

The search process There has been considerable theoretical discussion of optimal growth pattern and search strategies to capture resources (see broader, conceptual discussion in ►Chap. 3 and Hein et al. 2016). **Here we focus mainly on the challenge for a sessile organism in acquiring resources,** which involves modulating growth so as to strike a morphological compromise between optimal search and exploitation. In other words, the evolutionary goal in principle is to efficiently locate resources while minimizing overlap of RDZs. As noted in ►Chap. 3, how this is accomplished is as much a part of optimal foraging theory, broadly construed, as are strategies related to the active pursuit of prey. Most mobile, unitary animals actively search their environment in organized fashion using their neural/motor circuitry (e.g., by processing olfactory, visual, acoustic, and other cues); modular organisms tend to do so by positioning their modules at feeding sites. Clonal plants, for instance, may accomplish this by alternating feeding locations (e.g., leaves and roots at horizontal nodes) with spacers (rootless stems; non-photosynthetic tissue) (Bell 1984). Plant shape is drastically affected both above- and below-ground by sunlight distribution/intensity and soil nutrient patches, respectively (de Kroon et al. 2005; Chen et al. 2016). Bell visualizes four basic feeding behaviors among organisms: (i) a single (unitary) organism, mobile along some sort of search or foraging route (animal); (ii) a single, immobile organism feeding at one site (aclonal plant); (iii) a clonal organism feeding at many sites reached by growth of a branching system that potentially fragments; and (iv) a 'community' of the same organism foraging as a unit along a defined branching structure, each feeder contributing to the organized whole. Bell uses in this last case as examples social insects such as the army ant, although slime molds and similar microorganisms would be the microbial analog (◘Fig. 5.5).

Assuming some information is available to the forager on resource distribution and the foraging objective at one extreme is to leave no resource site unexplored, a spiral growth pattern without branching is theoretically the most economical of effort (Chaps. 2 and 4 in Stevens 1974; Harper 1985; Bartumeus et al. 2005). This is analogous to an efficient search strategy for locating an object such as a sunken ship when something is known about its presumptive location. Fossil evidence shows that the primitive sea-floor slug *Dictyodora* (a unitary, not a modular organism) fed in a spiral pattern—a foraging behavior that allowed it to cover the whole of an area of sediment systematically (Seilacher 1967; Sims et al. 2014). Similarly, a species of ginger, *Alpinia speciosa*, has a hexagonal branching pattern with characteristics of a tight spiral (Bell 1979). Simulations suggest that a spiral search also can result from random (Lévy) walks if certain conditions are imposed such as cueing or avoidance of self-trails (►Chap. 3 and Sims et al. 2014). This is more efficient than a simple 'scribble' pattern, but it covers an area less efficiently than tight 'meandering' because the spaces between the spirals are not exploited. In order of appearance among various worms, snails, and trilobites, scribbling appeared about 600 Mya and was replaced by a meandering and occasionally a spiraling mode between 425 and 500 Mya. Complex meanders and dense double spirals, considered the most efficient foraging patterns, did not appear until the Mesozoic (135-63 Mya) (Seilacher 1967; see also Seilacher et al. 2005).

At the other extreme, to exploit a pocket of an unlimited resource with minimal overlap, a linear growth pattern is best. The closest approximation to avoiding overlap while leaving no resource zone untapped is to branch and to change the branching pattern as growth proceeds (Harper 1985). Using a simple two-dimensional, hexagonal array of equi-spaced dots, Stevens (1974: Chap. 2) compared four basic patterns—the spiral, meander, explosion (center dot connected directly to each outlying dot), and various forms of branching—in terms of four geometric attributes: uniformity, space-filling, overall length, and directness. Each pattern has advantages. However, if the organism must both search and exploit with limited biomass at its disposal, branching, which is both short and direct, is the best compromise. (Similar patterns result in a three-dimensional array but it is noteworthy that the spiral becomes a corkscrew or helix and cannot completely fill space, unlike its two-dimensional counterpart.)

In interpreting theoretically optimal search strategies, the issue of compromises to multiple demands needs to be kept in mind (see ►Chaps. 1, 3, and see below). Perhaps key amongst these is that there are other forces shaping growth pattern than simply the search for

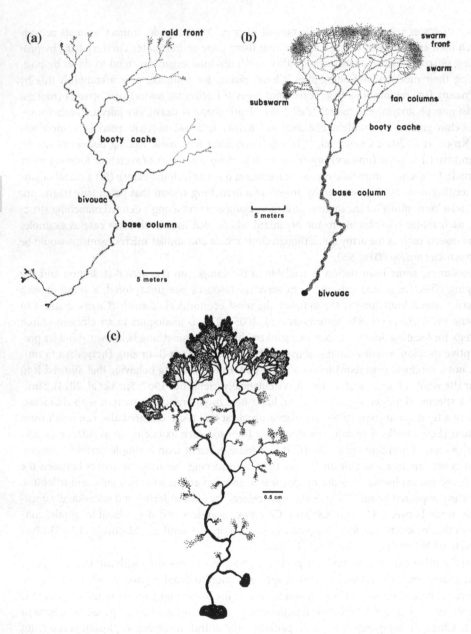

Fig. 5.5 Similarity in architecture of some growth network systems. **a, b** Raiding patterns of two species of army ants. The scale bars in A and B are 5 m. From Rettenmeyer (1963); reproduced by permission of The University of Kansas Science Bulletin, ©1963. **c** Foraging pattern of the slime mold *Physarum polycephalum*. The scale bar is 0.5 cm. Drawn from a photograph of a plate culture, courtesy of Tom Volk, University of Wisconsin-La Crosse

▣ **Table 5.3** Life history strategies of guerrilla and phalanx architectures (modified from Lovett Doust 1981a,b; Buss and Blackstone 1991)

Trait	Phalanx versus guerrilla
Size at first reproduction	Larger
Growth rate	Slower
Fecundity	Lower
Capacity for dispersal/sampling of environment	Lower
Regenerative potential	Higher
Competitive ability	Higher
Morphological stability	Higher
Frequency of interspecific contacts	Lower
Frequency of intraclonal contacts	Higher

and extraction of resources. And, with respect specifically to plants and resources, although at any one instant only one resource may be *the* limiting factor, in reality many resources will be limiting over time. Since RDZs are dynamic and vary with the resource, a strategy that may be optimal for capturing a particular resource (e.g., readily diffusible nitrate ions in soil) may not be necessarily so for another (e.g., weakly diffusible phosphate ions in soil). There are other considerations. The complex spatial pattern of a clonal plant in a field will reflect to varying degrees the search for light, water, soil nutrients and, especially, the positive and negative effects of neighbors (and neighbors here may be of three types: a different species; the same species but a different clone or genet; or the same clone but a different portion or member). Indeed, a detailed study over many years reaching conclusions emphasizing the first-order importance of neighbors was reported by Harper and colleagues in a Welsh pasture (see e.g., Turkington and Harper 1979a,b).

The guerrilla versus the phalanx habit Extending the 'explore versus exploit' argument shows that an open architecture (longer spacers) would promote rapid invasion of new habitat, while a closed architecture (shorter spacers) would favor consolidation of acquired habitat. (This is conventionally depicted in terms of horizontal expansion but is equally true for vertical growth as represented for example by tightly or loosely packed canopies of corals and trees.) For plants, the ecological implications have been recognized implicitly at least since Salisbury (1942, pp. 225–226) contrasted the aggressive vegetative spread of the creeping buttercup (*Ranunculus repens*) with the slow expansion of the figwort (*Scrophularia nodosa*). Lovett Doust (1981a) coined the terms 'guerrilla' and 'phalanx', reminiscent of military arrays, to describe distinctive clonal plant growth patterns (▣Table 5.3 and ▣Fig. 5.6). The **guerrilla** architecture is characterized by long internodes, infrequent branching, spaced modules, and minimum overlap of RDZs. While this is characteristic of the creeping buttercup, Lovett Doust showed (1981 a, b) that local populations in grassland vs woodland vary in the extremity of guerrilla-like form. Other examples include white clover and strawberry. According to Lovett Doust (1981a), the guerrilla strategy maximizes interspecific contacts and sampling of the environment that, in a woodland habitat for example, would be advantageous to locate sunflecks in the understory. Conversely, the **phalanx** arrangement is characterized by short

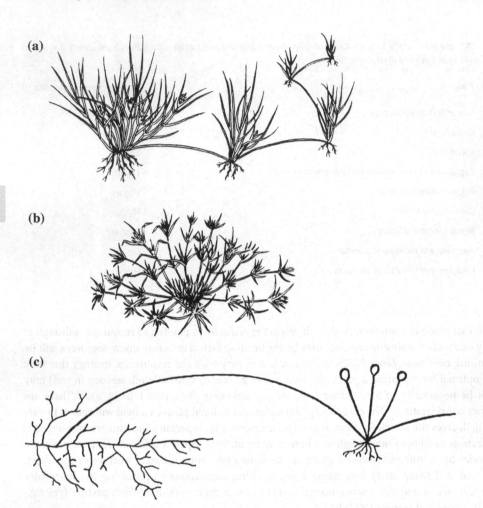

⬛ Fig. 5.6   Clonal morphology of sessile, branched, modular organisms as represented by plants and fungi. Compare with ⬛Fig. 5.7. **a** 'Guerrilla', or the extensive type with widely spaced ramets, shown by buffalograss (*Bouteloua dactyloides*). **b** 'Phalanx' or the intensive type with tightly packed ramets, shown by false buffalograss (*Munroa squarrosa*). From Silander (1985) based on terminology of Lovett Doust (1981a). Reproduced from: *Population Biology and Evolution of Clonal Organisms,* edited by Jeremy B.C. Jackson, Leo W. Buss, and Robert E. Cook, by permission of Yale University Press ©1985. **c** An aerial stolon of the zygomycete *Rhizopus stolonifera* ('black bread mold') has arisen from behind the advancing margin of a colony growing on nutrient agar. It has 'rooted' in an unexploited zone of medium and produced a new ramet distant from the original parental ramet. In this fashion and by the prolific production of asexual spores, *Rhizopus* rapidly colonizes bread and other substrata in guerrilla-like fashion. From Ingold (1965). Reproduced from: *Spore Liberation* by C.T. Ingold, by permission of Oxford University Press, ©1965

internodes, frequent branching, and closely packed modules. Consequently space is densely occupied, contact is mainly intraclonal, and RDZs overlap. Plant examples include various tussock grasses such as *Deschampsia cespitosa* (Harper 1985) and the common golden rod (*Solidago canadensis*) (Smith and Palmer 1976). As suggested above, both phalanx and guerrilla species can adjust their architectures as a result of competition or in response to resource-rich patches (Lovett Doust 1981b; de Kroon and Hutchings 1995; de Kroon et al. 2005) and there is evidence that such plasticity can be adaptive (Schmid 1986; van Kleunen and Fischer 2001; however, see Buss and Blackstone 1991; Ferrell 2008).

Phalanx and guerrilla microorganisms Clearly the descriptors phalanx and guerrilla also fit the growth patterns of other sessile, clonal organisms. Indeed, the specific epithets of some animals (e.g., the corals *Pectinia paeonia*, *Pavona cactus*) are derived from architecturally similar plants (Harper et al. 1986). Crustose (mat-forming) lichens grow as phalanxes as do some corals, bryozoans, and clonal ascidians; other corals and the foliose (aerially branched) bryozoans and lichens have, relatively speaking, a guerrilla-type habit (Harper 1981a, 1985). Marine benthic invertebrates and algae are shown in ◨Fig. 5.7. Of the six basic forms of sessile marine animals (Jackson 1979; see also Taylor and Wilson 2003), the 'runners' and 'vines' are essentially a guerrilla habit, while the 'trees', 'plates', 'mounds', and 'sheets' are to varying degrees phalanx in form (for terminology and architectural variation, see Ferrell 2008; Cherry Vogt et al. 2011).

The above nomenclature from studies on invertebrates and plants is clearly relevant also to certain microbial patterns. Because of the plasticity of microbial growth, the terms can be applied in two contexts. First, as with other modular organisms, the descriptors can be used in a relative sense to compare different taxa. For example, Kohli et al. (1991; see also Kohn 1995) contrasted the clonal, guerrilla-type growth and colonization pattern of the fungus *Sclerotinia sclerotiorum* with the phalanx characteristics of *Armillaria bulbosa*. Other fungi and fungal-like organisms, particularly the Zygomycetes and Oomycetes, extend rapidly in guerrilla-like fashion. This is probably facilitated by their coenocytic nature, which permits unimpeded cytoplasmic flow and rapid communication (at the cost to the thallus of vulnerability to damage). *Rhizopus stolonifer*, as its name implies, colonizes territory rapidly by stolon-like runners that develop rhizoids and tufts of sporangiophores at 'nodes' (◨Fig. 5.6 noted earlier). It is the creeping buttercup of the microbial world. At the other extreme, some fungi such as the apple scab pathogen *Venturia inaequalis* expand as a compact, scab-like mass, whether across a standard laboratory medium or in their natural substrata. Possibly the dimorphic fungi (which can alternate between a yeast and a hyphal phase) have the best of both worlds. *Saccharomyces cerevisiae*, for example, produces single cells in classical yeast-like fashion by budding, but under conditions of nutrient depletion forms filaments (pseudohyphae). The pseudohyphal pattern has been interpreted adaptively as a means to enhance foraging, i.e., in the current context, to escape from RDZs (Gimeno et al. 1992; see also Andrews 1995). This example and the case of dimorphic fungi in general are discussed with respect to phenotypic plasticity in ▶Chap. 7.

Second, the terms guerrilla and phalanx can also describe different growth phases of a particular organism (Andrews 1991). Basidiomycetes colonizing woody substrata in soil, or on the forest floor, commonly do so in exploratory, guerrilla-like fashion by producing elongated mycelial cord or branching rhizomorph systems (Thompson and Rayner 1982; Thompson 1984; Rayner et al. 1985; Boddy et al. 2009). When the fungus is within a nutrient cache a diffuse mycelial network tends to become established. This is strikingly similar to tropical lianas that initially produce rapidly growing 'searcher' shoots to find a suitable environment, which is then colonized by densely branching shoots (Strong and Ray 1975).

Branching networks Plants and fungi share many attributes as sessile, branching, modular organisms, but some differences exist in their processes of resource capture. In part this is because, for the former, some resources move to the organism. Sunlight impinges on leaves (although there is a race to avoid becoming shaded by one's neighbors), carbon dioxide and oxygen move by convective flow and diffusion, and inorganic nutrients tend to move quickly

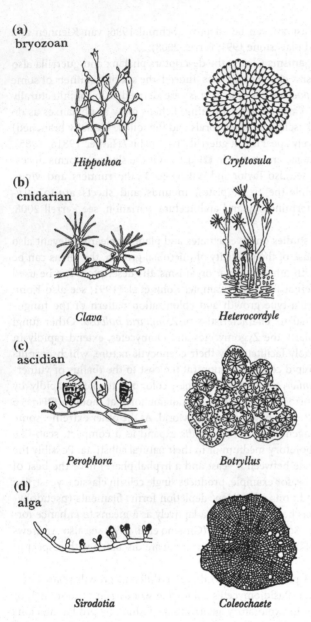

(a)
bryozoan

Hippothoa   *Cryptosula*

(b)
cnidarian

Clava   *Heterocordyle*

(c)
ascidian

Perophora   *Botryllus*

(d)
alga

Sirodotia   *Coleochaete*

▫ **Fig. 5.7**   Genera of sessile marine invertebrates (**a–c**) and algae (**d**) representing the loosely spaced, 'guerrilla' (*left column*) and tightly packed, 'phalanx' (*right column*) life forms. From Buss and Blackstone (1991). Reproduced from *Philosophical Transactions of The Royal Society of London B* by permission of The Royal Society, ©1991

by bulk flow driven by transpiration. In contrast, fungi typically move to their resource, both by linear growth and dissemination. Here, dispersal by asexual spores moves the clone to new, potentially uncolonized resource patches. The arrangement of leaves, stems, and roots tends to be relatively well organized and structured relative to hyphal ramification (as noted above; also see ►Chap. 7), which is much more plastic and involves mutual cooperation.

It is in this sense of communal effort that the fungi are like the social insects in resource capture division of labor, and genetic interactions (Rayner and Franks 1987).

Horn (1971, 2000) interprets the geometry of trees and the distribution and shape of leaves largely in terms of resource acquisition. His 1971 synthesis provides an explanation for why succession starts with trees intolerant of shade and proceeds to more tolerant species, rather than simply starting with the latter. The concept is based on calculations showing that net photosynthesis of leaves increases with light intensity up to about 20% of full sunlight. Beyond this, photosynthesis is not enhanced by further increases in light. Hence, strategies for light interception will be important whenever there is sufficient shade (in effect RDZs imposed by foliage) to reduce light to less than this 20% threshold value. Leaf placement in crowns of plants was viewed as either monolayered—leaves being concentrated at the periphery with few gaps and little overlap—or multilayered—leaves being randomly scattered vertically and horizontally throughout the canopy. Where light is abundant as in the open fields typical of early successional environments, the multilayer was considered to be adaptive. In shaded, late successional environments, the monolayer form can theoretically grow faster. This is because the multilayer can produce several tiers of leaves in the open habitat without reducing intensity on the lowermost to less than the critical 20% level. In the shade, however, photosynthetic gains of the lowermost tier may not compensate for respiratory losses, so the advantage would swing to the monolayer design.

In overview, the branching pattern of modular organisms, and especially those that are clonal, has generally been interpreted in the context of foraging for energy and nutrients. Most of the research has been done with plants because of their morphological diversity and accessibility and increasingly has involved complex modeling and computation (e.g., Wong et al. 2011; Campillo and Champagnat 2012). Architectures, represented in the extreme by guerrilla and phalanx forms, are usually assumed to be adaptive though this assumption has rarely been directly tested and the sparse evidence is mixed. Form could also result to varying degrees from, for example, a search for refuges or defense, and is inherently constrained by the principles of biomechanical design or possibly even shaped by a purely mechanical process akin to 'self-organized criticality' (the classic example of criticality is the 'avalanches in a sand pile' analogy—see Bak and Paczuski 1995; Sanchez et al. 2004). The architecture of sessile marine animals (Jackson 1979; McKinney and Jackson 1989) and seaweeds (Koehl 1986; Martone et al. 2012), particularly those in the intertidal zone, has been interpreted relatively less with regard to optimal nutrient acquisition and more in terms of resistance to hydrodynamic forces, substrate stability and characteristics, boring organisms and predators, and promotion of juvenile recruitment. However, comparison of a food acquisition model with a space limitation model has shown the former to predict better than the latter the morphology, distribution, and abundance of bryozoans under most conditions tested (Okamura et al. 2001).

The shape of plants must also be a compromise to multiple demands, of which light interception is only one, though admittedly an important one. Trees, because of their size, have to withstand strong winds, translocate fluids in a long plumbing system, disperse seeds, and in some climates shed snow loads, as well as display leaves efficiently for light interception and gaseous exchange (see ▶Chaps. 4 and 7 and Koch et al. 2004, 2015). Many influences are covariant and correlations may be direct or inverse. For example, adaptations for shade tolerance may conflict with those for drought tolerance (Valladares and Niinemets 2008) and functions such as mechanical stability and translocation efficiency impose engineering constraints (Niklas 1994a; Vogel 2003; Read and Stokes 2006). To use the words and metaphor of Wright (1932) and Niklas (1994b), the genetic analysis of such complicated adaptations resembles fitness walks over very complex landscapes. These morphological trade-offs by sessile organisms to multiple demands nicely illustrate the point made in ▶Chap. 1 that every phenotype is necessarily a compromise among different and often opposing selection pressures. Traits have no significance in isolation from the whole organism (Dobzhansky 1956). Natural selection acts on the aggregate of what is available; perfection in form or any other attribute is not necessary for reproductive success and indeed cannot be achieved.

5.5.5 When Should a Clonal, Modular Organism Divide?

Division has an obvious effect on the birth rate of a clonal organism that suddenly doubles its occurrence in a population. There is, however, a less apparent demographic consequence of clonal multiplication that should not be overlooked. This is the decline in probability of death that follows mathematically as a result of dividing oneself into two physiologically independent individuals instead of continuing to grow as a single entity. As Cook (1979, 1985) has pointed out, *the probability of genet extinction is the product of the probabilities of the separate chances of death of the ramets, i.e., the extinction of a genotype becomes up to half as likely if two independent lethal events are necessary.* Thus, an ecological strategy for some species of modular organisms may be to reproduce asexually, or to alternate occasional sexuality with recurring rounds of asexual reproduction. It should be emphasized that in many cases clonal units remain physically associated or within the same neighborhood, in which case mortality events are rarely independent. Also the mathematics assumes that the subdivided units do not suffer a higher individual probability of death than does the parental unit and indeed they may because fission typically is associated with an initial decrease in size and mortality is often size-dependent. Nevertheless, wind-dispersed seeds produced asexually (apomictically), such as occurs in some dandelion populations noted earlier (Janzen 1977b; van Dijk 2013) and the intercontinental movement of microbial clones (Goodwin et al. 1994; Brown and Hovmøller 2002; Gomez-Alpizar et al. 2007; Dutech et al. 2012), are examples of widespread dissemination of ramets. The persistence, size, and vast distribution of such clones imply that they are ecologically successful, at least in the relatively short term (see additional comments in ▶Chaps. 2 and 6).

Thus, speaking generally of modular organisms, we see two trends in life cycles. In aclonal species such as a maple tree, there is concurrence between the genetic and physiological individual, the two in effect coexisting in the same entity. In clonal species, such as the creeping buttercup or a bryozoan, it is adaptive for the two kinds of individual to be separated.

There are also physiological implications of a life history plan that involves either continuous growth as a singular individual or division. As summarized above in ◻Table 5.1, modular organisms characteristically have indeterminate growth. The adjective 'indeterminate' has been applied variously in the biological literature; here it will be used following Sebens (1982, p. 209) to mean *"the ability to increase and decrease size over a wide range as conditions vary, without an apparent genetically determined upper size limit."* Thus, the size of modular organisms, though ultimately set by mechanical constraints, depends in practice on biotic and abiotic attributes of the local habitat. For instance, it may be imposed in the case of animals by size-selective predators; by refuge size; and by prey abundance and prey quality, or more broadly, energy intake (Sebens 1982). In the case of plants, the factors influencing size include competition for light, water, or inorganic nutrients; pathogens; herbivores; and abiotic assaults from wind storms, lightning strikes, or snow loads.

Clonal, modular organisms can escape some of the allometric consequences of size increase as discussed in ►Chap. 4. One can hypothesize that as a result of natural selection adult size maximizes individual fitness and test whether growth stops, for example, at an energetically optimal size. This has been a common approach in research on marine sessile benthic invertebrates that are active or passive suspension feeders (Hughes and Hughes 1986; Sebens 1979, 1982, 2002). Such invertebrates include anemones, sponges, hydroids, soft and stony corals, bryozoans, and ascidians (Jackson 1985). Active feeders expend energy to draw water across a feeding surface; passive feeders rely on water currents to move food particles to the organism. In both cases, rates of prey capture depend in part on surface area of the feeding apparatus. These studies take on an additional dimension when the animals are clonal, colonial creatures where the question becomes when (at what size) should a colony divide asexually (McFadden 1986). In essence this tests the broad question of whether modularity can free an organism from metabolic allometry—the advantage of a fragmented genet being more than just in demographic terms.

McFadden (1986) studied particle capture rates in a species of small, alcyonacea soft coral that lives in the lower intertidal zone along the coastline of Western North America. The animal feeds passively by extending its polyps and tentacles to the current; feeding dynamics, particularly as influenced by water velocity, are complex and beyond the scope of this discussion. (Polyps are one body form of members of the Phylum Cnidaria, and each constitutes 'an individual'. In colonial species, as is the case here, a colony may consist of up to about 100 such feeding polyps, each approximately 6-mm long when fully extended. Tentacles protrude from the polyps and are the prey-capturing surfaces.) This coral is capable of localized movement across hard substrata and occurs as aggregations of colonies, each colony being typically up to about 1.5 cm in diameter, a size maintained by ongoing, endogenously controlled fission. Why are the colonies limited to this size?

Under controlled conditions in laboratory aquaria where particle capture rate could be measured as a function of flow speed and numbers of neighboring conspecific colonies McFadden found that, at all speeds tested, per polyp particle capture rates decreased as colony size increased. This is apparently due to physical interference among the polyps and the declining ratio of peripheral to interior polyps as the circular colony expands. Peripheral polyps are relatively more important than interior polyps at intermediate and higher flow speeds where most particles are trapped because of changes in colony shape accompanying colony expansion. Fission helps maintain a high ratio of peripheral polyps to overall colony area and therefore at first glance it should be best to have the smallest colony possible (recall area:volume relationships in ►Chap. 4). However, an increased rate of fission implies having more neighbors, which complicates things. Presence of neighbors decreased the per polyp capture rates of a colony at low water speeds but slightly enhanced capture rates at the highest speed. At low speeds this appears due to depletion of the nutrient flux by upstream colonies, decreased flow velocity, and lower penetration of particles into aggregated colonies. While these factors also play a role at higher speeds, the decreased velocity within an aggregation and reduced deformation of the tentacle crown (and therefore a beneficial reduction in drag forces) by neighbors act to enhance particle capture. On balance, it appears better to have relatively larger colony size at lower speeds while at intermediate speeds it is advantageous for colonies to divide at a smaller size. Feeding efficiency is maximized at high flow speeds by aggregations of very small colonies. Overall, in terms of feeding efficiency, though it is advantageous for colonies to be as small as possible there is a tradeoff at the point where the neighbor effect diminishes capture rate more than fission enhances it—as McFadden concludes (p. 15) *"fission should occur when the decrease in per polyp particle capture rate due to increasing colony size exceeds that due to the presence of neighboring colonies."*

Some caveats apply in interpreting the significance of these specific findings, and are summarized by McFadden (see also Sebens 1979, 1982, 1987, 2002 for general considerations). Beyond these it is worth noting that there are several optimum size models with varying assumptions, for instance about what function is being optimized. In McFadden's work, the premise is that selection is driving maximization of nutrient intake per unit biomass of the clone, which in turn is assumed to be directly related to growth rate, and ultimately clone size and sexual reproductive potential. More generally, it should be kept in mind that other hypotheses or models based on other selective forces might have produced similar results. Thus, because a model may be in accord with field observations does not mean that it is the correct explanation. Nevertheless, the energetics postulate provides general testable predictions, which can be further refined and tested should they be borne out in specific situations. Finally, with respect to broader implications of body architecture, Sebens (1982; see also Jackson 1977) makes the following interesting point about indeterminate growth. The size of a *solitary* indeterminate animal will be set by its shape and the dynamics of energy intake and energetic cost, so growth stops at some energetically optimal, habitat-dependent size. However, *clonal*, colonial organisms escape from such energetically imposed size limits by being able to undergo repeated rounds of fission into smaller daughter colonies. In related work, Hughes and Hughes (1986) have shown for the colonial cheilostome bryozoan *Electra pilosa* that the relationship between metabolic rate and biomass is isometric, i.e., not the negative allometric relationship that normally accompanies volumetric increase. Interestingly, even though zooid production is confined to the colony margin and therefore declines as peripheral area declines relative to volume, the animal compensates partially for this by increasing the budding rate of peripheral zooids and forming a protruding or lobate meristem.

The important point here is that the work described above on **optimal feeding with respect to colony architecture in benthic invertebrates surely has relevance to other clonal, modular organisms** including microorganisms such as fungi (Andrews 1994). Fungi, of course, do not obtain their nutrients like suspension feeders, but in many cases their colony dynamics, area:volume relationships, and growth patterns on hard substrata are closely analogous. The environmental triggers influencing fungal colony size and morphogenesis, including sporulation, are complex and involve cell density and nutrient depletion, among others (e.g., Chen and Fink 2006; Alkhayyat et al. 2015) and are discussed further with respect to phenotypic plasticity in ▶Chap. 7.

5.5.6 Competition or Complementation?

Because modular organisms are sessile and grow in indeterminate fashion, often by asexual increase as clones or colonies, they tend to contact parts of themselves and parts of other organisms in their immediate vicinity. On one hand, this situation presents the hazard of overgrowth or displacement by a competitor, or invasion by a foreign cell line (or pathogen). In other words, one can view such interactions purely in terms of two competitive pressures: at the level of the individual (colony) or at the level of the cell line (Buss 1990; Laird et al. 2005). On the other hand, the fusion of compatible conspecific genotypes offers potentially substantial advantages such as immediate increase in size. Another is the complementation of physiological repertoire between partners or even the emergence of novel traits. Such attributes have adaptive value in a contest for limited space as often occurs on the surface of, or within, a localized habitable substratum (by marine encrusting invertebrates or fungal hyphae, respectively). Thus, one would expect there to be strong selection pressure for the ability to determine self from non-self and that this discrimination likely would have arisen fairly early in geological time. Indeed, recognition ability transcends the issue of unitary/ modular design and seems to be a correlate of multicellularity: some manifestation of self/ non-self recognition occurs in essentially all multicellular organisms across all phyla. There are even suggestions of analogous recognition processes among some extant social bacteria that operate as collectives (Gibbs et al. 2008).

Here we will focus on issues directly relevant to modular micro- and macroorganisms and on allorecognition phenomena (operative among members of the same species) rather than xenorecognition (among members of different species). Incidentally, sessility implies that the effects of one neighbor on another can be specifically documented, whereas interactions among mobile (unitary) organisms usually have to be assigned abstractly to 'density effects' (Harper 1981a). Most of the attention has focused on genetic interactions between individuals, though even the most primitive modular organisms can respond to physical, chemical, and mechanical cues (▶Chap. 7 and Mydlarz et al. 2006). The formation of fusion chimeras has been most extensively documented in the fungi and the sedentary clonal benthic invertebrates, including sponges, bryozoans, cnidarians, and ascidians. Both groups (see following examples) share a mode of development known as somatic embryogenesis, i.e., they do not sequester a distinct germline and a given cell lineage can contribute to both somatic and germ cells throughout development (Buss 1987; see **Sidebar** and also ◻Table 5.1 and comments in ▶Chap. 2).

SIDEBAR: A Case Study: On Being a Neighbor

One consequence of a sessile, modular lifestyle is interactions among neighbors. The organism cannot escape from predators by running away or from competitors by moving to a different pasture to graze. Thus, one would expect to find evidence of various defensive mechanisms (and resource exploitation strategies) in such life forms. For instance, competition may be mediated by chemicals, a phenomenon commonly referred to as allelopathy by botanists, and as antibiosis by microbiologists. Sage plants of the genus *Salvia* in California evidently inhibit the growth of other vegetation at least in part by an allelopathic mechanism (e.g., Muller 1966; for the current status of allelopathy see Inderjit et al. 2011), although interactions are complex and alternative explanations have been put forward (Bartholomew 1970; Orrock et al. 2010). With respect to microorganisms, the production of antibiotics was formerly alleged to be an artifact of laboratory culture conditions. It is now clear that antibiotics are produced in nature and strong, multiple lines of evidence implicate their role in the antagonistic interaction among some microbes (Raaijmakers and Mazzola 2012). They also have multiple other and perhaps more prevalent roles, among them in intercellular signaling, biofilm ontogeny, virulence, and acquisition of trace elements.

Competitive (and other) interactions are also reflected by morphological behavior. In experiments on competition between the free-floating aquatic plants *Lemna* and *Salvinia*, the latter dominated because it overtopped *Lemna* by expanding its fronds above, and then

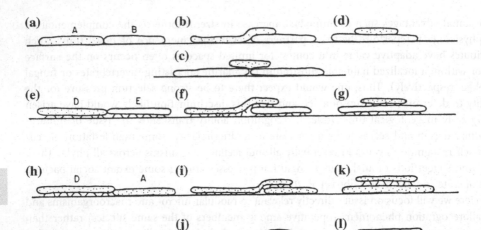

■ **Fig. 5.8** Schematic vertical sections through crustose (mat-forming; shown lying flat like pancakes) and foliose (on stilts) lichens, illustrating the potential range of interactions among these organisms. **a–d** Interactions between two crustose species, *A* and *B*. **e–g** Interactions between two foliose species, D and E. **h–l** Interactions between a foliose (*D*) and a crustose (*A*) species. Foliose forms can overtop most crustose forms and thus a stalemate (**h**) or overtopping by a crustose species (**j**) are uncommon. (**d, g, k, l**) Designate epiphytic growth which requires colonization of the underlying thallus by propagules containing both the algal and fungal components of epiphytic species, a rather uncommon event. From Pentecost (1980). Reproduced from *The Lichenologist* by permission of Cambridge University Press © 1980

down onto, the water surface (Clatworthy and Harper 1962). The outcome could not be predicted either by the carrying capacities or by the intrinsic growth rates of the respective species in isolation. When lichens of various morphologies meet on a rock surface, different outcomes are possible, but most commonly the foliose (upright) types override the crustose (appressed) forms (■Fig. 5.8; Pentecost 1980). However, the foliose forms are more vulnerable to being removed by abrasion.

In the hydroid *Hydractinia*, intraspecific competition between genetically unrelated colonies also takes on various phenotypic forms ranging from passive to aggressive (■Fig. 5.9; Rosengarten and Nicotra 2011). The former response involves the secretion of a fibrous matrix by both colonies and inhibition of growth at the interface. Active rejection is characterized by production of "a specialized organ of aggression," the hyperplastic stolon, at the tip of which nematocysts accumulate and discharge into the neighboring colony. So, as Buss and Grosberg (1990) comment, these simple invertebrates can display a range of aggressive reactions analogous to those of organisms that are much more developmentally and behaviorally complex.

In the fungi, intraspecific contact may be marked by tolerance, fusion (vegetative or somatic compatibility), and potentially the formation of a cooperative ecological unit (physiological individual). This phenomenon has become the standard test for self versus non-self in studies on the distribution of fungal genets. If the interacting colonies differ genetically at more than a few loci, however, the interaction is manifested by a localized darkened zone in which the hyphae are deformed (vegetative or somatic incompatibility). Interspecific responses may occur at a distance or following contact. In the latter case, several outcomes are possible (see Chap. 2 in Cooke and Rayner 1984). Interestingly, overall, the fungal interactions broadly parallel the interactions among hydroids.

An important practical significance of combative responses is that fungi and bacteria can be selected for their competitive, antagonistic ability and used as biological control agents to preempt or counteract plant pathogens. For further information, see Cook and Baker (1983). To cite but one example, the crown gall disease caused by *Agrobacterium tumefaciens* (overview in ►Chap. 3) is controlled by a closely related organism, *A. radiobacter* strain K84, which carries a plasmid coding for the production of an antibiotic that is toxic specifically to virulent *A. tumefaciens* (Burr and Otten 1999; Brencic and Winans 2005; Gelvin 2010). Similar principles are becoming actively researched and applied to manage the microbial ecology of the human gut (Costello et al. 2012; Donaldson et al. 2016).

Colonial invertebrates When two colonies of a benthic invertebrate fuse, stem cells can move throughout the chimera with the result that one partner can more-or-less overwhelm the other, thus becoming disproportionately represented in sexual or asexual propagules. In extreme forms of takeover the result is 'somatic cell parasitism' by the aggressor and genomic replacement of the victim (Buss 1982, 1987). Chimeras have been observed in detail in the field and laboratory (Stoner and Weissman 1996) and, in the case of *Botryllus* (below), invasion by relatively few stem cells is enough to accomplish a chimera (Laird et al. 2005). As discussed below and in ►Chap. 2, an analogous fusion/rejection interaction occurs in the fungi, although it is the nuclei rather than stem cells that move within the chimeric hyphal network, which in fungal semantics is generally referred to as a heterokaryon (Roper et al. 2011, 2013). In both situations there would be significant selection pressure to limit fusion to genetically compatible partners. Genetic surveillance systems oversee such interactions and are coupled to defensive response systems of varying sophistication that counteract an invader.

The genetic basis for allorecognition in colonial invertebrates has been studied for the most part in the ascidian *Botryllus* (Phylum Chordata) and the hydroid *Hydractinia* (Phylum Cnidaria). In both cases, at least one polymorphic fusibility/histocompatibility locus controls the outcome, where fusion depends on the sharing of one or two alleles (Grosberg 1988; Magor et al. 1999; Cadavid 2004; Rosengarten and Nicotra 2011). Studies on *Botryllus* show that randomly paired isolates rarely fuse and it is inferred that the fusibility locus is highly polymorphic with more than 100 alleles. However, Grosberg and Quinn (1986) showed that localized sites are settled by sibling larvae, which distinguish kin based on shared alleles at histocompatible loci. This means that developing colonies fuse with each other or with parental colonies at much higher rates than by random association, so potential benefits accrue to both kin members (see also De Tomaso 2006). *Hydractinia* colonies, shown in the **Sidebar**, grow as a surface encrustation on gastropod shells inhabited by hermit crabs. Allorecognition responses have been documented in multiple studies over more than a century and take three forms: (i) fusion (establishment of a common gastrovascular system within a few hours), (ii) passive or active rejection (the latter involving a specialized organ and tissue destruction mediated by stinging cells), and (iii) transitory fusion (Cadavid 2004) (◘Fig. 5.9). The *Hydractinia* allorecognition complex appears similar in some ways to *Botryllus*, but is believed to be more complex, at this point assumed to involve at least two linked loci (reviewed by Rosengarten and Nicotra 2011).

(a)

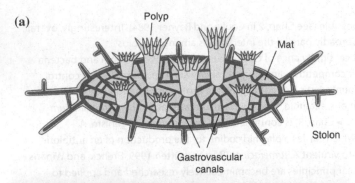

(b)

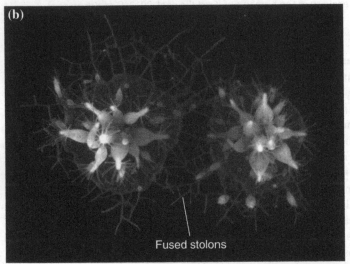

Fused stolons

(c)

Hyperplastic stolons

◄ ▣ **Fig. 5.9**   Allorecognition patterns in the cnidarian *Hydractinia symbiolongicarpus*, a colonial hydroid. **a** A colony showing the main parts. **b** Fusion between two histocompatible colonies. Note fused, small, web-like stolons barely visible in background. **c** Rejection between two incompatible colonies. Note enlarged, 'hyperplastic' stolons, typical of an aggressive response. From Rosengarten and Nicotra (2011), reproduced from *Current Biology* by permission of Elsevier©2011. Figure (C) is from Poudyal et al. (2007); reproduced from *Proceedings of the National Academy of Sciences* by permission of the National Academy of Sciences, USA, ©2007

Fungi Most fungi (primarily the ascomycetes and basidiomycetes) grow by an extensively radiating mycelium, noted earlier, that acquires nutrients from a patchy environment by ramifying upon, within, or at the interfaces between substrata, such as between soil and leaf litter (Andrews 1995; Bebber et al. 2007). Inevitably parts of the same and of different fungal individuals come into frequent contact. *Intra*-organism hyphal fusions are advantageous in terms of communication, homeostasis, transport, and repair. The overall network geometry of branching and fusions appears to be developmentally programmed and designed to optimize the conflicting demands of foraging with cytoplasmic mixing and transport (Roper et al. 2013). *Inter*-organism fusions can be advantageous in several ways as addressed below. Whether individuals fuse and go on to form a chimera is regulated by various incompatibility systems, including (i) somatic or vegetative incompatibility; (ii) sexual incompatibility; and (iii) intersterility. Of these, the most intensively studied and broadly analogous to macroorganisms (in some cases probably homologous, see below) is somatic incompatibility (▶Chap. 2 and Worrall 1997).

Where fusion between different genotypes occurs (or in the case of somatic mutation), two or frequently more, genetically different nuclei occupy the same cytoplasm (heterokaryosis; ▶Chap. 2). As noted in ▶Chap. 2, the extent of heterokaryosis in nature and its ecological advantages have been debated going all the way back to the original studies in the 1940s and 1950s (Pontecorvo 1946; Jinks 1952). In general the benefits relate to increased phenotypic plasticity, enhanced physiological capability and, in the case of pathogenic fungi, possibly increased virulence. Buss (1987) argues further that the filamentous fungi, by virtue of their cytology, are uniquely predisposed to both benefits from chimerism, as well as to be vulnerable to invasion. This sets up a situation where the different nuclei are potentially cooperators or competitors (a theme echoed by more recent investigators, e.g., Roper et al. 2011). Because of the porous, pipeline-like structure of hyphae, potentially thousands or even millions of genetically different nuclei can occupy the same interconnected network and travel quickly by bulk flow over relatively long distances (Roper et al. 2011, 2013). Many species, principally in the phyla Oomycota and Zygomycota, completely lack cross-walls (septa) so the cytoplasm is continuous, multinucleate (coenocytic), and the nuclei can readily move long distances relatively unimpeded (Roper et al. 2013). Species in the phyla Ascomycota and Basidiomycota, and their associated asexual states, are generally septate. These septa vary structurally, are to varying degrees porous to the passage of nuclei, and in some cases can even discriminate even among different nuclei. This appears to be one form of control to protect the individual thallus from systemic takeover by deleterious mutant or invader nuclei.

Aside from somatic fusion events and mutation, heterokaryosis also arises in other ways, the principal one being mating. Unlike the case with other macroorganisms, karyogamy in the fungi frequently does not immediately follow plasmogamy. In the ascomycetes, mating-type heterokaryons usually are physically restricted to certain cells (ascogonium and ascoge-nous hyphae) and the dominant life cycle phase, barring fusions among different genotypes, is as a haploid homokaryon (Worrall 1997). The success of hyphal anastomoses and heter-okaryon formation is regulated by heterokaryon incompatibility (*het*) loci that limit fusions to closely related homokaryons; if any of these typically multiple loci contain different alleles in the partners then fusion fails and lysis occurs.[5] Increasing numbers of *het* genes are being sequenced; they encode various polypeptides, including mating-type transcriptional regula-tors (for elaboration on the genetics see Glass et al. 2000; Glass and Kaneko 2003). The mat-ing-type locus can also function in some fungi as one of the somatic incompatibility loci (Worrall 1997).

In the basidiomycetes, typically the heterokaryon is the result of mating and is called a dikaryon (►Chap. 2) because each hyphal compartment has a pair of nuclei, one donated by each partner or mating type. However, unlike the case with the ascomycetes, the genomes are very different at other loci (James et al. 2008). The allocation of the two nuclei per cell, their synchronous division, transcription, and even their mutual spacing are exquisitely bal-anced and regulated (Gladfelter and Berman 2009; also see discussion of clamp connections in ►Chap. 2). Enforced dikaryotization prior to karyogamy ensures nuclear balance, though where the asexual propagules are monokaryotic the two nuclei may compete for representa-tion in the conidia. The nuclear situation has led to two competing views on basidiomycete heterokaryons: either they operate functionally as diploids with selection operating at the level of the heterokaryon individual; or as populations of nuclei where the unit of selection is the individual nucleus (Rayner 1991; James et al. 2008; Roper et al. 2011). Though recent advances in cytological technique enable nuclear dynamics to be assessed precisely (Roper et al. 2013), the conceptual issue is essentially as framed about 70 years ago by Pontecorvo[6] (1946). This returns us to the observations of Buss (1987, pp. 129–139), noted above, that the very structure of fungal hyphae that facilitates formation of genetic chimeras and result-ing organism vigor renders the fungi vulnerable to aberrant (mutant) nuclei or nuclear invasion.

From this welter of details the important broader point emerges that these various fungal processes regulate self/non-self discrimination in a manner directly analogous to the colonial invertebrates. The systems have evolved to enable the 'appropriate' fusions and the associated ecological advantages to occur, while protecting the individual from somatic cell or nuclear parasitism (Buss 1982, 1987). Allorecognition and rejection phenomena apparently reflect convergent evolution in each of the major evolutionary groups, though it would be interest-ing to know whether they have arisen from a common ancestral gene(s) basal to the clades. As more genes are sequenced and the products identified, the picture will come into focus.

---

5   Studies under laboratory conditions among strains of *Neurospora crassa* suggest that during mating the controls on heterokaryon compatibility are relaxed and the specialized hypha involved (trichogyne) evidently is able to accept nuclei from one or multiple hyphae of compatible mating type belonging to any *het* genotype. The gen-erality of this finding remains to be seen. For synopsis and references, see Roper et al. (2011); Roche et al. (2014).

6   In a remarkably prescient assessment, Pontecorvo (1946) stated (p. 199) … "*we may be justified in considering a hypha as a mass of cytoplasm with a population of nuclei. Such a population is subject to: (1) variation in numbers; (2) drift—i.e., random variation in the proportions of different kinds of nucleus; (3) migration—i.e., influx and outflow of nuclei, following hyphal anastomoses; (4) mutation; and (5) selection. Selection may act either on the nuclei themselves as proposed here, or on the hyphae carrying them.*"

But the important and even broader message here is that recognition phenomena are fundamental and transcend the arbitrary cleavage of the living world into microorganism and macroorganism.

5.6 Summary

Growth form, that is, the mode of construction and related morphology, sets fundamental opportunities and limits on the biology of organisms. All living things are basically either unitary or modular in design. It is argued here that, on balance, the modular versus the unitary distinction is probably more biologically significant than a demarcation based on size (microorganisms versus macroorganisms). To the list of modular macroorganisms, it is instructive to add fungi and bacteria, based on their morphology and ecological traits that align them with organisms such as plants and benthic invertebrates. Doing so places the emphasis on the organism as a whole and opens a vista of fruitful analogies for investigations in microbiology as well as in plant and animal biology.

Unitary organisms grow non-iteratively (not based on a repeated multicellular unit of construction), are usually determinate, and growth form is not highly influenced by the environment. Such organisms, represented typically by mobile animals, follow sequential lifecycle phases predictably and their number of appendages is fixed early in ontogeny. They display generalized (systemic) senescence and characteristically are unbranched. Reproductive value increases with age to some peak and then declines. The genetic individual (genet) and physiological or numerical individual are the same entity, which is repeated only by the start of each new life cycle.

In contrast, modular organisms grow by the reiteration of a basic multicellular unit of construction (module). Such organisms, characterized by plants, many invertebrates, bacteria, and fungi, grow indeterminately (no genetically fixed upper limit) and in a manner much more subject to environmental influences. Modules may remain attached to the body developing from the single-celled stage (zygote or asexual propagule) or become detached and function as separate physiological individuals (ramets), as occurs among clonal modular organisms. Thus, the ratio of genets:modules is 1:many. Modular organisms are frequently branched and are for the most part sessile or passively mobile. They typically display an internal age structure (e.g., a cohort of leaves is 'born', expands, and dies over a discrete block of time); absence of generalized senescence; a non-segregated germplasm; and often an increase in reproductive probability with age.

The modular paradigm was developed for plants and many clonal, colonial animals (primarily benthic invertebrates). Little attempt has been made to place bacteria and fungi within the unitary/modular context. Fungi are modular organisms architecturally by virtue of their extension of size by iteration of a hyphal growth unit as well as by their overall ecological and morphological characteristics. Furthermore, some fungi (e.g., those with a significant life phase as balanced dikaryons and in some cases as heterokaryons) are unique as modular organisms because a given physiological individual may become a true genetic mosaic in a way that other modular creatures cannot, even by somatic mutation (see also ►Chap. 2).

Bacteria grow in nature generally as microcolonies, aggregates, or as an extensive biofilm of cells. Architecturally and in many ways ecologically, bacteria operate as modular organisms because the individual bacterial cell (module) is iterated indefinitely, as are higher orders of organization. (Note, however, that in eukaryotes the term module is reserved for multicellular units.) Genetic recombination in bacteria (►Chap. 2) is non-meiotic but in effect sexual. However, because formation of the recombinant is not generally tied to a particular morphological structure, a divisional event, or a characteristic life cycle stage, the genet concept for bacteria is even more abstract than it is for the fungi. In a manner analogous to some clonal aquatic plants, in effect bacterial clones fall to pieces as they grow and frequently can be disseminated worldwide.

The key evolutionary implications of modular design stem directly or indirectly from growth by iteration and sessility. The former include (i) high phenotypic plasticity (shape; size; reproductive potential; growth as a population event); (ii) exposure of the same genet to different environments and selection pressures; (iii) iteration of germplasm and the potentially important role for somatic mutation; and (iv) potential extreme longevity of the genet.

Habitable space is for sessile organisms a critical resource that affects survival, and it can limit size and hence reproductive potential. The evolutionary implications of sessility include (i) adaptations for the individual to reproduce without a mate or to reach a mate and disperse progeny by growth or transport (pollen; spores; seeds; bacteriophage movement of bacterial genes); (ii) potentially strong interactions with neighbors; (iii) inability to escape from adverse environments, hence the means to adapt in situ such as by dormancy; and (iv) development of resource depletion zones and consequently resource search/exploitation strategies from relatively fixed positions. In clonal organisms, morphology and search are determined in large part by how the organism places its ramets, represented at the extremes by the guerrilla or the phalanx pattern. Ultimately, the architecture of modular organisms is a compromise response by the organism to multiple forces. Among these are energy and nutrient acquisition in the face of competition from neighbors; tolerance of abiotic conditions such as water turbulence, wind, or snow loading; dispersal of gametes and zygotes; mechanical support; and plumbing system constraints for translocation.

While the implications of modularity are profound, the correlates are inextricably linked to other related properties, namely the tendency for many modular creatures to grow as clones and as colonies. The latter attributes appear to have been in many cases the outcomes of a modular design.

Suggested Additional Reading

Campillo, F., and N. Champagnat. 2012. Simulation and analysis of an individual-based model for graph-structured plant dynamics. Ecol. Model. 234: 93-105. *This paper typifies the advanced computational analyses and simulations being applied in particular to growth and spatial distribution of clonal, modular organisms.*

de Kroon, H., and J. van Groenendael. 1997. The Ecology and Evolution of Clonal Plants. Backhuys Publishers, Leiden, The Netherlands. *The perspectives of multiple authors on the ecology and evolutionary biology of clonal plants. In many ways, this is the counterpart of the book on clonal animals by Hughes (1989), cited below.*

Harper, J.L. 1977. Population Biology of Plants. Academic Press, London. *This beautifully written, comprehensive synthesis remains the classic treatise on plant population biology and its lessons apply to ecology in general. There is extensive discussion of many topics raised in this chapter, such as the concept of genets and ramets, clonality, and modularity.*

Harper, J.L., B.R. Rosen, J. White (Eds). 1986. The growth and form of modular organisms. Phil. Trans. Roy. Soc. B 313: 1-250. Also, Proc. R. Soc. B 228: 109-224. *Perspectives of various authors on modular growth and its implications.*

Hughes, R.N. 1989. A Functional Biology of Clonal Animals. Chapman and Hall, London. *A thorough treatment of animal clones and a chapter on modular animals.*

Jackson, J.B.C., L.W. Buss, and R.E. Cook (eds). 1985. Population Biology and Evolution of Clonal Organisms. Yale Univ. Press, New Haven, Conn. *Proceedings of a symposium. Interesting perspectives; inconsistent terminology; absence of microbes from the synthesis.*

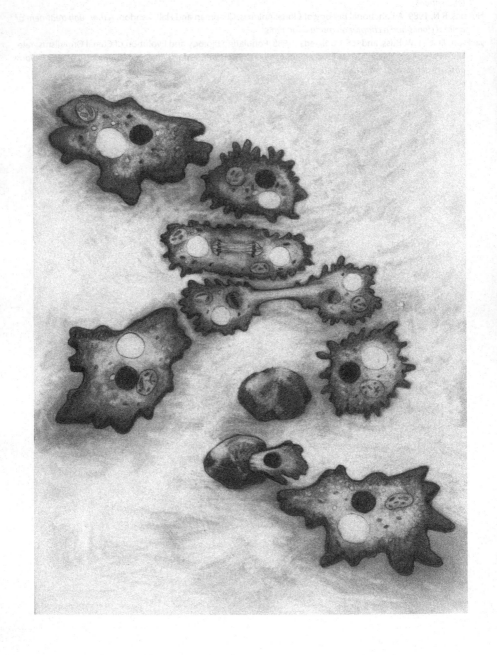

The Life Cycle

© Springer Science+Business Media LLC 2017
J.H. Andrews, *Comparative Ecology of Microorganisms and Macroorganisms*,
DOI 10.1007/978-1-4939-6897-8_6

The 'survival machines' of genes are the life cycles that link parental zygotes to progeny zygotes.

Harper and Bell, 1979

6.1   Introduction

The life cycle is the standardized, cyclic sequence of physiological, genetic, and morphological events that mark the passage of one generation to the next. Its normal starting and ending points are marked by birth and death of an individual, respectively, or the span from a parental zygote to a progeny zygote, as reflected in the quotation above. The life cycle unfolds, usually from the single-celled zygote, with the new individual being literally 'reproduced' from information encoded in that cell (as worded by Harper 1977, p. 27). Bonner (1965, 1993) argues that the organism should be considered to be an integral of its life cycle phases (not merely visualized in the adult stage, as is typically the case) and emphasizes the importance of the cycle by calling it (1965, p. 3) *"the central unit in biology"*. He examines many of its interesting attributes and their ramifications. Here, we consider the diversity of life cycles among both macro- and microorganisms, explore a few of them, in particular those ecologically deemed 'complex', and conclude by examining senescence as a phenomenon and as manifested among different phyla.

6.2   Origins and General Considerations

Historically speaking, the original and simplest life cycles probably arose in unicellular organisms concomitantly with cellular division of labor and the onset of sequential events at the molecular and physiological level. A contemporary analogue would be the cell cycle in undifferentiated, unicellular prokaryotes. That cycle begins with a single cell and ends with two separate, superficially identical, daughter cells (in fact they are not strictly identical, as will be developed later). It should be emphasized in passing that **the cell cycle is distinct from the life cycle of the prokaryotic clone, which continues indefinitely** (more on this later).

With respect to prokaryotic life cycles and unlike the case with eukaryotes, reproduction (fission) in bacteria and gene exchange are different processes as was reviewed previously. The cycle is typically haploid, since most (not all; see ▶Chap. 2) prokaryotes contain only a single chromosome and therefore a single copy of each gene. Notwithstanding little conspicuous morphological change prior to the actual divisional event, underpinning the process is a regulated temporal pattern in gene expression, biosynthesis, etc. At a slightly more complex level, discrete morphological changes occur in the cell cycle of bacteria that form spores as a resistant state (e.g., *Bacillus*), or that alternate between a sessile, feeding stage and a motile, reproductive state of two or more cells. This latter group includes, for example, *Caulobacter* and relatives (Brun and Shimkets 2000; Laub et al. 2000) (◻Fig. 6.1). The molecular genetic underpinning of this *Caulobacter* life cycle and the morphological transitions, particularly with respect to regulatory proteins, are known to some extent and are probably quite analogous in both prokaryotes and multicellular eukaryotes (Jiang et al. 2014). Other prokaryotes may exhibit multiple alternative states (myxobacteria discussed in ▶Chap. 4) or differentiate specialized cell types (e.g., akinetes and heterocysts in the filamentous cyanobacteria) (Dworkin 1985).

Thus, 'development' in the bacterial cycle consists largely of significant transitions in function and, in some species, the morphological changes between alternative cell types in

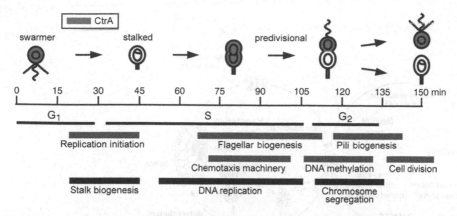

◻ Fig. 6.1 A simple dimorphic life cycle represented by the aquatic bacterium *Caulobacter crescentus*. Morphogenesis, cell cycle stages (G_1, S, G_2), and certain associated biochemical activities are shown as a function of time. (The *shaded* cells contain a response regulator, CtrA.) During each cell cycle a flagellated, swimming cell (swarmer) and an ultimately stalked, sessile cell are produced. Following a motile phase, the swarmer sheds its flagellum and an adhesive holdfast is produced cementing the cell to the substratum, following which a stalk is synthesized at the holdfast pole. Interestingly, through its role in cell elongation the stalk enhances nutrient uptake, advantageous in oligotrophic waters (see Wagner and Brun 2007). From Laub et al. (2000); reproduced from *Science* by permission of The American Association for the Advancement of Science, ©2000

response to environmental signals. This sequencing and the diversity of prokaryotic life cycles are discussed by Dworkin (1985, 2006) and Brun and Shimkets (2000). These seemingly simple changes are the prokaryotic counterpart to morphological complexity and developmental staging in eukaryotes. Remarkably, however complex the life cycle ultimately becomes in multicellular macroorganisms, it characteristically if not universally incorporates, sooner or later, at least one single-cell stage (typically the spore and/or zygote; apparent exceptions to this principle are discussed by Grosberg and Strathmann 2007). The significance of passage through a single-celled stage is discussed in a later section, below.

Embellishment on the rudimentary life cycle described above evolved in tandem with increasing complexity accompanying multicellularity. For example, in the case of plants, distinct phases include the zygote; post-zygotic development; seed dormancy; germination (often equated with birth); juvenile (vegetative) development; staging of reproduction and longevity as an annual, biennial, or perennial; senescence or lack thereof; and death (Harper and White 1974). Harper (1977, p. 519; see also Niklas and Kutschera 2010) further identifies 10 life cycles among flowering plants with respect to their fertility schedules. Animals display similarly diverse cycles, among the most dramatic being those of insects with complete metamorphosis. Buss (1987, pp. 165–167) developed seven sets of traits (based on such factors as the nature of the cell enclosure, ploidy, presence of germline sequestration, etc.). If all permutations were possible some 972 life cycles would result; in practice there are specific conditions that influence others and whether or not there can be compensations. This reduces the feasible number to a much smaller, though still sizable, subset of a couple of dozen or so and these are based only on the specific conditions chosen by Buss. Clearly, the actual number of life cycles, though numerous, is a subset of the very many theoretically possible, suggesting constraints. A generalized life cycle for multicellular organisms is shown in ◻Fig. 6.2 though there are many variations on the basic theme (e.g., Bell 1994).

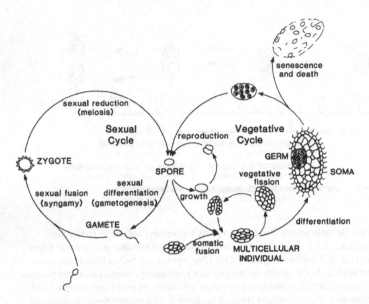

□ **Fig. 6.2** A simplified eukaryotic life cycle, typified here by a protist, showing a sexual phase beginning with a single-celled, diploid zygote and a vegetative phase beginning with a single-celled, haploid spore that links the two phases. Note the seminal role of the unicellular state. From Bell and Koufopanou (1991); reproduced from the *Philosophical Transactions of the Royal Society of London B* by permission of The Royal Society, ©1991

6.2.1 **Alternation of Generations**

With the evolution of sexuality in eukaryotes, the regularized haploid and diploid phases that are morphologically striking in land plants and many algae (Niklas and Kutschera 2010) became dominant elements of the conventional life cycle. To vastly simplify, plasmogamy and karyogamy, typically involving gametes, establish the diploid phase, which is followed more or less rapidly by meiosis, restoring the haploid condition. However, karyogamy may not closely follow plasmogamy and there is a tremendous phylogenetic range in the number of mitotic cell generations separating the two events of meiosis and syngamy (Valero et al. 1992; Coelho et al. 2007): The fungi, for instance, are almost exclusively haploid (meiosis rapidly follows syngamy); the n and $2n$ phases are about equally prominent in many algae such as sea lettuces, *Ulva* spp. (meiosis and syngamy are separated in time and space); while in most animals the haploid condition is relatively brief, inconspicuous, and represented only by the gametes (syngamy rapidly follows meiosis). Terminology varies, but generally speaking the three types of cycles are referred to as haploid, haploid/diploid, and diploid, respectively. As discussed in ►Chap. 2, some fungi are unique in having an extended dikaryotic ($n + n$) condition. Broadly speaking, the eukaryote life cycle is one in which growth and reproduction alternate with sexual fusion and meiosis; vegetative processes occur in the haploid phase, the diploid phase, or both. Growth during haploidy is a primitive condition, with diploid growth having evolved independently in numerous clades (Bell 1994). Representative organism life cycles and the nuclear states (n, $n + n$, $2n$) of key basic stages are summarized in □Table 6.1.

In modern land plants (**embryophytes**) the life cycle is particularly remarkable (Walbot and Evans 2003; Niklas and Kutschera 2010). Here the products of meiosis are not gametes but spores that undergo mitosis to produce a haploid, multicellular body termed a gametophyte, which produces the eggs and sperm. The gametophyte alternates with a multicellular diploid sporophyte (spore-producing) generation that originates upon fusion of the gametes and ends with the production of haploid meiospores. All land plants (as well as some algae) display a conspicuous alternation of generations and the sporophyte is the dominant form among the more modern (vascular) lineages (**tracheophytes**: ferns, lycopods, horsetails, and seed plants). Among the higher seed plants (angiosperms), the nonphotosynthetic gametophyte typically is reduced to only 2–7 cells that rely on the sporophyte for nutrients. Conversely, among the earliest land plants (**bryophytes**: liverworts, hornworts, mosses) the gametophyte generation dominates, with the diploid phase characteristically being inconspicuous. These fundamentally different life cycles are called diplobiontic (alternation of two multicellular phases, one diploid and one haploid) and haplobiontic (only one multicellular generation), respectively (◘Fig. 6.3).

Interestingly, the land plants evolved from the **charophyte**-like (stoneworts) ancestor within the green algae (see ►Chap. 4), i.e., from a condition where the gametophyte dominates and the single-celled diploid phase is only transient, to one where the sporophyte is ascendant (Niklas and Kutschera 2010; Evert and Eichhorn 2012). This ancient transformation probably resulted from a delay in meiosis possibly orchestrated by genes encoding regulatory homeoproteins that maintained sporophytic function and suppressed gametophytic development. A similar argument pertains to extant plants that undergo regular alternation of generations in their life cycles (Sakakibara et al. 2013). Indeed, Niklas and Kutschera (2010) suggest that the reproductive structures of the extant multicellular charophytes and embryophytes are homologous and that the embryophytes largely redeployed the ancient algal homeodomain gene network. An intriguing debate centers on whether the shift to a diplobiontic cycle was essential for transition from an aquatic to the terrestrial environment (i.e., a developmental constraint) or merely correlative (i.e., a phyletic legacy) with the successful colonization of land by plants (Friedman 2013).

Across the diversity of living organisms, life cycle patterns show an evolutionary trend to a preponderantly diploid condition, evidently a correlate of individuality and complexity. Numerous theoretical hypotheses based on genetic or ecological interpretations have been put forward in favor of diploidy (summarized by Hughes and Otto 1999; Coelho et al. 2007). In the main they relate to: (i) the masking of deleterious mutations by the normal allele; (ii) retention of the masked mutations as a genetic repository available for future use; (iii) the twofold greater amount of DNA enables new beneficial mutations to accumulate at a greater rate; (iv) the 'additional' allele can evolve new functions while the other copy is providing for ongoing function. However, the advantages of diploidy have been challenged and may be situation-specific (e.g., Zeyl et al. 2003); also, the trend to diploidy does not explain the maintenance of haploid and biphasic haploid/diploid cycles (Hughes and Otto 1999).

The green plant lineages beautifully illustrate the spectrum ranging from an essentially haploid gametophyte body, through a relatively balanced condition, to complete sporophyte dominance in the most advanced embryophyte phylum (for details and terminology see Graham 1993; Graham et al. 2009). By virtue of having the gametophyte of varying size and duration, plants can impose stringent selection during the genetically active haploid phase and likely have a lower genetic load than animals; this may be the reason that, unlike many animals, they do not have to designate a germ line early in development (Walbot and Evans 2003; however, see Szövényi et al. 2013). Under some conditions, genes promoting haploid selection may be favored but not if such mutations are selected against in the diploid phase—a condition referred to as 'ploidally antagonistic selection' (Otto et al. 2015). The balance in such pressures and the broader genetic and ecological interactions governing biphasic cycles are explored by Rescan et al. (2016).

6

◻ Table 6.1 The life cycles of some representative micro- and macroorganisms (modified from Fincham 1983)

Organism	2n; diploid phase	n; products of meiosis	n; haploid phase	n; gametes	n + n; dikaryon	2n; zygote
Escherichia coli (bacterium)	–	–	Entire life cycle	Direct gene exchange	–	–
Neurospora crassa (filamentous fungus)	Ascus (within fruit body perithecium)	Ascospores, four-spore pairs	Filamentous mycelium bearing asexual spores (conidia)	Ascogonium (in protoperithecium); conidium (opposite mating type)	Ascogenous hyphae within perithecium	Ascus initial
Schizophyllum commune (bracket fungus)	Basidium (borne on gills of fruit body)	Basidiospores, borne externally on basidium in tetrads	Monokaryotic mycelium	Mycelial cells of compatible mating types	Mycelium—the major growth phase—forming fruit bodies (brackets)	Young basidium
Ulva (multicellular marine green alga; two-layered sheet of cells)	Sheet-like thallus	Zoospores formed in tetrads	Sheet-like thallus (similar to diploid)	Motile cells—opposite mating types, morphologically identical	(Karyogamy immediately follows gamete fusion)	Germinates to give haploid thallus
Any moss—bryophyte, land plant	Capsule borne on haploid plant	Spores, initially in tetrads, liberated from capsule	Leafy plant bearing archegonia and antheridia at fertile shoot apices	Egg (in archegonium); sperm (liberated from antheridium)	(Karyogamy immediately follows gamete fusion)	Fertilized egg develops into diploid capsule
Zea mays (angiosperm, seed plant)	Maize (corn) plant	Megaspores in ovules; microspores (pollen) in anthers of flowers	Embryo sac (eight nuclei including egg nucleus); pollen tube (three nuclei including two gamete nuclei)	Egg (in embryo sac); pollen tube nucleus	(Karyogamy immediately follows gamete fusion)	Fertilized egg develops into dormant embryo in maize kernel
Drosophila melanogaster (fruit fly)	Larval stages leading through pupal stage to adult fly	Eggs; spermatozoa	–	Eggs; spermatozoa	(Karyogamy immediately follows gamete fusion)	Fertilized egg develops into larva

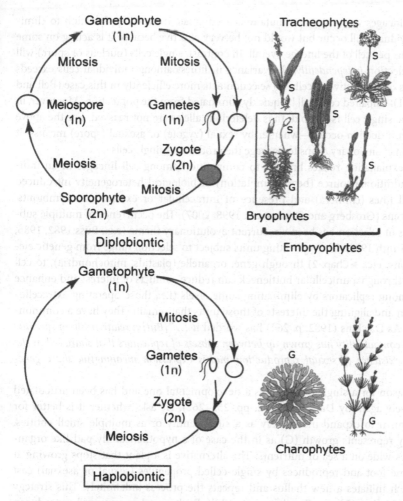

□ Fig. 6.3 Attributes of basic life cycles in the plant world. The modern land plants or embryophytes have a diplobiontic cycle (*upper figure*) where both the haploid gametophyte (G) and diploid sporophyte (S) are multicellular and more-or-less conspicuous. In contrast, the green algae, here represented by two charophytes (likely the group from which the land plants emerged), have a haplobiontic cycle (*lower figure*), where either only the haploid phase is multicellular (haplobiontic-haploid) as illustrated here, or only the diploid phase is multicellular (haplobiontic-diploid) as is the case for animals. See text for taxonomic nomenclature. From Niklas and Kutschera (2010). Reproduced from *New Phytologist* by permission of Karl Niklas and John Wiley and Sons, Inc. ©2009

6.2.2 The Centrality of a Single-Cell Stage

The life cycle of all organisms sooner or later passes through at least once single-celled stage. Surely this is more than just a coincidence. There are at least three biological hypotheses as to why a single-cell bottleneck would be advantageous. First, Crow (1988), among others (e.g., Bell and Koufopanou 1991), has argued that the single-cell sets essentially a tabula rasa for a new generation. The alternative of reproducing by vegetative (multicellular) propagules or organs leads to lineages where somatic mutations accumulate irreversibly. Selection to elimi-nate loaded individuals will occur but would not be very effective because it is acting on some *mean* of mutations per cell of the lineage overall. In contrast, single cells (nucleus or spore) will be exposed to selection *independently*; the variance in fitness among individual cells exceeds that of the means of a collective of cells, so selection acts more efficiently in this case (Bell and Koufopanou 1991). Impaired cells will be quickly eliminated from the population. Moreover, in many lineages the single cell is haploid, thus deleterious alleles are not masked. For the above reasons, when reproduction occurs—whether by sexual (zygote) or asexual (spore) means—it is advantageous in evolutionary terms to package the products as single cells.

The second explanation relates broadly to competition among cell lineages in an indi-vidual and the additional source (beyond mutation) of biological heterogeneity introduced by fusion of cell lines (chimera) and presence of intracellular or extracellular symbionts, including pathogens (Grosberg and Strathmann 1998, 2007). The occurrence of multiple sub-organismal levels of selection has been a recurrent evolutionary theme (e.g., Buss 1982, 1985, 1987; Maynard Smith 1988). The replicating units subject to selection range from genetic ele-ment (transposons, etc.; ►Chap. 2) through gene, organelle (plastids, mitochondria), to cell lineage. Passage through a unicellular bottleneck can reduce conflicts of interest and enhance compatibility among replicators by eliminating some levels (i.e., those operating extracellu-larly) of variation and aligning the interests of those units that remain. They have a common-ality of descent. As Dawkins (1982, p. 264) has worded it … "*[But] a relationship of specially intimate mutual compatibility has grown up between subsets of replicators that share cell nuclei and, where the existence of sexual reproduction makes expression meaningful, share gene-pools.*"

The third reason for a single-cell stage is a developmental one and has been articulated most imaginatively again by Dawkins (1982, pp. 253–262). He asks whether it is better for a growing organism to expand indefinitely as a single entity or as multiple small entities. The first strategy represents growth (G) as in the case of a hypothetical lily pad-like organ-ism several miles wide on a sea of nutrients. The alternative is a plant that stops growing at a diameter of one foot and reproduces by single-celled propagules (sexual or asexual) cast to the wind. Each initiates a new thallus and repeats the process indefinitely. This strategy represents reproduction (R). So G only grows, while R alternately grows and reproduces. Which is better developmentally? Dawkins argues that descendants of R can evolve complex features in a manner that those of G cannot. Each young, expanding R unit can differentiate and potentially outperform its parental colony, and at each turn of the cycle new variations can be introduced. **This is a specific illustration of the principle that a complex state must evolve from a less complex condition**. In contrast, each cell of G is destined to assume only a small role subordinated amidst the massive, expanding lily pad. Novelties can be made at the cellular but not the multicellular level because there is no passage through a structurally simple stage that would enable more complicated, organ-level innovation to occur. Dawkins' metaphor is that whereas it takes only a little tinkering to transform a Bentley into a Rolls Royce late on the assembly line, to go from a Ford to a Rolls Royce you have to start with new blueprints. Indeed Dawkins argues that **the large, complex multicellular organism is pos-sible because of recurrent life cycles that include a single cell where mutations can effect wholesale change, a process that works only by** *"repeated returns to the drawing board dur-ing evolutionary time"* (p. 262).

6.2.3　Phylogenetic Constraint

While every life cycle eventually passes through a single-celled stage and may have in essence a tabula rasa, this does not mean that the slate is completely blank. The influence of ancestral legacies has been recounted in compelling detail by Buss (1987), which can only be touched on here. A first broad limitation relating to animals stems from their origin in a unicellular protistan flagellate, which imposed a constraint that cells cannot divide while ciliated (related to concurrent need both in mitosis and flagella of the microtubule organizing center; this is recounted in ▸Chap. 4). Some protists overcame the impasse but the line giving rise to metazoans did not. In turn, some of the early metazoans never did resolve this limitation and abandoned dispersal, but for most lineages the solution was gastrulation. For example, sometimes an early embryonic stage involved a two-layered entity capable of both movement and of further development because ectoderm cells were ciliated whereas the endoderm was not (an analogous example pertains to the motile alga *Volvox* discussed in ▸Chap. 4). In other words, one evolutionary solution to the conflict between increasing cell complexity and continued division was resolved by allowing some cells to become elaborated and specialized while retaining others unspecialized, capable of ongoing division, and hence in the creation of germ layers. The synergistic and competitive interactions among these embryonic cell lineages, together with selection operating at the level of the individual, have shaped both early and late ontogeny, the resulting diversity among phyletic blueprints, and ultimately the corresponding life cycles (Buss 1987).

Second, the ancestors of the major multicellular clades (animals, plants, fungi) also began with different starting materials that set the particular boundary conditions and shaped the possible, presenting both evolutionary opportunities and constraints (Buss 1987). The starting condition for animal cells was a flexible membrane boundary, which enabled them to move. Accordingly, the zygote could undergo multiple rounds of division giving rise in ontogeny to differentiated lineages destined for much later positions and varied functions. This, however, rendered animals vulnerable to 'somatic cell parasites' that abandoned their regulated somatic role and reverted to uncontrolled mitoses. A counter to this threat is early (within the first few cell divisions of the zygote) germline sequestration. In its earliest ontogenetic form, this occurs only in animals with the preformistic mode of development; ▸Chap. 5 and Buss 1987, 1985). Such development precludes asexuality because ramet production depends on the presence of a totipotent lineage. Thus, animals have either the capability for cloning at the expense of vulnerability to mutant lineages, or ontogenetic controls that limit such variants at the expense of losing asexuality, but not both.

In sharp contrast to animals, plants have rigid walls and consequently their cells cannot move. Somatic variants, though they still occur, are therefore not a systemic threat. Embryogenesis is not marked by early germline determination. Plants preserve totipotent lineages represented by their various meristems (see ▸Chap. 5) that produce somatic structures and eventually flowers; accordingly, plants have the potential for asexual reproduction. They contend with deviant lineages in unique ways, including stringent selection in the haploid gametophytic phase, noted above (see ▸Chap. 5 and Walbot and Evans 2003). Considerable intercellular communication remains possible in plants via plasmodesmatal channels between adjoining cells.

Fungi, in turn, have certain attributes of both plants and animals. Fungal cell walls are rigid but, unlike plants, septation in the end walls of adjoining hyphal compartments is generally incomplete so they are to varying degrees coenocytic. Certain processes and cytological features such as specialized septa, cell synchrony, balanced dikaryosis, and cellularization at sites of reproduction, regulate systemic mixing. Their largely haploid condition but varied nuclear states involving dikaryosis and heterokaryosis (►Chaps. 2 and 5) are related to the exceptional and frequently complex life cycles of fungi, as developed below.

6.3   Complex Life Cycles

6.3.1   Concept and Definition

The terminology and theory of complex life cycles (CLCs) were formalized largely by zoologists to describe life cycles in which there is *"an abrupt ontogenetic change in an individual's morphology, physiology, and behavior, usually associated with a change in habitat"* (Wilbur 1980; p. 67). An organism that passes through two or more such (irreversible) distinct phases was considered to have a **complex life cycle** (Istock 1967). In contrast to these complicated cycles, if the offspring of a species are born into essentially the same habitat occupied by the adults, and do not undergo sudden morphological or ontogenetic change, the organism is said to have a **simple life cycle** (SLC). Examples of the latter include birds, humans, and other mammals.

Animals The classic example of a CLC is the frog, many species of which occupy two distinct ecological niches: that of an aquatic herbivore in the tadpole phase, followed by life as a terrestrial carnivore in the adult stage. As Wilbur notes, it is interesting that two such species may be ecologically and morphologically more similar as tadpoles than either is to the adult frog it will become. CLCs are represented in most animal phyla, have evolved independently many times, and have persisted through geological time (Moran 1994). Examples include in general the amphibians, insects (especially aquatic insects, which live for months or years as immature forms in streams, followed by an ephemeral adult phase of hours to weeks on land), and most marine invertebrate species such as barnacles, starfish, snails, etc., that have a tiny larval phase that feeds in coastal waters followed by an adult phase on rocks or in mud (◨Fig. 6.4a and Roughgarden et al. 1988). Where the individual organism passes sequentially from larva to adult by metamorphosis the type of CLC is single-generational (as opposed to multigenerational, below).

Algae While zoologists played a seminal role in developing the conceptual framework of CLCs, clearly organisms other than animals exhibit an analogous pattern of drastic morphological and ecological change during the course of their life histories. The seaweeds and bryophyte land plants typically have a pronounced alternation of generations consisting of a prominent gametophytic phase contrasting with morphologically distinct sporophytes, discussed earlier. The heteromorphic algae, of which the red, brown, and green seaweeds are good examples, have two separate, ecologically distinct phases that may be so dissimilar morphologically that the phases have even been classified as separate species. For instance, in the red alga *Porphyra*, a haploid thallus of leafy sheets about 3–15 cm long growing on rocks in the intertidal zone alternates with an inconspicuous, shell-inhabiting sporophyte, the *Concho-*

celis phase. Frequently, the morphological and ecological divergence within the algal life cycle parallels the alternation of haploid with diploid generations. In some cases, the phases are self-perpetuating and neither morphological nor genetic alternation is obligatory (Lubchenco and Cubit 1980). Even in cases where the phases are morphologically similar, ecological differences likely occur and may explain in part the origin and maintenance of biphasic life cycles (Hughes and Otto 1999) noted previously. Instances of such isomorphy technically do not rise to the level of a strictly defined CLC. The heteromorphic algae, certain insects such as aphids that display different phenotypes during a seasonal cycle, many parasites, and the fungi discussed below, are examples of multiple generation type CLCs.

Fungi A situation analogous to the bryophytes and heteromorphic algae is the filamentous fungi. Members of the largest group of fungi (at least 30,000 species in the phylum Ascomycota) have diverse lifestyles but typically have two distinct, alternating forms, a perfect or teleomorph (sexual) state and an imperfect or anamorph (asexual) state. Each has its own characteristic structures, spore type and frequently, habitat (◘ Fig. 6.4b). Many ascomycete species are associated with living plants both above and below ground as parasites, mutualists, or commensals. In terms of their ecological activities, these and other filamentous fungi could be conceptualized as passing through two life phases in which biomass is allocated primarily to either foraging or reproduction. The former, a juvenile vegetative condition, is characterized usually by a diffuse hyphal network adapted for exploration and nutrient acquisition upon or within a substratum. The latter, a mature reproductive condition, involves initially specialized hyphae and ultimately propagules produced asexually, usually through repetitive cycles and released into the medium (air, water, or soil). Upon death of the plant or plant organ, the fungus frequently continues to grow as a saprophyte, undergoes sexual reproduction, and enters a relatively dormant condition until meiospores are released upon the appropriate environmental cues. In summary, although there is no host alternation here (unlike the rust case below), with this basic type of cycle there are morphologically and functionally distinct phases and two distinct niches. This pattern is consistent with Wilbur's and Istock's criteria for organisms with a CLC.

Animal and plant parasites Among the most complex life cycles are those of many animal and plant parasites. These include the rust fungi reviewed at length below and many of the nematodes, trematodes, protozoa, and phytophagous insects that must use two or more hosts in sequence (Chaps. 2 and 3 in Price 1980; Thompson 1994). Usually the sexual phase of the parasite life cycle is associated with a particular host or habitat while the asexual stage, functioning mainly to increase and disperse offspring, occurs with a different host. The case for parasites can also be equivocal as one must distinguish complexity in the epidemiological sense from that of the life cycle of the causal organism. Many parasites attack a single host species. Although the epidemiology of the resulting disease may be complex, it does not follow that the incitant has a CLC. In other cases, several species may potentially be hosts, yet neither is their sequential involvement mandatory for completion of the parasite's life cycle nor does the parasite have to undergo an irreversible ontogenetic change. The epidemiology of scrub typhus presents a fascinatingly intricate case that illustrates the above points. The concurrence of events culminating in human infection depends on: (i) presence of the bacterial pathogen *Rickettsia tsutsugamushi*; (ii) the vector, chiggers (trombiculid mites); (iii) small mammals, particularly rats, on which the chiggers feed; and (iv) transitional or secondary forms of vegetation ('scrub'), through which humans chance to pass. The intimate relationship of these four factors led to the descriptor 'zoonotic tetrad' (Traub and Wisseman 1974). However, chiggers are the primary reservoir of the causal organism and the related disease is not passed through either rodent or human in sequence to complete their cycle. This parasitic cycle is not complex in the ecological sense of Istock and Wilbur.

(a)

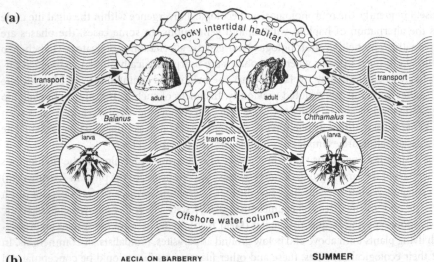

(b)

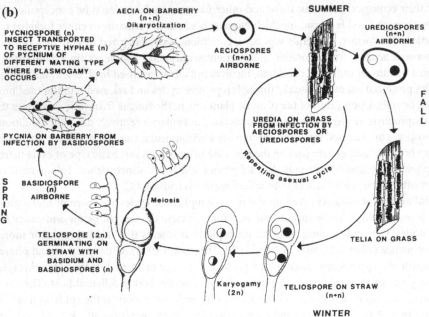

◻ **Fig. 6.4** The complex life cycle as it pertains to macroorganisms and microorganisms. **a** The tiny, motile larval phase of barnacles (e.g., *Balanus* and *Chthamalus*) lives in offshore waters; the sessile adult phase occupies the rocky intertidal zone. Figure redrawn, based on Roughgarden et al. (1988); barnacles reproduced from Darwin (1859). **b** The rust fungus *Puccinia graminis* alternates between the stems of grasses (cereals) and the leaves of barberry, with multiplication phases on each. The five spore stages clockwise from the bottom right-hand corner are the teliospore, basidiospore, pycniospore (spermatium), aeciospore, and urediospore (urediniospore) with associated nuclear stages shown as *n*, *n* + *n* (dikaryotic), and 2*n*. For contemporary terminology and life cycle details, see Webster and Weber (2007, Chap. 22). From Roelfs (1985). Reproduced from *The Cereal Rusts. Vol. II. Diseases, Distribution, Epidemiology and Control*, A.P. Roelfs and W.R. Bushnell (Eds.), by permission of Alan Roelfs (Univ. Minnesota) and Academic Press, ©1985

However, chiggers are the primary reservoir of the causal organism and the rickettsia need not pass through either rodents or humans to complete their life cycle. This rickettsial cycle is not complex in the ecological sense of Istock and Wilbur.

In contrast to typhus, the malarial parasite *Plasmodium* must pass through both the mosquito and human to complete its CLC, during which it undergoes alternation of generations and several morphological states. Similarly, the liver fluke (class Trematoda) has a CLC by virtue of obligately passing through snails and humans (or usually another vertebrate such as dogs, cats, pigs, or sheep) to complete its life cycle. In the process, the parasite passes sequentially from zygotes to ciliated larvae (miracidia), sporocysts, nonciliated larvae (rediae), tadpole-like larvae (cercariae), metacercariae, young and finally adult stages from which are produced the zygotes.

6.3.2 The Rust Fungi: Extreme Specialists with a Complex Life Cycle

These remarkable plant parasites (>7,000 species in the monophyletic order Pucciniales, Basidiomycota) illustrate nicely some general attributes of the CLC. As a group they are among the most highly evolved fungi and may have undergone as many as 300 million years of coevolution with their hosts since emerging from a common ancestor. From the inception of rust evolutionary studies in the early 1900s, this storied ancestor has broadly assumed to have been a parasite of tropical ferns (very early vascular plants, member of pteridophytes)[1] in the Carboniferous Period of the Paleozoic era (e.g., Savile 1955; Leppik 1959). From these hosts arguably the rusts diverged through geological time and changing climate to become parasites successively of gymnosperm trees (ancient and then more recent conifers in the Abietaceae, Cuppressaceae, and Taxaceae), then the arborescent angiosperms, and ultimately the modern, herbaceous angiosperms (Leppik 1959). Many if not most of the major host groups likely served as 'gene centers' or stepping stones in the further diversification of the rust clade on successor host taxa (Leppik 1959). In the case of the extant heteroecious grass rusts (discussed later), the herbaceous hosts originally were members of the Berberidaceae (barberry family), and ultimately that rust group extended to the sedges and grasses worldwide. Rust evolution as recounted in the many papers of Leppik and Savile was construed as being one of continual adaptation to phylogenetically younger hosts (parallel cladogenesis or co-speciation), with only occasional jumps geographically associated but taxonomically distinct hosts. Alternatively, however, it may have been primarily the latter (Hart 1988).

Significantly, the rusts appear to have been derived exclusively from parasitic ancestors (i.e., not secondarily from free-living forms as is believed to be the case for many parasitic animals), though this cannot be stated with assurance. Parasitism appears at the base of the Basidiomycota, including the subphylum Pucciniomycotina (which contains the rusts),

---

1   The origin of rusts is highly controversial and speculative. The so-called 'ancient fern hypothesis' advanced by several classical rust taxonomists evidently was based largely on the tenuous assumption that ancient hosts should harbor ancient parasites, in contrast to those found on evolutionarily younger hosts such as the gymnosperms and angiosperms. If correct, this fern association implies a vast geological time frame of coevolution between host and parasite. However, phylogenetic studies by Aime and colleagues (2006, 2014) show that the most ancient (basal) of the rusts in her phylogram of 46 species is *Caeoma torreyae*, originally found on California nutmeg (Bonar 1951), a member of the Taxaceae (i.e., the evolutionary much more recent yew family). Thus, the fern hypothesis has been substantially weakened but not yet entirely disproven by Aime's work, as well as by other molecular phylogenetic studies (see Sjamsuridzal et al. 1999; Wingfield et al. 2004), which are generally consistent with the earlier cladistic analysis of Hart (1988) that challenged the dogma. Hart's thorough work was based on an analysis on 28 morphological or character states. For further details, see footnote 2.

although the mycorrhizal basidiomycetes in the sister subphylum Agaromycotina have several independent origins from saprophytic ancestors (James et al. 2006; Kohler et al. 2015). Unlike most fungi, the rusts do not grow saprophytically in nature and thus are intimately coevolved with the living host. **This obligately biotrophic nature has set the stage for specialization and the most complex life cycle among fungi and probably in the entire living world.**

Let us now turn to the details of why this is the case for the rusts. During the course of their life cycle, the rusts as a group are distinctive in exhibiting up to five morphological stages (and correspondingly diverse nuclear, structural, and ecological states), each typified by a characteristic spore form (◘Fig. 6.4b). Moreover, these stages occur in a definite sequence and typically on two taxonomically unrelated plant hosts. In the basic life cycle a complete sequence of spore types—which characterizes a so-called **long-cycled** or **macrocyclic** rust—consists of teliospores, basidiospores, spermatia, aeciospores, and urediniospores (older terminology varies, see Chap. 22 in Webster and Weber 2007). So, unlike the complex life cycle of the frog, rust multiplication occurs at many points in its cycle. Apart from the basidiospores, which are borne naked on modified basidia extending from the germinating teliospores, all spore stages are produced within specialized structures (sori) called, respectively, spermogonia, aecia, uredinia, and telia. This basic macrocyclic life cycle is modified (abbreviated) in various ways. For example, species lacking the uredinial stage are termed **demicyclic**; those lacking aecia and uredinia are called **microcyclic or short-cycled**. Two significant consequences of a microcyclic life cycle are that many such rusts do not undergo normal sexual recombination (spermogonia are absent) and all such rusts are **autoecious** (basidiospores reinfect the same host species). In contrast, there is a conventional sexual cycle in macrocyclic rusts and these parasites can be autoecious or **heteroecious** (requiring two dissimilar hosts to complete the life cycle) (Terminology varies here as well and some authors consider microcyclic rusts to produce only teliospores and basidiospores; e.g., Ono 2002; Agrios 2005).

Puccinia graminis, the black stem rust fungus, is distributed worldwide and attacks hundreds of species of cereals and grasses in at least 54 genera (Leonard and Szabo 2005). Like almost all the other rusts, this species consists of highly specialized forms each attacking certain host genera and frequently only specific cultivars (at which level they are specific enough to be termed races; ►Chap. 2). For example, a specialized form, *P. graminis* f. sp. *tritici*, attacking wheat causes devastating epidemics attributable in part to the rapid evolution of races virulent only to specific lines of wheat (see gene-for-gene interactions in ►Chap. 3 and Singh et al. 2011). Not surprisingly, this pathogen is the focal point of rust investigations and is one of the most intensively studied of all organisms. The fungus is a macrocyclic, heteroecious species that forms uredinia and telia on the Gramineae and alternate stages only on various native *Berberis* (barberry) or rarely on *Mahonia* hosts. While most of the contemporary attention is on the graminaceous hosts because of their economic importance, in an evolutionary context alluded to above, barberry is the primary host and grasses the secondary hosts. The stem rust life cycle contrasts with other cycles represented in the *Puccinia* genus. For instance, *P. asparagi* (asparagus rust) is a macrocyclic species but autoecious, while *P. malvacearum* (hollyhock rust) is microcyclic. For *P. graminis*, or any other heteroecious rust to complete its life cycle, geographic overlap of the alternate hosts (or effective dispersal of appropriate infective forms) is obviously a necessity. Details of the life cycles of various rusts can be found in Bushnell and Roelfs (1984), Roelfs and Bushnell (1985), and Webster and Weber (2007). Here we focus on the evolution of complexity or simplicity in the life cycle.

The early dogma is confused as to the respective origins of macrocyclic, typically heteroecious, rusts and the microcyclic, autoecious, category (Jackson 1931; Leppik 1959, 1961).[2] Hart's cladistics analysis (1988) shows that heteroecism is well nested within his cladogram, i.e., heteroecism is a relatively recent, derived condition that evolved several times. There is no evidence that it is the primordial state. Based on the 30 rust genera examined, Hart concluded that several short cycle, mainly tropical and warm climate rusts are the most basal (i.e., 'primitive') as indicated below in footnote 2. This contradicts much of the early work cited above that argued the heteroecious long cycle is primitive, based largely on the faulty premise that since such cycles occur on supposedly ancient hosts they must also be the incipient form. (Of course, until Hart's cladistical treatment, the same 'logic' was used by others to argue the opposite case, that the microcycle was primitive.)

Heteroecism, perhaps having arisen fortuitously fairly early in rust evolution if not as the primordial state, was maintained arguably because susceptible tissue was scarce when basidiospores were being discharged. Inspection of the macrocycle suggests that heteroecism may have arisen from mating between rusts individually pathogenic to one of the two unrelated host species, with the resulting dikaryon able to infect both (▶Chap. 2 and Buss 1987, p. 158). Whatever its origin, the advent of host alternation must have carried considerable benefit because the increased distance of travel between two hosts involves higher costs in population mortality than travel between tissues of the same plant. (The practical implication of host alternation is that frequently an effective disease control strategy for heteroecious rusts is to eradicate the alternate host.) Heteroecism presumably fostered evolution of uredinia. The incipient uredinium would have provided for multicycles of urediniospores to offset population mortality in travel to a new host. Evolution of spermogonia and aecia would have facilitated outcrossing, nuclear transfer, and dikaryotization in a much more efficient manner than the former process of hyphal fusion from germinating basidiospores.

Just as environmental conditions perhaps selected for heteroecism and the associated life-cycle stages, so likely did the environment set the stage for evolution of secondary autoecism and the microcyclic rusts (Jackson 1931; Savile 1976; Anikster and Wahl 1979). How this occurred also has been actively debated, but there may have been an immediate conversion to the microcycle on the aecial host, where telia take the place of aecia on the alternate host of the ancestral rust. In fact, this pattern exists so consistently in nature that it has become known as Tranzschel's Law after the Russian mycologist who formalized the generalization. He inferred that when the alternate host for the aecial stage of a suspected heteroecious rust

---

2   Savile (e.g., 1953, 1955, 1971a, b, 1976) and Leppik (e.g., 1953, 1959, 1961), among others since the early 1900s, reconstruct the evolution of the rusts based on evidence derived from comparative morphology, host relationships, and biogeography. The synopsis in the following paragraphs is based mainly on their work together with the pioneering synthesis by Jackson (1931). A careful cladistics analysis by Hart (1988) reaches different conclusions as to the likely evolutionary history. As discussed in footnote 1, in the ensuing years the advent of molecular phylogenetics has refined the story (e.g., Maier et al. 2003, 2007; Aime 2006; Aime et al. 2006; van der Merwe 2007, 2008; Duplessis et al. 2011), in some cases (Sjamsuridzal et al. 1999; Wingfield et al. 2004; Aime 2006) undermining the early work and strengthening the arguments of Hart. Importantly, Hart's work showed that the most *primitive* (basal) clade he studied was tropical short-cycle rusts on *angiosperms*. Extant rusts on conifers and ferns highlighted by the early workers were actually a nested *terminal (i.e., most recent)* clade. Thus, the larger picture evolutionary syntheses characteristic of Savile and Leppik, now considerably dated, are probably best regarded as hypotheses given the rudimentary fossil evidence and constraints in dating fungal divergences (e.g., Lücking et al. 2009; Berbee and Taylor 2010; see Thompson's [1994] comments on difficulties of reconstructing phylogenetic events after the fact, pp. 60–63).

is unknown it should be sought on plant species infected by microcyclic rusts with morphologically similar teliospores to the heteroecious rust (Shattock and Preece 2000; Webster and Weber 2007, Chap. 22). Of course, the search will be fruitless if the alternate host is extinct. And, this raises the important caveat of Hart (1988): while many rust researchers have doggedly assumed that the absence of stages is explicable simply by ignorance of the 'missing' host, in reality these taxa may be primitively short cycle.

Whatever the contested primacy of the microcycle versus macrocycle in geological time, the adaptive benefit of a microcycle is strikingly apparent among extant species in arctic, alpine, or desert areas. Here the growing season for the host and hence the parasite may be only a few weeks (Savile 1953, 1971a,b, 1976). Convergent similarities among rust species include not only autoecism and short-cycling, but also self-fertility and suppression of spermogonia. Savile (1953, 1976) contrasts the life cycles of heteroecious rusts along a latitudinal gradient from temperate zones to the arctic. In the former, most of the species have a macrocycle similar to that described above for the temperate *Puccinia*. Numerous uredinial generations occur and the rusts can persist with the alternate hosts relatively far (many meters or more) apart. As one moves northward, time available for the repeating stage decreases progressively; near the tree line the identities of the plant species involved become obvious because the rust cannot survive unless the alternate hosts are within about 50 cm of each other. Just inside the tree line the plants must be contiguous. Beyond the tree line the only heteroecious rusts are those that can abbreviate the life cycle, for example, by self-fertility or by producing dispersible (diasporic) rather than sessile teliospores, or by 'double-tracking' (Savile 1953, 1976) the life cycle: *Chrysomyxa*, for example, overwinters as a dikaryotic perennating mycelium rather than as a telium in the leaves of its evergreen host, *Ledum*. In the spring this mycelium produces uredinia and telia nearly simultaneously. These sori mature to produce their respective spore forms. The urediniospores infect *Ledum*; the teliospores germinate in situ to produce air-borne basidiospores that infect the alternate host *Picea*. The life cycle thus is abbreviated because gene recombination (teliospore/basidiospore phase) and dispersal/multiplication (urediniospore phase) proceed concurrently. (For an analogous discussion of environmental constraints as they influence amphibian life cycles, see Wilbur and Collins [1973]. It is notable that the life cycles in closely related species of sea urchins and relatives may run the full gamut of stages or involve loss of larval stages and accelerated development [Raff 1987]. Moran and Whitham [1988] discuss analogous life cycle reduction in the aphid *Pemphigus*.)

So, what do the welter of details and the diversity of life cycles among the rusts tell us about evolution of CLCs? First, by their adaptive adjustments in the cycle, these fungi exemplify the postulate that natural selection acts to adjust the length of time spent in a particular life stage to maximize lifetime reproductive success. 'Telescoping' of the rust life cycle by the dropping of one or more stages is analogous to progenesis, a term used in developmental biology to refer to abbreviated ontogeny by accelerated (precocious) sexual maturation. Gould (1977: Chap. 9) reviews some of the many examples among animals, including animal parasites (e.g., the cestodes and certain copepods; see also Moran and Whitham 1988). Thus, there are numerous instances where a parasitic species is progenotic with respect to free-living relatives or related parasites. This has been accomplished apparently by the adoption of a simplified life cycle from what was originally a CLC by deletion of the alternate host. Why heteroecism is *retained* seems a more challenging question. It has been traditionally interpreted adaptively, though retention may reflect constraint; this aspect is considered further in the next section.

Second, what we see in the complex cycles of the rust fungi may reflect an historical legacy—an extreme means to regulate nuclear access to the germ line. As we saw earlier, animals, plants, and fungi all accomplished this regulation differently. The rust fungi appear to have done so in particularly extreme fashion. An interesting and imaginative interpretation has been advanced by Buss (1987 p. 158) and is worth quoting at length below (emphasis added):

» *The forces which spawned the unusual life cycles of fungi are the forces which challenged metazoans to develop elaborate mechanisms of historecognition. Competition within organisms for access to the germ line arises not only through mutation, but also through fusion between conspecifics. The potential consequences of fusion for integrity of the individual varied as a function of the primitive conditions of the plant, animal, and fungal kingdoms with respect to cell membrane architecture and cellularization. Plants were not challenged. Those animal groups that were challenged responded by evolving sensitive mechanisms of self-recognition.* **The fungi turned the challenge to their own advantage inventing from it novel life cycles. A life cycle trait, fusibility, challenged the different clades differently, producing a phylogenetic restriction in those groups which allow fusion. One group, the fungi, responded with a unique developmental innovation which subsequently fueled the evolution of complex, idiosyncratic life cycles."** [emphasis added]

6.3.3 Optimization and Constraint in Molding Life Cycles

The CLC has been explained in terms of both adaptation and developmental constraint (Moran 1994; Chap. 9 in Raff 1996). According to the former, it has arisen and is maintained as a means of accommodating the conflicting needs of competitive growth at a fixed site or time with genetic exchange and dispersal of offspring to new sites or to better conditions at a later time. According to the latter, it is a developmental constraint, nothing more than a historical legacy, retained because there is no alternative ontogenetic pathway.

The adaptive interpretation The CLC allows important activities to be segregated by specialized stage, the organism moving progressively from one functional compartment into another, broadly speaking by metamorphosis. In this manner it can exploit transient opportunities (Wilbur 1980), efficiently executing one task rather than compromising among many as do organisms with a SLC. As we have seen in the multigenerational type of CLC typified by the rusts, the uredinium and its associated spores evolved as the long-range, repeating, dispersal stage; the spermogonium is specialized for nuclear transfer and fertilization; and the telium is the main resistant (overwintering) stage associated with survival and karyogamy. Fleeting opportunities—for example, in the form of an accessible alternate host at precisely the appropriate stage of susceptibility and in an environment conducive for infection—are clearly just as apparent for parasites with CLCs as they are for, say, anuran larvae in resource-rich but temporary ponds, or insect larvae that feed on carrion or seeds.

It is apparent that where a CLC includes heteroecy, host shifting has usually been interpreted as currently adaptive. Because organisms with CLCs exhibit 'decoupling' or compartmentation of stages, the opportunity arises for independent evolution in the different phases. The extent to which this may occur is contested (Istock 1967; Strathmann 1974; Bonner 1982b, pp. 226–229; Petranka 2007). Based on a life table model, Istock (1967) proposed that the adaptiveness of the two stages of a hypothetical organism with a larval and an adult form is largely independent. He suggested further that it is this independence that usually makes CLCs unstable over evolutionary time because the flow rates of individuals between the phases are unbalanced (e.g., flow of larvae from the larval to the adult environment; flow of progeny from the adult to the larval environment). In Istock's CLC model, the ecological advantages are fully exploited only where both the larval and adult compartments are demographically saturated (Istock 1967). Even if achieved in nature, he notes that this condition is very difficult to maintain over evolutionary time, leading eventually to reduction or loss of one of the stages. Strathmann (1974), on the other hand, using marine invertebrates as an example, emphasized that evolutionary feedback acts in regulatory fashion to integrate the various phases. Interaction between the two phases is described for marine intertidal invertebrates by Roughgarden et al. (1988).

In the microbial context, it would be interesting to know to what extent the various stages (such as the oversummering parasitic phase in contrast to the overwintering, saprophytic phase) in plant pathogen life cycles can be modified independently or whether there is close overall evolutionary coordination. (This is roughly analogous to the cycle of a songbird that spends its summers in Canada or the northern U.S. and winters in the Caribbean or Mexico, adjusting to the markedly different habitat in each place.) In other words, how plastic is the overall life cycle? Regarding the rusts in particular, evolution of the CLC and particularly host alternation arguably was one mechanism whereby certain fungal parasites were able to advance from their early hosts, possibly in the primeval swamp, into otherwise inhospitable environments represented in part by new host taxa. As noted earlier, the extraordinary complexity of the rust cycle is unique; if assessed on its prevalence among taxa, it must therefore be judged a relatively *unsuccessful* evolutionary experiment. Nevertheless, the rusts persist, just as do the amphibians with their CLC.

The developmental constraint interpretation The constraint hypothesis emphasized by developmental biologists (see Moran 1994; Chap. 9 in Raff 1996) interprets CLCs as persisting because the stages are inherent parts of a developmental program largely beyond the reach of natural selection. Because of linkage among traits, modification of a stage may be restricted or precluded without adverse impact on others. Moran (1988) reviews the evidence for aphids that switch hosts. She concludes that the cycle represents a historical constraint, being an evolutionary dead-end rather than a beneficial adaptation to optimize resource use. For such aphids, according to Moran, host alternation is a relict, maintained because a particular aphid morph (the fundatrix) became 'over-specialized' to the ancestral (woody) host, precluding the cycle moving in its entirety to the herbaceous host. In this case sequential specialization may be a trait difficult to jettison.

But, what of the rusts? Clearly many species have shown evolutionary plasticity in eliminating the alternate host and in some cases adopting a sharply abbreviated cycle in some populations but not in others (as in arctic vs. temperate latitudes, for example). As such they appear to be more malleable than do the aphids. Actually, some aphids evidently are more malleable than assumed by Moran and the adaptiveness of the CLC may be lineage-specific (Hardy et al. 2015). Have the rusts become too specialized? Do they exemplify Harper's (1982) observation that, far from representing the perfect match between organism and environment, partners in a coevolutionary spiral may drive each other into "ever deepening ruts of specialization"? (see also ►Chap. 3 and Huffaker 1964). Coevolutionary specialists do not necessarily endure such fates and a march from generalist to specialist is not supported by the broad phylogeny of specialization (Chap. 4 in Thompson 1994), ... "*nor is there any reason to expect that specialization is an evolutionary dead end from which species cannot escape, or that specialists, as a rule, show sustained phylogenetic tracking of host taxa*" (Thompson 1994, p. 75).

Overall, the complexities of CLCs, at least among parasites, are probably best interpreted in a situation-specific manner reflecting both constraint and natural selection: constraint on the life histories that lead to host alternation; on the adaptive elimination of a host after it has become a part of the life cycle; how that host is used in the relationship; and selection pressure for retention (Chap. 6 in Thompson 1994).

6.4 Senescence

» *It's good to be immortal.*
New York Times Magazine article headline, June 2008

We consider the issue of senescence in the life cycle context for a couple of reasons. First, while zygote-to-zygote cycles continue indefinitely, senescence and death are an inevitable end stage in the life cycle for the individual. Or are they? That in part philosophical question is the second reason for our attention: Is it possible that microorganisms as opposed to macroorganisms escape senescence? Is any organism that grows as a clone 'immortal' or 'potentially immortal' as is routinely claimed and does this descriptor even have any useful biological meaning? Why has senescence evolved?

At the level of the individual, **senescence is generally defined as a decline in physiological function with age manifested by an increasing probability of death and declining fecundity** (Finch 1990; Rose 1991; Ricklefs 2008). It has been argued persuasively that *both* attributes need to be considered (though commonly they are not) because they are interdependent and selection acts on the product of survival and fecundity (Partridge and Barton 1996; Pedersen 1999).

What is termed senescence of individuals in botany is often called aging in medicine and zoology, although usage varies and some authors (e.g., Stearns 1992; Baudisch 2008, 2011) do not use the terms synonymously. Clearly, the many physiological changes that accompany increasing age span the gamut from being innocuous correlates (such as changes in color and density of body hair in humans); to changes rendering the individual increasingly vulnerable to extrinsic sources of mortality (e.g., decrease in auditory or visual acuity, reaction time, flexion, endurance, and bone strength); and ultimately to those that are pathophysiological and potentially lethal in themselves (e.g., cancer) (◘Fig. 6.5 and Williams 1999). Although senescence is popularly assumed to depict the senility and frailty accompanying old age, in an evolutionary context it is classically taken to begin much earlier, at about the time of sexual maturity in humans and some animals. Of course, in actuarial terms, increasing death rates may not become striking until later in life. The reasons for this are discussed later when we take up the evolutionary aspects of senescence.

6.4.1 Some Demographic Considerations

Senescence *At the population level, senescence is quantifiable within a cohort as an increasing mortality (decreasing survivorship) rate with age as physiological decline takes its toll.* The factors responsible have been categorized broadly as inherent (*intrinsic*), i.e., attributable to disability and disease (e.g., cancer, stroke, cardiovascular malfunction), or as increased vulnerability to mortality from **extrinsic** sources (e.g., accidents; extreme weather events). The resulting hypothetical plots (◘Figs. 6.6 and 6.7) with more-or-less extreme variation, are descriptive of the process in perhaps most multicellular animals (indeed of unitary organisms generally) and, importantly, are the form on which the classic theories of senescence are based.

Non-senescence *If the mortality rate is constant with age*, however, as noted in one case in both ◘Figs. 6.6 and 6.7, *the number of survivors declines exponentially with time.* Obviously this may occur rapidly or very slowly depending on the coefficient of the exponent. The demise of glass tumblers in a cafeteria (Brown and Flood 1947) or Medawar's famous test tube analogy, described later, follows an approximate non-senescent decline as do some organisms in nature. Such plots have given rise to the ambiguous adjective '**potentially immortal**', also discussed later.) In other semantics, Baudisch and colleagues (e.g., Baudisch 2008; Vaupel et al. 2004; Baudisch et al. 2013) refer to organisms like *Hydra* as having '**negligible senescence**' (cf. 'negative senescence', below and G. Bell 1984; Chap. 4 in Finch 1990; Pedersen 1999; Finch 2009; Munné-Bosch 2015). Non-senescent curves also could, in principle, imply simply that populations in the wild are pruned by high mortality early in life before senescent decline can appear significantly in actuarial terms. Therefore, such plots may not necessarily be evidence against intrinsic senescence of the individual, simply that non-senescence cannot necessarily be inferred from the demographic data.[3] However, the notion that wild animals do not survive long enough to show senescence, once dogma, appears to be largely discounted at least probably for most animals by relatively recent research (Gaillard et al. 1994; Nussey et al. 2013; plants do survive long enough for senescence to appear in at least some species, discussed later; see Roach 1993; Munné-Bosch 2015). It is noteworthy and counterintuitive that short-lived organisms (by our standards as humans), such as many small birds, may still experience senescence and, conversely, that a constant rate of mortality (if low enough) can still result in very old members of a cohort. Thus, **non-senescence does not imply longevity and the occurrence of very long-lived exemplars of various taxa** (▶Chap. 5 and this section) **is not, of itself, evidence for absence of senescence** (◘Fig. 6.6 and Pedersen 1999). As Caswell (1985) noted, **the alternative to senescence is not immortality but a mortality rate that does not increase with time. Indeed there is considerable confusion in the literature between longevity and senescence.** The ramifications are discussed later when we take up the evolutionary theories of senescence.

Negative senescence Finally, for many organisms there may be a generally *increasing expectation of life over time* (Pearl 1940; Haldane 1953; Chap. 1 in Comfort 1979; Jones et al. 2014) *evidenced by declining mortality rates and increasing reproductive output with age.* This pattern is now popularly referred to as '**negative senescence**' (Munné-Bosch 2015). Apparently this is the case for many plants and possibly certain animals (among unitary organisms this may possibly include some reptiles, fish, and amphibians; however, see Warner et al. 2016 on turtles). Herbaceous perennials and many shrubs and trees appear to belong to this group or the 'negligible senescence' category above (Harper 1977; Watkinson and

---

3   To the extent to which increasing age-related mortality reflects primarily intrinsic physiological decline as opposed to increasing vulnerability to extrinsic mortality factors, the two sources have been controlled as an incidental effect of raising animals in zoos (though such 'experiments' have their own shortcomings; see discussion in Ricklefs 2008). These manipulations have shown reduced mortality for some protected species compared to their counterparts in the wild, suggesting that older individuals are indeed more vulnerable to extrinsic sources of mortality; there are also cases of similar patterns of mortality in zoos and in the wild, implicating primarily intrinsic factors for senescence (Ricklefs and Scheuerlein 2002; Ricklefs 2008).

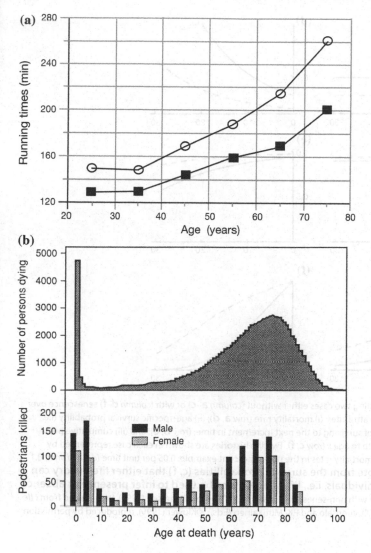

■ **Fig. 6.5** Manifestations of senescence: **a** Declining functional capacity with age: the mean running times of the top 10 male (*black boxes*) and female (*open circles*) athletes in 65 marathons. From Leyk et al. (2007). Reproduced from *International Journal of Sports Medicine* by permission of Thieme Publishers (▶www.thieme. com) ©2007. The distinguished evolutionary biologist G.C. Williams presented similar data (1999) for a single individual, presumably himself, from age 52 to 63, as his capability declined. **b** Distribution of deaths from all causes (*top*) is similar to pedestrian deaths in road accidents (*bottom*), showing a peak at the outset of life and increasing with advanced age. Redrawn from Comfort (1979) based on data of DeSilva (1938) and Lauer (1952). Reprinted from *The Biology of Senescence, 3rd Ed.* by A. Comfort. Reproduced by permission of Elsevier Science Publishing Co, Inc. ©1979

White 1985; Petit and Hampe 2006; Baudisch et al. 2013; Munné-Bosch 2015). Indeed, this pattern appears to typify many modular organisms (▶Chap. 5), in particular those that are clonal. The demographic theory for plants is more complex and unlike that for unitary animals primarily because of differences in their structure, developmental biology, reproduction,

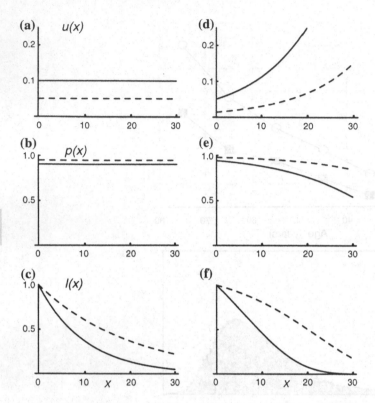

■ **Fig. 6.6**   Models contrasting two cases either without (*column* **a–c**) or with (*column* **d–f**) senescence over time characterized by three attributes: (i) mortality rate (*row* **a, d**); (ii) age-specific survival probability (i.e., probability at a given time of surviving to the next increment in time; (*row* **b, e**); and (iii) cumulative probability of surviving from birth to age *x* (*row* **c, f**). Two life histories are shown in each case represented by two different but constant mortality rates in the non-senescent example: 0.05 per unit time (*dashed line*); 0.1 per unit time (*solid line*). **Note from the survival probabilities (c, f) that either life history can result in long-lived individuals, i.e., life span cannot be used to infer presence or absence of senescence.** Compare with non-senescence in ■Fig. 6.10. From Pedersen (1999). Reproduced from *Life History Evolution in Plants*, T.O. Vuorisalo and P.K. Mutikainen (Eds.). ©Kluwer (1999), reproduced by permission of Springer ©2001

and environmental relationships (see section below on macroorganisms and Harper 1977; Caswell 1985; Roach 1993; Roach and Carey 2014). Their life events are frequently more size- and stage-dependent rather than age-dependent. *A constant or declining mortality rate would be anticipated in general for modular clonal organisms at the level of the genet whether animal, plant, or microbe.* These points and the evidence are examined in the following sections.

Mathematical models on demography Various mathematical functions have been used to interpret actuarial senescence, the most prominent of which are the Gompertz and the Weibull. These differ primarily in the manner in which age-independent (often called extrinsic) and age-dependent (intrinsic) death events are handled (Ricklefs and Scheuerlein 2002; Ricklefs 2008), though these sources are not as cleanly separated as may be implied. Both models begin with an assumed basal or initial mortality rate common to relatively young, healthy members of a population before senescent decline begins. As noted above, such mortality events are viewed as being essentially accidents occurring independently of age caused by such factors as predation, starvation, and adverse weather. In the Gompertz, the relative physiological decline increases and the mortality rate increases exponentially as: $m_x = m_o e \gamma^x$ implying that senescence is a *multiple* of initial mortality. The constant γ is the exponential rate of increase in mortality rate with age; m_x (mortality rate at age x) and m_o (initial mortality rate) are instantaneous rates and expressed as time^{-1}. The mortality rate can range between 0 and infinity (in both models).

In contrast, in the Weibull the age-dependent component is *added* to the initial mortality rate as: $m_x = m_o + \alpha x \beta$ implying that the causes of death in a young population are different and independent from those affecting older members. The initial or extrinsic factors are the same as in the Gompertz; this rate is supplemented by a separate category (represented by the term $\alpha x \beta$) of age-dependent deaths due to intrinsic dysfunction and disease (cancer, strokes, etc.), as well as deaths from increased vulnerability to the extrinsic factors by virtue of the disabilities of old age (ineffectual escape from predators, etc.; see examples above). Thus, here the relative rate of increase in mortality rate is age-dependent and slows with increasing age (for details on both models, see Ricklefs and Scheuerlein 2002).

These Gompertz and Weibull are examples of relatively simple models and have their proponents and critics. They need to be interpreted in the context of the usual simplifications, caveats, and assumptions inherent in model building (see e.g., Levins 1966; Chap. 1 in Levins 1968; Abrams and Ludwig 1995; Pedersen 1999; Williams 1999).[4] **Clearly, these models as applied to senescence are being used to represent the demography of discrete, unitary individuals such as humans and birds. They do not lend themselves well to the population biology of clonal or modular organisms** where what constitutes 'the individual' is more-or-less abstract and where it is operationally difficult or impossible to separate parents from offspring. *Furthermore, the models focus on survival and ignore the fecundity implications of senescence.*

In summary, whereas the only complete actuarial data are for humans, substantial demographic evidence exists for certain other higher organisms such as zoo and domestic animals, many annual and some perennial plants (such as a few of the shorter lived trees), and for selected species used in research. These cases are discussed in the following section. Elsewhere, data are few and conclusions have been drawn largely by extrapolation and inference, occasionally based only on age of an organism at death. Life span is not the same as senescence. Although some actuarial evidence can be compiled from existing records or by inference from mortality tallied across all age groups at one time (Comfort 1979 p. 53), the best method would be, where possible, to follow over time survivorship and fecundity of a particular age cohorts under controlled and uncontrolled conditions (however, see Rose 1991, pp. 21–28 on laboratory artifacts). Usually this is technically difficult and sometimes operationally impossible to do. As will be apparent from the following sections, survivorship and fecundity over time are tracked in some plant studies, but, with notable exceptions (e.g., Gustafsson and Pärt 1990), rarely for animal populations (or microbial clones) in nature. Variation among and within taxa in the occurrence of senescence is considerable, suggesting genetic variation plays a significant role.

4　Here and elsewhere in our discussion of models in this text, the famous and succinct words of Box and Draper (1987, p. 424) should be kept in mind ... **"Essentially, all models are wrong, but some are useful".**

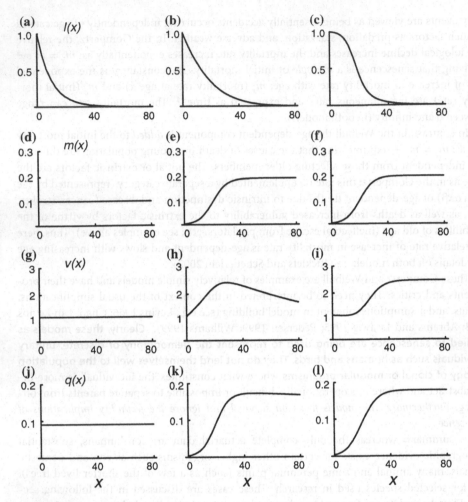

Fig. 6.7 Models illustrating demographic aspects of senescence as a function of time with respect to four attributes: (i) probability of survival from birth to age *x* (*top row* of figs. **a–c**); (ii) fecundity at age *x* (*second row*, **d–f**); (iii) reproductive value (*third row*, **g–i**); and (iv) the product of age-specific survival probability and fecundity (figures **j–l**). Three life histories are shown: the *first column* (figures **a, d, g, j**) depicts constant death and fecundity; the *second column* (**b, e, h, k**) shows increasing mortality according to the Gompertz model (see text); the third column (**c, f, i, l**) shows the case of constant mortality at advanced age. **Note that if senescence is defined based on reproductive value, only the second life history depicted in b, e, h, k shows senescence**. From Pedersen (1999). Reproduced from *Life History Evolution in Plants*, T.O. Vuorisalo and P.K. Mutikainen (Eds.). ©Kluwer (1999), reproduced by permission of Springer ©2001

6.4.2 Evidence for Senescence Among Macroorganisms

Animals It appears that most if not all animal taxa that reproduce exclusively or primarily sexually senesce (the so-called ovigerous or egg-bearing organisms; Bell 1984; Finch 1990; Ricklefs 2008). This includes the vertebrates and many invertebrates such as the nematodes, rotifers, crustaceans, and insects (unitary organisms in general, ▶Chap. 5). According to Rose

(1991, p. 86) *"No one has yet found a vertebrate that does not senesce under laboratory conditions when the relevant demographic parameters are measured."*

Interestingly, non-tumorigenic, somatic cells from various animal species (most evidence is from human fibroblasts) grown in vitro have a finite divisional or doubling life span (the famous 'Hayflick limit') that correlates loosely in a relative sense with senescence in vivo (Hayflick 1965; Fridman and Tainsky 2008). This appears attributable at least in part to progressive loss of DNA from the specialized termini (telomeres) of eukaryotic chromosomes (see non-evolutionary senescence hypotheses, later). Limited but intriguing evidence suggests that cells undergo fewer doublings when taken from older as opposed to younger organisms, as they do from organisms with shorter as opposed to longer life spans (Goldstein 1974; Röhme 1981). Nevertheless, this correlation is controversial (Rubin 2002) and allegedly 'neither quantitative nor direct' (Campisi 2001).

Among invertebrates where clonal growth (asexual reproduction) is prominent in the life cycle, data are either strongly against or equivocal for senescence, at the clonal lineage (genetic individual level; see section Clonal Organisms, below). The strength of the case against senescence in animals appears to depend largely on whether the products of clonal division are essentially morphologically indistinguishable (binary fission) or not. As will become evident later, where a somatic lineage cannot be distinguished from a germ line, or parent from offspring, senescence should not evolve (the classic example is bacteria; among animals the closest parallel would be among certain protozoa; both are discussed later). Slightly more complicated are cases where division results in products relatively equal in size and requiring only limited development to reach reproductive maturity (**paratomical fission**; examples include sea anemones, hydroids, planarians, and various oligochaetes). The next step in the gradient is where the asexual offspring are markedly smaller and much less differentiated than the parent, requiring development before becoming an adult (**architomical fission**; examples include sponges and ribbon worms). This progression implies that senescence should accompany **ovigerous reproduction** (i.e., where fertilization of eggs is involved), but be relatively negligible for architomical taxa, and absent in paratomical organisms (Bell 1984). Bell provided limited evidence from the culture of six freshwater invertebrates that survival decreased significantly with age in the ovigerous animals (two rotifers plus an ostracod and a cladoceran), but did not change for the two paratomical oligochaetes. This is consistent with predictions. However, to differentiate clearly in general between paratomical and architomical categories seems somewhat subjective and arbitrary; it may also be the case that having a soma, however rudimentary, may be sufficient to trigger senescence (Buss 1987). Why the early developmental segregation of germ from soma appears to doom an organism to senescence is not clear. A further complication is that the frequency, timing, and overall importance of asexual reproduction in the life cycle can be expected to vary within a particular taxon depending on the environment (e.g., for marine invertebrates see Hughes and Cancino 1985; Hughes 1989).

The clonal animal, like all clonal organisms, grows indeterminately, in other words without definite restrictions or limits. The ultimate size of the **clone** is potentially immense as has been discussed previously (▶Chap. 5). **It is the fate of the clone as the genetic individual that is of evolutionary significance. Here, as is the case with microorganisms discussed below, 'the individual' (implying a physiologically independent individual or ramet) is often confounded with the clone.** For example, following on from Bell's study, Martinez and Levinton (1992) ostensibly report senescence in the oligochaete *Paranais litoralis*, which apparently reproduces almost exclusively in nature by asexual fission. However, the survivorship that they studied was of individual worms, *not* clones. They also reassess the data of Sonneborn (1930) on survival of the flatworm *Stenostomum*, where again the focus is on individual, *not* clonal, survivorship. These and related comments are expanded later, where clonal organisms are discussed as a group.

Plants Harper's seminal work on the population biology of plants (e.g., Harper and White 1974; Harper 1977, 1981a) set the stage for all subsequent demographic studies and interpretations of senescence. In pointing out the unique attributes of plants (see also Roach 1993; Munné-Bosch 2015), he questioned Hamilton's conclusion (1966, p. 12) that "senescence is an inevitable outcome of selection" (see extensive discussion in Evolutionary Theories section, later). Among other things, Harper noted that the reproductive activity of plants can be dictated more by population density than by age of an individual; thus, the age-structured demographic tables so important in animal demography (e.g., Charlesworth 1980, 1994; Caswell and Salguero-Gómez 2013) have limited utility for plants. In general, however, being an older plant means being either the successful larger clone or a bigger tree dominating a canopy. The difference in seed production between large and small plants (of the same species) may be several orders of magnitude, with a small proportion of plants thus making a grossly disproportionate contribution to fecundity (Levin 1978). While precocious reproduction contributes to reproductive value and is thus favored by selection in animals, it may be so in plants but there are cases where this does not apply and what is important is massive reproduction in later years or just before death (see following comments for trees and semelparous plants). While the old (unitary) animal usually contributes relatively little or nothing directly to population increase, "*the oldest individuals in a plant population may have the greatest reproductive output, control the largest fraction of the resources, and control the recruitment of new seedlings*" (Harper 1977, p. 700). Thus, it is not surprising that evidence on the occurrence of senescence in plants is mixed and is influenced significantly by factors such as the architecture and life history of a particular species.

The senescence of determinate annual plants after fruit maturation has long been recognized though until recently poorly understood mechanistically (Noodén and Penney 2001; Noodén 2004; Wuest et al. 2016). After a variable period of vegetative growth these plants typically undergo one round of reproduction followed by death within a season, that is, they are monocarpic or semelparous in zoological semantics. Their fecundity rises steeply from zero to a peak and then drops precipitously (◻Fig. 6.8a). In contrast, the reproductive schedule of almost all perennials is polycarpic (iteroparous; i.e., they undergo multiple rounds of reproduction) because the genet can continue to form new meristems, which themselves reproduce, and so on. The fecundity schedule depends on whether the perennial has a single meristem or multiple apical meristems. In the former case, as represented by the coconut palm *Cocos nucifera*, seed production can be expected to rise and then level off for an indefinite period (◻Fig. 6.8b) (Watkinson and White 1985). Where there are multiple meristems, the outcome depends on whether the plant functions as one physiological individual (aclonal) or several (clonal). In the former case fecundity increases more steeply than for plants with a single meristem; in some systems this increase may be curtailed (Watkinson and White 1985), but most commonly fecundity increases exponentially with size (possibly also with age) as is the case for the clonal plants (◻Fig. 6.8c, d). Clonal (and to a large degree aclonal perennial) plants are in this sense directly analogous to the clonal marine invertebrates discussed above, and in distinct contrast to the fecundity patterns of unitary organisms that rise early to a maximum and then gradually decline (Charlesworth 1980; Finch 1990). Similarly, unlike the Hayflick limit, which determines the number of divisions of cultured somatic animal cells noted above, cells taken from plants continue to grow indefinitely in vitro provided that they are subcultured at regular intervals and that growth conditions remain adequate (Murashige 1974; d'Amato 1985). Plant cells are also totipotent (provided that they retain a nucleus and living protoplast), that is, each retains the full genetic/developmental potential of the organism and differentiation is entirely reversible.

The implication of the foregoing is that survivorship rates of genets of semelparous plants are qualitatively similar to those of many unitary organisms (Watkinson and White 1985). In contrast, the indefinite and in general exceptionally long-lived nature of iteroparous clonal and at least many aclonal plants is akin to that of clonal marine organisms. Evidence for such plants is mixed for occurrence of senescence (summarized by Munné-Bosch 2015). Silvertown et al. (2001) examined 65 species from 24 families of herbs and 12 families of woody plants. Senescence rates, unlike the case for many animals, were independent of the initial mortality rate so the authors used the Weibull function instead of the Gompertz to fit their data. Some 55% of the plants showed an increase in or maximal value of age-specific mortality late in life that they attribute either to inherent dysfunction or environmental factors, notwithstanding the fact that 61 of the 65 species showed an increasing trend in fecundity. However some caution is warranted in interpreting the synopses. For example, the data for several of the herbs and tree species with increasing age-specific mortality late in life pertained to ramets and not genets. Also, in broad trends, most of the herbs and almost all of the woody species showed a pattern of either asymptotic decrease with time, or a minimum death rate between the youngest and oldest ages. In the latter case, the pattern of higher mortality in latter stages of life was dominated by the longer lived woody plants (>100 years; see Watkinson and White's Fig.1). The authors speculate that the short-lived iteroparous perennials die from extrinsic (environmental) causes and the long-lived perennials from physiological deterioration later in life. They conclude that a clonal growth habit is necessary (but insufficient) to prevent the evolution of senescence and that fragmentation of a clone, rather than retention of the clone as an intact unit, is the key determinant. In a fragmented clone, any later disadvantageous trade-offs associated with advantageous early growth or reproduction would tend to be associated with the fragment, i.e., be ramet-directed mortality rather than clonal mortality) rather than the clone as a whole (Silvertown et al. 2001; see later discussion of disposable soma theory).

The work by Silvertown et al. (2001) was extended by Baudisch and colleagues (2011, 2013) in a large study of senescence in 290 angiosperm perennials spanning the gamut in architecture and global distribution (multiple ecoregions). They consider in terms of mortality metrics both patterns in pace and shape. *Pace* refers here to speed of life measured by life expectancy and categorized as fast, moderate, or slow. This results in short, moderate, or long life, respectively. *Shape* refers here also to three categories: increasing mortality, designating senescence; decreasing, designating negative senescence; or constant, designating negligible senescence). In principle, species showing any pace of life can exhibit any shape of age-dependent mortality. The five plant architectures were: (i) cryptophytes (shoot meristems belowground); (ii) hemicryptophytes (shoot meristems near the surface; typically rosette-type plants); (iii) chamaephytes (shoot meristems <25 cm aboveground); (iv) phanerophytes (shoot meristems >25 cm aboveground, typically shrubs and trees); and (v) epiphytes (height determined by their position on the plant, frequently on branches). Pace was related to form, with woody species typically living longer, whereas shape was mainly a function of phylogenetic relatedness. Senescence was observed only among the trees but even here, in the cate-

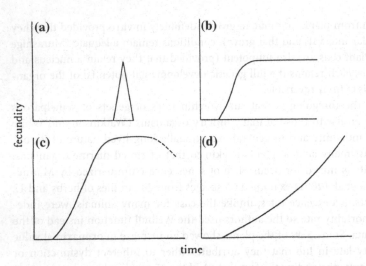

◨ **Fig. 6.8** Reproductive schedules (relative fecundity over time) approximated for **a** semelparous (mono-carpic) plants (i.e., those with one round of reproduction, typically annuals); **b** iteroparous (polycarpic) plants (those with multiple rounds of reproduction, typically perennials) with a single shoot; or with multiple shoots that are **c** aclonal, or **d** clonal. From Watkinson and White (1985); reproduced from *Philosophical Transactions of the Royal Society of London B* by permission of A.R. Watkinson and The Royal Society, ©1985

gory of phanerophytes, 81% showed negligible or negative senescence. *In all, 93% of the plant species did not demonstrate senescence.*

The case for trees, broadly speaking, is complicated, with various authors arguing for senescence or absence of senescence. In general, it appears that most species do not senesce, at least not from intrinsic causes the way animals do, but the paucity of long-term demographic studies hampers interpretations, especially for long-lived (hundreds of years) trees (see non-evolutionary theories, below and Roach 1993; Pedersen 1999; Larson 2001; Mencuccini et al. 2005; Petit and Hampe 2006; Stephenson et al. 2014). Even where large and ancient trees become moribund or break in storms, they frequently resprout from adventitious buds (see the extraordinary case for coastal redwoods described in ►Chap. 7) so life of the genetic individual continues. Meristems (►Chap. 5) are the 'life-blood' of juvenility for plants (see ►Chap. 5 and Bäurle and Laux 2003; Heidstra and Sabatini 2014; Klimešová et al. 2015; Morales and Munné-Bosch 2015).

Clonal organisms The early interpretations of senescence and most of the subsequent dogma are based on populations of age-structured, unitary, aclonal animals, i.e., situations in which an individual is obvious. The complications that arise where multiple and in many cases separate, physiologically independent subunits (modules operating at the level of ramets) constitute in aggregate one genetic entity (genet) warrant special comment (cf. ►Chap. 5). The clone is thus the uppermost organizational level in a hierarchy of units where one or multiple types of asexual reproduction at any lower level constitutes growth of the clone (whether intact or disaggregated), whereas sexual reproduction produces new clones (genets). As discussed in ►Chap. 5, clones arise in various ways such as by polyembryony, or by budding, fission, or fragmentation of the parental unit. As discussed previously (►Chaps. 2 and 5), most microbes are clonal, as are many plants (e.g., buttercup, clover) and animals (e.g., corals, anemones). *While it is clear that senescence in all such organisms occurs at the* *level of the* ramet, *the key issue is whether it occurs also at the level of the genet.* One must also be cautious in interpreting the literature as some authors take 'the individual' with respect to senescence to mean the physiologically independent ramet (e.g., Orive 1995), and others take it to be the entire clone or genet (Pedersen 1999). This is particularly important in the case of microorganisms (discussed in following subsection) where senescence at the ramet level is routinely misconstrued to imply clonal senescence. A second level of complexity is added where, as is commonly the case, clonal organisms are also reproducing sexually, i.e., producing new genets (clones) sexually as well as new ramets asexually (see Caswell 1985; Gardner and Mangel 1997).

Little can be said authoritatively about the full extent of clones in nature because their entirety has rarely been mapped exhaustively in the field by any method. With respect to animals, genets of clonal benthic invertebrates survive substantially longer than their constituent ramets (Cook 1985; Hughes and Cancino 1985; Hughes 2005), and fecundity and survivorship generally increase with size (of the physiological individual and, where known, of the genet; Jackson 1985). Sponges and corals, for instance, have been estimated at ages of from one to several centuries (Jackson 1985). This has been taken by some authors to imply absence of senescence, but as noted above and in ◨Fig. 6.6, longevity, in itself, is not evidence for or against senescence. With time, encrusting organisms may die locally, become subdivided, and thus come to exist as clonal fragments over an unknown area. Jackson's review shows that sponges, hydrozoans, bryozoans, ascidians, and corals periodically degenerate ("regress") locally, but regenerate from other areas and so may not senesce as a clone (Hughes 1989; Babcock 1991). This is in striking contrast to the rotifers, for example, where the zygote produces a species-specific number of cells after which subsequent cell division ceases (Buss 1985), the powers of regeneration are negligible, and senescence is well documented (Bell 1984; Chap. 2 in Comfort 1979).

With respect to plants, the shoot and ramet dynamics of many clonal perennials such as clover or woodland violets, noted above under 'Plants', are well known (Harper 1977; Cook 1983, 1985; Sackville Hamilton and Harper 1989; Roach 1993). Localized death, i.e., at the ramet level, may not decrease survival probabilities of either aclonal or clonal iteroparous plants (Watkinson and White 1985; however see conclusions from theoretical studies, below). Watkinson and White attributed both the relative absence of senescence and "considerable longevity" of iteroparous plants to the capacity for continuous activity of their apical meristems. They concluded in part (p. 31) that … "*insofar as they retain the capacity for rejuvenescence from apical meristems, genets of modular organisms do not senesce*" and that whereas the longevity of aclonal plants may be limited by the problems associated with large size or the accumulation of dead matter, "*clonal plants are, in contrast, potentially immortal*" (see caveats and discussion by Klimešová et al. 2015).

As discussed in ▶Chap. 5, clones of *Populus tremuloides* (trembling aspen) in British Columbia were thoroughly mapped by Ally and colleagues in a molecular study involving microsatellite marker loci over several years. This species forms clones composed entirely of male or female ramets (trees), which can reproduce sexually (beginning at 10–20 years) and asexually (after 1 year). The clones are potentially extremely large, extending up to about 44 ha and arguably as old as one million years. In one aspect of this multifaceted study the issue of senescence was considered (Ally et al. 2010). They found a decline in sexual fitness (significant reduction in average number of viable pollen grains per catkin per ramet) with age. This implies that long-lived clones might senesce as a result of accumulating somatic mutations, possibly because the deleterious mutations are recessive and masked in the diploid condition but revealed in the haploid pollen. In other words selection on sexual fitness does not occur during clonal growth. There was no strong correlation between male fertility and various measures of clonal fitness, hence no evidence, such as higher asexual fitness of ramets associated with lower sexual fitness, that would imply involvement of trade-offs (negative pleiotropy). Technically, the affected clones are senescing if one adopts the attribute of the definition pertaining to declining fertility; however, this is really a study in clonal longevity and there is no evidence for or against age-specific mortality at the genet level. The impaired sexual reproduction does threaten the population (sexual) *lineage* and eventually the rate of asexual clonal expansion also may be impeded locally by environmental constraints.

Numerous theoretical analyses of clonal senescence have appeared since the 1990s. Among the most detailed are those by Pedersen (1995, 1999) who modeled clonal dynamics where ramets increase by asexual reproduction and new genets (clones) are produced sexually. He found that as ramets and clones aged the force of natural selection declined. This implies that natural selection could not prevent accumulation of deleterious genes with time, or that there may be selection for genes conferring beneficial effects early on with deleterious effects later (negative pleiotropy; see evolutionary theories later). Thus, Pedersen inferred that clones will senesce and cannot escape senescence by asexual reproduction alone. Sexual reproduction provides an escape by, in his terminology (1995, pp. 306–308), "*resetting the clonal level age clock*" (as well as the ramet age clock); hence parental effects do not accumulate over generations. Of course, sexual reproduction marks the end of a genet and the origin of a new genet in the zygote. Pedersen's model is based on somewhat different assumptions than the earlier one by Caswell (1985) though their conclusions are in general accord. Caswell argued that clonality does not necessarily prevent senescence but that it *may* do so by altering the intensity of selection acting on the various stages of the life cycle.

Orive (1995) applied a model to the demographic data of Babcock (1991) on the life history of three Scleractinian corals. Babcock had argued that there was no evidence that the older corals in his study underwent physiological senescence, though their clonal expansion was probably ultimately limited by factors such as colony morphology and availability of favorable terrain. The *youngest* genets and smallest ramets had the highest mortality. Orive concludes that clonality by itself retards but does not preclude the evolution of senescence. Gardner and Mangel (1997) reach essentially the same conclusion while focusing on the trade-offs between clonal and sexual reproduction by the same individual.

The generality emerging from the clonal models is that clonality, though admittedly important, is but one factor among many in the life history of a macroorganism that determine whether senescence will evolve.

6.4.3 Evidence for Senescence Among Microorganisms

Certain attributes of the population biology of microorganisms need to be kept in mind when one attempts to interpret whether senescence occurs. First, although vital stains are available for determining the live/dead status of microbes (Ericsson et al. 2000; Nelson et al. 2002) viability, at least for bacteria, is defined operationally as the ability to grow to detectable levels in or on a recovery medium (Postgate and Calcott 1985). Survival is thus equated in practice with the ability to divide repeatedly. By this criterion, all aclonal organisms would be dead! In fact, as judged by other criteria, noncultivable bacterial cells can be alive (e.g., Roszak et al. 1984), though replication potential may be lost at least in a fraction of such populations (Ericsson et al. 2000; see also the bacterial 'persisters' phenomenon, ▶Chap. 7).

Second, importantly and as noted earlier, it is the clonal lineage, not the individual cell that is of ecological and evolutionary interest. This is comparable to the genet/ramet distinction in macroorganisms.[5] Strictly speaking, for mortality statistics, the appropriate comparison would be *among* numerous clones, over time, as opposed to the population dynamics *within* a single 'clone'. The latter, however, is what bacteriologists typically study, as in growth curve experiments. Usually microbe cultures have an age structure and consequently cells will not be physiologically uniform throughout the population at any given time as well as over time (Yanagita 1977), unless precautions have been taken to do the experiments in 'continuous culture' (e.g., Dykhuizen 1990). Such procedures maintain a tightly controlled environment with replenishment of nutrients and removal of staling products. Otherwise, potential senescence effects, i.e., those due to some intrinsic organismal dysfunction, are confounded with those arising from externally imposed stress due to a drastically changing environment.

Third, estimating the distribution of clones of microorganisms in nature, other than within an arbitrarily delimited local area such as a field, is of necessity grossly incomplete due to the typically vast dissemination of clonal fragments (typically propagules such as spores, bacterial cells, etc.). This is much more difficult operationally than tracking plant and animal clones discussed above, but for different reasons. The problem in microbial ecology surrounds the unexcelled dispersal powers of bacteria and fungi (as well as some of the protists), not the lack of sophisticated identification or tracking methods. Although it is rarely possible to census an entire clonal population, in some situations it is possible to estimate the extent of emigration (Lindemann et al. 1982) or immigration (Andrews et al. 1987) from a source.

Finally, the occurrence of dormancy or quiescence, frequently under starvation conditions, together with various physiological states of activity and growth rate within a clonal population (▶Chap. 7 and Kester and Fortune 2014; Palkova et al. 2014), complicate the interpretation of senescence. This is somewhat analogous to modifications in the life cycle of reptiles and amphibians caused by diapause, diet, and temperature.

Bacteria To explore the phenomenon of senescence among prokaryotes we might begin by asking whether it occurs among the relatively few differentiated species, i.e., those that have more than one morphotype. To do so we return to the example at the outset, *Caulobacter crescentus* (◻Fig. 6.1), which lives in aquatic environments. As noted earlier, the organism consists of a flagellated swimming cell called a 'swarmer' that eventually ejects its flagellum, settles, and differentiates into an elongated, sessile, 'stalked cell' tipped with an adhesive holdfast. Undifferentiated swarmer cells cannot replicate but stalk cells produce successive swarmer cells asexually, operating essentially like stem cells (Curtis and Brun 2010). Ackermann et al. (2003) used microscopy to follow age-specific output by the stalked cells in three replicated in vitro experiments (actually subpopulations) of the same strain of *C. crescentus*. Some cells produced up to 130 progeny in the approximately 300 h monitored, but division rate decreased or halted over time in many other cells. From this result they concluded that "*senescence can indeed evolve in bacteria if there are systematic differences between the two cells*

---

5   The term 'clone' has been used variously by bacteriologists and, depending on the author, may or may not be equivalent to 'strain' (see e.g., Selander et al. 1987 and previous discussion of clones in Chaps. 2 and 5). 'Strain' normally refers to descendants of a given isolation that frequently but not necessarily arise from a single colony, which in turn is assumed to have arisen from a single cell.

emerging from division". However, what is supposedly 'senescing' here is a particular cell type, effectively a ramet, *not* the bacterial clone. Assuming this phenomenon is real, i.e., occurs in nature, what the results imply is that the genetic individual will continue indefinitely by the released swarmers, which in turn will differentiate, etc., as ongoing cycles repeat themselves, even though there may be some attenuation in the rate of clonal expansion.

Stewart et al. (2005) examined cell division through time in the well-known bacterium *Escherichia coli*. The rod-shaped cell divides in the middle forming two seemingly identical daughter cells. However, they find that the fission products are physiologically different: one new end ('pole') per cell is produced during division, implying that one of the ends of each progeny cell is preexisting from a previous division ('old pole', distal from the axis of division) and one is newly synthesized ('new pole', along the division axis). Old poles persist through multiple cell cycles and can be discriminated microscopically from new poles. At each division the cell inheriting the old pole is somewhat slower to divide and over time the effects on this lineage are cumulative. The older a pole cell is, the slower its growth rate and offspring production; there was some evidence that the older lineage also has an increased incidence of death. This physiological and reproductive asymmetry is associated with the polar localization of cell components, including proteins and peptidoglycan, known to accompany bacterial growth (Saberi and Emberly 2013; Kysela et al. 2013; for related studies on another rod-shaped bacterium and possible medical implications, see Aldridge et al. 2012). However, whether the inequality is a cause or consequence of bacterial cell aging is unknown (Stewart et al. 2005). Wang et al. (2010), using different experimental conditions, reached somewhat different conclusions with respect to the nature of population dynamics in *E. coli*; Rang et al. (2011) reanalyzed the data from both studies and provided a reconciliation.

The foregoing studies (as do those of microorganisms generally as recounted in other examples below) elucidate the process of <u>cellular</u> 'aging' and are <u>not</u> evidence for clonal senescence. This situation recounted above is analogous to the aging and death rates of individual ramets in an aspen clone. Importantly, the bacterial genetic individual continues indefinitely, as does the aspen genetic individual and fitness may actually improve with time. Thus, for comparisons with senescence in unitary organisms, the closest approximation is *among clones, i.e., how different microbial clones behave over time*. The trajectory of a given clone should also be followed, as judged by its replicative potential at various ages (as evidenced by generation time or in competition assays between representatives of younger vs. older populations). For example, the long-term evolutionary dynamics of 12 subpopulations of *E. coli* cultured from a common ancestor have been studied since 1988 by Lenski and colleagues who find, among other things, that fitness apparently increases 'without bound' (Wiser et al. 2013). To date this remarkable experiment has proceeded over some 30 years, >50,000 bacterial generations, and is ongoing. There is no evidence from such work that these bacteria senesce.

Yeasts and filamentous fungi The classic phenomenon of replicative senescence in some yeasts is, at least superficially, similar to so-called 'aging' in bacteria noted above. Some 60 years ago, Mortimer and Johnson (1959) determined that although the budding yeast *Saccharomyces cerevisiae* continues to divide indefinitely when cultured under favorable conditions, the individual 'mother' cells have a finite 'replicative life span'. Many ensuing studies have embellished this basic point (see Steinkraus et al. 2008; Henderson and Gottschling 2008).

In budding yeasts (as opposed to fission yeasts, below), the result of division is two morphologically different cells, a mother and a daughter that originate from the bud (i.e., replication is clearly asymmetrical, though aberrant division occurs in the oldest mothers resulting in indistinguishable mothers and daughters; Steinkraus et al. 2008). Mortimer and Johnson (1959) followed the replicative history of 36 mother cells by micromanipulation and microscopy. They reported a mean life span of ~24 (±8 S.D.) generations and also noted a progressively longer generation time as mothers aged, with early generations taking 60–100 min and late ones up to 6 h. When budding ceased there was visible evidence (granularity, lysis) that most mothers died (Subsequent work has shown that such post-replicative cells can remain in a viable and metabolically active state for days and given rise to the term 'chronological life span'; Fabrizio and Longo 2003; Steinkraus et al. 2008). Interestingly, the replicative age of the mother is not passed to daughters early in her life span though in the latter half of their lives mothers produce daughters with progressively shorter replicative life spans. There are numerous hypotheses to explain such observations and to account for a presumed 'senescence factor' (Henderson and Gottschling 2008).

Using a different yeast model wherein the rod-shaped organism divides into two superficially identical daughters much as does E. coli, Coelho et al. (2013) report the absence of aging in the fission yeast Schizosaccharomyces pombe under standard culture conditions. Lineages followed by time-lapse microscopy showed no progressive increase in replication time or mortality rate. A mutant cell line that divided off-center into larger and smaller cells likewise showed no increase in division time. Cell death was random, not preceded by aging phenomena, and correlated with inheritance of protein aggregates that possibly interrupt cytokinesis or formation of the cell walls in the daughters. Cells exposed to heat or oxidation treatments to simulate environmental stress underwent asymmetry in aggregate segregation whereby the lineage inheriting the large aggregates aged whereas their sisters inheriting few such aggregates did not age. The authors suggest that asymmetrically induced segregation of damage has evolved to partition damage into a cell line that is sacrificed so that the other escapes.

In the filamentous fungi, intraclonal 'aging' or localized regression or senescence of ramets analogous to that in degenerating and regenerating benthic invertebrate clones unquestionably occurs (Trinci and Thurston 1976). On balance the evidence is against senescence of an entire clone. In the absence of robust demographic data, inferences must be made from the characteristics in culture or estimated age and size (terrain occupancy) in nature. Extensive and extremely old clones of various fungi have been mapped (see ►Chaps. 4 and 5 and Dickman and Cook 1989; Smith et al. 1992; Bendel et al. 2006). However, as noted at the outset of this section, increasing evidence shows that length of life and senescence are poorly correlated (Baudisch et al. 2013; Jones et al. 2014). Furthermore, in many if not most circumstances the vast and occasionally global dispersal of fungal clones, typically by asexual spores, means that clones are discontinuous and their full extent impossible to estimate accurately (Kohli et al. 1992; Goodwin et al. 1994; Anderson and Kohn 1998). It was established long ago from continuous growth experiments in so-called racing tubes (Ryan et al. 1943; Gillie 1968) that clones can grow indefinitely (Fawcett 1925; Perkins and Turner 1988; Gow and Gadd 1995), although not necessarily continuously (Bertrand et al. 1968). There is evidence from laboratory culture that strains of some fungi senesce (Griffiths 1992; Griffiths and Yang 1993); in Podospora anserina the phenomenon has been attributed to mitochondrial DNA instability (Albert and Sellem 2002) and similarly aberrant lines have been associated with cytoplasmically transmissible factors (Bertrand 2000). Senescence in this coprophilic (dung-inhabiting) fungus has been ascribed possibly to its colonization of an ephemeral resource (Geydan et al. 2012; van Diepeningen et al. 2014).

Diatoms The clonal dynamics of diatoms resemble bacteria and are very interesting. Unlike bacteria, however, each diatom cell is covered by a rigid silica wall (frustule) formed in two components, the slightly larger 'lid' (epitheca) overlapping the smaller diameter 'bottom' (hypotheca) somewhat like a petri dish. At division, one progeny cell receives the parental epitheca and regenerates a new hypotheca. Hence this cell is the same size as the parent, as will be one of its descendants at each subsequent cell division. The other daughter cell receives the smaller hypotheca, which becomes its epitheca, and a new hypotheca is generated. This cell is thus smaller than the parent and will give rise to a lineage of sequentially smaller cells (Yanagita 1977). The consequence of this asexual division protocol is the rapid dilution of the larger daughter lineage by the successively smaller. Eventually a threshold in the lineage of progressively smaller cells is reached that triggers either sexual reproduction, whereupon cell size is restored, or the cells become critically small and die (Chepurnov et al. 2004). Of course either event marks the end of that clonal progression. Time-lapse imagery of a strain of the centric diatom *Ditylum brightwellii* showed that the smaller (hypothecal) lineage actually inherited more and possibly 'better' parental material at each division than the epithecal lineage, and that it divided about 4% faster (Laney et al. 2012). This shortens the average interval between rounds of sexual reproduction and the authors suggest may be a strategy to increase representation of the ancestral genome in a population. The generality (other strains and species) and realism (applicability in nature) of these intriguing findings await further work. Cultural studies suggest that some diatoms exist for exceptionally long periods as asexual populations; however, the population dynamics and rates of mortality over time of such clones await further study. Therefore, there is no evidence for clonal senescence, though such diatom data are embedded in the senescence literature. Even if there were, this form of 'senescence' would appear to be unique, unlike senescence as construed in unitary organisms.

Protozoa The amoebozoa and alveolates such as *Paramecium* reproduce asexually by dividing in half, as well as sexually. The evidence for senescence in this group is mixed. Some, perhaps most, species display abnormalities that approximate senescence if cultured over long periods of time, though lineages are rejuvenated by sexual reproduction. Other species persist with little if any evidence for declining rates of fission. Woodruff (1926), in what is now a classic experiment, grew a single culture of *P. aurelia* asexually for >11,000 generations. He sampled the culture periodically and found no evidence for a sustained decline, rather (p. 437) "that there are inherent, *normal* [his emphasis], minor, periodic accelerations and depressions of the fission rate due to some unknown factor in cell phenomena." Other ciliates such as *Tetrahymena* are referred to as 'immortal'. Bell (1988b) in a definitive assessment of the early work in protozoology concluded that the general trend for species in fission rate was negative but that some cultures can be propagated for thousands of generations without perceptible decline and that "*although very general, senescent decline is not universal*" (p. 43). He (and others) interpret metazoan senescence as being mechanistically different from protozoan senescence, the former ascribed to an unavoidable consequence of selection for an optimal life history, and the latter a nonadaptive consequence of accumulating deleterious mutations reflecting Muller's ratchet (►Chap. 2).

6.4.4 Non-evolutionary and Evolutionary Interpretations

Non-evolutionary 'theories' (actually, hypotheses) The non-evolutionary interpretations view senescence strictly in physiological, mechanistic terms of causality. It has been clearly established and known for decades that the rate of senescence has a genetic basis (Rose 1991; Ricklefs 2008). Undoubtedly, length of life in general is influenced by multiple genes whose effects on extending or shortening life influence numerous pathways and processes. Many of these mechanisms are conserved across phyla. The literature in this rapidly advancing frontier driven by research in medical gerontology is vast and details are beyond our scope here; some examples will be noted only in passing with commentary on similarities or differences, where known, among animals, plants, and microorganisms (for reviews see e.g., Finch 1990; Rose 1991; Kirkwood 2005; Ricklefs 2008; Roach and Carey 2014).

Research with animals including humans has demonstrated the effect on aging of single-gene mutations, many of which affect the insulin-like growth factor pathway, which is ubiquitous (Junnila et al. 2013). Growth hormone (somatotropin) and insulin-like growth factor (IGF-1) have multiple pleiotropic effects but the hormones act differently on glucose and lipid metabolism. Mutations affect longevity in mice; similar mutations have been observed in humans though their effect on longevity has not yet been ascertained. Considerable attention is focused on transcription factor p53 for its role as a tumor suppressor and as part of a generalized response to stress and cell death (Berkers et al. 2013; Salama 2014). Oxidative damage (Haigis and Yanker 2010; Selman et al. 2012) and injury associated with inflammatory responses (Ashley et al. 2012) have long been associated with *cellular* senescence.

An active area of biomedical research is directed at the role of telomeres and telomerase in cell biology and aging (Signer and Morrison 2013; Pfeiffer and Lingner 2013). Telomeres are nucleoprotein caps that protect the tips of chromosomes from end-to-end fusions, DNA degradation, and recombination. They contain short, noncoding sequences of DNA repeated for up to thousands of base pairs—for example, in humans the nucleotides 5′-TTAGGG-3′ are reiterated about 10,000 times. Telomeric repeats are conserved across lineages as diverse as protists, mammals, plants, and fungi, and are probably common to all eukaryotes. Cell division tends to shorten telomeres because RNA polymerase cannot fully replicate the lagging strand and DNA is lost at each replicative cycle. While the caps prevent coding sequences from attrition, eventually a critical threshold may be reached where DNA can no longer replicate. At this point chromosomes become physically and functionally aberrant, and the affected cells enter a 'replicative cell senescence' state or undergo programmed cell death. Thus the replicative life span of certain types of cells is finite, declines with age, and is regulated by telomeres together with the enzyme telomerase, which acts in part to synthesize telomeres. Telomerase enhances cellular replicative capacity and has been associated, as has telomere length or rate of erosion, with life span in various animal species (e.g., Heidinger et al. 2012; however see Monaghan and Haussmann 2006).

Interestingly, Skold et al. (2011) report that old (7–12 yr.), asexually reproducing strains of the colonial ascidian *Diplosoma listerianum* have lower telomerase activity and somewhat shorter telomeres and impaired asexual propagation rates compared with their recent, sexually produced progeny. So here is an example of a clonal, colonial, modular organism that may *not* have escaped senescence. The authors speculate that because these metazoans have evolved generally from sexual ancestors they may have retained the constraint of somatic senescence as an evolutionary legacy necessitating periodic sexual reproduction for rejuvenation.

In plants, the mechanistic basis of 'senescence' of *cells* and *organs* in the physiological context of normal cell differentiation and organogenesis is relatively well documented (Pedersen 1999; Noodén 2004; Thomas 2013). The process typically follows an orderly (coordinated) programmed cell death or apoptosis-like process under the control of hormones and frequently nutrient source–sink relationships and remobilization of nutrients (Schippers et al. 2015; Van Durme and Nowack 2016). For example, in tissues destined to be the water-conducting (vascular xylem) elements, precursor cells that are living and meristematic undergo complex ontogenetic changes to become elongate, interconnected, dead cells specialized for water transport. Other examples of coordinated events include the staging of leaf and flower senescence and possibly even the rapid death ('whole-plant senescence') that follows immediately upon the flowering stage in monocarpic (semelparous) plants, mainly annuals and biennials. However, unlike the case in animals, these senescence events do not appear to result from the action of deleterious genes or somatic mutation. Such episodes of plant 'senescence' seem fundamentally distinct from the life history or evolutionary context of senescence of the animal individual, though some molecular causal mechanisms may be shared.

Telomere attrition and telomerase activity have received limited attention in plants; as a senescence mechanism they have not risen to the level of potential importance in plant senescence that they have in animals. This appears due in large part to the greater tolerance of telomeric damage in plants, perhaps attributable to developmental plasticity, including the organization of their various meristems (Riha et al. 2001; Thomas 2013; Amiard et al. 2014). Thus, where gradual, whole-plant senescence occurs among polycarpic plants, the causal mechanisms remain unclear and are usually attributed to multiple, fairly general problems related to accumulating live and dead biomass, size in the case of trees, transport, respiratory tissue burden, and increased susceptibility to herbivores and pathogens (Pedersen 1999).

To summarize, there is intense focus on mechanisms of **cellular** senescence in animal models and humans driven largely by research in medical geriatrics. In plants, senescence research has emphasized physiological mechanisms associated with nutrient remobilization and developmental staging of such relatively local events as leaf or flower senescence, rather than whole-plant senescence. The latter has been considered with respect to systemic senescence of annual plants, with *Arabidopsis thaliana* being a favorite model (e.g., Morales and Munné-Bosch 2015; Wuest et al. 2016). In the case of both plants and animals, it is not clear whether cellular senescence is the cause or consequence of senescence at the organismal level or, more broadly speaking, how fruitful such reductionist approaches ultimately will be in explaining organismal senescence mechanistically. Similarly, it is not clear whether results from research models can be extrapolated to other organisms.

Evolutionary 'theories' (hypotheses) The classic so-called theories reviewed briefly below were developed largely with humans and other animals in mind. They attempt to explain the paradox of why selection has not removed a conspicuously disadvantageous trait lowering organism fitness. Seminal early work on the evolutionary biology of senescence argues that the phenomenon reflects the unavoidable consequence of the timing of gene action and declining force of natural selection (Medawar 1946, 1952; Williams 1957; Hamilton 1966; Kirkwood and Holliday 1979). Stearns (2000, p. 482) summarizes the evolutionary interpretations as consolidating to senescence being "… *a byproduct of selection for reproductive performance*".

Fisher's seminal observation The early studies built upon the now famous insights on population increase and reproductive value by Fisher (1930, pp. 27–30), whose aim was to understand how rates of death and reproduction of a population of organisms aligned with the force of natural selection. With respect to persons of any age he asked specifically (p. 27) … "*what is the present value of their future offspring? … To what extent will persons of this age, on the average, contribute to the ancestry of future generations? … the direct action of Natural Selection must be proportional to this contribution.*" For humans, reproductive value rises with age to puberty and then declines to a minimum beginning in about mid-life. Fisher went on to note (p. 29), prophetically but seemingly in passing, that the mortality rate in humans was generally the inverse of the plot of reproductive value, "... *points which qualitatively we should anticipate⌊,⌋ if the incidence of natural death had been to a large extent moulded* [sic] *by the effects of differential survival*" (◘Fig. 6.9).

Thus, reproductive value is the product of the probability of survival to a particular age and the rate of production of offspring (fecundity) (for synopsis see Begon et al. 1996; pp. 530–532). If we ignore for a moment the numbers of offspring and look at just the likelihood that an organism such as a human will reproduce at any given age, the probability distribution rises sharply from zero prior to sexual maturity to a peak followed by steep decline as potentially reproductive individuals are removed by death from the population, as Fisher had noted. If we consider next the fecundity component of reproductive value, this will decline with adult age largely if not entirely because of senescent changes. So it cannot be invoked without circular reasoning to explain why the phenomenon occurs. The key, however, is that over time, even *without* senescence, there will be a cumulative probability of death simply from random events. This dictates the decline in reproductive probability illustrated because the likelihood of reproducing at an age clearly depends on the chances of surviving to that age (◘Fig. 6.9).

Medawar's insight By such reasoning, it fell to Medawar to avoid the pitfall that caught his predecessors. His early ideas (1946), influenced by Fisher's reasoning, were presented in a lecture at University College, London, in 1951, and published as an essay shortly thereafter (1952). He circumvented the tautology by not first assuming that senescence was due to wear and tear and then explaining it in those terms, but rather asked *why* older organisms should be decrepit and worn out. His reasoning used the analogy of a population of 1000 'potentially immortal' glass test tubes and followed their fate over time. The population declined progressively due (only) to random breakage, i.e., it did not become increasingly prone to 'death' (in other words, Medawar initially ignored any inherent imperfections in the glass, i.e., he took as a premise that it did not intrinsically 'senesce'). The attrition rate therefore was exponential, the chances of dying did not change with time; for illustration he used a random attenuation rate of 10% or 100 tubes per month. The broken tubes were replaced with 100 new ones each month so the population size remained constant, and after a number of years of losses and replacements the age distribution stabilized (◘Fig. 6.10). Thus, regardless of age, the probability of survival for any tube from one month to the next is the same (0.9) and for any month (x) in question can be given as 0.9^x. Obviously, the older the tubes are, the fewer there will be in that age group simply because they will have been exposed longer to random breakage.

In the next step, Medawar allowed the test tubes to maintain their own numbers rather than being replaced by the investigator or, in Medawar's words, to reproduce themselves ('no matter how' pp. 16–17) at an average rate of 10% per month, with each remaining tube contributing an equal share. This means that in aggregate 100 tubes would be added each month and since there are 900 remaining, each tube, regardless of age, contributes 1/9 of a tube to the population monthly. This is the same as saying that fecundity on a per capita basis is not changing with age. **The key insight here is that although the average death rate and birth rate per tube does not change among the age classes, the contribution to the population of the older classes does decline, not because the tubes are more fragile but because there are fewer of them. This is comparable to saying that the older the age group the lower its overall reproductive value.**

As a final step, Medawar imposed mortality on the tubes by allowing some to disintegrate spontaneously and asked what the effect the timing of such lethal 'genetic' events would be on the population. Clearly the earlier such a disaster occurs, the greater is its impact on the pop-

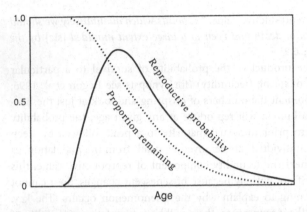

1.0

0.5

0

Reproductive probability

Proportion remaining

Age

⬛ Fig. 6.9   Relationship between age and probability of reproduction (for some organisms—see text; theory and graph based on Williams 1957). The *solid line* is the reproductive probability distribution curve, which measures the expectation that the organism will reproduce at any given age. Area under this curve thus reflects the total expectation for reproduction. The *dotted curve* is the proportion of the total probability for reproduction remaining at any given age. It is also a measure of natural selection in force at any age for such organisms because selection acts through reproduction and will be highest when the potential for reproduction is highest. The graph is similar but not identical to Fisher's (1930) concept of *reproductive value* (see text), and to a plot of *rate* of reproduction versus age as described by Kirkwood and Holliday (1979). Plot redrawn from Williams (1957); reproduced from *Evolution* by permission of The Society for the Study of Evolution and Allen Press, Lawrence, Kansas ©1957

ulation: at the advanced age of 5 years it would be virtually negligible because only a tiny fraction of that age group (probability of survival to that age =0.0018 or <1 in 500 total population) would be still 'alive'. At 12 months (an age reached by about 28% of the test tubes), the destruction would be severe but not disastrous because that group would have previously contributed most of its offspring to the population. Medawar's important conclusion (1952, p. 18) was "*… that the force of natural selection weakens with increasing age—even in a theoretically immortal population, provided only that it is exposed to real hazards of mortality*" and further (p. 19) "*… the selective advantage or disadvantage of a hereditary factor is rather exactly weighted by the age in life at which it first becomes eligible for selection. A relatively small advantage conferred early in life of an individual may outweigh a catastrophic disadvantage withheld until later*".[6]

Expansion by Williams, Hamilton, and others   If the potency of natural selection is maximal near the age of puberty and begins to decline subsequently in populations such as humans, a key implication of Medawar's analogy is that the timing in expression of genes with a beneficial effect will tend to move forward to youth whereas those with deleterious effect will move backward to older age. Those genes will remain in the gene pool, however, because selection at that point is insufficient to remove them and they will tend to accu-

---

6   Buss (1987, pp. 142–144) extends Medawar's logic to populations reproducing both sexually and asexually. Asexual individuals begin with a *physiological age* of zero but a *genetic age* equivalent to that of their parent, going back to the last zygote. Buss's argument is consistent with conclusions of the theoretical models of cloning noted earlier: senescence, though much attenuated in its appearance, will nevertheless occur eventually in asexually reproducing species. Furthermore, a key evolutionary ontogenetic development was that "*germ-line sequestration not only closed off asexuality as a developmental alternative, it limited organisms possessing it to only a brief span of life*" (p. 144).

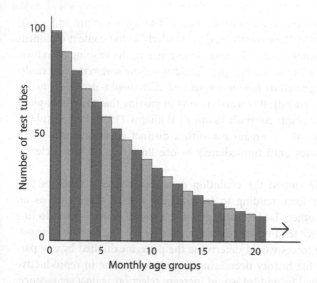

□ Fig. 6.10 Stable age distribution after many months in a population of 1,000 test tubes 'dying' randomly at a constant rate of accidental breakage and replacement of 10% per month. Thus, in month class 0–1 there are 100 tubes; then 90 aged 1–2 months; et seq. This pattern gave rise to the terminology 'potentially immortal' in the senescence and longevity literature to describe organisms where the chances of dying do not change with time. Such descriptors date back to the work of Medawar (1952) and extend to this day. Note that if the tubes are allowed to 'reproduce themselves' at some constant rate per tube regardless of age, the younger age classes will nevertheless contribute more offspring to the population purely because there are more such tubes than in the older classes. Plot and theory based on Medawar's lecture delivered in 1951, published in 1952 for University College, London by H.K. Lewis & Co., 1952

mulate in the population (the special case of antagonistic pleiotropy is discussed later; Williams 1957). Candidate late-manifesting human genes include those involved in Huntington's chorea, neurodegenerative diseases, immunological dysfunction, cancers, cardiovascular problems, stroke, etc., that in aggregate are associated with the characteristic physiological decline and dysfunction of older age. It bears repeating that Medawar's insight was that such a sequence of events could arise even in his 'theoretically immortal' non-senescent population subject only to the hazards of extrinsic accident.

Although Medawar alluded to the possibility of pleiotropy and linkage in discussing the implications of his scheme (1952, see p. 19), he had in mind mutation accumulation and it remained for Williams (1957) to take up pleiotropy at length. Williams emphasized the role of genes with pleiotropic effects at different stages of life, drawing on Medawar's work to show that those with beneficial early effects would tend to spread in a population even if causing comparable reduced survival later (the 'antagonistic pleiotropy' model). Williams emphasized the importance of the separation of germ line and soma, arguing that senescence is an evolved characteristic of the soma. Thus, absence of a clear soma meant absence of senescence.[7] Returning to our earlier examples, organisms reproducing by fission such as

---

7   Bell (1984, 1992), among others, emphasizes that the real criterion is whether there is a distinction between parent and offspring or, at the extreme, whether reproduction produces a symmetrical or asymmetrical result. Martínez and Levinton (1992; see also Roach 1993) argue that the evolution of somatic differentiation, and not germline sequestration per se, was the critical precursor to senescence.

bacteria, many protozoa and yeasts, and a few animals, lack a clear age structure and soma; theoretically they should not senesce. Importantly, and particularly in the context of confusion between individual (ramet) senescence and clonal senescence in the ensuing literature, Williams observed (p. 403) that *"while asexual <u>clones</u> should not show senescence, asexually reproducing <u>individuals</u> may be regarded as having somas and they should, according to the theory, show senescence"* [emphasis added]. It is worth noting in passing that the whole-plant senescence typified by monocarpic plants contradicts one of Williams' (1957) key postulates that senescence should be found only in organisms with a distinct germ line and soma. Plants do not segregate such tissues until immediately before flowering yet some clearly senesce.

In overview, Williams broadly viewed the evolution of senescence as a result of two opposing forces: one, an indirect force tending to increase the rate of senescence as an adverse side effect of early-acting, otherwise favorable genes, the cost of which is later decline (senescence); the other, a direct force tending to reduce the price by slowing the rate of senescence. The balance between these forces would determine the pattern exhibited by any particular species. Any factor in the life history decreasing the rate of decline in reproductive probability distribution (survival and fecundity) would increase selection against senescence. Subsequently, Charlesworth (1980, pp. 206–209; 1994, pp. 198–200) described Williams' idea in population genetics terms.

Rather than adopting Fisher's (1930) somewhat nebulous ('indirectly relevant' according to Hamilton 1966) concept of reproductive value in interpreting the force of natural selection as Medawar and Williams had accepted intuitively, Hamilton used a different criterion in his tightly argued mathematical analysis. This influential paper became a benchmark in the assessment of senescence. He advocated that the Malthusian parameter (equivalent to r) be used to assess age-specific survival and fecundity and noted that the change in reproductive value over time differed from the sensitivity of fitness to changes in survival and fecundity. Among other things, this led to different predictions with respect to the timing of gene action and onset of senescence in some situations (pp. 21–25 in Hamilton and pp. 190–198 in Charlesworth 1994; Charlesworth 2000). Indeed, in cases where fecundity increases exponentially with adult age (as is the case for many trees), reproductive value as a gauge of selection intensity does so as well and predicts the *reverse* of senescence (see Baudisch, below). Reproductive value depends on the individual having survived to a given age x discounting population growth to that age, whereas Hamilton's measure of r is an assessment of the fitness effect of change in death at age x for individuals *at the time of conception* (Chap. 5 in Charlesworth 1994; see also Chap. 1 in Rose 1991 and Charlesworth 2000). Reproductive value generally reaches its maximum shortly after the age of first reproduction, whereas Hamilton's measures of the sensitivity of fitness to changes in survival and fecundity decline throughout reproduction. Hamilton therefore emphasized and Charlesworth later clarified the importance in Hamilton's calculations of early reproduction, and accordingly the role of mutations increasing early survival compared to those acting later. This effect would lead to life histories with lower age-specific mortality early in adulthood and higher mortality rates later.

Hamilton took as the limiting case for potential senescence a hypothetical organism like *Volvox* (described in ▶Chap. 4) in which all cells divided synchronously and fertility increased exponentially with age indefinitely. At the tetrad stage one cell is ejected to become a spore and the other three separate, position themselves in the enlarging sphere, and continue growing in geometric progression at the ratio of three per every two cell generations. No limits to growth are imposed in this extreme case so growth, expansion, fertility, continue exponentially ... *"so that all its members are immortal"* (p. 25). Yet even Hamilton's 'Utopian population' is not immune from the workings of natural selection and genetic variation. He accepts Williams' (1957) idea of antagonistic pleiotropy as well as noting that a mortality factor imposed early, however small, shows that selection acts to resist it at the cost of increased vulnerability later. Thus, Hamilton's prophetic words have echoed down through the decades: *"senescence will tend to creep in"* (p. 25); and (p. 26) ... *"for organisms that reproduce repeatedly, senescence is to be expected as an inevitable consequence of the working of natural selection"*.

Baudisch (2005, 2008; Vaupel et al. 2004), however, argues that Hamilton's conclusions are a result of his restrictive modeling assumptions and parameterization. Different assumptions and parameters lead to different conclusions, even within Hamilton's original model (Baudisch 2005). Instead of emphasizing the impact of mutation accumulation on age-specific fitness, she and colleagues have focused on optimization of trade-offs in shaping mortality and fertility schedules, arguing that the force of natural selection actually can *increase* with age in some organisms such as plants as alluded to earlier. Caswell and Salguero-Gomez (2013) use different terminology but reach similar conclusions in a model that considers plants both in terms of age and stage of development.

Kirkwood and colleagues (e.g., Kirkwood 1977; Kirkwood and Holliday 1979, 1986; Kirkwood and Rose 1991; Abrams and Ludwig 1995) explained senescence in terms of physiological ecology, specifically pertaining to optimal resource allocation (e.g., see ►Chap. 3 and Partridge and Barton 1993). Kirkwood asked how the life history of an organism should be shaped by allocating resources and energy optimally between somatic upkeep and reproductive function. Obviously the production of progeny requires some somatic investment, but how much? They viewed maintenance costs as accrued from: (i) construction of nonrenewing parts such as teeth; (ii) cell renewal, such as those of the skin and immune system; and (iii) intracellular upkeep and traffic. The soma is merely a means of transmitting genes to one's progeny. A close to error-free (non-senescent) soma would necessitate a very high level of maintenance at the expense of input to progeny; conversely, too little invested in a soma could result in no or negligible reproduction. In their model, the investment that maximizes fitness (rate of increase, r) for an iteroparous species is a strategy that allows some unrepaired somatic defects (senescence). This 'disposable soma' theory is actually a variant of Williams' model, a potential explanation for one way it might operate. Even in the absence of a soma, a single-celled organism like a bacterium faces trade-offs between maintenance (survival) functions and reproduction (Nyström 2002).

In overview, it appears that where organismal senescence occurs in various phyla it can reflect the declining force of natural selection with age. Theoretical models support such an interpretation, and the limited experiments to date on model organisms such as *Drosophila* that vary the force of natural selection at different ages, support the evolutionary interpretation of aging. The underlying mechanisms remain unclear and may include the accumulation of deleterious genes, the pleiotropic impact of the timing in gene action, soma/gene tradeoffs, or other factors as yet unidentified.

6.5   Summary

The life cycle has been called the central unit in biology. In this chapter the origin and attributes of life cycles in general are discussed, and two of the many ecological questions are explored: why complex cycles as opposed to simple cycles have evolved, and whether all organisms are doomed to undergo senescence.

The inception of the life cycle is traceable to the simple cell cycle of prokaryotes. With the origin of multicellularity and eukaryote sex, life cycles became more expansive and identified with distinct phases related to developmental stage or nuclear condition. Apparently all organisms ultimately pass through a single-cell stage, which is advantageous in many ways: as a tabula rasa for implementing developmental novelties; in purging mutations from loaded individuals; and in aligning sub-organismal levels of selection.

When offspring are born into essentially the same habitat as the adults and do not undergo sudden ontogenetic change, the organism is said to have a simple life cycle (e.g., mammals). Where organisms have two or more ecologically distinct phases separated by an abrupt ontogenetic change, the life cycle is considered complex in ecological semantics (e.g., the frog, many species of which metamorphose from a tadpole living in a pond to an adult living on land. Other examples include toads, the holometabolous insects [i.e., those having a complete metamorphosis], various animal and plant parasites, and algae). An adaptive interpretation of the complex life cycle is that it appears to be (or at least to have been at the many points in geological time when it evolved) a specialization that allows the organism to exploit certain opportunities, and forgo compromises. Organisms with such cycles are adapted for particular activities such as feeding or reproduction in a particular stage.

The rust fungi, many of whose parasitic life cycles consist of a sequence of five morphological states on two distinct hosts, illustrate the intricacies and principles of the complex life cycle. CLCs in general and the rust cycle in particular usually have been interpreted adaptively. That reasoning follows in the rusts in part because of the seeming ideal match between pathogen stage and environment, and the truncation of the cycle in short-season environments somewhat analogous to progenesis (precocious sexual maturity in animals). However, the CLC may be an evolutionary dead-end and a prime example of how evolution can drive organisms to greater degrees of specialization. In the rust case, the driving force would be an ever-deepening spiral of host–parasite coevolution. CLCs present a paradox because in theory they should be unstable over evolutionary time but have remained fixed in major taxa, including the rusts, even over geological time. Thus, an alternative interpretation of the CLC is that it persists because the stages are inherent parts of a developmental program largely beyond the reach of natural selection. Because of linkage among traits, modification of a stage may be restricted or precluded without adverse impact on others.

Senescence is the manifestation of various deteriorative effects that decrease fecundity and the probability of survival with increasing age. It has been depicted most clearly at the population level by actuarial statistics showing an increasing mortality rate over time. If mortality is constant with age, the number of survivors declines exponentially and there is no basis for inferring from the data that senescence occurs. Non-senescent curves mean either that the organism inherently does not senesce or possibly that insufficient numbers of the population in nature pass through early- to mid-life for senescent effects to be detectable numerically.

Among macroorganisms that reproduce largely or exclusively sexually, particularly the unitary organisms, a good case can be made for the occurrence of senescence. Examples include the vertebrates and many invertebrates such as the nematodes, crustaceans, insects as well as some plants such as the determinate annuals. On the other hand, evidence for senescence appears limited or nonexistent generally for modular organisms, including benthic sessile invertebrates, most perennial plants, and microorganisms. Indeed, there is evidence for some organisms that the probability of survival actually increases, along with fecundity, later in life. The terms 'negligible senescence' and 'negative senescence' have been used to describe such population dynamics. The ultimate demise of clonal organisms, particularly clones of bacteria and fungi, is speculative. As genetic individuals, they can clearly be exceptionally long lived and of indeterminate size and duration, but there are virtually no actuarial statistics on the age-specific mortality or fecundity for such clones. Terms such as 'immortality' and 'potential immortality' are better suited to abstract mathematics, metaphysics, and religion than as biological descriptors in the senescence literature.

Senescence probably results mechanistically from multiple causation. In vertebrates and particularly in humans, these include decline of the immune and key organ systems leading to numerous forms of dysfunction manifested in diseases such as arthritis, neurological degeneration, cancer, and cardiovascular impairment. Among the specific causal factors implicated are faulty DNA repair, inflammation, crosslinking of macromolecules, and oxidative injury. Such progressive deterioration is attributable in part to cumulative insult and in part to defective repair. In an evolutionary context and as a first approximation, senescence is predicted to occur wherever the reproductive value of the individual diminishes with increasing age. Thus, a gene with positive effects early in the reproductive period, that is, when reproductive value is high, will tend to be selected even if it may later have deleterious effects. Likewise, because of the declining strength of selection with age, pressure to remove harmful genes that are expressed late in life would be lower than for those acting early. The major evolutionary postulates relating to mutation accumulation, antagonistic pleiotropy, and a disposable soma are intuitive though backed up by relatively little convincing data. It remains generally unclear whether 'senescence' effects documented at the cellular level typically in research models are the cause or consequence of senescence at the organism level; likewise it is not clear or even likely that results from a few model organisms can be extrapolated to other taxa.

The occurrence of senescence among unitary organisms and its general absence or exceptionally delayed expression at the level of the genet among modular organisms is yet one more manifestation of the marked ecological difference between these two major groups of life forms. Most of the population dynamics and evolutionary theory have been developed for unitary organisms. Much remains to be done in sorting out the distinctive behaviors of these two classes and the biological research needed transcends the domain of microbiologists, botanists, or zoologists.

Suggested Additional Reading

Baudisch, A. 2008. Inevitable Aging? Contributions to Evolutionary Demographic Theory. Springer-Verlag, Berlin. *A refreshing perspective on senescence dogma, emphasizing organisms where senescence is delayed or absent.*
Bonner, J.T. 1965. Size and Cycle: An Essay on the Structure of Biology. Princeton Univ. Press, Princeton, N.J. *The case for considering the whole life cycle, not just the adult, as the organism.*
Charlesworth, B. 1994. Evolution in Age-Structured Populations. Cambridge Univ. Press, Cambridge, U.K. *Mathematical treatment of the evolution of life histories and senescence.*
Finch, C.E. 1990. Longevity, Senescence, and the Genome. University of Chicago Press, Chicago. *This comprehensive treatise, though now dated, remains a benchmark in the literature.*
Rose, M.R. 1991. Evolutionary Biology of Aging. Oxford Univ. Press, NY. *A concise, well-argued interpretation of senescence in an evolutionary context. This book is reviewed by Bell (1992). Evolution 46: 854-856.*

The Environment

© Springer Science+Business Media LLC 2017
J.H. Andrews, *Comparative Ecology of Microorganisms and Macroorganisms*,
DOI 10.1007/978-1-4939-6897-8_7

The known facts of development and of natural history make it patently clear that genes do not determine individuals nor do environments determine species.

R. Lewontin, 1983, p. 276.

7.1   Introduction

This book started with a review of how organisms change genetically, because genetic variability provides the raw material for evolution. Since natural selection always occurs in a setting, we now conclude with the other side of the coin, namely, how an organism's surroundings influence the evolutionary process. We proceed, first, by reviewing the issue in terms of what is meant by 'the environment', broadly construed. The remainder of the chapter then contrasts how different kinds of organisms experience, respond to, and shape their environments.

Environments are often conceived only in terms of the purely abiotic elements in which the organism is immersed. It is important to remember that *the setting includes both physical and biological components*. The latter includes surrounding individuals of the same and different species. Some organisms, most obviously symbionts or predators and their prey, obviously have a direct influence on one another. The association among others may be only indirect or sporadic. The point is, first, that the biotic and abiotic environmental complex drives natural selection by influencing which organisms survive and reproduce, how many offspring are produced, and of these, how many in turn survive and reproduce; and second, that natural selection is itself altered by the changing composition of the survivors. Throughout the chapter examples will be given of the influence of organisms on one another, i.e., how important and often far-reaching is the biotic component of an organism's environment. Here, a couple of instances suffice. Johnson and Agrawal (2005) dissect the contributions of an organism's genotype and environmental factors operating at large and small spatial scales in determining the multi-trophic arthropod community on evening primrose. Plant genotype is particularly important, largely determining the variation in arthropod diversity, evenness, abundance, and biomass on particular plants, but the effects vary across habitats. Second, Meinhardt and Gehring (2012) show how an invasive tamarisk adversely affects associated native cottonwoods in Arizona, not directly, but (rather amazingly) by disrupting the cottonwoods' mycorrhizal associations. These two combating plant species are major components of the landscape and the outcome reverberates throughout the whole ecosystem.

To the extent possible, the environment should be considered from the **organism's viewpoint**, because what is biologically pertinent is determined by the organism, not by the observer. Von Uexküll (1957) recounts with fascinating examples how different organisms partition the same environment into their own, closed worlds as they perceive them (his term to describe this is "*Umwelten*")—for example, how creatures as different as a human, a fox, an owl, or an ant, view and respond to the same oak tree. Some inferences can be made by humans in attempting to discern the behavior of other organisms, but they always entail interpretive risk. In this context, Harper (1982, p. 19) has argued against the use in ecology of emotive terms such as "stress" (e.g., in the context of "environmental stress"), which tend to be anthropomorphic and ambiguous. Many people would consider the tropics clement and the polar latitudes inclement or "stressful" to biota, yet an emperor penguin, well adapted to Antarctica, presumably would find life on the equator intolerable. Nevertheless, the term remains widely embedded in various disciplines, and when carefully qualified such as with respect to water potential or frost tolerance can be useful.

7.2   The Environment and Organism Are Tightly Coupled

Continuous, reciprocal involvement between organism and environment means that the environment does more than merely set the evolutionary stage for the organism. The environment and organism together establish the 'fitness function', in other words, the success or failure of the phenotype within a selective milieu. The environment plays a role in developmental processes thereby influencing the phenotype itself (◘Fig. 7.1 and Scheiner 1993).

Lewontin (1983, 2000) has made the essence of the points shown in ◘Fig. 7.1 clear in two forceful analogies as well as several corollaries (◘Table 7.1). First, the developmental paradigm that genetically predestined 'normal' ontogeny simply unfolds against a conducive environmental backdrop—much as photographic film would be converted from latent image to image in the appropriate developing solution—is wrong. Rather, the organism is both cause and consequence of a particular ontogenetic sequence. There is no overall best phenotype; what is best depends on the environmental context. This is dramatically shown in the classic experiment of Clausen et al. (1948) of differential performance of clonal transplants of *Achillea* grown at three elevations in California. The stature of each [genetically identical] clone varied more-or-less extensively as a function of elevation. Second, Lewontin argues that the phylogenetic paradigm wherein the environment poses successive 'problems', which each species 'solves' adaptively, also is mistaken. Each species, indeed each organism, is a unique product of a genotypic and environmental history, not to mention random events (later). In Lewontin's words (1983, p. 276), noted at the outset of this chapter, "...*genes do not determine individuals nor do environments determine species.*" By extrapolation this implies that evolution is the concerted change of both organism **and** environment. There is probably no more elegant example of the organism creating its own environment than that of the tumor-inducing plant bacterium *Agrobacterium tumefaciens*. As part of the complex signaling process between the pathogen and host, and ensuing pathogenesis, the bacterium, which resides in the root, genetically engineers the plant to produce novel, specific metabolites (opalines)

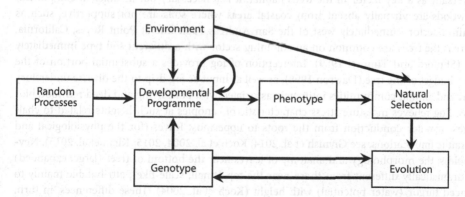

◘ **Fig. 7.1**   Model illustrating how environment interacts with genotype (and random events) to influence the developmental program and, ultimately, to shape the variable or plastic phenotype. The environment also drives natural selection by determining in part the survivors and is, in turn, affected by the composition of the resulting population and community. Note also that genetic effects on phenotype can extend *beyond* the traditional phenotype level to become manifest at higher organizational levels (Whitham et al. 2006), producing what have become known as 'community and ecosystem phenotypes'. From Scheiner (1993). Reproduced from *Annual Review of Ecology and Systematics* by permission of Annual Reviews Inc. ©1993

◻ **Table 7.1**   The interaction of organism with environment: Some postulates (Lewontin 1983, 2000)

1. The organism and its environment are an integrated unit.

2. The organism determines what is relevant in its environment

3. The organism alters the external world through its activities

4. The organism transduces physical signals from the external environment

5. The organism constructs a statistical pattern of its environment distinct from that of the external world

6. The ability of the organism to change is constrained in part by physical relations in the external world

useful to itself (see discussion in ►Chap. 3 and McCullen and Binns 2006; Platt et al. 2014). In effect, it has obtained the keys to the pantry and has locked itself inside.

Recognizing that the association between organism and environment is dynamic, reciprocal, and continuous invigorates the longstanding dogma that environment and genetics together determine the individual. As an example, we might consider some interesting aspects of the life history of the remarkable coastal redwood tree (*Sequoia sempervirens*) in its many phenotypic representations (◻Figs. 7.2 and 7.3). This redwood species has been measured at heights commonly of 100 m or more and extreme longevity, the oldest confirmed tree being at least 2200 years (Noss 2000). Present-day stands of the species are confined by the environment to a narrow strip of the Pacific coastline extending some 700 km from southwestern Oregon to Monterey County south of San Francisco (Noss 2000). These favorable conditions result from a set of unique circumstances, principally high summer humidity and free moisture associated with frequent oceanic fogs, continuously cool temperatures, low evapotranspiration, and deep, moist, especially alluvial soils. The atmospheric conditions in turn hinge largely on the broader climatic setting, which features a coastal upwelling system driven by wind (The California Current; see Black et al. 2014). While fog is rightly implicated universally as a key factor in the tree's habitat, it is a necessary but insufficient component. Redwoods are virtually absent from coastal areas where soils are not supportive, such as within a sector immediately west of the San Andreas Fault line at Point Reyes, California, whereas the trees are common on an adjoining sector with a different soil type immediately east (Shuford and Timossi 1989). Interception of fog provides a substantial portion of the water budget of the tree (Dawson 1998), as well as input as fog drip to the diverse understory plant and animal communities intimately associated with both living and dead redwood biomass. Fog relieves moisture stress characteristic of canopies at such heights related to challenges of water conduction from the roots to uppermost leaves (for the physiological and hydraulic implications, see Givnish et al. 2014; Koch et al. 2004, 2015; Klein et al. 2015). Nevertheless, the morphology and anatomy of leaves near the bottom of trees (large, expanded) are dramatically different from those near the top (small, scale-like), attributable mainly to reduced turgor (water potential) with height (Koch et al. 2004). These differences, in turn, affect many processes, including photosynthesis. In other words, the leaves have substantially different ecophysiology and effectively live in dramatically different environments.

Older coastal redwoods typically have a complicated canopy structure marked by 'reiterations' (multiple trunks arising when the main axis breaks in storms and sprouts new trunks).

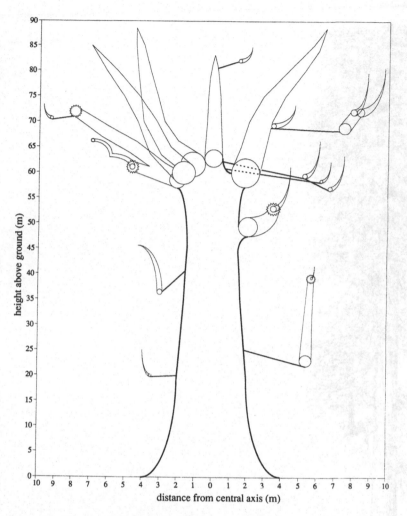

◘ Fig. 7.2 Phenotypic model at a moment in time of a remarkable organism as shaped by the environment: The 'Atlas Tree', the fifteenth largest known living coastal redwood at 88.6 m high, 7.1 m diameter at breast height (dbh), and estimated wood volume of 791 m³, located in Prairie Creek Redwoods State Park, Humboldt County, California. Note expanded scale of x-axis. The irregular conical lines represent branches; only branches giving rise to 'reiterated' trunks (those arising after breakage of a main axis) are shown. There are 21 such trunks. *Circles* denote girth of trunks and branches and are drawn to same scale as the x-axis. Jagged-edged circles represent broken trunks. See ▶Chap. 4 in Noss (2000). From *The Redwood Forest* (2000) edited by Reed F. Noss, copyright ©2000 by Save-the-Redwoods League. Reproduced by permission of Island Press, Washington, DC. Compare with ◘Fig. 7.3

Multiple fusions also occur among branches and from trunk-to-trunk and branch-to-trunk from branch-to-branch; such fusions can themselves resprout (as do fallen branches or trunks on the ground) with each reiterated trunk supporting its own branch system. Remarkably, even non-fused and completely severed branches lodged in the crown can remain alive and resprout (see ◘Figs. 7.2 and 7.3 and Noss 2000). This aerial scaffold in turn

7

◘ Fig. 7.3   A scale drawing of the 'Atlas Tree' shown in ◘ Fig. 7.2 (drawing by R. Van Pelt). Note the complexity of the crown featuring the multiple trunks. The specks at the base of the tree represent humans and a dog at the same scale as the tree. From *The Redwood Forest* (2000) edited by Reed F. Noss, Copyright ©2000 by Save-the-Redwoods League. Reproduced by permission of Island Press, Washington, DC

supports a complex community of epiphytes, including microorganisms, lichens, bryophytes, and vascular plants, as well as animals (Noss 2000; Williams and Sillett 2007). The base of the tree also provides a complex habitat for multiple organisms, as does dead wood on the forest floor. In short, clearly the coastal redwood is the epitome of an organism that both shapes and is itself dramatically shaped by its environment.

Finally, it is noteworthy that organisms can differ in phenotype for reasons beyond both conventional genetics and environment. This random component of variation is known as **noise** (Lewontin 2000; Griffiths et al. 2015). For example, the number of eye facets in *Drosophila* is typically different between the left and right eye. As the eyes of the same individual are genetically the same and their environments are effectively identical, such differences almost certainly result from stochasticity operating at the molecular level during ontogeny. Even clonal bacterial cells growing in homogeneous (shake-culture) conditions display nongenetic individuality attributable to fluctuation of key molecules in such low quantities that they are subject to statistical variation (Spudich and Koshland 1976; Avery 2006). The phenomenon is typically referred to as **"phenotypic heterogeneity"** in bacteriology (Ackermann 2015). In fact, noise arises from stochastic events both extrinsic and intrinsic to a given cell (Elowitz et al. 2002). Although it has been recognized for decades, only relatively recently has the far-reaching impact of noise and stochasticity become more fully appreciated. This is attributable to expanding interest in epigenetics and the rapidly advancing technology to dissect genetic and metabolic circuits, often at the level of individual molecules in living cells, (e.g., Losick and Desplan 2008; Eldar and Elowitz 2010). We discuss several examples and implications in the ensuing sections. It should further be noted that, notwithstanding the vagaries and versatility imposed by noise and the environment, ontogeny can be strongly genetically driven and adaptive. For example, Holeski et al. (2009) show in a common garden experiment with ramets propagated from the top, middle, and bottom of cottonwood trees that resistance to aphids is significantly higher in the top than bottom, i.e., that it is genetically programmed and not a function of environmental factors that vary with height such as sunlight or water relations.

7.3 How Organisms Experience Environments

7.3.1 The Influence of Life Span and Growth Form

In biological terms, there are no absolutely constant environments either in nature or even in the laboratory. Differences occur in space from minute increments to vast reaches ($<\mu$m to >km); likewise, cyclic and noncyclic changes occur through time (milliseconds at one extreme to eons on a geological scale). For instance, with respect to space, soil properties vary over several orders of magnitude influencing microbial distribution and activity at one extreme, to plant distribution patterns over regional scales at the other extreme. A similarly vast spectrum is found with respect to micro- and macro-variation in nutrients within the ocean extending from the microhabitat surrounding particles of marine detritus (Alldredge and Cohen 1987; Poindexter 1981a, b) to the nutrient upwellings along coasts or impacts of the Gulf Stream. Temporal fluctuations range from the recurrent seasonal and diurnal cycles, to the highly irregular and episodic. Conducive environments may be fleeting, as in a shifting sand bar in the Mississippi River on a scale of years or decades for plant colonists; or phases in host susceptibility that change over minutes, hours, or days for plant and some animal pathogens, or on a geological timescale if we are considering continental drift and ancient phylogenies or global warming/cooling cycles. Such vastly different terms of reference frame the important question: At what threshold does environmental fluctuation become significant to the organism and what life history attributes most determine the response?

Each creature experiences the environment differently. A major part of this difference relates to growth form and longevity, including their correlates. A shorter lifespan, generally speaking, means a greater acuity to short-term fluctuations. Variability occurring at intervals beyond the typical lifespan is beyond the frame of reference of the organism (although obviously not of the lineage). Clonal organisms, as noted previously (▶Chap. 5), persist and expand indefinitely. The clonal unit (ramet) may only experience a particular environment over a short period (possibly minutes or hours in the case of a single bacterial cell; see example below), whereas the clone as a whole may persist for decades or centuries. Hence, as a genetic individual, the clone typically experiences over time greater ranges in environmental variables than does the genet in most unitary organisms. Also, by virtue of their size and whether aggregated or disaggregated, clones frequently experience multiple environments simultaneously.

Among unitary organisms, for a rotifer with a life of 10 days, weekly changes are probably about comparable in periodicity to yearly changes for a bird (Chap. 1 in MacArthur and Connell 1966). The life cycles of animals in the Florida Everglades are coupled to the annual periodicity of the wet and dry seasons. The wood stork breeds when water levels are falling and it can easily obtain fish in the receding pools. The bird will not nest if this water cycle is disrupted (Kahl 1964). Analogously, in the annual cycle for many fungal pathogens of plants, the apple-scab fungus *Venturia inaequalis* overwinters as a saprophyte in apple leaf litter on the orchard floor where it undergoes sexual recombination; it then oversummers as a parasite, undergoing repeated rounds of asexual fragmentation in association with the living plant. Both phases are intimately tied to the seasonal activities of the host as well as the orchard environment (Andrews 1984, 1992; Aylor 1998). In terms of a shorter time scale, leaf-associated bacterial populations respond quantitatively and qualitatively on the order of minutes to hours as physical conditions change (Hirano and Upper 1989, 2000). Growth rates over brief periods in nature can even be on an order of magnitude (doubling times ca. 2–3 h) approaching those under optimal laboratory conditions.

There are various biological recorders of ambient conditions. Perhaps the best known of these is variation in the width of annual growth rings of trees, which reflects the impact of many environmental variables, particularly water. This living chronograph can extend over thousands of years in the case of extraordinarily long-lived species such as bristlecone pine (*Pinus longaeva*), among others (Pilcher et al. 1984). Thus, long-term oscillations in climate also are captured. A remarkable example is that growth rings of the blue oak (*Quercus douglasii*) projected back to the year 1428 have been used to infer climate pattern in the southwestern U.S. (Black et al. 2014). Such exhaustive reconstruction is possible because the same atmospheric conditions that control winter-time coastal upwelling (and associated marine life) in the California Current, noted earlier, also affect precipitation and tree growth.

For species comparison purposes, one can array various life spans or generation times against a set of hypothetically important environmental temporal variables (◨Fig. 7.4 and Istock 1984). Every organism experiences many variables concurrently, and differences in life history mean differences in aggregate experience with the environment. Organisms such as annual plants with a short life span relative, say, to a yearly pattern of seasonal change, are restricted to one season and become specialized to it such that they could not develop at other seasons. Species that live over many seasons could be generalists—doing more or less well throughout the year, or specialists—flourishing in one season while remaining dormant or migrating in others (see Dormancy below).

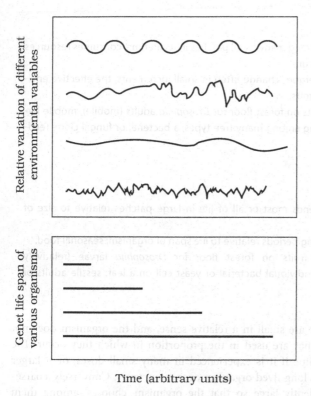

Y-axis labels: Relative variation of different environmental variables; Genet life span of various organisms

X-axis label: Time (arbitrary units)

◘ **Fig. 7.4** Hypothetical periodicities of various abiotic environmental variables (*top*) arrayed against life spans or generation times of organisms. Variation also occurs in three-dimensional space as well as in time (not shown)

7.3.2 **The Concept of Environmental Grain**

The most historically prominent effort to portray surroundings from the organism's standpoint is the notion of 'grain' (MacArthur and Levins 1964). This attempts to relate environmental variation to an individual's size, longevity, and activity, especially as regards foraging. In other words, the idea involves more than simply a matter of relative scale of the organism. Not only is the environment per se important but also how the organism experiences or samples it. This is influenced in turn by various attributes such as the kind of organism, its size (state of maturity), and growth form (which includes mobility or sessility) (►Chap. 5). Depending on how resource units or 'grains' of varying size are presented to an individual or species, environments can be classified as being either fine- or coarse-grained in space or time. The idea is summarized as follows[1]:

---

1 Extended mainly from MacArthur and Levins (1964), Levins (1968). Grain refers to size and/or duration of environmental patch relative to size and activity of the organism or species. With increasing size and motility of organism, coarse-grained environments may become fine-grained, and some fine-grained environments may become inconsequential (limiting case: all individuals experience same variation, no uncertainty; Levins (1969, p. 3). See also McArthur and Connell (1966), McArthur and Pianka (1966).

- **Fine-grained environment**

Space: Mobile individual moves among many small patches in its lifetime, consumes resources in proportion to which they occur.
Time: Since the environment undergoes change often in small increments, the effective environment is the average of the units.
Examples: Differences among fruits on forest floor for *Drosophila* adults (mobile); mobile larvae of barnacles; a bird foraging among many tree types; a bacterial or fungal clone feeding as a generalist.

- **Coarse-grained environment**

Space: Nonmobile organism spends most or all of life in large patches relative to size of organism.
Time: Environment varies over long periods relative to life span of organism; seasonal food.
Examples: Differences among fruits on forest floor for *Drosophila* larvae (relatively immobile, cf. adults above); individual bacterial or yeast cell on a leaf; sessile adults of barnacles.

Fine-grained patches in space are small in a relative sense, and the organism does not really distinguish among them: they are used in the proportion in which they occur. The environment is fine-grained in time if it is experienced in many small doses, or if larger fluctuations are encountered by a long-lived organism over many years. Conversely, coarse-grained environments are sufficiently large so that the organism 'chooses' among them (space) or spends its life in a single environment (time). An oak-hickory forest appears fine-grained to a scarlet tanager, which forages in both oak and hickory trees, but as coarse-grained to a defoliating insect, which, as a specialist, attacks only the oaks (Chap. 1 in MacArthur and Connell 1966). There are instances for sessile or sedentary organisms (e.g., trees; MacArthur and Levins 1964) where an environment appears coarse-grained to individuals but fine-grained to the widely distributed species. In general, microscale heterogeneity increases as organism size decreases; small creatures, especially if they are relatively sedentary, experience the environment in coarse-grained fashion, living out their lives on a leaf or under a rock.

How organisms might experience the grain in their environments relates to competition among related species, the evolution of specialization in resource use, and how environmental instability affects the degree of specialization. MacArthur and Pianka's (1966) theoretical analysis predicts that specialist feeders are favored over generalists in fine-grained systems and the converse in coarse-grained systems, where the generalist can compensate for its less efficient feeding with lower hunting time (see also ▶Chap. 3).

Mathematical models based on the grain concept have been developed to describe selection in heterogeneous environments (see especially Levins 1968) and some experimental tests have been reported. Baythavong (2011) used the logic that plant lineages in fine-grained environments (spatial scale of environmental variation less than dispersal distance of a species) experience habitats different from their maternal parents and selection should favor phenotypic plasticity (discussed later). In coarse-grained environments (spatial variation on scales exceeding dispersal) the dispersing progeny likely experience environments similar to their parents; selection should favor genetic differentiation more so than plasticity (see ▶Sect. 7.5).

Data on attributes of the annual plant *Erodium cicutarium* (redstem filaree, Geraniaceae) growing in closely adjacent serpentine and nonserpentine soil patches in California supported the hypothesis. Such soils vary considerably in edaphic and biotic properties within spatial scales of 0–10 m.

Notwithstanding the intuitive appeal of Levins' model and others like it, there are some caveats worth mentioning. First, it is unclear what is being maximized here by natural selection (Hamilton 1970). For example, at the population level, this could be the *mean* of individual fitness (as usually stated) over all environments. Selection in this case eliminates extreme individuals, preserving those near the population mean, and is said to be '*stabilizing*'. Alternatively, it could be acting directionally to elevate the minimum of individual fitness in the population over all environments. This is the so-called '*Maximin strategy*' (Templeton and Rothman 1974). To quote the authors in part (p. 425), "*under this strategy the optimum population is that population which maximizes its minimum fitness over all environments instead of maximizing its average fitness. By adopting this strategy, the population further insures its survival by letting the worst conditions it experiences dominate in importance.*" Levins (1968) evidently assumes the former mode of natural selection with respect to grain and his predictions have been challenged (Hamilton 1970; Strobeck 1975; Templeton and Rothman 1974).

Second, the grain concept is somewhat abstract and would seem useful mainly conceptually. The level at which a resource or any other environmental attribute appears fine- or coarse-grained is arbitrary; also, whether a species is specialized relative to another in a meaningful way can be in the eye of the beholder (Futuyma and Moreno 1988). Levins' model was designed to isolate and address specifically the resource component of the environment. Given the simplifying assumptions in a modeling context, this can be a powerful investigative tool. Yet, in reality, multiple life history facets interact among themselves, with the resource, and with the forager, in complex fashion. Furthermore, as is apparent from the examples summarized above, how any particular organism views the surroundings will vary with its life cycle stage. The sluggish caterpillar sees life in coarse-grained fashion, while the butterfly it will become flits about a fine-grained environment. Finally, different parts of the same modular organism (modules or in some cases ramets; ►Chap. 5) are exposed to potentially quite different environments. The mathematical models of Levins (1968) and others as applied to hypothetical situations are, of necessity, simplified abstractions. To reemphasize the introductory remarks to this chapter: One should not overlook the fact that we do not know, and can never really know very accurately, how any other organism experiences the world.

7.4 Organism Size and Environmental Variation

Much of what follows addresses how size differences affect environmental relationships (extending comments in ►Chap. 4), but it is worthwhile to start with some commonalities. Monod has famously said (personal communication reported by Koch 1976, p. 47) that "*what is true of* E. coli *is also true of the elephant, only more so,*" by which he probably meant that they had many biochemical reactions in common (this likely was from Monod's closing conference synopsis at Cold Spring Harbor, see Monod and Jacob [1961]). Koch, however, continues in his own words (p. 47) about their ecological parallels…"*There never was a single* E. coli, *nor a single elephant, on which Darwin's law has not operated separately, and equally; it has done so on them and on all their ancestors. The law of survival of the fittest has been obeyed, and the little* E. coli *has survived.*" Both organisms have passed the screen of natural selection within the context of what is possible for each of them. For the elephant, coping has involved… *"many more cells, much more DNA, more neurons, and the ability to walk and do other things that E. coli cannot do."* For the bacterium, it has involved extensive phenotypic plasticity and extreme genotypic versatility.

It should be added that what the elephant does as a genetic individual comprising one huge mass of coordinated, differentiated cells, the *E. coli* genetic individual does as a diffuse, essentially undifferentiated clone. Most of the elephant's cells, being internal, are buffered from exterior fluctuations, but they do have to contend with such things as pathogens. The cells react in unison at the tissue or organ level under centralized control by neurons and chemical signals such as hormones. Homeostasis is most apparent in the exquisite mechanisms that mammals have for balancing temperature and blood and tissue chemistry. As discussed earlier for unitary organisms (▶Chap. 5), the physiological and the genetic individual are the same entity. How well the entire corpus responds to diverse environmental stimuli determines whether the elephant is sick or robust, whether it dies young or as an old matriarch, and whether it will contribute significantly to the population gene pool. Being by far the largest land animal, the healthy adult elephant has no natural predators. Evidently the main environmental challenge it faces is to find food, a process that takes about three quarters of the animal's time (▶Chap. 5 in Eltringham 1982). Each cell is a party to this venture and if the functional unit dies, all components die. In short, the animal as an integrated unit has a complex neural network with which it can modify its behavior to stimuli such as hunger, thirst, or temperature extremes to maintain homeostasis.

In contrast, while cells of an *E. coli* clone may occur in aggregates, each is relatively more exposed than is an elephant cell to oscillations in the external environment. Each responds, and lives or dies, largely as a physiological individual (though coordinated responses such as quorum-sensing are known among some bacteria; Miller and Bassler 2001). Serological, electrophoretic, and molecular typing reveals the population structure of this species and shows that resident *E. coli* strains may persist for weeks or longer in a healthy human or other host despite loss of cells en masse when the host defecates (Selander et al. 1987; Touchon et al. 2009). (Note we are considering here events on a relatively local scale; the widespread and in some cases global distribution of microbial clones has been discussed previously; see ▶Chaps. 4 and 5.) There is also turnover and sporadic reappearance of strains, the inoculum originating from a few cells that evidently persist in protected intestinal sites. Intra- as well as inter-specific competition is presumably extreme. The environment is highly variable from the bacterial cell's perspective both in time and space. Koch (1971, 1987) postulates that this is a "feast and famine existence"—brief periods of glut alternating with chronic malnutrition. In spatial terms, the clone is subject to the environmental vagaries of, say, the colon as opposed to the ileum, the intestinal wall or the lumen, the intestinal milieu of different hosts, and possibly to sporadic doses of antibiotics (variation in time and space). Finally, there is life, albeit in a declining phase, outside the host associated with soil, water, plants, or feces. Savageau (1983; see also Gordon 2013) estimates roughly that an average *E. coli* cell spends about half its life in the intestine and about half on the surface of the earth, drastically different environments! It manages environmental challenges by metabolic versatility through vast, intricate systems of biochemical reactions and metabolic regulation to reach some overall optimal compromise between rapid growth when conditions permit and persistence when they do not. For *E. coli*, this coordinated biochemical network is the counterpart to the elephant's neurological network which, through natural selection, bestows upon the bacterium an appearance of similar predictive behavior (Tagkopoulos et al. 2008).

Of course, size and complexity inevitably impose many differences. Environmental signals of interest to E. coli are obviously different from those important to the elephant. To the bacterium, gravity is of no consequence, but Brownian motion, Reynolds number, and molecular diffusion coefficients are important (▶Chap. 4). Natural selection will be primarily for growth rate. To grow in the highly competitive environment of the intestine means in part being able to remove nutrients, often at very low levels, before the host or competing microflora do, and to efficiently convert these metabolites into cellular macromolecules. Efficiency of nutrient removal (defined as the equivalent number of volumes of medium that can be cleared of nutrient by a unit volume of cytoplasm per unit time; Koch 1971) is increased within limits by a decrease in size, asymmetry in shape, increase in the number of transport units per unit membrane, and increase in the capability of the transport mechanism. Transport systems in E. coli work typically at about 1000 times lower concentrations than do those of yeast, algae, and the epithelium of macroorganisms (Koch 1976). Koch calculates (1971) that E. coli could theoretically clear 2800 times its own volume per second in growth media at 37 °C. He observes that one reason, if the bacterial cell were elephant-sized, it would starve to death in the midst of plenty is because of its inability to take up nutrients fast enough. The transport mechanisms evidently have evolved to a peak where further refinements would be useless, because the bacterium is constrained by viscosity of the fecal environment, which limits diffusion rate and, in turn, growth (Koch 1971, 1976; also, recall from ▶Chap. 4 that viscosity is the denominator in the equation for Reynolds number. Therefore, an increase in viscosity will cause a decrease in Reynolds number, i.e., reduction in relative velocity of cell movement).

Efficient conversion of nutrient to biomass depends in part on the processes of protein synthesis, that is, on the efficiencies of transcription and translation. As an example consider protein synthesis. The rate of protein synthesis is directly proportional to the number of ribosomes, and each ribosome functions at a constant biosynthetic rate, regardless of the nutrient environment (Koch 1971; Koch and Schaechter 1984). Thus, regardless of the rate of cell division, each ribosome will wait the same length of time for a mRNA strand and take the same time to add an amino acid. While the cost to E. coli is that there is an excess of poorly utilized or nonfunctional ribosomes in very slowly growing cells, the benefit is that the bacterium is well equipped with ribosome machinery to get a head start for fast growth when a pulse of nutrients appears (Koch 1971; Koch and Schaechter 1984). Generation time is inversely proportional to size and, for E. coli in its intestinal environment, is about 40 h (Savageau 1983; Hartl and Dykhuizen 1984), while for the African elephant it is about 1 generation per 12 years (birth to puberty; Chap. 4 in Eltringham 1982).

The preceding paragraphs consolidate to this: E. coli cannot control its environment, whereas an elephant, by virtue of its bulk and related complexity and homeostasis mechanisms, can modify its environment markedly (Smith 1954; Bonner 1965, pp. 194–198). However, the bacterium can much more rapidly accommodate to changing environments and natural selection has shaped it to do so extraordinarily well in its dynamic conditions. This tracking of environments entails many genotypic and phenotypic adjustments, including ultrastructural and morphological changes, enzyme inhibition or stimulation, and induction or repression of protein synthesis, metabolic adjustments that may affect entire pathways (▶Chap. 3 and Harder et al. 1984; Forage et al. 1985). So the issue of size as it relates to environment is in large part one of being either a well-buffered, homeostatic individual destined to change relatively slowly in the face of adversity, or being exposed and vulnerable but capable of fast adjustment.

7.5   Genotypic and Phenotypic Variation

Sessile organisms cannot escape environmental vicissitudes. One would expect to see very good examples of adjustments to stimuli and extremes among these organisms. Focusing on plants, Givnish (2002) reviews the basic responses in various contexts to spatial and/or temporal variation, summarized below and with elaboration in the subsequent sections.

First, when individuals of a species experience relatively the same environment throughout their lifespan, locally adapted genotypes producing relatively invariant (canalized) phenotypes (ecotypes or races, below) are expected. A classic example is *Achillea* adapted to different elevations in California (Clausen et al. 1948), noted at the outset of the chapter. Another example is the intensive research over many sites and years on plants tolerant to heavy metals growing on mine tailings (e.g., Antonovics et al. 1971). Under such pervasive and relatively extreme conditions, plasticity would *not* be advantageous. Invariant phenotypes (phenotypic generalism) would also be the optimum strategy when the environment varied rapidly relative to the potential response time of an individual, or where a generalist phenotype conferring intermediate fitness across multiple environments outperforms, overall, that of alternative specialist phenotypes (Levins 1968, pp. 21–22; Moran 1992). A somewhat analogous example from microbial ecology pertains to bacteria specialized for growth in various extreme environments ('extremophiles' in bacteriology jargon), such as acid mine drainage, which is characterized by high concentrations of toxic metals, heat, and extreme acidity (e.g., with pH < 1.0; Baker and Banfield 2003). These and other *Bacteria* and *Archaea* in extreme habitats prosper at the physiological limits of life (Madigan et al. 2015).

Second, where individuals are exposed to an environment that varies principally spatially, phenotypic plasticity is expected. Many excellent examples are found among clonal organisms that encounter patchy habitat horizontally or vertically as they expand (see ▶Chap. 5 and van Kleunen and Fischer 2001; de Kroon et al. 2005).

Third, where the environment varies temporally, such as in the seasonal drying of a pond or onset of winter, plasticity should also be favored. This assumes that the adaptable phenotypes can respond effectively, producing competitive advantages over invariant phenotypes.

Finally, when environments vary significantly in space as well as time—a frequent occurrence—plasticity should be favored over invariance, again subject to the caveat of the ability to effectively track or anticipate environmental change. Phenotypes may vary as a continuous function of the environment, or they may alternate in a switch-like fashion as a critical threshold is reached (note later discussion of dimorphic fungi). Some organisms may deploy multiple phenotypes in the absence of environmental information, thereby increasing the likelihood that at least one is suited to any prevailing environment (this is one form of bet-hedging; see later discussion of this topic and Moran 1992; Gremer et al. 2016). Givnish (2002) emphasizes that the important criterion in assessing the costs and benefits of plasticity is relative competitive superiority over a range of environments, not the absolute extent of plasticity. His and the related conceptual work of others builds on the foundation of theoretical genetics laid by Levins. In a series of now classic papers and a book in the 1960s, Levins (e.g., 1965, 1968) laid out the genetic and phenotypic basis underpinning population dynamics in fluctuating environments, alluded to at various points in this chapter.

7.5.1 Genetic Differentiation

In terms of genetic variation among populations within a species, **ecotypes or races** are local matches between organism and environment. Among macroorganisms, they were first described in plants and subsequently have been well documented (Bradshaw 1972). Among microorganisms, races of fungal plant pathogens virulent at the level of "gene-for-gene" for specificity to host cultivars were described in the classic research by Flor (1956) beginning in the 1930s (recall Sidebar in ▶Chap. 3 and rusts in ▶Chap. 6). Pathogen races are discriminated on the basis of their ability to infect one or more host genetic lines or cultivars. In plants, the ecotype/race form of genetic variation has traditionally been separated from phenotypic plasticity by various approaches, principally by either interchanging representatives of local populations or transplanting them to a common environment ('common garden' experiments; Clausen et al. 1948; Bradshaw 1959; Holeski et al. 2009), or by various in situ manipulations (Turkington and Harper 1979a,b; Linhart and Grant 1996). If the major attributes of such transplants are maintained when they are grown in a new situation, then the evidence is for the genotypic component of variation. The most common observation is that phenotypic and genotypic variation, acting together, produces some intermediate result.

There is extensive evidence for localized, genetically determined variation in many traits, for example with respect to heavy metal tolerance noted above, as well as to natural soil mosaics (e.g., Baythavong 2011), flowering time, nutritional and physiological responses, parasite resistance, competitive ability, and growth form along altitudinal gradients (examples summarized in Begon et al. 1996, pp. 39–46; Linhart and Grant 1996). Selection can be strong for such local specialization at spatial scales down to a few cm, and hence margins of the populations can be sharply drawn even in cases of substantial gene flow across them (Becker et al. 2008). Temporal differentiation also occurs among hosts and pathogens, as in the oscillation of plant subpopulations carrying resistance genes against certain pathogen races and conversely, the local distribution and prevalence of races in the pathogen population able to overcome the resistance (e.g., Burdon 1987a,b; Tack et al. 2012), though the temporal component in general has not been studied as extensively as spatial variation.

Some important generalities emerge from multiple studies on such genetic differentiation in plants. The first is that it arises from biotic and abiotic environmental heterogeneity that generates selection pressures (Linhart and Grant 1996). These different environments also can cause barriers to gene flow. The combined effects of heterogeneity and restricted gene flow tend to accentuate differentiation, particularly among relatively isolated subpopulations (an example being where exposed sites at higher altitudes affect phenology).

Can the environment directly impose heritable genetic change on organisms? This idea sounds heretical and a throwback to Lamarckism and the Lysenko era, but there is strong evidence that indeed this can happen in some cases. A good example is flax, where exposure of certain varieties to suboptimal growth conditions within a generation can result in progeny with properties distinct from their parents (Durrant 1962; Cullis 1983, 2005). Notably, the phenomenon in certain flax varieties is stable across generations (transgenerational), unlike many well-known epigenetic changes in plants that are expressed only within an individual (intraorganismal; see later discussion of epigenetics and Whittle et al. 2009; Ceccarelli et al. 2011). The differences include height and biomass, total nuclear DNA, and the number of genes coding for various RNAs (Walbott and Cullis 1985). Frequently, seeds from such plants breed true, producing stable lines (termed genotrophs by Durant 1962), regardless of their growth regimen. These transitions are associated with changes in DNA evidently resulting in

◘ **Table 7.2**   Major forms of phenotypic plasticity (Piersma and Drent 2003, except where noted otherwise)

Plasticity category	Change is reversible?	Variability occurs within single individual?	Change seasonally cyclic?	Examples
Developmental plasticity	No	No	No	Barnacle snail predator (Lively 1986)
Polyphenism	No	No	Yes	Spring versus summer caterpillars (Greene 1989)
Phenotypic flexibility	Yes	Yes	No	Size shift in sea urchins (Levitan 1989)
Life cycle staging	Yes	Yes	Yes	Plumage soiling in male ptarmigan (Montgomerie et al. (2001)

part from activation of transposable elements and can result in copy number variations. How widespread the flax phenomenon is phylogenetically remains unclear, as are all the genes and possibly epigenome involved, the underlying mechanisms, and evolutionary implications. Plants do seem to be genetically much more plastic than animals (Walbot and Cullis 1983; Heslop-Harrison and Schwarzacher 2011) and could change potentially in response to environmental stimuli in ways that animals cannot.

7.5.2 Phenotypic Variation

Any kind of environmentally induced phenotypic variation in behavior, physiology, or morphology is termed **phenotypic plasticity** (Stearns 1982; West-Eberhard 1989; DeWitt and Scheiner 2004). When the phenotype varies as a continuous function of an environmental parameter the relationship is known as a reaction norm. Such changes can be complex functions of the environment and may involve delays in expression, reversibility, and occur at different developmental stages, in addition to wide variation in degree of expression (Schlichting and Pigliucci 1998; Givnish 2002). The topic as well as the biological phenomena underlying it are indeed vast as West-Eberhard (1989) has emphasized—'phenotype' including all aspects of an organism other than genotype, and 'environment' including both the external surroundings as well as the internal cellular milieu that influences gene expression.

Piersma and Drent (2003) subdivide phenotypic plasticity into developmental plasticity, polyphenism, phenotypic flexibility, and life cycle staging, expanded briefly below (see also ◘Table 7.2 with the caveat that terminology varies in the literature and is used inconsistently). The table and examples cited give some impression of the pervasiveness and range in plasticity possible.

Developmental plasticity refers to the irreversible variation in some trait(s) arising during development in response to one or more environmental signals. For example, the acorn barnacle *Chthamalus anisopoma* responds developmentally by changing the architecture of its external calcareous plates in the presence of a snail predator (Lively 1986). Dramatic changes in phenotype as a result of morphogenesis are also seen in plants, many benthic marine invertebrates, and the holometabolous insects (where the same genetic entity may at one moment be a caterpillar and then next a butterfly). Some authors, however, consider such varied forms produced in standardized, ontogenetic sequence to be plasticity and others do not.

Polyphenism is the ability of an organism to change developmentally but in a seasonally cyclic fashion (Moran 1992). An extraordinary example is developmental polymorphism in caterpillars of a geometrid moth of the southern US and northern Mexico (Greene 1989). The spring brood mimics oak catkins on which they feed; the summer brood develops into mimics of oak twigs. This switch is triggered evidently by diet, as all caterpillars fed experimentally on catkins, which are low in tannin, developed into catkin morphs whereas those raised on leaves, which are high in tannins, resembled twigs. How such striking morphological responses at the organism level result from a cascade of physiological and genetic reactions at the subcellular level, initiated by an environmental signal, is a fascinating question.

Phenotypic flexibility refers to the noncyclic, potentially reversible variation within a single individual in behavior, physiology, or morphology in response to environmental signals. Although sea urchins have a hard exoskeleton, they are able to change body size by positive or negative growth in response to food abundance (Levitan 1989); plants, because of their modular nature, are highly morphologically plastic (▶Chap. 5) in response to local stimuli such as shade, nutrient availability, and moisture (see below); many organisms, and in particular the fungi and numerous bacteria, form spores. As any weightlifter knows, a plastic trait in humans is muscle mass.

Where flexibility is seasonally cyclic it is termed **life cycle staging**. Rock ptarmigan in the Canadian arctic camouflage themselves from predators by changing their plumage from white to dark to mottle in concert with the seasons. Where this form of plasticity differs from conventional polyphenism is in the further, reversible phenotypic variation that occurs *within* a male bird. When snow melts in the spring, females molt into their summer cryptic plumage, whereas males remain white for some time. As such, they are more conspicuous to females in competitive displays, evidently enhancing reproductive success. The cost, however, is that they are also more conspicuous to predators. When females begin egg-laying, the males start to soil their plumage, and are maximally dirty when incubation begins (and their mates no longer fertilizable). Shortly thereafter the males molt to their cryptic summer plumage. This interpretation of events was developed from a careful, long-term study by Montgomerie et al. (2001) that systematically eliminated alternative hypotheses. In passing it is worth noting that here, and in related cases of seasonal coat coloration, one would expect strong selection for synchrony between organism and environment because mismatches (e.g., change in color of an animal to white in absence of snow) are striking and render the compromised individuals highly vulnerable. In those animals where cyclical polyphenism may be regulated by photoperiod, mismatches are a particular issue of increasing concern given prospective decreased snow duration as climates warm (Mills et al. 2013).

As suggested by its various manifestations summarized above and exemplified in ◻Table 7.2, phenotypic plasticity offers the potential major advantage of dampening the effects of selection by uncoupling the phenotype from the genotype (Stearns 1982). In so

> **⬛ Table 7.3   Key potential benefits, costs, and limitations of phenotypic plasticity**
>
> **Benefits:**
> - Adaptation to temporally or spatially variable environments (at the individual level)
> - Acceleration of novelty, speciation, macroevolution (at lineage level)
>
> **Costs:**
> - Maintenance: sensory and regulatory mechanisms
> - Production: inducible costs in excess of those paid by fixed genotypes
> - Information acquisition: energy expended in sampling environment
> - Developmental instability: imprecise development or fluctuating asymmetry
> - Genetic: developmental instability; pleiotropy; epistasis
>
> **Limitations:**
> - Information reliability: maladapted phenotypes
> - Lag-time: time interval between environmental change and response
> - Developmental range: fixed phenotypes possibly better than plastic in extreme conditions
> - Epiphenotype: plastic, 'add-on' phenotypes may be inferior to those inherently constructed
>
> For benefits see West-Eberhard (1989); costs and limitations are from DeWitt et al. (1998) and references therein

doing it provides for adaptation to variable environments. While plasticity is generally viewed as expanding the capability of a given genotype, it also allows different genotypes to assume the same phenotype. At the population level, such plasticity may allow retention of genetic variation, analogous to dominance (Bradshaw 1965). Plasticity may also facilitate the origin of novel changes leading to speciation events and macroevolution (i.e., above the level of species, such as those affecting multiple families within a lineage; see examples in West-Eberhard 1989 and Pfennig et al. 2010).

It is generally assumed intuitively that plasticity is adaptive but there have been relatively few direct tests (for one example, see Dudley and Schmitt 1996). There are costs and constraints to plasticity (see ⬛Table 7.3 and DeWitt et al. 1998; Givnish 2002; Valladares et al. 2007) and quantitative methods for assessing these in principle have been proposed (DeWitt et al. 1998). Plasticity cannot be invariably advantageous: No phenotype is favored in all environments and natural selection may favor fixed, plastic, or entirely novel phenotypes depending on the environmental context (Bradshaw 1965; Givnish 2002). The potential range in phenotypic variation is set genetically and is trait-specific (Bradshaw 1965). Hence, plasticity in some attributes is associated with and may enable stability in others. Certain taxa, especially plants as already noted, and fungi (below), are more morphologically plastic than others, and modular organisms more so than unitary creatures. Within a taxon, seed size in plants and spore size in fungi are relatively fixed: A large as opposed to a small plant does not produce larger seeds, but rather more of them (for caveats see Thompson and Rabinowitz 1989). Within lineages the most common ultimate constraints on the extent of developmental polymorphisms probably are allometric (▶Chap. 4). For example, upward plasticity in birth weight cannot exceed a limit without increase in size of the adult; if adult size increases, an increase in birth weight will necessarily follow, whether it is adaptive or not (Stearns 1982).

Where selection is directional (i.e., favoring phenotypes at one end of the range), plasticity can extend the response of a population with limited genetic variability. However, it seems to be more important where selection is disruptive (i.e., favoring individuals at both extremes of a range). Whenever there is noncyclic, recurrent variation in time at intervals less than the generation time of the organism, adaptation can only be by plasticity (however, see later comments about mobile genetic elements; also comments later on environmental grain with respect to phenotypic or genotypic responses). Environmental conditions may change rapidly for desert plants exposed to sudden rainfalls or for microbes exposed to sporadic pulses of nutrients. Disruptive selection can also occur over short distances, as for the clonal plant whose stolons encounter varying degrees of competition, or the aquatic plant that roots in the littoral zone, grows up through the water column, and finally produces leaves and flowers above the surface (below).

Plants Aquatic or desert plants in particular are noted for their conspicuous phenotypic plasticity (Bradshaw 1965; Valladares et al. 2007; Nicotra et al. 2010), perhaps most strikingly shown by environmentally determined leaf form called heterophylly (☐Fig. 7.5 and Hutchinson 1975, pp. 157–196; Bradshaw 1965). For instance, a single aquatic plant may bear two or more kinds of leaf, typically a dissected form in the water column and a laminar form above it. The finely dissected leaves increase area-to-volume ratio. This means that tissues are more accessible to dissolved CO_2 for photosynthesis underwater (Hutchinson 1975, pp. 146–148). The triggering factor, depending on species, may be the prevailing conditions such as water level; in other cases it may be a seasonal cue such as temperature, photoperiod, or light intensity when the leaf is initiated. The molecular mechanisms underpinning heterophylly are not known in detail. The developmental switch in one case appears to be the regulation of the hormone gibberellin by expression levels of homeobox *KNOX1* genes (Nakayama et al. 2014). Dimorphism among desert plants is related to water economy (Orshan 1963; Westman 1981). Plasticity is also apparent in the resource allocation patterns common to all plants, either to vegetative biomass for foraging purposes, or to reproduction, depending on the prevailing environment (Abrahamson and Caswell 1982; Grime et al. 1986; Bazzaz et al. 1987). Other common manifestations of the enormous morphological plasticity of plants include the 'shade avoidance syndrome' of shade-intolerant plants (Pierik and de Wit 2014) and the variation among flowers of the same individual (e.g., adaptations of some plants for open pollination in early-season flowers versus adaptation by way of closed flowers produced later to ensure self-pollination).

How do plants sense and respond to environmental variation? A striking example is the report by Braam and Davis (1990) that *Arabidopsis* can detect numerous experimental stimuli such as sprayed water, touch, sub-irrigation, wind, darkness, and wounding. The expression of at least four touch-induced (*TCH*) genes is affected by these stimuli and, additionally, growth is inhibited by touch. The touched plants have shorter petioles, begin to bolt later, and develop shorter bolts. This implies that plants can detect, transduce, and respond to environmental signals such as wind and rain, as well as mechanical loading and above-or belowground obstacles. Interestingly, subsequent work shows that the *TCH* genes are inducible by nonmechanical stimuli such as darkness and extreme temperatures (Braam 2005; see also Monshausen and Haswell 2013). The growth response in *Arabidopsis* is gradual and differs from the immediate morphological reaction of such plants as the Venus fly trap (*Dionaea muscipula*) or the sensitive plant (*Mimosa pudica*). Nevertheless, *Arabidopsis* responds very quickly at the molecular level (within 10–30 min the amount of mRNA increases up to 100-fold). Three of the *TCH* genes encode calmodulin and related proteins, which implies that Ca^{+2} is involved in signal transduction and response. The fourth is a cell-wall modifying enzyme, a glucosylase/hydrolase. Other molecules implicated in the signaling network include reactive oxygen species, ethylene, and octadecanoids. Braam (2005) showed by gene-expression profiles that

◨ **Fig. 7.5** The water marigold (*Megalodonta beckii*) exhibits a type of phenotypic plasticity known as het-erophylly, illustrated by the broad aerial leaves and the finely dissected leaves below the water line. Dissection of the leaf lamina enhances the rate of photosynthesis underwater by increasing the area:volume ratio, which increases CO_2 assimilation from the surrounding water. Note that this rooted aquatic plant occupies three environments: sedimentary, aqueous, and aerial. This type of environmental polymorphism is controlled by a condition-sensitive switch (West-Eberhard 1989; Nakayama et al. 2014). From Fassett (1957), reproduced by permission of the University of Wisconsin Press, ©1957

there are 589 genes or more than 2.5% of the *Arabidopsis* genome upregulated at least two-fold in expression within 30 min of touch stimulation. Among the most highly represented functional classes involved are genes involved in defense responses. Recent evidence suggests that plants can respond systemically with chemical defenses to leaf vibrations induced by the feeding activity of insects (Appel and Cocroft 2014). Thus, we now have the foundation for a mechanistic interpretation as to how plants may be able to react rapidly and sensitively to environmental cues.

Microorganisms To what extent are microbes phenotypically plastic? Indeed, they are extraordinarily malleable though one might not draw this conclusion because much of the relevant literature is presented in molecular or physiological semantics in the specialist journals, not within the conventional ecology mainstream. While most attention to bacterial evolutionary ecology has been paid to their genetic dynamics (▶Chap. 2), phenotypic plasticity is probably as important or arguably more so to their adaptive success. Bacterial plasticity is expressed mainly at the physiological level illustrated by an ability to adjust growth optimally as conditions warrant (previous discussion above in ▶Sect. 7.4 and ▶Chap. 3). One way bacteria do this is by producing spontaneously and reversibly a slow-growing cohort of the population that is resistant to unfavorable conditions such as the presence of antibiotics (Kussell and Leibler 2005; Helaine and Kugelberg 2014; see discussion of bistability, bet-hedging, and the bacterial persister phenomenon under Dormancy, later). Similar switching mechanisms are also common in the fungi and slime molds (below).

Another bacterial metabolic tactic is to alter the structure of certain macromolecules for metabolic optimization. This can be important in habitats where nutrient conditions are not simply always constraining but rather fluctuate irregularly from non-limiting to limiting. The case of adaptation to S-limitation by cyanobacteria is instructive. These bacteria are photoautotrophs (▶Chap. 3) that harvest light with chromophoric proteins called phycobiliproteins in a manner not too dissimilar from plants (Madigan et al. 2015). Such protein complexes may account for 60% of the soluble proteins of cyanobacterial cells, with sulfur represented in the amino acids cysteine and methionine. It follows that the phycobilisome proteins would be a likely target for selection pressure to economize cellular sulfur to avoid impairment of protein synthesis in habitats where that element is periodically limiting. Mazel and Marlière (1989) showed that *Calothrix* compensates in such situations by encoding S-depleted versions of its most abundant phycobilisome proteins, which it specifically expresses under S-deprivation. Remarkably, the bacterium in effect carries the genetic equipment (operons) for two complete, functional sets of these proteins—one set containing the sulfur amino acids as usual and the other the methionine and cysteine-depleted set (for background information and analogous examples, see Merchant and Helmann 2012; Flynn et al. 2014). In habitats such as these where a resource supply oscillates erratically between sufficiency and insufficiency, it can make better adaptive sense to carry the capability to function in both circumstances rather than rely on mutation to provide the requisite set whenever it is needed. (In a similar context, Levins [1968, pp. 11–12] discusses potential adaptive strategies for bacteria carrying an inducible enzyme under various scenarios.)

Being of relatively simple and more-or-less fixed shape (Young 2006), bacteria do not appear to vary growth form as dramatically as do the fungi (below), though there are exceptions. For example, the bacillary shape of *E. coli* cells can become filamentous, a transition that is adaptive in uropathogenic strains as the latter phenotype resists phagocytosis (Justice et al. 2008). An initial, transient, resistant response to the antibiotic ciprofloxacin, also involving *E. coli* cells, is formation of extensive filaments containing multiple chromosomes (Bos et al. 2015). Bacterial filamentation, long considered to be a 'sick' phenotype, is increasingly viewed as an adaptive response at least in some situations. A more subtle but ecologically and medically critical modification is the difference in external structure between attached and planktonic cells of biofilm-forming bacteria (Costerton et al. 1995; Hall-Stoodley et al. 2004; for mechanisms underpinning sessile versus motile phenotypes, see Norman et al. 2013). The two cell types are functionally distinct though genetically identical, as is the case for differentiation of cell types within a biofilm in response to environmental as well as cell–cell signals (Vlamakis et al. 2013).

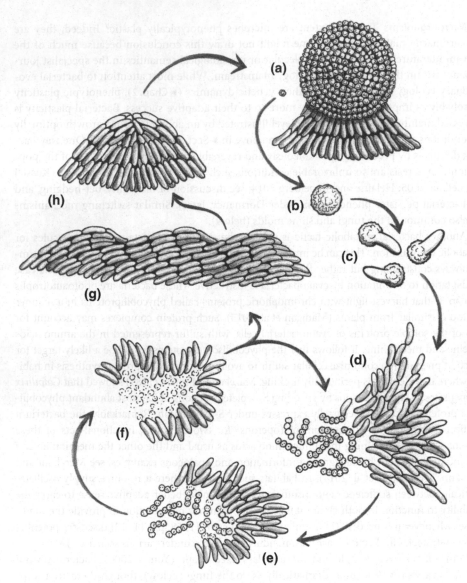

◧ **Fig. 7.6**   The life cycle of *Myxococcus xanthus,* a predatory, social, soil-dwelling bacterium, shown here engulfing micrococci (chains of spherical cells in **d, e**). It transitions among various phenotypically plastic morphological states by integrating many cues from its external environment with physiological signals from within. Nutrient limitation triggers cellular aggregation (**g**), fruiting body development (**h**), and ultimately cell differentiation and sporulation (**a, b**). Nutrient sufficiency signals spore germination (**c**) with cells congregating into a feeding 'swarm' (**d**) that moves en masse by gliding, surrounding (**d, e**) and digesting (**f**) prey cells. Cooperative interactions are orchestrated by cell-to-cell signaling. From Goldman et al. (2006). Reproduced from *Proceedings of the National Academy of Sciences,* by permission of the National Academy of Sciences ©2006

Furthermore, there are a few extraordinary instances of morphological plasticity among bacteria. The myxobacteria (◧Fig. 7.6), noted previously (▶Chap. 4), are fundamentally single-celled, but manifest multiple phenotypic forms, changing among them based on intracellular and extracellular signals (Goldman et al. 2006; Madigan et al. 2015). They inhabit mainly topsoil but also diverse other interesting sites such as the bark of some trees (Reichenbach 1993). Individual cells move by gliding and can aggregate into swarms and mounds. This stage typically progresses eventually to a multicellular, differentiated fruiting body of varying, species-specific complexity, within which some cells form myxospores that germinate to produce vegetative cells. The fruiting bodies are macroscopically visible, often brightly colored, and may consist of a billion cells. The onset of these and the various social aggregates is controlled mainly by environmental nutrient status, to which the bacterial cells respond by cell-to-cell exchange of soluble signals likely involving quorum-sensing or chemotactic mechanisms. These organisms present a fascinating example of an entity at the threshold of multicellularity and 'collective individuality' in its numerous phenotypic states from unicellular to multicellular, and prompt introspection as to how such lifestyles have evolved. Sequencing of the *M. xanthus* genome shows it to be comparatively large (9.14 Mb), an increase of some 4–5 Mb relative to other sequenced relatives (δ-proteobacteria) (Goldman et al. 2006). After ruling out other possibilities, the authors suggest that gene duplication and divergence were major contributors to separation from the most recent common ancestor; moreover, the genes were not duplicated at random but at specific sites such as for cell–cell signaling and small-molecule sensing. These social bacteria represent evidently a very successful evolutionary experiment as judged by the large numbers of known myxobacterial species (~50), high densities of cells (millions) per gram of many kinds of soils, and worldwide distribution (Reichenbach 1993; Kaiser 2003; Goldman et al. 2006).

Fungi These organisms have shorter generation times than plants and are composed of fewer cell types (▶Chap. 4), all of which are modifications of the basic hyphal unit. Not surprisingly, then, the fungi are even more plastic than plants (see e.g., Jennings and Rayner 1984; Gow and Gadd 1995). Some may generate complex organs such as rhizomorphs from diffuse or compact mycelium, which is itself variable in many features such as internode length, branch angle, and growth pattern (▶Chap. 5 and Cooke and Rayner 1984; Rayner 1991; Andrews 1991, 1995; Riquelme et al. 2011). Some organisms can alternate between a harmless, commensal state and a virulent, pathogenic condition depending on the environment. Certain fungi take this to the extreme by undergoing a dimorphic transition, changing from a filamentous (hyphal) form typically found in an environmental reservoir such as soil or vegetation, to a single-celled (yeast) state in an infected human host (Nemecek et al. 2006; Klein and Tebbets 2007). The hallmark environmental trigger for this phase transition is temperature (e.g., *Histoplasma capsulatum* grows in a mycelial form at 25C, but as yeast at 37C), although other host factors such as nutrients and hormones are to varying degrees involved.[2] This extreme phenotypic shift is associated with differential expression of hundreds of genes; it is concurrent with and may be mandatory

---

2   Interestingly, some plant pathogens are also dimorphic and at least one of these is very closely related to a human pathogen. Strains of the fungi *Ophiostoma novo-ulmi* and *O. ulmi* cause the famous Dutch elm disease, the characteristic wilt symptoms of which are attributable in part to masses of budding, yeast-like spores produced by the pathogen in the water-conducting vessels of the tree (see Sidebar in ▶Chap. 4 and ◧Fig 4.13). The fungus in this phase is morphologically very similar to members of the genus *Sporothrix*, so much so that decades ago this state came to be known informally as the "Sporothrix stage" of Dutch elm disease (Agrios 2005, pp. 528–532). The dimorphic human pathogen causing sporotrichosis is *Sporothrix schenckii* (Dixon and Salkin 1991). Berbee and Taylor (1992) show by 18S rRNA sequence methods that, phylogenetically, *S. schenckii* is a member of the *Ophiostoma* genus and very closely related to the Dutch elm disease fungus (as well as to *O. stenoceras*).

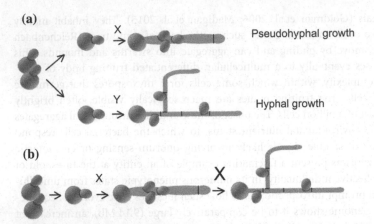

(a)

X → Pseudohyphal growth

Y → Hyphal growth

(b)

X → X →

▣ **Fig. 7.7** Phenotypic plasticity as represented by morphological transitions of dimorphic or trimorphic yeasts represented by the commensal and human pathogen *Candida albicans*. The fungus can undergo dimorphic transitions from budding cells to either hyphae (a) or an extended, intermediate form known as pseudohyphae (also a) under the control of different factors (X, Y). In (b) there is continuous transition from yeast to pseudohyphae to hyphae triggered by a dosage-dependent factor (x, X). For elaboration of this model, see Carlisle et al. (2009). From Bastidas and Heitman (2009). Reproduced from *Proceedings of the National Academy of Sciences*, by permission of the National Academy of Sciences ©2009

for virulence (Klein and Tebbets 2007, but see Sudbery 2011). Some of the human pathogens are actually trimorphic, having the capacity to grow as yeasts and hyphae as well as an intermediate form, pseudohyphae (Bastidas and Heitman 2009; see Carlisle et al. 2009 for morphological differences between pseudohyphae and hyphae, and role of gene expression in morphological switching). In *Candida albicans*, for example, the yeast phase is characteristic of its benign life as a commensal associated with the human mucosa, skin, and the gastrointestinal tract. However, in the pathogenic state of the fungus, the yeast form is adapted for systemic dissemination within the host, while the hyphae are responsible for infection and death of host cells (Bastidas and Heitman 2009). The role of pseudohyphae remains controversial as is the evolutionary origin and development of the dimorphic phenomenon. It is notable that some nonpathogenic fungi, such as the common yeast *Saccharomyces cerevisiae*, can display a dimorphic (yeast to pseudohyphae) state that tends to be overlooked because it is unique to certain strains and culture conditions (Gimeno et al. 1992). In *Saccharomyces*, pseudohyphae are triggered by nitrogen limitation and may be adaptive in nature, enabling the wild yeast to forage for nutrients by expanding radially thereby advancing into uncolonized habitat (Gimeno et al. 1992; see also Brückner and Mösch 2012) (▣Fig. 7.7).

Unfortunately, relatively little is known about the ecology of the many human pathogenic fungi that spend most of their life histories in environmental reservoirs and for which association with a human host is largely accidental. Attention has focused instead on clinical aspects. Many if not all of these fungi probably grow as commensals or saprophytes of living or dead plant matter or other biomass and have adapted only secondarily as pathogens.

7.6 The Environment and Life Cycle Changes

7.6.1 Pauses: Dormancy

In the life cycles of most organisms there are one or more reversible stages characterized by reduced metabolic activity and related changes. These periods of delay in development are variously termed dormancy (plants), 'shut-down' (microbes), diapause (insects), hibernation (certain mammals) or estivation (certain birds, mammals, and lungfishes). While obviously each is unique in detail, all share the attribute of quiescence and for simplicity will be construed here broadly as dormancy. Entry into such a phase is typically associated with a morphological transition such as the dropping of leaves by deciduous plants, or the formation of resistant propagules such as spores, seeds, or cysts by numerous taxa. In a life cycle context, dormancy frequently coincides with the dispersal stage and is often closely associated with genetic recombination. This is unlikely due to happenstance, as discussed later. A major consequence is that one or more of the emerging products of meiosis and genetic fusion (new genets) are suited to the new environment. Dormancy may also appear as an interruption of the growth phase, as in the annual cycle of vegetative expansion, growth cessation, and bud set in trees.

Dormancy is usually though not always (see e.g., Simons 2014) interpreted as a **bet-hedging strategy**[3] to adverse conditions (Childs et al. 2010; Scott and Otto 2014; Rajon et al. 2014). Levins (1969) points out that it may arise instead as an adaptation simply to extreme environmental uncertainty. A developmental delay is in effect a 'cost-averaging' mechanism that spreads reproductive output over time and across environments. Adversities may occur regularly or irregularly and can involve abiotic factors such as drought and temperature extremes, or biotic factors such as extreme predation (for an unusual example with bdelloid rotifers, see Wilson and Sherman 2013) and scarcity of resources. Microbial life in oligotrophic environments such as the open seas, deep sub-seafloor, and soil is largely one of varying degrees of starvation, dormancy, and physiological quiescence where metabolic rates and turnover times are several orders of magnitude lower than in nutrient-rich environments (Morita 1997; Hoehler and Jørgensen 2013). Relevant timescales there are centuries to millennia. Attempts to interpret the evolution of dormancy should recognize that while in some cases it may have *arisen* in response to a particular adversity or uncertainty, dormancy may now be *maintained* by selection for other reasons. For example, the internal physiological environment during insect diapause is complex and, where diapause is obligate, the phenomenon may have become a developmental necessity and operate under the control of factors different from those having driven its inception (Levins 1969).

Thus, if environments were always benign and predictable, would quiescence as a trait be maintained? This is difficult if not impossible to test because other than within the artificial confines of a laboratory setting (e.g., unrestricted bacterial growth in a chemostat) these situations are rare, though occasionally we see such glimmers (Wolda 1987, 1989). Harper (1977, p. 74)

---

3   What has become known as bet-hedging in effect was outlined originally in mathematical, population genetics terms by Gillespie (1974). He drew attention to the importance of spreading offspring across clutches in a breeding season (thereby reducing variance in offspring number) as opposed to putting all of them in the same large clutch. Slatkin (1974) then used the term in generalizing Gillespie's model in a brief, conceptual paper. Bet-hedging has since been embellished extensively and used more-or-less strictly by various authors (see e.g., de Jong et al. 2011). For relevant details see the cited papers in the text and references therein.

points out that seed development and germination are continuous in some tropical maritime mangroves. The prominence of physical factors as potential phenological cues is attenuated in lowland tropical rain forests, where leaf production and leaf senescence of most tree species may *"appear to follow a free-running periodicity (i.e., no annual biological clock)* [sic] *based on internal regulation"* (Reich 1995, p. 168). However, even the moist tropics have some modest seasonality, for example with respect to irradiance and the amount of precipitation. Moreover, changes in a physical cue tend to be transmitted and frequently amplified among community members because of species interactions in food and energy webs (►Chap. 3 and Wolda 1987; van Schaik et al. 1993). For example, phenological changes in the plant community imposed by the abiotic environment are often reflected in their consumer populations, which switch their diet, change breeding pattern, or their range (van Schaik et al. 1993).

In most if not all organisms, including even at least some prokaryotes, numerous biological physiological processes such as patterns of gene expression oscillate with a standardized periodicity of activity. (Note that these cyclic changes are in contrast to plasticity resulting from noise and stochastic switching—as exemplified by bacterial persisters, see later discussions and Ackermann 2015.) When the cycles occur in about 24-h increments they comprise what is known as **circadian rhythm**. The circadian clock may be set by an environmental cue, such as the well-known light:dark cycle, or it may run irrespective of an environmental signal (free-running rhythm) (Bell-Pedersen et al. 2005). Though varying extensively in complexity, all circadian systems are based on an internal oscillator system of negative and positive components that compose an autonomous, internal regulatory feedback loop used to generate the periodicity (for mechanism and examples, see Bell-Pedersen et al. 2005). Among the processes regulated by a circadian clock, depending on the organism, are photosynthesis (even in cyanobacteria; Cohen and Golden 2015), solar-tracking (heliotropism) in sunflowers (Atamian et al. 2016), cell division, uptake and other metabolic activities, locomotion, and melatonin levels. The unicellular cyanobacterium *Prochlorococcus* (discussed in ►Chap. 3) lives throughout the euphotic zone in tropical and subtropical oceans (Zinser et al. 2009; Biller et al. 2015). It is an oxygenic phototroph and so, not surprisingly, its daily activities are closely coordinated with the light:dark cycle. In a detailed study, Zinser et al. (2009) report on patterns of cell physiology (e.g., cell cycle; C-fixation; nutrient acquisition and assimilation, etc.) and gene expression patterns. They examined the entire transcriptome and found that 80% of the annotated genes exhibited cyclic expression over the diel cycle in tightly scripted fashion, with expression activity for most peaking at sunrise or sunset. New cells were born usually late in the day or at night; when sunlight appeared at dawn they were fully metabolically primed for maximum photosynthesis and biomass increase. This tiny unicellular creature illustrates beautifully how the ability of organisms to coordinate biological processes with daily environmental cycles has adaptive value.

Where there are seasonally recurring hazards, dormancy tends to be initiated synchronously with and typically somewhat in advance of the changes based on an external clock such as photoperiod. In the case of seeds, germination is more reliably triggered by photoperiod than by temperature, which, because of its vagaries, can be an unreliable predictor (e.g., hazards of a late frost). There is considerable theoretical work on optimal clock timing, presumably set by natural selection (Chap. 3 in Harper 1977; Scott and Otto 2014). A photoperiod-based clock may not allow the organism to take advantage of an extraordinarily early season but would prevent a catastrophic loss associated with being unprepared for an exceptionally late lethal event. Using photoperiod rather than temperature as the cue means that efficient, advance preparation is possible rather than having to adjust to events themselves and usually being too late. Thus, the physical form of the signal and the response may bear no relationship (Levins 1968; Harper 1977) and the less seasonal is the environmental change, the less useful is an external clock (e.g., unpredictable disasters such as hurricanes; see comments on adaptive phenotypic plasticity in Simons [2014] and imposed stochastic switching related to bacterial bet-hedging by Beaumont et al. [2009]). *While the cyclical synchrony between environmental adversity and organism preparations may give the appearance of prediction, it is important to remember that the pattern is ingrained because of the actions of natural selection acting over countless past generations* **on the individual's ancestors** (recall discussion on adaptation in ▶Chap. 1).

Some form of developmental delay is phylogenetically widespread and continues to be maintained by natural selection in organisms that physiologically could have shorter generation times than they in fact have (Levins 1969). This suggests that dormancy is beneficial especially as it is not a cost-free transition. Organisms usually have to make investments in switching to a resistant form; in the maintenance of the dormant state where, for example, environmental signals must be transduced; and in the process of resuscitation (Dworkin and Shah 2010; Lennon and Jones 2011). These costs vary substantially with the degree of shut-down and other internal and external factors such as temperature (Price and Sowers 2004), but are rarely if ever nonexistent. They have been examined in numerous papers by Maughan and colleagues (e.g., 2009) for the spore-forming microbe *Bacillus subtilis*. In principle, costs are theoretically lower for adaptation based on a purely stochastic switching mechanism (no sensing; involves population heterogeneity) than a responsive switching mechanism (involves direct response to an external cue) (Kussell and Leibler 2005). This leads to the conclusion that the cost of a responsive mechanism is justified where environmental uncertainty is higher, whereas the stochastic switching is favored where change is relatively infrequent.

Dormancy also comes with the ecological cost that by becoming quiescent the organism is opting out of active competition for resources, territory, and contributing its genes to the gene pool. As such, Harper (1977, p. 62) views dormancy, at least in plants, as a *"weak solution to the problem of adaptation to a changing environment"* (for one example of such costs, see Janzen and Wilson 1974). An alternative in a cyclically unfavorable environment is potentially to migrate. To a degree, organisms adapted for dispersal combine both options because dispersed propagules typically are dormant. At the extremes, Harper (1977, p. 82) views two strategies for dealing with a deteriorating local habitat: *"escape to somewhere else"* (i.e., migration), or *"wait until the right habitat reappears"* (dormancy). An intriguing idea is that mobility and the more versatile physiological adaptation of many animals to changing environments have largely replaced the need for dormancy (Bonner 1958). An intermediate option apparently is to transition to a different but metabolically still relatively active phenotype. A case in point is the alternation between large- and small-leaved forms in certain plants in seasonal wet and dry seasons; this is one example of **seasonal dimorphism** (▶Chap. 3 in Harper 1977; Palacio et al. 2006).

Bacterial 'persisters' In meeting potential environmental adversity, clonal organisms, by virtue of their growth form (▶Chap. 5), can do things that unitary organisms cannot: They put a fraction of their ramets on a standby condition, while others continue active growth. This would be effectively a form of bet-hedging and there is some evidence from clonal plants for this strategy (de Kroon et al. 2005; Magyar et al. 2007). This option does not involve sensing but rather simply population heterogeneity and has been discussed by Kussell and Leibler (2005) under the semantics of stochastic switching. There is even better evidence for such a strategy in the microbial world, particularly for bacteria. Certain bacteria exhibit standard forms of dormancy closely analogous to seeds based on production of a resistant, frequently long-lived propagule (e.g., endospore; McKenney et al. 2013). Of more interest, however, is that **all or virtually all bacteria produce at very low frequency a subpopulation of slow- or nongrowing cells called 'persisters'.** Other species segregate a population into 'sporulators' and 'diauxic growers' (Veening et al. 2008a). Not only is the persister phenomenon ecologically significant from the standpoint of survival, it has important medical implications because traditionally defined persisters are metabolically more-or-less inactive and as such are usually tolerant to antibiotics that kill replicating cells.[4]

Persistence has been recognized almost from the inception of the antibiotic era (early 1940s) and a classic example is the protracted regimen of multiple antibiotics necessary to control diseases such as tuberculosis caused by *Mycobacterium tuberculosis* (Hu and Coates 2003; Zhang 2014; see further comments on mycobacteria below). Persisters are a specific example of the ability of bacterial populations to exhibit **'bistability'**, which refers to the bifurcation of a system into two dynamically stable states, in this case phenotypically distinct subpopulations (Dubnau and Losick 2006; Veening et al. 2008b). The cells in the original metapopulation are clonal, effectively genetically identical, and grown under effectively homogeneous, 'identical' environments. Bistability can arise from stochastic, epigenetic alteration in gene expression (generally referred to as 'noise', noted at the outset and with further discussion later).

The persister phenomenon was investigated elegantly by Balaban et al. (2004) who directly monitored the time course of population dynamics in wild-type (*wt*) and persister-enhanced mutants of *E. coli* cells. Growth was followed microscopically in an ingenious microfluidic device where narrow grooves provided a system whereby the descendants of each bacterial cell formed a separate, linear microcolony. The authors found that the *wt* population consisted of three components: (i) persisters that were continuously generated by spontaneous, stochastic phenotype-switching; (ii) persisters that were triggered to form by responsive switching during stationary phase (for bacterial growth phases, see Finkel 2006); and (iii) typical, rapidly growing cells. This implies that the original clonal bacterial population became at some level heterogeneous and cells could switch reversibly between persister and normal growth. (The broader issue of individual cell *phenotypic* heterogeneity is discussed by Avery 2006 and Hashimoto et al. 2016.) Persisters also can be induced by various triggers. In numerous recent reports in this rapidly emerging field, several endogenous and exogenous 'stress' cues have been identified, among them starvation, DNA damage, and antibiotic treatment (Cohen et al. 2013; Amato et al. 2014; Zhang 2014).

It now appears that some forms of persistence may not be attributable to simply nondividing persisters, but rather to balanced division and death in this subpopulation; or to quiescence in most cellular activities but selective activity in others such as antibiotic efflux

---

4   Note that persistence is a transient phenotype and distinct from conventional drug resistance attributable to genetic mutation discussed at length in ►Chap. 2. Persisters regrow when the antibiotic is withdrawn and the resulting population consists of approximately the same fraction of persisters as did the original population and is of identical antibiotic sensitivity. Also note that persistence is related to but distinct from the 'viable but uncultivable' bacterial condition described in ►Chap. 6: Persisters recover and resume growth from a dormant state when the stressor is removed; noncultivable microbes by definition at least presumptively cannot be cultured.

systems (Pu et al. 2016; however, this seems to confuse conventional resistance mechanisms with the persister phenomenon. See ►Chap. 2 Sidebar and Nikaido 2009). With respect to balanced division and death, using similar microfluidic devices and time-lapse microscopy, in this case of *Mycobacterium smegmatis,* Wakamoto et al. (2013) describe stochastic pulses of the enzyme catalase-peroxidase (KatG) that activates a drug (isoniazid), which kills the bacterium. Hence, such bursts are negatively correlated with cell survival. Pulsing and death are not a function of single-cell growth rates but apparently are determined by molecular noise (see following section). Persistent subpopulations (termed cell lineages) of *M. smegmatis* may survive because of infrequent pulses of the enzyme, which in turn fail to activate isoniazid to lethal levels. Notable also is that mycobacteria are distinctive in several attributes: their unique cell-wall characteristics; their ability to extensively restructure the cell envelope in deterministic fashion based on environmental triggers; and their enlargement and division processes that result in asymmetric daughter cells (Aldridge et al. 2012; Kieser and Rubin 2014). The populations resulting from asymmetry have different growth rates and antibiotic sensitivity (Richardson et al. 2016). In other words, in mycobacteria these basic phenomena contributing to substantial phenotypic plasticity are supplemented by stochastic changes in gene expression.

To summarize, the persister phenomenon, whether arising from a truly quiescent or a dynamically maintained cell birth:death state, has important survival and evolutionary implications. It may actually be an 'evolutionary reservoir' for emergence of resistant cells (Cohen et al. 2013). It appears to be one of several examples of a bet-hedging strategy by bacteria (e.g., Veening et al. 2008b). While in origin, persistence, like mutation, may be inevitable and nothing more than the result of various errors (Johnson and Levin 2013), it is intuitive that the consequences have survival value. A clone possessing such a property would have an advantage over one that did not or that produced persisters less frequently. This remains to be seen; presently, some work supports such an interpretation (Kussell et al. 2005) and some authors argue otherwise (Johnson and Levin 2013).

7.6.2 Epigenetics, Bistability, Noise, and Development

Bacterial persisters and the role of stochastic gene expression as a putative mechanism in at least some cases of persistence (Wakamoto et al. 2013) raise the broad issue of gene regulation in development and evolution in changing environments. Gene regulation was alluded to earlier (see ►Chap. 2) and in many respects is similar in prokaryotes and eukaryotes. In bacteria, a fundamental regulatory mechanism is the **operon**, a cluster of functionally related structural genes under the control of a promoter, which together operate as a single transcriptional unit (Chap. 8 in Madigan et al. 2015). Enzymes for an entire pathway can be synthesized concurrently under the control of a single 'on/off' switch (operator). This design allows bacteria to adjust their metabolism rapidly to environmental fluctuations (►Chap. 3). Multiple operons, in turn, may be controlled at a higher level by a functional unit called a **regulon**. Control mechanisms exist at even higher organizational levels as well as at multiple lower levels. In eukaryotes, genes are packaged in chromatin and the transcription machinery is significantly more complex. As noted in ►Chap. 2, eukaryotic gene regulation can occur at any or several of six steps from gene to protein (◘Table 2.4 and Alberts et al. 2015). Of these, control at the transcriptional level (including chromatin changes as well as the initiation of transcription) is the most efficient and presumably the most important. Even in *Archaea* and *Bacteria*, expression is coordinated at the transcriptional level by activator and repressor proteins coded for by a regulatory gene outside a particular operon. These proteins bind to *cis*-acting regulatory sequences (so-called **cis-regulatory elements**) on the DNA, as they do in eukaryotes, where they regulate the degree of transcription by controlling the attachment of RNA polymerase. In multicellular eukaryotes, a given cell expresses only a small fraction of its genes and gene regulation is a central feature of cell differentiation, as well as in the reception/transduction of external or internal signals (Doebley and Lukens 1998; Carroll et al. 2005; Peter and Davidson 2015).

Virtually from the outset of genetics, the dogma became entrenched that phenotypic variability is attributable to mutation (implicitly, mutation of a structural gene). It was not until the classic work of Jacob and Monod on bacterial genetics in the 1960s that changes in gene *expression* became increasingly recognized as being evolutionarily important. Expression could change either directly, through alteration of what we know today as the *cis*-regulatory DNA of the gene, or indirectly, by changes in the upstream transcription factors that regulate the gene. The Jacob-Monod model established the important regulatory principles that gene expression is managed by an 'on/off' switch controlled by a DNA-binding protein and that this protein recognizes a particular DNA sequence in the vicinity of the gene. These concepts were extended by numerous theoretical and experimental papers in the 1970s on gene expression in eukaryotes (e.g., Britten and Davidson 1971; King and Wilson 1975). *When nongenetic, chemical changes to histones or to DNA (i.e., without affecting DNA sequence) are heritable, they are referred to in current semantics as* **epigenetic** (Griffiths et al. 2015) and the phenomenon is well documented both in bacteria (Casadesús and Low 2006) and eukaryotes (Gilbert and Epel 2009). Examples include some forms of phase variation in bacteria; prion-associated phenotypic change (below); genomic imprinting (gene inherited from one parent is not expressed because DNA is methylated); and various forms of histone modification. **At the population level, epigenetically inherited traits may be functionally indistinguishable from allelic alterations**; in fact, in the case of plants, several instances of heritable variation now attributable to chromatin architecture were originally interpreted as DNA variants (Johannes et al. 2008; He et al. 2011).

Some forms of epigenetically acquired traits may be relatively fluid. In interpreting the epigenetic influence of a prion-like protein, Jarosz et al. (2014a) noted two differences between this inheritance mechanism and conventional nucleotide variation: First, because cells can acquire and lose prions relatively quickly, such elements could provide for a more rapid and dynamic means of adjusting to environmental fluctuations than does the process of mutation and reversion to wild-type in acquiring complex traits; second, environmental stresses can actually increase the rate of acquisition or loss of these proteins. In the case of *Saccharomyces cerevisiae*, acquisition of the protein-based epigenetic unit allows the yeast to overcome glucose repression (i.e., the inability to use even trace amounts of most other carbon sources in the presence of glucose; see ►Chap. 3) (Jarosz et al. 2014 a, b). Certain bacteria associated with the fungus secrete a chemical factor that induces the epigenetic element. The resulting yeast phenotype becomes capable of using alternative carbon sources, becoming metabolically a carbohydrate generalist rather than a specialist, as it has come to be known from laboratory culture. This is a survival trait arising spontaneously in some members of a yeast population and likely of metabolic importance in the wild. Of particular interest and presumed ecological importance, the prion-like element was experimentally induced in wild strains but poorly or not at all in laboratory strains of *S. cerevisiae* by a factor secreted by wild but not by almost all laboratory strains of *E. coli* and *B. subtilis* tested. Presence of the protein element in evolutionarily diverse fungi showed that this and similar epigenetic mechanisms are conserved over at least 100 million years. As the authors point out (Jarosz et al. 2014b), the phenomenon really is a heretical example of 'Lamarckian evolution': a heritable, epigenetic trait elicited by a chemical signal secreted into the environment. As prions have usually been assumed to be pathogens, this work also indicates how they may have originated or been conserved by evolution (see similar remarks regarding pleiotropy, later).

The fundamentals of gene expression in prokaryotes and eukaryotes summarized above are relatively well known in very broad terms and comprise standard fare in biology textbooks. However, historically, the biochemical and molecular genetic processes have been inferred from studies on large populations of cells and macromolecules. Events are viewed as deterministic and interpretations are based on averages compiled across countless individual events. Only recently with the advent of techniques to examine living, single cells has it become possible to examine transcriptional and translational mechanisms at the single-molecule level (Eldar and Elowitz 2010; Xie et al. 2008; Li and Xie 2011; Chong et al. 2014). The biochemistry of individual cells is quite unlike that of mass solution chemistry of cell populations. The former is characterized by macromolecules in low copy numbers, stochastic reactions, nonequilibrium conditions, and holistic complexity rather than reductionist simplicity (Xie et al. 2008). In bacterial cells in particular there are only one to a few copies of a particular gene, few of a particular mRNA (due to short lifetimes), and few of many important proteins such as those regulating gene expression (Li and Xie 2011). Events such as gene regulation therefore need to be viewed as single-molecule processes and as such are inherently noisy (stochastic), potentially leading to phenotypic heterogeneity within an otherwise identical, isogenic cell population.

Noise Various types of dynamic behavior in eukaryotic, multicellular macroorganisms as well as in single-celled microorganisms occur as a result of stochasticity (Eldar and Elowitz 2010). These include the role of noise: (i) at the shortest timescales, in enabling certain physiological regulatory mechanisms such as the coordination of gene networks; (ii) in permitting a range of stochastic differentiation options not available in a deterministic system; and (iii) over the longest timescales, in facilitating adaptive evolution (some examples follow and for others see Eldar and Elowitz 2010). A specific case with a microorganism noted above is persistence of certain cell lineages of *Mycobacterium smegmatis* in the presence of the drug isoniazid (Wakamoto et al. 2013). Adaptation to the drug could potentially take the form of a lineage of cells characterized by infrequent stochastic pulses of the enzyme KatG. Although the pulsing mechanism is not clear, it appears due to noise-induced transcriptional bursts of *KatG*. Here the pulses, in turn, evidently result from reversible gyrase association and dissociation with a DNA segment, which changes the degree of supercoiling (Chong et al. 2014), though sources of molecular noise are various (Maheshri and O'Shea 2007; Potoyan and Wolynes 2015).

Other well-known bacterial phenomena ascribed to phenotypic bifurcation of a population (bistability) have been revealed in studies of the soil bacterium *Bacillus subtilis*, including genetic competence, spore formation/cannibalism, and swimming/chaining (Dubnau and Losick 2006). Noise can also lead to coexistence of distinct, bimodal states without strict bistability as such (To and Maheshri 2010). Transcriptional noise has been investigated in eukaryotes of varying complexity such as *Saccharomyces* (Blake et al. 2003) and mammalian cells (Suter et al. 2011), where it can be a significant variable influencing differentiation and development. Noise, while unavoidable and of functional value (Eldar and Elowitz 2010), may be subject to natural selection for minimization above a baseline level (Blake et al. 2003; Fraser et al. 2004) such as by modifications to promoter architecture (Kaern et al. 2005; Sanchez et al. 2013).

Molecular changes in multiple transcriptional control mechanisms undoubtedly played a major role in the evolution of metabolic pathways in bacteria and in the evolutionary divergence of bacterial species. Analogously, the importance of changes in regulatory as opposed to coding DNA has increasingly been emphasized in plant (Doebley and Lukens 1998) and animal (Peter and Davidson 2015) developmental biology. Carroll and colleagues (see e.g., Carroll et al. 2005; Prud'homme et al. 2007; Carroll 2008) attribute the greater role of regulatory DNA in morphological evolution to three factors: (i) the higher 'degree of freedom' in cis-regulatory sequences to mutational change (no need to maintain a particular reading frame); (ii) modularity of the cis-regulatory elements (individual elements can evolve independently); and (iii) combinatorial action of the transcriptional factor repertoire in animal cells. Doebley and Lukens (1998) view plant development as a cascade of events beginning with internal signals at embryogenesis, followed by both internal and external environmental signals conveyed through a hierarchically arranged modular system that is progressively activated. The components include transducers of the signals (e.g., ligands, receptor kinases); transcriptional regulators such as the cis-regulatory sequences and associated proteins noted above; and the target genes at any point in the relay system.

Depending on location, some target genes have only a local impact while others near the start of the cascade (e.g., signaling) can have far-reaching effects. An allele affecting several organism properties is generally termed pleiotropic. **Pleiotropy** is a good example of the developmental category of evolutionary constraints discussed in ▶Chap. 1. Pleiotropy constrains the kinds of changes possible in morphological evolution (Prud'homme et al. 2007). A bacterium can be assembled faster and more easily than can an elephant. Therefore, the constraints of informational relay and sequential gene action in a cascade are presumably much less for bacteria than for elephants. Importantly, however, **pleiotropic and related effects can nevertheless in some cases be beneficial in facilitating coordinated rather than individualized change of multiple phenotypic traits.** Thus, despite its potentially adverse consequences, and depending on trade-offs, pleiotropy may be favored overall by natural selection (Cheverud 1984; Guillaume and Otto 2012). Changes in cis-regulatory regions rather than in coding sequences tend to circumvent or minimize pleiotropic effects (Prud'homme et al. 2007).

7.7   Habitable Sites and the Evolution of Dispersal

》 .. a *a species* of plant is an island in evolutionary time to the insect *species* that feed on it and .... the *individual* plant may also be analyzed as an island in space and contemporary time to the *individual* insects that feed on it.
>   Janzen, 1968 p. 592, commenting on MacArthur and Wilson's theory of island biogeography [with word emphasis as per Janzen]

Much of the conceptual framework of organism and environment can be consolidated into a model that interprets the distribution and abundance of organisms in terms of habitable sites. Aspects of the habitable site idea were developed mathematically by Gadgil (1971) and extended conceptually by Harper (1977, 1981b), among many others (e.g., Johnson and Gaines 1990; McPeek and Holt 1992; Travis and Dytham 1999; Levin et al. 2003; Ronce 2007). What follows here builds on Gadgil's notion and depicts it descriptively and graphically, rather than mathematically. In the ensuing decades since his conceptual paper on the evolution of dispersal, the issue has been explored in depth, as have the closely related topics of colonization, landscape heterogeneity, and metapopulation biology (see e.g., Hanski and Gilpin 1997; Turner et al. 2001), as well as the life history correlates of dispersal (Stevens et al. 2012). With some exceptions (e.g., Andrews et al. 1987; Taylor and Buckling 2010) this body of theory has not been integrated into microbial ecology (or vice versa), where the focus has been on bacterial and fungal dispersal, host range of pathogens, cropping patterns, and epidemiology (Baker and Cook 1974; Burdon 1987a).

Habitable space can be defined as a zone in which biotic and abiotic conditions allow an organism to become established and survive competitively. Obviously the clearest example of a habitable site is the place where an organism can complete its life cycle (i.e., including a reproductive phase). A site is not habitable to a species simply because it contains viable propagules, such as seeds or spores, of that organism. Habitable sites are thus distinct from 'sink' sites, which may contain individuals—perhaps even consistently and in relatively high numbers—but only because they are supplied as immigrants from a source site nearby (Pulliam 1988). The human colon but not a soil particle is a habitable site for *E. coli*. Thus, habitable space consists of localized, favorable conditions that occur as islands or patches within a biologically and physically hostile sea. Distribution and abundance will then be determined in large part by the match between characteristics of the organism (e.g., dispersal ability; intrinsic population growth rate) and environmental pattern (e.g., size, number, distance between, duration, and carrying capacity of habitable sites). As seen in the preceding sections, chances for a match are improved by phenotypic plasticity of individual colonizers, and by genotypic variation within the colonizing population to cope with the variability of sites. Occasionally, habitable sites used alternately by a particular colonist, can have very different characteristics (see **Sidebar, A Case Study**).

SIDEBAR: A Case Study: The American Legion, Cooling Towers, and a Novel, Phenotypically Plastic Bacterium

There are probably few organisms with a more interesting ecology (not to mention notoriety) than *Legionella pneumophila*, the bacterium that causes what became known as Legionnaires' disease or Legionellosis. This inconspicuous microbe achieved international fame when 149 of some 4400 delegates attending a convention of the American Legion in Philadelphia in the summer of 1976 became ill. All developed, to various degrees, similar symptoms including fever, cough, muscle aches, chills, diarrhea, and pneumonia. More cases were discovered within the local populace. Eventually 221 people were affected (including those not directly associated with the Legion event) and 34 died. Interestingly, it was shown in due course that many hotel employees carried antibodies to the pathogen, implying that it had been present for some time. Reinvestigation of earlier cases attributed to nonspecific pneumonia showed that the Legionnaires' pathogen was infecting people at least as early as 1947 (reviewed by Gomez-Valero et al. 2009). Pathogenesis of the disease is reviewed in detail by Swanson and Hammer (2000).

The Philadelphia mystery was unraveled through intensive, persistent epidemiological and etiological studies (Fraser et al. 1977; McDade et al. 1977; Fraser and McDade 1979). The story is a fascinating case study in insightful medical sleuthing (see excellent synopsis by Fraser and McDade 1979). An early, significant finding was that the victims had only one thing in common: they lived in, or spent time in or near, the hotel where the convention was held. Epidemiological work by the CDC did not support person-to-person transmission and there was no evidence from surveying subsequent hotel guests that the epidemic continued after the Legion event. Eventually, the causal organism was cultured by an unorthodox procedure for bacteria, identified—again by unusual methods in bacteriology—and shown to be a new species. In particular, the culturing procedure was a pivotal development, not only in demonstrating the cause of the disease, but in establishing one aspect of the complex environmental relationships of the causal organism. *L. pneumophila* was found to have many distinctive properties, including fastidious nutritional requirements such that it grew in the laboratory only if culture media contained unusually high amounts of both cysteine and iron. (Subsequent work has elucidated over 50 species of *Legionella*. Some are closely related to *L. pneumophila* and about half are clinically important; Lau and Ashbolt 2009.)

Epidemiological evidence pointed to an airborne pathogen transmitted from a point source and eventually suggested strongly that the singular feature of that particular hotel was the cooling towers for the air conditioning system, which provided a mechanism for concentrating and distributing the bacterium. In later outbreaks (as well as in past episodes, known retrospectively from serological evidence to have been caused by *L. pneumophila*), such sources have been implicated unequivocally (as have other water systems subsequently, including domestic water supplies; Stout et al. 1985; Lau and Ashbolt 2009). The evidence includes patients known to have been exposed to drift from the source and culture of the pathogen from the cooling towers or evaporative condensers.

So, in terms of the environmental biology of this organism, we are faced initially with an intriguing paradox: How does a nutritionally demanding pathogen, which can be cultured in the laboratory only under the most exacting conditions, and that in its human host multiplies primarily in lung tissue, maintain substantial populations in water? A key later discovery was that *Legionella* could grow in ciliated protozoa and amoebae, which are the hosts in a commensal or parasitic, intracellular relationship (reviewed in Abu Kwaik et al. 1998; Swanson and Hammer 2000; Steinert et al. 2002). This aspect of the life history of the pathogen is complex though epidemiologically important because it is an environmental reservoir for survival and multiplication. The bacterium exhibits several distinct phenotypes within its protist hosts (Gomez-Valero et al. 2009). In the aquatic environment, *Legionella* not only obtains nutrients from its associated microorganisms within biofilms, but is protected from desiccation while in transit. Fliermans et al. (1981) isolated the bacterium from diverse aquatic habitats (e.g., temperature range 5.7–63 °C; pH 5.0–8.5) although the concentration in lakes and rivers appears low other than in association with the protozoal host and biofilms.

L. pneumophila is now considered to be a thermotolerant component of freshwater microbial communities (Steinert et al. 2002). It also grows well in hot water tanks and distribution systems, particularly in areas where sediment (scale and organic floc) accumulate (Lau and Ashbolt 2009). Interestingly, the sediment seemingly acts by promoting growth of other bacteria, which in turn enhance growth of *Legionella* (Stout et al. 1985). (The occurrence of *Legionella* spp. in various water systems including drinking water is discussed by Lau and Ashbolt [2009].) When encased within amoebic cysts or protozoal vesicles, the bacterium resists desiccation and biocides. It seems more than coincidental that amoebae share many attributes of alveolar macrophages, a key site of pathology in humans (Swanson and Hammer 2000), who evidently are incidental habitats in the life history of this opportunistic pathogen. The infection cycle is very similar in both amobae and human macrophages (Gomez-Valero et al. 2009; Escoll et al. 2013). Its long coevolutionary relationship with a eukaryotic, protozoal host likely preconditioned it via numerous virulence factors for pathogenesis in humans. The genomic sequence of the pathogen revealed genes for unexpected pathways and other attributes that in retrospect help explain the organism's ecology, including its broad host range (Chien et al. 2004; see also Burstein et al. 2016).

Tolerance of the pathogen for thermal aquatic environments, among other ecological attributes, explains why air conditioning systems can be ideal reservoirs and distribution systems for airborne infection. Their associated water cooling devices provide warm, wet habitats and often receive makeup water from sources containing protozoa and multiple other microorganisms. The cooling towers or, in other cases, evaporative condensers, also act as efficient scrubbers of microbes from the air. A typical tower of the size for a large hotel handles water at a rate of about 3800 L per minute. This hot water from the compressor unit is sprayed over splash bars, cooled by evaporation, and returned to the compressor. Fresh air is pulled through the spray with a fan to enhance evaporation. Airborne microbes drawn in by the fan are entrained in the spray, as are any microbes originating from the water system itself. Many are expelled in minute water droplets and aerosols as drift in the discharge airstream (Fraser and McDade 1979).

In overview, as is true of the Case Studies in previous chapters, this example presents an interesting problem in basic ecology with obvious practical implications. The organism grows in more-or-less habitable sites as diverse as portions of the human body (even though the human host is a dead-end), in protists, and also in various forms principally as biofilms in natural or artificial water systems. *Legionella* clearly has the ability to dramatically change its environment; in turn, survival of the bacterium in nature seems to depend on a complex association with other microorganisms that provide the appropriate nutritional and protective milieu. In its various habitats and growth forms it exhibits multiple states and extreme phenotypic plasticity. Knowledge of the environmental biology of *Legionella* provides a basis for the intelligent application of controls, such as elevating the temperature of domestic hot water supplies, or chlorination, which can be directed at the localized inoculum reservoirs involved in amplification and dispersal of the pathogen.

Habitable sites also have variable temporal and spatial components. This is perhaps best illustrated by considering how environments could appear in time and space to an organism (Southwood 1977). With respect to time they can be: (i) constant (favorable or unfavorable indefinitely); (ii) predictably seasonal; (iii) unpredictable; or (iv) ephemeral (predictably short, favorable conditions followed by unfavorable conditions for an indefinite time). In space, habitats can be: (i) continuous (favorable area exceeds area that organism can reach regardless of dispersal mechanism); (ii) patchy (unfavorable areas surrounding favorable islands which can be reached by dispersal); or (iii) isolated (favorable but unreachable). Consider a migratory bird that breeds over the summer in northern Canada and passes the winter in the southern United States. The two sites are predictably seasonal and the bird's life cycle is adjusted to exploit both. On a local scale, the relative area occupied at each site and population density will fluctuate annually depending on factors such as pressure from competitors and predators, which also occupy the site. The life cycles of plant and certain animal pathogens that alternate habitats is directly analogous. The remarkable intersection in time and space between habitable sites and appropriate growth stage to colonize them is best seen among species with complex life cycles (▶Chap. 6; see also Chap. 3 in Price 1980). So, sites are highly plastic in size, may be temporally discontinuous, and are not species-specific. They can be depicted simplistically in three-dimensional form through time as meandering tubes of irregular diameter (◻Fig. 7.8). Individuals in contemporary time or lineages in evolutionary

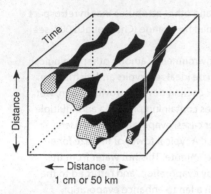

← Distance →
1 cm or 50 km

◘ Fig. 7.8 The habitable space concept. Favorable sites, projected through time, appear as continuous or discontinuous patches (irregular tubes). Species cohabiting a site may or may not directly interact. In response to biotic or abiotic factors, any given species may increase, become temporarily dormant, or decline possibly to local extinction pending recolonization

time can escape extinction by surviving competitively in a particular habitable site, reproducing sexually or asexually, and then dispersing to a former or different habitable site.

The dynamic aspect of habitable space results not only from changes in the organism or the site, but also from interactions between organism and site. It was noted at the outset of the chapter that, strictly speaking, separation between organism and environment is impossible. As the organism moves through its life cycle it selects those parts of the site that are **relevant** to itself (Lewontin 1983) and **reorganizes** them: Twigs become nesting material for birds; trees are felled to become food and shelter and dams for beavers; lignin and cellulose in wood are degraded by fungi and the monomers incorporated into fungal protoplasm while fungal metabolites are also extruded and further modify the habitat and influence succession (Lonsdale et al. 2008); the DNA of a plant host is reprogrammed by a tumor-inducing plasmid of the crown gall bacterium to encode unique molecules (opalines and nopalines) used as nutrients by the pathogen and not the host (▶Chap. 3 and Platt et al. 2014). Organisms modify a site further by such activities, in the process often making it more-or-less habitable by certain other species. In the extreme case, a successional sequence develops, as consortia of organisms are displaced by successors better adjusted to the dynamically evolving environment-community matrix (Weiher and Keddy 1999). A classic example is the loss of a major 'foundation species', chestnut, from the southern Appalachians where it has been a co-dominant with oak for thousands of years, as well as in the Northeast (Paillet 2002). This was caused by the introduction from Asia of a virulent fungal pathogen in the early 1900s. The result was both the devastation of an entire tree species throughout its range and an entirely changed forest ecosystem. This is but one example among many of the far-reaching ramifications surrounding loss of dominant, influential species (for examples and terminology see Ellison et al. 2005).

Distance between habitable sites is arbitrary and related to distribution of favorable conditions relative to the size and growth form of the organism, which in turn determine other significant correlated attributes. For instance, colonizable patches may be separated by distance on the order of micrometers for bacteria or yeasts growing on a leaf surface; by meters for earthworms; or by up to kilometers for the Glanville fritillary butterfly in the Åland

Islands of Finland. That habitable space may exist unfilled is clear from the frequent success of intentional species introductions or inadvertent[5] invasions (Mooney and Drake 1986).

Distance between real or virtual islands is a factor driving dispersal and molding life cycle features. The probability of colonizing any site is a function of several factors, among them the probability of propagule survival and the number of arriving propagules. Gadgil's (1971) models show that, as a result of dispersal, isolated, poorly accessible sites should be less crowded than an average site. A single episode of dispersal can produce appreciable differences in the extent of crowding at the various sites. A phenomenon that might be called 'immigrant amplification' can occur in which, even though dispersal results in no net additions or losses of individuals, it can still increase mean population densities if the episodes of net migration are positive when populations are growing and negative when they are declining (Ives et al. 2004). When migrants have high rates of mortality, the dispersal event could reduce crowding at all sites, effectively lowering the total population size of the species. Likewise, the greater the variation in carrying capacity among the different sites, the greater the impact on over- or under-crowding. This will tend to depress the metapopulation size over its entire range.

Species have different tendencies to disperse and different dispersal characteristics. This will affect gene flow (used here to mean gene movement among different populations of the same species). For instance, dispersal is generally strong and rigidly programmed in insects, evidently as an adaptive response to their short life cycles and ephemeral breeding sites. Migration is less programmed in vertebrates, but still predictable based on certain factors such as sex and age (e.g., young males are the most prone to disperse in baboon troops and lion prides; Wilson 1975). Plant seeds and microbes even more so are the ultimate nomads; the old adage *"everything is everywhere, the environment selects"* is commonly applied to microbial biogeography and powers of dispersal.[5] Historical, geographical, or geological constraints are less important considerations in assessments of distributional patterns than they are for macroorganisms (▶Chap. 4 in Brock 1966; Hanson et al. 2012). Small size facilitates long distance movement by wind, water, or vectors across barriers which would be insurmountable for macroorganisms. A short generation time and high fecundity, which imply potentially high population densities, favor colonization. Fungi particularly seem to be close counterparts to insects in programmed dispersal characteristics (timing in life cycle; specialized migratory forms), although there are obvious differences in that fungal dispersal is largely passive, whereas insect dispersal is frequently directed. Dispersal characteristics and colonization dynamics have been examined at length elsewhere (MacArthur and Wilson 1967; MacArthur 1972). Where dispersal is passive, as is typical in plants and microbes, the number of individuals declines exponentially (e^{-x}) with distance x from the source. Where dispersal is directed, as in the case of many animal search patterns, the decline follows a normal distribution pattern (Wilson 1975, pp. 104–105). The two dispersal types will thus have very different effects on the rates of gene flow.

How has dispersal evolved? Since travel is risky, migration will only be favored by natural selection if the chances for finding a better site exceed those of colonizing a worse site and of death in transit. A strategy of partial population migration may be the fittest (Gadgil 1971). Harper (1977) has added that for plant seeds it may occasionally be better to remain dormant indefinitely in situ without dispersal. A case would be where there is a higher prob-

---

5   Not surprisingly, this simplistic assertion, still hotly debated, grossly overstates the case. While the environment undoubtedly selects, microbial distribution is nonrandom over all levels of scale. For expansion of these points, see Martiny et al. (2006), de Wit and Bouvier (2006), Hanson et al. (2012).

ability of a conducive environment reappearing that that of dissemination to a distant habitable site. In general, however, dispersal must be advantageous because it is exhibited in some form in all phyla and is typically marked by adaptations that promote survival and dissemination. Dispersal can also entail considerable cost to the organism in expended time and resources. Birds migrate only after physiological preparation and training flights, and lose massive amounts of stored energy reserves en route. In the slime mold *Dictyostelium*, up to 90% of the formerly free-living cells are assigned to act as stalk cells to elevate the fruiting body. They die as they are incorporated into the stems (Whittingham and Raper 1960), so the benefits of getting spores to a favorable 'launch site' and a new, potentially habitable site must outweigh these costs (Bonner 1982a). The benefit of dispersal to the individual (as opposed to the weaker argument for group selection at the species level) is the chance to colonize an empty or rarefied habitat, and the initial mating advantage over the locals upon entering the new population (part of the process of so-called 'migrant selection'; Wilson 1975).

In many if not most unicellular organisms, dispersal may be incidental and the cells involved essentially no different from those remaining at the site of production. Incipient forms of dispersal were likely rudimentary, haphazard processes. It is easy to see how chances for successful migration could be improved by modest physiological changes, such as evolution of a shut-down phase (Lennon and Jones 2011), or by cells that became less adherent than their counterparts. The population would thus become partitioned into members suited for active growth or for survival and transport. Among bacteria, several extant species, such as among the myxobacteria and *Caulobacter* spp. discussed previously, have stages in their life cycles morphologically adapted for dispersal. This is not unlike the segregation of insect populations into sedentary and mobile forms (winged versus non-winged aphids; *solitaria* and *gregaria* phases of the African locust). A later evolutionary development than simply shut-down, perfected by the fungi, would have involved cells (spores) packaged for transport with a nutrient reserve and thick walls and, in many cases, propelled on their way by elaborate release mechanisms (Ingold 1971). Wind dispersal of bacteria, fungal spores, and pollen has been studied for decades since the classic volume of Gregory (1961) on aerobiology. The importance of dispersal for the fungi is strikingly revealed in the elaborate adaptations for spore production, discharge, and dispersal. This includes the shape of the mushroom fruiting body (including the bell-shaped cap and height of the stalk), mechanisms to forcibly discharge spores or promote entrainment in air currents, and the evolution of spore shape to minimize drag (Deering et al. 2001; Trail 2007; Roper et al. 2008, 2010).

In plants and fungi there is obviously a close linkage between dispersal and sexual reproduction. Possibly selection acting on the latter led to adaptations also fostering the dispersal process. The container enclosing the products of meiosis, such as pollen, or of genetic recombination such as seeds, sexual spores, cysts, might have been an easily transported unit (Bonner 1958). Organisms reproducing largely by clonal means in effect sample new environments by growth (Jackson et al. 1985; Hughes 1989). It was noted previously (▶ Chaps. 5 and 6) that while such organisms as strawberries and corals can in theory grow in unlimited fashion, in practice it is only a matter of time before clonal individuals reach the boundaries of their habitable sites (Chap. 3 in Williams 1975). Dispersal by clonal fragmentation is merely an extension of the growth process whereby the genet can move farther afield, freed from the handicaps related

to physiological transport and allocation problems, or systemic toxins and infections, of oper
ating as a single, massive physiological unit. Mobile, unitary organisms typically select sites in
an active, controlled manner, whether this be accomplished by relatively primitive chemotaxis
mechanisms (nematodes with their plant hosts), or by sophisticated olfactory, visual, and acoustic cues (vertebrates and higher invertebrates). Hence, a fundamental distinction in dispersal
seems to be drawn along lines of growth form (►Chap. 5), rather than kingdom.

Whatever its origin, dispersal is likely best interpreted within the context of challenge and
opportunity met *gradually* and improved upon over geological time by different organisms
interacting with their environments. This returns us to the opening remarks of the chapter: In
evolutionary terms it seems no more likely that movement by one means or another to habitable sites suddenly became a 'problem' to be solved by terrestrial species than did swimming
in water become a 'problem' that seals 'solved' by losing their legs. Lewontin (1983) observes
that seals developed flippers and in so doing likely incorporated water as a progressively
greater component of their environment. Analogously, almost from the inception of life, a
migratory phase joined a sedentary phase as an increasingly important and programmed
component of the life cycle.

7.8   Summary

The environment includes the physical and biological setting of an organism with which it
is coupled and reciprocally interacting. This external milieu is superimposed on the internal
environment in which the genes and other cellular machinery function. Environments both
drive natural selection by their differential impact on survivorship and are in turn altered by
the changing composition of the survivors. The developmental programs of organisms neither simply unfold in mechanical fashion against a placid environmental backdrop, nor do
organisms simply create solutions to problems posed by the environment.

Organisms experience the same absolute fluctuations very differently depending on their
lifespans. For a bacterial cell (as opposed to the bacterial evolutionary individual or clone)
with a generation time of hours to days under natural conditions, absence of nutrients or a
slight drop in temperature for several hours would have major physiological and growth rate
implications. Either change over the same absolute time frame for an elephant with a generation time of 12 years would be negligible.

How organisms experience and respond to environmental fluctuations also is affected by
their size and growth form. Increasing size of the physiological individual is related directly
to increasing complexity manifested in various ways: increase in number of cell and tissue
types, and hence in the interactions among them; division of labor among cell types; formation of support structures; insulation and homeothermic mechanisms; centralized hormonal
and neural control. Size is inversely proportional to generation time. While a bacterial cell
is not as well insulated from its environment as an elephant cell (or, correspondingly, at the
genet level, the bacterial clone or the whole elephant), the bacterium can track environmental
fluctuations much more rapidly by phenotypic and genotypic changes. The macroorganism
responds by virtue of a complex neural network; the counterpart of the neural network for

the microorganism, enabling in effect predictive behavior, is the network of regulated bio-chemical pathways and metabolic controls. Modular organisms, composed of iterated parts and being usually sessile (►Chap. 5), respond differently to the environment than do unitary organisms: For example, resistant propagules (seeds, spores, bulbs, etc.) may outlast local adversity in situ or, frequently, be transported to new sites, while the soma changes by addi-tion or subtraction of modules. Unitary organisms, which are typically mobile, adapt primar-ily by migration, and by physiological and behavioral mechanisms.

Environments have been classified from the organism's perspective with respect to resources as fine-grained (experienced in many small doses and not actively sought out or avoided) or coarse-grained (sufficiently large so that the individual chooses among them or spends its whole life in one patch). Though conceptually appealing and useful in a modeling context, the theory of environmental grain is abstract, restricted to behavior in the context of resources, and delimited by the assumptions of the model.

When individuals of a species experience relatively the same environment throughout their lifespan, locally adapted genotypes producing relatively invariant phenotypes known as ecotypes or races, are expected. Where individuals are exposed to an environment that var-ies principally spatially or temporally, phenotypic plasticity—defined as any kind of environ-mentally induced phenotypic variation in behavior, physiology or morphology—is expected. Phenotypic plasticity offers the major advantage of dampening the effects of selection by uncoupling the phenotype from the genotype and in so doing provides for adaptation to vari-able environments. Such plasticity, frequently called 'phenotypic heterogeneity' in bacteriol-ogy, can occur within a genetically homogeneous cell population in a constant environment triggered by stochastic molecular noise. Noise enables types of dynamic behavior in eukary-otic, multicellular macroorganisms as well as in single-celled microorganisms.

Dormancy, apparent as various manifestations of quiescence, is usually interpreted as a bet-hedging strategy to adverse conditions. Organisms receive environmental cues and transduce these from a physical or chemical mode into an appropriate biological response. Where environmental fluctuations are irregular, the response needs to occur directly as the environment changes. Where fluctuations are predictably cyclic, it is advantageous for organ-isms to 'anticipate' them by recognizing some form of early signal. For example, both mac-roorganisms and microorganisms exhibit circadian rhythms. In seasonal terms, the message "day length is changing" is transduced via hormones to activate flowering in some plants, or reproductive activity in certain animals. Correlations between environmental parameters allow one form (e.g., photoperiod) to act as a predictor of another (e.g., temperature). Par-ticular phases of life cycles appear to match those environmental conditions for which they are suited but this is a result of natural selection acting on the individual's ancestors over gen-erations.

Certain bacteria and most fungi exhibit standard forms of dormancy closely analo-gous to seeds based on production of a resistant, frequently long-lived propagule such as an endospore. Of more interest is that many, if not all, bacteria produce at very low frequency a subpopulation of slow- or non-growing cells called 'persisters'. Such bifurcation (bistabil-ity) of populations has survival value in that the quiescent cell fraction tolerates antibiotic or other environmental adversities that kill active, dividing cells.

Beginning with the classic work of Jacob and Monod on bacterial genetics in the 1960s, changes in gene *expression* became increasingly recognized as being as evolutionarily impor-tant, if not more so, than mutation of structural genes. When changes in gene expression are heritable, they are referred to in current semantics as epigenetic and the phenomenon is well established both in bacteria and eukaryotes. At the population level, epigenetically inherited traits may be functionally indistinguishable from allelic alterations. Events such as gene regulation should be viewed as single-molecule processes and as such are inherently noisy (stochastic), potentially leading to phenotypic heterogeneity within an otherwise essentially identical, isogenic cell population. Molecular noise enables dynamic behaviors in eukaryotic, multicellular macroorganisms, as well as in single-celled microorganisms.

The distribution and abundance of species can be interpreted in large part by the match between organism and environmental pattern. Habitable sites are patches where the organism can develop competitively. Sites are dynamic in time and space due to changes in the organism, the environment, or the organism–environment interaction. Species invasions and successful introductions show that habitable space may exist unfilled. Distance between habitable sites is a factor driving dispersal and molding life cycle features. Gene flow patterns are quite different for organisms that disperse propagules passively, which decline in numbers logarithmically from a source pool (such as plant seeds, microbes, and clonal benthic invertebrates), and organisms that disperse progeny in directed fashion (most unitary animals), represented generally by a normal distribution pattern of decline. Despite the high mortality that typically occurs during a migratory phase, dispersal is evidently favored by natural selection because in some form it occurs universally and is often facilitated by intricate adaptations. Dispersal is an ancient evolutionary development that has been progressively improved upon by various mechanisms.

Suggested Additional Reading

Bonner, J.T. 1988. The Evolution of Complexity by Means of Natural Selection. Princeton Univ. Press, Princeton, NJ. *How increase in size and complexity of organisms has evolved, and implications for environment-organism interaction.*

Carroll, S.B., J.K. Grenier, and S.D. Weatherbee. 2005. From DNA to Diversity: Molecular Genetics and the Evolution of Animal Design, 2nd ed. Blackwell, Malden, MA. *A clearly written, beautifully illustrated book on animal development, including the seminal role of gene regulation as the major creative force underlying morphological innovations.*

Lewontin, R. 2000. The Triple Helix: Gene, Organism, and Environment. Harvard University Press, Cambridge, MA. *A reminder that organisms both make and are made by their environments. Organisms are not simply constructed from a DNA recipe.*

Peter, I.S. and E.H. Davidson. 2015. Genomic Control Process: Development and Evolution. Academic Press (Elsevier), NY. *A comprehensive, readable, timely overview of genomic regulatory systems in animal development.*

Conclusion: Commonalities and Differences in Life Histories

© Springer Science+Business Media LLC 2017
J.H. Andrews, *Comparative Ecology of Microorganisms and Macroorganisms*,
DOI 10.1007/978-1-4939-6897-8_8

*The most important feature of the modern synthetic theory of evolution is its foundation upon
a great variety of biological disciplines.*

G.L. Stebbins, 1968, p. 17

8.1   Levels of Comparison

Because the life history of every species is at some level unique, the challenge in attempting to
forge a comparative synthesis is finding instructive (as distinct from trite) commonalities. On
one hand, if important differences are overlooked, the analogies are misleading and forced.
Indeed, this must happen occasionally because what is truly ecologically important may
become known only in hindsight. Francis Crick has said ... *"There isn't such a thing as a hard
fact when you're trying to discover something. It's only afterwards that the facts become hard"*
(Judson 1979, p. 114). Boulding (1964) has cautioned about the dangers of false analogies in
a world driven by the "rage for order", where order or pattern is perceived but in reality does
not exist. Importantly, however, he says further that (p. 36) ... *"the remedy for false analogy is
not no analogy but true analogy."*

A good case in point of seemingly empty coincidence is what became of Hutchinson's
(1959) famous observation that average individuals from sympatric, congeneric species
showed regular differences in various morphological features such as length of proboscis
in bumblebees or length of skull in squirrels (the so-called 'Hutchinsonian ratio' or 'rule').
This was a potentially important contribution to the concept of character displacement, or
the shift that occurs when two partly allopatric (nonoverlapping) species become sympatric (overlapping) in portions of their range. The data he presented (see Hutchinson's Table 1,
p. 153) from various sources referred to skull dimensions in some mammals and the cul-
men length (upper ridge of the bill) in birds. Hutchinson's brief comment on this specific
subject in the context of an extensive conceptual paper ended with the summary and conjecture that in general ... *"where the species co-occur, the ratio of the larger to the small form
varies from 1.1 to 1.4, the mean being 1.28 or roughly 1.3. This latter figure may* **tentatively
[emphasis added]** *be used as an indication of the kind of difference necessary to permit two
species to co-occur in different niches but at the same level of a food-web"* (p. 152). In an era
when the role of competition as a force shaping communities was in its ascendancy, and
despite Hutchinson's caveats, these ratios quickly became elevated to the stature of a rule
embedded in the literature, and supplemented by numerous subsequent reports from other
authors. The ratios frequently were seized on ... *"as prima facie evidence that communities
were organized according to the principles of limiting similarity"* (Eadie et al. 1987, p. 1). Horn
and May (1977) showed later, however, that the rule also applied to such inanimate objects as
an ensemble of musical instruments and the wheel sizes of children's bicycles and tricycles.
The ratios are now generally regarded as being artifactual (Roth 1981), an inevitable outcome
simply of a lognormal distribution in animal size with small variance in distribution (Eadie
et al. 1987). The conditions apply to many animate and inanimate objects. While the ecological principles of character displacement and limiting similarity undoubtedly are valid, the
underlying processes cannot meaningfully be captured in a size ratio. Likewise, the similarity
between animal size and abundance patterns and those of car 'species' also raise a cautionary
note with respect to conclusions that can be validly drawn (see discussion in ►Chap. 4 and
Gaston et al. 1993).

☐ Table 8.1 Some levels at which all biota can be compared, the bases for comparison, and the anticipated similarities

Level	Comparative basis	Expected commonalities
Cell	Mechanics; physiology; function; biochemistry; genetics	Numerous: energetics; genetic code and its replication and translation; metabolic cycles; cell division
Organism	Mobility; homeostatic ability; size; shape; mechanical design; life history patterns, incl. resource allocation; modular versus unitary	Each organism unique in detail; many analogous[a] (e.g., foraging, reproductive, dispersal mechanisms); and some homologous[a] features (e.g., forelimbs of chickens, humans, dogs, whales; cellulose in cell walls of members of Kingdom Plantae)
Community	Major organizing forces (e.g., competition, predation; mutualism; abiotic factors); recruitment (open vs. closed); equilibrium versus nonequilibrium	Few: patterns such as species abundance distribution may be similar but for different reasons; organizing forces may be the same but relative roles differ; size–frequency distribution for unitary but not modular organisms

[a]Analogous here implies functional and/or morphological parallels among taxa not resulting from a common descent; homologous is similarity resulting from common ancestry. See ▶Chap. 1. "**Organism**" is highlighted because it is the level of focus in this book

On the other hand, sweeping generalities set at a level to avoid peculiarities of specific circumstances are usually so bland as to be trivial and uninformative. This dilemma is shared by model builders who compromise between accuracy and generality (see earlier quotation from Box and Draper 1987 footnoted in ▶Chap. 6). As seen above, the problem is confounded by the intermingling of fact and dogma.

At what biological level does it make sense to compare the ecology of microorganisms and macroorganisms? As noted in ▶Chap. 1, all organisms are quite similar in terms of their basic cellular function. Similarities among taxa diminish, as they do among other entities, as one moves through higher levels of systems organization (☐Table 8.1 and Slobodkin 1988; see also Zamer and Scheiner 2014). This principle is but one of many examples of the general laws of integrative levels (Feibleman 1954): Each succeeding systems level embodies the attributes of its predecessors as well as imparting some of its own (emergent) properties. Hence, organisms can be expected to be more complex and distinctive as entities than the molecules of which they are mutually composed, and populations and communities, though sharing broad attributes, are yet more complex as systems than are the component organisms. For these and other reasons related to tractability and the levels which natural selection effectively acts (see Preface and Introduction), the comparisons drawn in this book have been focused at the level of the individual. The complexities of ecological systems in many contexts, including the important implications of scale, are considered insightfully by Allen and Hoekstra (1992).

8.2   On Being a Macroorganism or a Microorganism

The major frame of reference for this book, as the title implies, is size differences across major groupings of taxa, that is, microorganisms versus macroorganisms. As noted in ►Chap. 1, however, there is obviously a continuum between the two categories for several reasons; in other words, the demarcation is one of convenience and is not a simplistic binary choice. Indeed, the impetus for this book is the arbitrary distinction, reinforced by disciplinary isolation, between microbial ecology and plant/animal ecology. Instead it is possible to compare all cellular life essentially along six axes as we have done here with respect to: genetics, nutrition, size, growth form, life cycle, and environment.

Be that as it may, size is clearly an ecologically important attribute of an organism and influences profoundly its interactions with the biotic and abiotic environment. Microbes are affected primarily by intermolecular forces while large creatures live in a world governed by gravity. The direct consequences of this were discussed in ►Chap. 4 and were developed, along with many indirect consequences, in each of the other chapters. Most if not all attributes of an organism are influenced by its size, and much of the variation in life histories is correlated with size. One cannot compare organisms strictly on the basis of size, however, because this factor is not independent of others. First, organisms are locked into developmental and ecological channels established by phylogenetic differences, the most obvious of which are design constraints (►Chaps. 1 and 4). A wolf and a fish of equivalent weight have very different life histories **and evolutionary potentials.** Second, even among geometrically similar organisms of a common phylogeny (such as different species of lizards), shape within limits tends to change with increasing volume leading to proportionate rather than progressively diminishing changes in surface area of many structures. Efficiency in biological (exchange) processes dictates an approximately constant ratio of surface:volume. The same relationship means that supporting elements (bones, stems) must be proportionately thicker as size increases if the organism is not to fall under its own weight.

Within the scale from microorganism to macroorganism, increasing size also means increasing complexity (more cell types) and concomitantly increasing division of labor and centralized control; increasing independence of the external environment (homeostatic ability); and increasing chronological time to maturity and between generations (although ratios of life history features to *physiological* time seem to be constant at least in unitary organisms; ►Chap. 4). Population densities (and intrinsic growth rates) of microorganisms are consequently much higher than those of macroorganisms. Since for a given nucleotide sequence the rates of base substitution are approximately constant per unit time across lineages (molecular evolutionary clock; ►Chap. 2), favorable mutations will spread more rapidly in a bacterial than in an elephant population. Put differently, this means that microorganisms, though more exposed to environmental variation (►Chap. 7), can evolve more rapidly in response to it.

8.3   Natural Selection as the Common Denominator

As noted in ►Chap. 1, a general property uniting organisms is that all are shaped by evolutionary pressures. The major evolutionary process underpinning comparisons in this book has been natural selection. This is as opposed to other forces or phenomena such as founder effects, archetype (phylogenetic) effects, genetic drift, or pleiotropic effects (for insightful discussion of these and others, see Harper 1982). Many genes or gene families (e.g., histone, cytochrome, etc.) have been conserved in widely divergent organisms through a vast evolutionary history. Thus, it should not surprise us that analogous responses or 'strategies' to similar fundamental challenges or 'problems' also have emerged in the ecology of organisms, regardless of their size (◘Table 8.1). By virtue of sharing fundamental biological attributes and being extant, all have passed thus far the endless sifting and winnowing of natural selection. It is therefore an unfortunate mistake that demarcation based on size, implicitly or explicitly, has acted to compartmentalize so much of the thinking in ecology (and biology in general).

8.3.1 The Principle of Trade-Offs as a Universal Currency

The general model presented at the outset (◘Fig. 1.2 and Chap. 1) as a common basis for ecological comparisons is the organism as an input/output system. The organism is viewed as a black box with an input and an output: it acquires various resources and produces progeny. The challenge for all organisms is, and always has been, the same: Each life history is of necessity a compromise to multiple demands on limited time, resources, and options. Stearns (1992, his Chap. 4) identifies "at least 45 trade-offs" (p. 72), particularly those between current reproduction and survival, current reproduction and future reproduction, reproduction and growth, number and quality of offspring, among others. In a world where no compromises were necessary, under natural selection every organism would operate at its optimum phylogenetic limits—an observation obviously inconsistent with a wealth of evidence from both microbial ecology and plant and animal ecology (see especially ▶Chaps. 1, 3, 5, and 7; also Stearns 1989, 1992). This, arguably the most fundamental ecological tenet at the organism level, might be called the Principle of Trade-Offs.

Natural selection should favor organisms that are cost-efficient **overall** in resource acquisition, allocation, and expenditure. As such, they should be competitively superior and leave more descendants. (Note that while the principle of allocation is intuitively acceptable, it is usually not clear to what extent any particular allocation pattern translates ultimately into numbers of **descendants**; ▶Chap. 3.) The activities of an organism on which selection acts can be subdivided for convenience into five arbitrary categories (Southwood 1988): (i) tolerance of the physical environment; (ii) defense; (iii) foraging; (iv) reproduction; (v) escape in time or space. Although the input/output scheme, as a simplistic model, does not explicitly recognize the functional categories (i), (ii) and (v), they are implicitly a part of it. Costs assignable to these categories will have a negative impact on foraging and a positive or negative effect on reproduction. For instance, with respect to defense, time spent by a bird avoiding predators or, with respect to escape, time spent by a fungus ensconced as a resistant, quiescent sclerotium, in both instances is time lost, for example, to feeding. Modifications in these diversions may or may not prove to increase the number of descendants (▶Chaps. 3, 7, and Andrews 1992). Natural selection is the ultimate arbiter.

Various levels of ecological generality and comparability can be recognized for each of the above categories in comparisons among taxa. Take, for example, adjustment to the physical environment. The most general comparison entails only a slight restatement of the principle of trade-offs, viz., microorganisms as well as macroorganisms must compromise allocations to physiological adjustment with other demands on their time and resources. This is a valid generality, but being a universal statement conveys nothing of specific value in any particular context. A second, more specific level of comparison could be that, in the short term, bacteria respond rapidly to a rapidly changing environment by altering metabolic pathways, whereas plants respond primarily by a system of cues and hormonal controls triggered by environmental change. The two organisms do so in part by analogous mechanisms of phenotypic plasticity (►Chap. 7). In fact, as we saw, at least some bacteria even exhibit circadian rhythm. For both creatures this implies costs as well as benefits. Being a more specific statement than the first, it carries useful predictions, though these predictions may mostly pertain to bacteria and plants. Finally, at the third and most specific level, one could postulate that bacterium species A copes with stochastic adversity by segregating a tolerant subpopulation (for example, of 'persisters'; ►Chap. 7) prepositioned on 'standby' as it were, while plant species B responds to seasonal adversity by shedding its leaves. At this level a given response does not necessarily apply even to fairly closely related taxa, much less from microorganism to macroorganism. A different genus of bacterium may form spores; a different plant species might cope with the environmental change by employing apical dormancy. In our context and across this spectrum from very general to quite specific, a very interesting and productive analogy is at the level of evolution of phenotypic plasticity: what forms does it take in microorganisms versus macroorganisms; to what extent does plasticity shape the respective life cycles; what is the genetic/epigenetic basis; and do some of the genes involved even have a common evolutionary origin? Parenthetically, it is noteworthy that such plasticity is among the most ancient of organism traits, having evolved in primitive bacteria, and in some form being phylogenetically universal. Clearly it appears to have been a seminal evolutionary event.

8.3.2 Growth Form as the Integrator of Life Histories Across Size Categories

If size establishes fundamental differences at the level of the individual, then growth form is arguably the best unifier. Instead of segregating the biological world by size into micro- and macroorganisms, it can be cut along the axis of growth form into modular or unitary organisms. Such a scheme constitutes a rational as opposed to an arbitrary demarcation and is more biologically justifiable than is one based on size.

The unifying feature of growth form is that microorganisms share with a whole set of macroorganisms a modular life style. Foremost among such characteristics are sessility, indefinite growth by iteration, and a developmental mode that is effectively based on totipotency or somatic embryogenesis (►Chap. 5). As developed earlier, the main ecological and evolutionary implications include: (i) replacement of active mobility with growth by extension, passive dispersal mechanisms, and dormancy; (ii) exposure of the same genetic individual to different environments (hence to different selection pressures); (iii) close interaction among subsets of a population (neighbor effects); (iv) high phenotypic plasticity; (iv) the potentially significant role of somatic mutation in evolution; and (v) substantially attenuated or absence of senescence at the level of the genet.

By varying their growth form directly in response to environmental variation, modular organisms make trade-offs visible in a way that their unitary counterparts do not. Consider a fungus that can either forage for nutrients by a diffuse mycelial network, or extend rapidly away from resource depletion zones into new terrain by aggregating its hyphae into telephone cable-like strands (rhizomorphs), or divert biomass into wind disseminated spores and the specialized structures built to broadcast them. Dispersal in some form is evident in all organisms, but the costs are shown most dramatically within the modular category. As addressed in earlier chapters, mortality may exceed 90% of the cell mass in dictyostelid cellular slime molds, where the stalk of the fruiting body consists of dead cells 'sacrificed for the cause'. Among the fungi (e.g., Aphyllophorales of the Basidiomycetes), resources are diverted into massive bracket conks, structures that exist purely for dispersal and are ultimately destined to die. Moreover, the wood-rotting fungus may produce conks prolifically, thus expending its finite resource quickly 'in brief but riotous living'. Alternatively, it may opt to stagger reproduction over time, thereby prolonging the food base but risking loss to competitors or by mishap. Among plants, the compromise is between the competing needs of photosynthetic and non-photosynthetic tissue. Among sessile marine invertebrates, it is between biomass allocated for support against turbulence, for food interception, or for production of progeny. While the modular life style makes the concept of trade-offs especially compelling, sorting out the forces involved in any situation and the relative roles of each in the evolution of form is challenging.

What limitations does growth form impose on the extent to which basic ecological concepts can be generalized? A good example is the logistic or sigmoid equation developed by Verhulst (1838) to describe growth of human populations. The sigmoid model and variations of it have been used almost universally as descriptors of population increase (►Chap 4; for microbial analogues see Andrews and Harris 1986). In terms of a unifying ecological principle, logistic theory at the population level is as basic a tenet as is that of trade-offs at the level of the individual. It is the foundation for fundamental sequels such as the Lotka–Volterra model of interspecific competition, and r- and K-selection (►Chap. 4). Although density-dependent regulation of populations in some form is generalizable, as is r/K theory, the predicted outcome may vary depending on growth form. Sackville-Hamilton et al. (1987) propose that the effects of r- and K-selection may not be the same on unitary as on clonal (modular) organisms. For instance, whereas genets with a phalangeal growth form (►Chap. 5) are expected to behave according to the theory, those with the guerrilla habit are not. In the latter case, r-selection favors nonreproductive (nonsexual), so-called immortal genets that spread clonally unless there is a lethal condition that kills all stages in the module-to-module cycle. K-selection acts similarly, provided the environment is reasonably stable over many generations (if it is not, then K-selection should favor reproduction). These predictions are of course opposed to convention, which associates greater reproduction with r-selection and competitive growth with K-selection (►Chap. 5). The authors observe that the genets of some higher plants are 'immortal' and never reproduce. It is also interesting that numerous fungi appear to grow indefinitely as clones. The significance of the paper by Sackville Hamilton et al. (1987) in our context is that growth habit, not size per se, may greatly influence response to selection pressure. Hence, it would be informative to align evolutionary predictions with respect to growth form, rather than segregated by size to microorganism as opposed to macroorganism.

8.4 Recapitulation of Some Major Points

For readers interested in a brief synopsis of some themes to take away from the book, the following list may be useful. If somewhat more elaboration is desired, a review of the summaries at the end of each chapter is recommended.

▬ Every organism is unique in detail, but all share fundamental physiochemical and cytological properties and all have been shaped by evolution operating through differential reproductive success. The latter attribute provides a common basis for comparing analogous strategies among organisms.

- The term 'individual' has various connotations: numerical, genetic, physiological, and ecological. An informative way to view an individual is in the ecological context as the entity through its entire life cycle.

- Microorganisms have essentially all of the capability of macroorganisms in generating genetic variability, though adaptive evolution proceeds rather differently in the two groups. Variation specifically in the prokaryotes is transmitted in dynamic, unordered fashion, as opposed to the orchestrated manner, characterized by meiosis and gametogenesis, characteristic of the eukaryotic microorganisms and macroorganisms. In prokaryotes, the gene pool of distant relatives is tapped by horizontal transmission that introduces fundamentally new traits in a manner that apparently occurs rarely in eukaryotes. The fungi, as eukaryotic microorganisms, have a fairly fluid genome as exhibited by such processes or conditions as parasexuality, dikaryosis, and heterokaryosis.

- The occurrence of different ontogenetic programs among taxa means that in some cases somatic variation can be transmitted to the germline. This major implication, together with the different and in some cases multiple nuclear conditions during the life cycle, and the ubiquity of mobile DNA, mean that the concept of the genetic individual (genet), though remaining instrumental in terms of assessing how evolution acts, must be viewed as more fluid than as originally conceived.

- The quest for nutrients and energy is universal and the evolution of search strategies is a universal basis for comparisons. Optimal foraging theory is a cost/benefit analysis in energetic terms developed primarily as an economic optimization model to interpret the foraging behavior of certain animals. It can be construed broadly, merged with optimal digestion theory, and applied informally, conceptually, and empirically to all organisms. Broadly speaking, in terms of foraging, bacteria appear to do largely by metabolic versatility what animals accomplish by mobility and behavior, and plants, fungi, and other sessile organisms do by morphology.

- The major multicellular lineages evidently all arose from different unicellular progenitors. These transitions apparently occurred many times and in all three major domains of life, sometimes through a colonial intermediary and with occasional reversals to unicellularity.

- For about 2–3 billion years, life on Earth was microscopic and simple, preponderantly if not exclusively prokaryotic, with only basic forms of differentiation occurring, if at all. Conventional multicellularity was ultimately an evolutionary dead-end for prokaryotes; eukaryotes exploited growth form, in large part enabled by a fundamentally different cell structure as the building block, in a way that prokaryotes could not.

- Physical and chemical laws ultimately set the theoretical lower and upper size constraints on life: The low end, approached by some very small bacteria, is established by the minimum package size to contain cell machinery and associated molecular traffic; the upper limit on cell size is set mainly by the declining surface area-to-volume ratio as size increases.

- The small size of prokaryotes (and microorganisms in general) means that their world is dominated by molecular phenomena such as diffusion, surface tension, viscosity, and Brownian movement. Macroorganisms, in contrast, are conditioned primarily by gravity and scaling relationships. Extremely large size can be attained in some circumstances where gravitational and other constraints can be mitigated, such as by clonal growth on land or partial buoyancy in water.

- Allometric scaling denotes regular changes in certain proportions or traits as a function of size according to the general relationship of $Y = aW^b$ where exponent b is the scaling constant that defines the general nature of the relationship. A broadly based and biologically important example is the surface area law or the 2/3-power law (exponent b is 2/3) dictating the relative decline in surface area as volume increases (Principle of Similitude).

- Both the mean and upper size limits among biota as a whole have increased over geological time. The most important event spurring evolution of greater complexity arguably was the transition from unicellularity to multicellularity. This enabled organisms to increase in size, differentiate to segregate function by specialized cell type, and to develop increasingly sophisticated forms of intercellular communication and division of labor.

- Growth form, that is, the mode of construction and related morphology, sets fundamental opportunities and limits on the biology of organisms. All living things are basically either unitary or modular in design. The modular versus the unitary distinction is probably more biologically significant than a demarcation based on size (microorganisms versus macroorganisms). Bacteria and fungi are inherently modular in design and should be considered analogous with modular macroorganisms.

- The key evolutionary implications of modular design stem directly or indirectly from growth by iteration and sessility. The former include: (i) high phenotypic plasticity (shape; size; reproductive potential; growth as a population event); (ii) exposure of the same genet to different environments and selection pressures; (iii) iteration of germ plasm and the potentially important role for somatic mutation; and (iv) potential extreme longevity of the genet. The evolutionary implications of sessility include: (i) adaptations for the individual to reproduce without a mate or to reach a mate and disperse progeny by growth or transport (pollen; spores; seeds; bacteriophage movement of bacterial genes); (ii) potentially strong interactions with neighbors; (iii) inability to escape from adverse environments, hence the means to adapt in situ such as by dormancy; (iv) development of resource depletion zones and consequently resource search/exploitation strategies from relatively fixed positions.

- The inception of the life cycle is traceable to the simple cell cycle of prokaryotes. With the origin of multicellularity and eukaryote sex, life cycles became more expansive and identified with distinct phases related to developmental stage or nuclear condition. Apparently all organisms ultimately pass through a single-cell stage, a key event with profound genetic and developmental implications.

- When offspring are born into essentially the same habitat as the adults and do not undergo sudden ontogenetic change, the organism is said to have a simple life cycle (e.g., mammals). Where organisms have two or more ecologically distinct phases separated by an abrupt ontogenetic change, the life cycle is considered complex in ecological semantics (CLC). The rust fungi, many of whose parasitic life cycles consist of a sequence of five morphological states on two distinct hosts, illustrate the intricacies and principles of the complex life cycle. CLCs usually have been interpreted adaptively. However, the CLC may be an evolutionary dead-end and a prime example of how evolution can drive organisms to greater degrees of specialization.

- Senescence reflects deteriorative effects that decrease fecundity and the probability of survival with increasing age. At the population level, actuarial statistics showing an increasing mortality rate over time are indicative of senescence. If mortality is constant with age, the number of survivors declines exponentially and there is no basis for inferring from the data that senescence occurs.

- Vertebrates and many invertebrates such as the nematodes, crustaceans, insects, as well as some plants such as the determinate annuals, senesce. Evidence for senescence appears limited or nonexistent generally for most modular organisms, including benthic sessile invertebrates, most perennial plants, and microorganisms. For some organisms the probability of survival actually increases, along with fecundity, later in life. The terms 'negligible senescence' and 'negative senescence' have been used to describe such population dynamics. In an evolutionary context and as a first approximation, senescence is predicted to occur wherever the reproductive value of the individual diminishes with increasing age.

- The environment includes the physical and biological setting of an organism with which it is coupled and reciprocally interacting.

- How organisms experience and respond to environmental fluctuations are affected by their size and growth form. The macroorganism responds by virtue of a complex neural network; the counterpart of the neural network for the microorganism is the network of regulated biochemical pathways and metabolic controls. Modular organisms, composed of iterated parts and being sessile, respond differently to the environment than do unitary organisms

- When individuals of a species experience relatively the same environment throughout their lifespan, locally adapted genotypes producing relatively invariant phenotypes known as ecotypes or races, are expected. Where individuals are exposed to an environment that varies principally spatially or temporally, phenotypic plasticity—defined as any kind of environmentally induced phenotypic variation in behavior, physiology or morphology—is expected. Plasticity dampens the effects of selection by uncoupling the phenotype from the genotype, providing for adaptation to variable environments. Such plasticity, frequently called 'phenotypic heterogeneity' in bacteriology, can occur within a genetically homogeneous cell population in a constant environment triggered by stochastic molecular noise.

- Dormancy, apparent as various manifestations of quiescence, is usually interpreted as a bet-hedging strategy to adverse conditions. Particular phases of life cycles appear to match those environmental conditions for which they are suited but this is a result of natural selection acting on the individual's ancestors over generations.

8.5   On the Comparative Ecology of Microorganisms and Macroorganisms

To conclude, we return to the contention (see Preface) that the ecology of macroorganisms is essentially phenomena in search of mechanistic explanation, whereas microbial ecology is experimentation in search of theory. Slobodkin (1988, p. 338) has commented that *"ecology may be the most intractable legitimate science ever developed"*. Despite the rich theoretical base of plant and animal ecology, few clear generalizations or specific predictions applicable to reality emerge that are other than trivial, and many of the precepts are contentious. Macroorganism ecology remains a field constrained by correlation. This is not surprising, given the complexity of higher organisms and their habitats, the interactions among traits, and the consequent difficulty of explaining phenomena mechanistically. The consequence is that controlled experiments frequently are difficult, if not impossible, and conclusions are often not robust (for criticism of ecological experimentation see Hairston 1990; however, for a good example of how critical field experimentation in macroecology *is* possible, see Daan et al. 1990 on optimal clutch size in birds; for a good discussion of the utility and limits of different types of experiment, see Diamond 1986). Long-running debates on such topics as niche theory; resource generalists versus specialists; adaptive radiation; and the role of contingency in evolution over either contemporary or geological time, are but a few conspicuous examples.

Microbial ecologists, meanwhile, have focused historically on autecological studies of selected species generally under laboratory conditions. Increasingly, such entities are known only by their nucleotide sequences, from which a phenotype is inferred. Broadly speaking, while microbial systems exist that could be used to answer rigorously many ecological questions that cannot be addressed mechanistically by plant and animal studies, they still remain comparatively unexploited. Conversely, the principles and concepts of macroecology, which could serve to guide and underpin microbial ecology, remain generally unintegrated as a body of work. There have been some rare, shining exceptions, such as the insightful approach of Alexander (1971) about 50 years ago in his classic text on microbial ecology whose foundation was ecological theory. Nevertheless, the still common academic practice of sequestering microbiologists in departments unto themselves (and of placing plant and animal ecologists in botany and zoology departments, respectively), and of structuring curricula accordingly, have promoted isolationism. The upshot is that the artificial distinction between microorganism and macroorganism is amplified, different concepts of what ecology supposedly is have emerged, and different vocabularies exist to describe fundamentally the same phenomena. This is but one manifestation of the march to specialization in science, a broader point addressed eloquently by Greene (1997). Finally, the gulf between microbial and plant and animal ecology has been widened by the historical difference in methodological approach, alluded to in the Preface and ►Chap. 1. The new, ever-expanding world of genomics (and numerous related "... omics") and molecular systematics offer powerful, potentially revolutionary insights into the evolution of organism function and design. Examples include 'soil metagenomics', now some decades old and, of much more recent origin, the so-called 'phytobiome' and the 'gut microbiome'. Despite some sanguine progress reports (Stearns and Magwene 2003), whether these novel subdisciplines serve to really bridge the gulf or merely extend reductionist thinking further into the macroorganism world, remains to be seen.

What does experimentation with microorganisms hold of conceptual value for the ecology of macroorganisms? How might such studies inform the discipline of ecology as a whole? One answer is that by virtue of being more amenable as research subjects, microbes provide excellent models to test ecological theory. There is a direct parallel here between such ecological exploration with the use by geneticists since the 1960s of microbial research models such as *E. coli*, *Saccharomyces cerevisiae*, *Neurospora*, and *Caenorhabditis* to answer fundamental questions in physiology and genetics applicable to macroorganisms. Though hardly free from their own limitations and artifacts, the numerous attributes of microbes and their potential utility as research models were alluded to in ►Chap. 1: easy clonal propagation of effectively a single genotype; fast generation times; capability for prolonged storage under conditions of suspended animation; high degree of control over environmental conditions.

Several examples of the insightful application of microbes to test ecological theory are noteworthy: Microbes have been used to address optimality concepts (e.g., the metabolic burden or maintenance costs associated with 'excess' gene functions can be tested with prototrophic and auxotrophic forms—see ►Chap. 3 and Zamenhof and Eichhorn 1967). Also as recounted in ►Chap. 3, Dykhuizen and colleagues have used bacterial systems to effectively address major ecological topics such as metabolic constraints, competition, and the evolution of specialism versus generalism in resource use (e.g., Dykhuizen 2016). In research extending over several decades, Graham Bell and colleagues at McGill University have focused on relatively simple model organisms in controlled culture, principally the unicellular alga *Chlamydomonas*, bacteria, and domestic and wild species of the yeast *Saccharomyces* (e.g., Replansky et al. 2008) to address fundamental ecological issues such as adaptive diversification, 'evolutionary rescue', historical contingency, the fitness of long-term sexual and asexual populations, and trade-offs. Paul Rainey and colleagues in New Zealand have used bacteria to examine numerous aspects of evolutionary processes, among them the role of environmental heterogeneity in adaptive radiation (Rainey and Travisano 1998), niche theory, and the origins of individuality and multicellularity. Interestingly, the same forces (environmental heterogeneity) driving Rainey's static bacterial culture to diverge genetically and ultimately morphologically are driving morphological divergence in sticklebacks that inhabit different environments (limnetic vs. benthic) in the same lake (Rundle et al. 2000). Richard Lenski and collaborators at Michigan State University, in a mammoth experiment begun in 1988 and ongoing, are studying fitness and evolutionary processes as revealed under controlled conditions in *E. coli*. Samples are taken every 500 generations from 12 clonal populations originating from a common ancestor, frozen, and compared through time. As of 2013 the subpopulations had been evolving for >50,000 generations. Among the Lenski group's many fascinating insights over the long course of this work is that average fitness continues to increase (and is projected to continue to do so by their model), although apparently at a decelerating rate (e.g., Wiser et al. 2013). Most of the 12 cultures behaved similarly in this regard, suggesting that evolution, at least among clonal populations in standard conditions, is broadly reproducible. Nevertheless, with respect to some key traits, historical contingency may be critical, a conclusion made elegantly by virtue of the power of their sequential sampling regimen that allows retrospective examination of population change through time in each line; e.g., Blount et al. 2008). Lenski's former student, Brendan Bohannan, currently at the University of Oregon, has been at the forefront of applying ecological theory to microbial ecology in many contexts (e.g., see Costello et al. 2012 and Horner-Devine et al. 2004), below. Aspects of our own work in ecological theory and microbes began in the early 1980s and have been noted in various chapters in this book (see example below on island biogeography).

Two areas of major convergence between microbial and plant and animal ecology are biogeography, and community ecology and community assembly, which have received significantly increased attention in the past two decades (Horner-Devine et al. 2004; Martiny et al. 2006; Vellend 2010; Costello et al. 2012; Nemergut et al. 2013). The stages, forces, and processes involved appear to be closely analogous for both microorganisms and macroorganisms despite some obvious differences in attributes such as dispersal, dormancy, and plasticity discussed in earlier chapters. Indeed, even within the microbial world, there are likely differences in the specific colonization dynamics between bacteria and fungi (Schmidt et al. 2014). Notwithstanding such idiosyncrasies, the simplicity and experimental tractability of microbial systems have and will continue to provide an instructive model for the experimental complexities of the macroorganism counterpart. Aspects of the colonization (invasion) process, particularly when a new habitat patch occurs, appear to be especially analogous (e.g., Gjerde et al. 2012; Tan et al. 2015). Notwithstanding close theoretical as well as practical similarities between the subdisciplines of invasion biology in ecology and biological control in plant pathology, there has been to date minimal crossover of literature or cross-fertilization of researchers and research ideas. This is ironic and regrettable.

As early as 1967, Ruth Patrick at the Academy of Natural Sciences in Philadelphia used the conceptual framework established in various indices of diversity, Preston's models of commonality, and MacArthur and Wilson's theory of island biogeography (1967), all developed for macroorganisms, to rigorously test area relationships and colonization dynamics of diatoms on glass slides immersed in flowing water from a spring or streams of different characteristics (e.g., species source pool size). Some of her experiments varied the invasion rate by controlling the flow rate of the water, and the size of the 'island' (slide). Her work not only established the importance of all three factors in the number of species and diversity of a community but highlighted the role of invasion rate in maintaining relatively rare species in a community and the implications of crowding to local extinction of those species represented by very small populations. Tom Brock and his students at Madison, Wisconsin, extended Patrick's work with studies on the population biology of aquatic bacteria (reviewed in Brock 1971). They ingeniously differentiated immigration from growth processes by pulsing the evolving population at short intervals by UV irradiation. Some years later, we implemented a conceptually similar disinfestation approach to test island biogeography and community assembly precepts with respect to microbial colonization of leaf surfaces. The microbes were filamentous fungi and wild yeasts, and the leaf islands were those of apple trees under manipulated (leaves initially effectively sterile; Kinkel et al. 1987; 1989a, b) or unmanipulated, orchard conditions, where species and population dynamics were followed from the incipient stage of bud break in the spring (surfaces relatively uncolonized) to leaf senescence and abscission in the fall (surfaces extensively colonized) (Andrews et al. 1987).

Another advantage of microbes in evolutionary ecology is they are unique in offering a relatively close, direct linkage between genotype and phenotype. Their developmental and structural simplicity means that the problems of pleiotropy and epigenetic effects (the so-called 'gene net' of Bonner 1988, pp. 144 and 174–175; or 'informational relay' of Stebbins 1968), though real, are comparatively small relative to those of macroorganisms. For instance, imagine that we are interested in whether a particular phenotypic trait is responsible for the competitive dominance of species A over species B in a community. To take a practical example from plant pathology, suppose further that organism A is seen to inhibit B in an agar plate assay, ostensibly by production of an antibiotic. We might notice also that under field conditions, application of populations of A also controls a disease incited by B. At this observational level there is merely a *correlation* between antibiotic production and competitive suppression, just as among macroorganisms there may be a correlation between, say, body size and spatial location of species along a resource gradient. In the latter case, interspecific competition has often (and occasionally erroneously) been inferred as the responsible mechanism (see examples cited by Roth 1981; Connell 1983; Price 1984; Eadie et al. 1987).

The cause for such an observed pattern typically can be determined directly and often at the level of the gene in microorganisms, however, in a way that it cannot for the macroorganism analog. First, the antibiotic from A could be purified, tested for an effect on B, and demonstrated to be present at inhibitory levels at the field microsites where antagonist A and pathogen B interact. Second, two complementary mutational analyses could be performed to implicate antibiosis as the mechanism. Mutants of A lacking inhibitory activity to B should not produce the antibiotic and should fail to prevent disease development. Complementation of the mutants to restore antibiotic production should also restore competitive dominance and disease suppressing ability. Second, mutants of the pathogen B could be made and tested in the same manner as those of antagonist A. B mutants insensitive to the antibiotic should cause disease in the presence of A; restoration of sensitivity should coincide with failure to cause disease. Finally, any remaining remote possibility of a correlative rather than a causal relationship between antibiotic production and biocontrol activity can be effectively eliminated by examining many mutants (for examples, see Hirano et al. 1997; Raaijmakers and Mazzola 2012).

The degree to which studies in microbial ecology will prove to be of heuristic value in macroecology, and vice versa, remains to be seen. But, clearly there are precedents and the prospects are encouraging. The fundamental commonalities as well as the practical examples reviewed in this book mark the progress of multiple research groups and provide many intriguing points of departure for further research. Nevertheless, it is evident that there will always be examples (and always a level) of comparison that just 'don't fit'. Nature displays an astounding richness of expression, but is surprisingly conservative in underlying pattern. Adopting a broad conceptual perspective in our individual studies is enlightening and intellectually gratifying.

8.6   Summary

This chapter recapitulates the central themes of the text. Although the life history of every species ultimately is unique, homologies and analogies exist at some levels of comparison. The latter are the focus of the book. Microorganisms and macroorganisms interact with their environments differently, for the most part as a result of size differences over orders of magnitude. Because of their small sizes, microorganisms are governed largely by intermolecular forces, whereas macroorganisms are affected mainly by gravity (▶Chap. 4). With increasing dimensions inevitably come increasing complexity (number of cell types; division of labor) and other correlates of size, among them increasing generation time, hence decreasing rates of genetic change at the population level (▶Chap. 2) but increasing homeostatic ability of the individual with respect to environmental change (▶Chap. 7). Nevertheless, use of size as a primary discriminator of the living world is artificial and largely an historical anachronism. A more biologically defendable basis is to integrate across size by delimiting with respect to growth form into modular and unitary organisms (▶Chap. 5).

A universal basis for comparison is the individual as an input/output system wherein trade-offs are inherent because of conflicting allocation demands among growth, maintenance, and reproduction for finite time and resources (▶Chaps. 1, 3, 5). One would expect that analogous strategies would be manifested across the domains of life in response to the sieve of natural selection. Many are exemplified herein. The life cycle itself, like all other attributes, is a compromise response by the organism (▶Chap. 6). Within phylogenetic constraints, natural selection should lead to increasing overall cost efficiency. Cost-efficient creatures should be competitively superior and should thereby leave more descendants. Within this general principle, progressively more specific statements or postulates can be made as bases for comparisons across taxa. Eventually a point is reached where contrasts are meaningful only at the intrageneric or intraspecific level. The body of theory on optimal foraging and optimal reproductive tactics, though introduced by ecologists in specific contexts with specific organisms in mind, are broadly relevant conceptually and take on different connotations depending on whether one is considering bacteria or elephants. Thus while r- and K-selection, for instance, is a general ecological concept (still highly contentious some 45 years after its introduction!), the predictions may vary with respect to growth form of the individual, and within the grouping of growth form the mechanistic basis for an r or K strategy may well differ at the guild or at the species level.

The ecology of macroorganisms remains, in part of necessity, an endeavor dominated by correlation rather than definitive causation. In contrast, the ecology of microorganisms lacks to date the strong theoretical basis and conceptual integrity of plant and animal ecology. Its reductionist focus does offer the potential for highly controlled experimentation and for dissecting life history features at the mechanistic and molecular genetic level not possible with the more complex eukaryotes. A broader perspective applied to studies with *any* organism would promote insight acquired by seeing things from a novel vantage point. It would also foster an appreciation that beneath a bewildering array of expressions there is fundamental integrity in nature. This would advance the whole discipline of ecology.

Literature Cited

Aarssen LW, Schamp BS, Pither J (2006) Why are there so many small plants? Implications for species coexistence. J Ecol 94:569–580

Abedin M, King N (2010) Diverse evolutionary paths to cell adhesion. Trends Cell Biol 20:734–742

Abrahamson WG, Caswell H (1982) On the comparative allocation of biomass, energy and nutrients in plants. Ecology 63:982–991

Abrams PA, Ludwig D (1995) Optimality theory, Gompertz' Law, and the disposable soma theory of senescence. Evolution 49:1055–1066

Abu Kwaik Y, Gao L-Y, Stone BJ, Venkataraman C, Harb OS (1998) Invasion of protozoa by *Legionella pneumophila* and its role in bacterial ecology and pathogenesis. Appl Environ Microbiol 64:3127–3133

Ackermann M (2015) A functional perspective on phenotypic heterogeneity in microorganisms. Nature Rev Microbiol 13:497–508

Ackermann M, Stearns SC, Jenal U (2003) Senescence in a bacterium with asymmetric division. Science 300:1920

Adl SM, Simpson AGB, Farmer MA, Andersen RA, Anderson OR et al (2005) The new higher level classification of eukaryotes with emphasis on the taxonomy of protists. J Eukaryot Microbiol 52:399–451

Agrios GN (2005) Plant pathology, 5th edn. Elsevier, NY

Aime MC (2006) Toward resolving family-level relationships in rust fungi (Uredinales). Mycoscience 47:112–122

Aime MC, Toome M, and McLaughlin DJ (2014) Pucciniomycotina In: McLaughlin DJ and Spatafora JW (eds) The mycota. VII. Part A. Systematics and evolution, 2nd Ed. Springer, Berlin, pp 271–294

Aime MC, Matheny PB, Henk DA, Frieders EM, Nilsson RH et al (2006) An overview of the higher level classification of Pucciniomycota based on combined analyses of nuclear large and small subunit rDNA sequences. Mycologia 98:896–905

Albert B, Sellem CH (2002) Dynamics of the mitochondrial genome during *Podospora anserina* aging. Curr Genet 40:365–373

Alberts B, Johnson A, Lewis J, Morgan D, Raff M et al (2015) Molecular biology of the cell, 6th edn. Garland Science, NY

Aldridge BB, Fernandez-Suarez M, Heller D, Ambravaneswaran V, Irimia D et al (2012) Asymmetry and aging of mycobacterial cells lead to variable growth and antibiotic susceptibility. Science 335:100–104

Alexander M (1971) Microbial ecology. John Wiley & Sons, NY

Alkhayyat F, Kim SC, Yu J-H (2015) Genetic control of asexual development in *Aspergillus fumigatus*. Adv Appl Microbiol 90:93–107

Alldredge AL, Cohen Y (1987) Can microscale chemical patches persist in the sea? Microelectrode study of marine snow, fecal pellets. Science 235:689–691

Allen TFH (1977) Scale in microscopic algal ecology: a neglected dimension. Phycologia 16:253–257

Allen TFH, Hoekstra TW (1992) Toward a unified ecology. Columbia University Press, New York

Ally D, Ritland K, Otto SP (2008) Can clonal size serve as a proxy for clone age? An exploration using microsatellite divergence in *Populus tremuloides*. Mol Ecol 17:4897–4911

Ally D, Ritland K, and Otto SP (2010). Aging in a long-lived tree. PLoS Biol 8:el000454. doi:10.137.journal.pbio.1000454

Amaral PP, Dinger ME, Mercer TR, Mattick JS (2008) The eukaryotic genome as an RNA machine. Science 319:1787–1789

Amato SM, Fazen CH, Henry TC, Mok WEK, Orman MA, and Sandvik EL et al (2014) The role of metabolism in bacterial persistence. Frontiers Microbiol. 5:doi: 10.3389/fmicrobiol.2014.00070

Amend JP, Shock EL (2001) Energetics of overall metabolic reactions of thermophilic and hyperthermophilic archaea and bacteria. FEMS Microbiol Rev 25:175–243

Amiard S, Da Ines O, Gallego ME, White CI (2014) Responses to telomere erosion in plants. PLoS ONE 9:e86220. doi:10.1371/journal.pone.0086220

Anderson JB, Kohn LM (1995) Clonality in soilborne, plant-pathogenic fungi. Annu Rev Phytopathol 33:369–391

Anderson JB, Kohn LM (1998) Genotyping, gene genealogies and genomics bring fungal population genetics above ground. Trends Ecol Evol 13:444–449

Anderson JB, Kohn LM (2007) Dikaryons, diploids, and evolution. In: Heitman J, Kronstad JW, Taylor JW, Casselton LA (eds) Sex in fungi: molecular determination and evolutionary implications. ASM Press, Washington, DC, pp 333–348

© Springer Science+Business Media LLC 2017
J.H. Andrews, *Comparative Ecology of Microorganisms and Macroorganisms*,
DOI 10.1007/978-1-4939-6897-8

Anderson JB, Ullrich RC, Roth LF, Filip GM (1979) Genetic identification of clones of *Armillaria mellea* in coniferous forests in Washington. Phytopathology 69:1109–1111

Anderson SH, Kelly D, Ladley JL, Molloy S, Terry J (2011) Cascading effects of bird functional extinction reduce pollination and plant diversity. Science 331:1068–1071

Andersson DI, Hughes D (2010) Antibiotic resistance and its cost: is it possible to reverse resistance? Nature Rev Microbiol 8:260–271

Andersson DI, Jerlström-Hultqvist J, and Näsvall J (2015). Evolution of new functions de novo and from preexisting genes. Cold Spring Harbor Perspect Biol 7:a017996. doi:10.1101/csperspect.a017996

Andersson JO (2005) Lateral gene transfer in eukaryotes. Cell Mol Life Sci 62:1182–1197

Andersson JO (2009) Gene transfer and diversification of microbial eukaryotes. Annu Rev Microbiol 63:177–193

Andrews JH (1984) Life history strategies of plant parasites. In: Ingram DS, Williams PH (eds) Advances in plant pathology, vol 2. Academic Press, NY, NY, pp 105–130

Andrews JH (1991) Comparative ecology of microorganisms and macroorganisms, 1st edn. Springer-Verlag, NY

Andrews JH (1992) Fungal life-history strategies. In: Carroll GC, Wicklow DT (eds) The fungal community: its organization and role in the ecosystem. Marcel Dekker, NY, pp 119–145

Andrews JH (1994) All creatures unitary and modular. In: Blakeman JP, Williamson B (eds) Ecology of plant pathogens. CAB International, Wallingford, UK, pp 3–16

Andrews JH (1995) Fungi and the evolution of growth form. Can J Bot 73(Suppl. 1):S1206–S1212

Andrews JH (1998) Bacteria as modular organisms. Annu Rev Microbiol 52:105–126

Andrews JH, Harris RF (1986) *r*- and *K*-selection and microbial ecology. Adv Microb Ecol 9:99–147

Andrews JH, Harris RF (2000) The ecology and biogeography of microorganisms on plant surfaces. Annu Rev Phytopathol 38:145–180

Andrews JH, Rouse DI (1982) Plant pathogens and the theory of *r*- and *K*-selection. Am. Nat. 120:283–296

Andrews JH, Kinkel LL, Berbee FM, Nordheim EV (1987) Fungi, leaves, and the theory of island biogeography. Microb. Ecol. 14:277–290

Angert ER (2005) Alternatives to binary fission in bacteria. Nature Rev Microbiol 3:214–224

Angert ER, Clements KD, Pace NR (1993) The largest bacterium. Nature 362:239–241

Anikster Y, Wahl I (1979) Coevolution of the rust fungi on Gramineae and Liliaceae and their hosts. Annu Rev Phytopathol 17:367–403

Antonovics J, Bradshaw AD, Turner RG (1971) Heavy metal tolerance in plants. Adv Ecol Res 7:1–85

Appel HM, Cocroft RB (2014) Plants respond to leaf vibrations caused by insect herbivore chewing. Oecologia 175:1257–1266

Arakaki Y, Kawai-Toyooka H, Hamamura Y, Higashiyama T, Noga A et al (2013) The simplest integrated multicellular organism unveiled. PLoS One 8(12):e81641. doi:10.1371/journal.pone.0081641

Armitage JP (1999) Bacterial tactic responses. Adv Microb Physiol 41:229–289

Arnaud-Haond S, Duarte CM, Alberto F, Serrao EA (2007) Standardizing methods to address clonality in population studies. Mol Ecol 16:5115–5139

Ashley NT, Weil ZM, Nelson RJ (2012) Inflammation: mechanisms, costs, and natural variation. Annu Rev Ecol Evol Syst 43:385–406

Atamian HS, Creux NM, Brown EA, Garner AG, Blackman BK et al (2016) Circadian regulation of sunflower heliotropism, floral orientation, and pollinator visits. Science 353:587–590

Avery SV (2006) Microbial cell individuality and the underlying sources of heterogeneity. Nature Rev Microbiol 4:577–587

Aylor DE (1998) The aerobiology of apple scab. Plant Disease 82:838–49

Azam F, Malfatti F (2007) Microbial structuring of marine ecosystems. Nature Rev Microbiol 5:782–791

Babcock RC (1991) Comparative demography of three species of Scleractinian corals using age- and size-dependent classifications. Ecol Monogr 61:225–244

Babin J-S, Fortin D, Williamshurst JF, Fortin M-E (2011) Energy gains predict the distribution of plains bison across populations and ecosystems. Ecology 92:240–252

Bachmann H, Fischlechner M, Rabbers I, Barfa N, and Branco dos Santos F et al (2013). Availability of public goods shapes the evolution of competing metabolic strategies. Proc Natl Acad Sci USA 110:14302–14307

Bada JL (2004) How life began: a status report. Earth Planet Sci Lett 226:1–16

Bak P and Paczuski M (1995) Complexity, contingency, and criticality. Proc Natl Acad Sci USA 92:6689–6696

Baker BJ, Banfield JF (2003) Microbial communities in acid mine drainage. FEMS Microb Ecol 44:139–152

Baker KF, Cook RJ (1974) Biological control of plant pathogens. Freeman, San Francisco

Balaban NQ, Merrin J, Chait R, Kowalik L, Leibler S (2004) Bacterial persistence as a phenotypic switch. Science 305:1622–1625

Baldauf SL (2003) The deep roots of eukaryotes. Science 300:1703–1709

Barbara GM, Mitchell JG (2003) Bacterial tracking of motile algae. FEMS Microbiol Ecol 44:79–87

Barrett LG, Kniskern JM, Bodenhausen N, Zhang W, Bergelson J (2009) Continua of specificity and virulence in plant host-pathogen interactions: causes and consequences. New Phytol 183:513–529

Barrett SCH (1989) Waterweed invasions. Sci Am 261:90–97

Bartholomew B (1970) Bare zones between California shrub and grassland communities: the role of animals. Science 170:1210–1212

Barton NH, Briggs DEG, Eisen JA, Goldstein DB, Patel NH (2007) Evolution. Cold Spring Harbor Laboratory Press, Cold Spring Harbor, NY

Bartumeus F, da Luz MGE, Viswanathan GM, Catalan J (2005) Animal search strategies: a quantitative random-walk analysis. Ecology 86:3078–3087

Bascompte J (2009) Disentangling the web of life. Science 325:416–419

Bastidas RJ. and Heitman J (2009) Trimorphic stepping stones pave the way to fungal virulence. Proc Natl Acad Sci USA 106:351–352

Bates M (1960) The forest and the sea. Random House, NY

Baudisch A (2005). Hamilton's indicators of the force of selection. Proc Natl Acad Sci. USA 102:8263–8268

Baudisch A (2008) Inevitable aging? Contributions to evolutionary-demographic theory. Springer-Verlag, Berlin

Baudisch A (2011) The pace and shape of aging. Methods Ecol Evol 2:375–382

Baudisch A, Salguero-Gomez R, Jones OR, Wrycza T, Mbeau-Ache C et al (2013) The pace and shape of senescence in angiosperms. J Ecol 101:596–606

Baulcombe D (2004) RNA silencing in plants. Nature 431:356–363

Baum DA, Smith SD (2013) Tree thinking: An introduction to phylogenetic biology. Roberts & Co., Greenwood Village CO

Bäurle I, Laux T (2003) Apical meristems: the plant's fountain of youth. BioEssays 25:961–970

Baythavong BS (2011) Linking the spatial scale of environmental variation and the evolution of phenotypic plasticity: selection favors adaptive plasticity in fine-grained environments. Am Nat 178:75–87

Bazzaz FA, Chiariello NR, Coley PD, Pitelka LF (1987) Allocating resources to reproduction and defence. BioScience 37:58–67

Beaumont HJE, Gallie J, Kost C, Ferguson GC, Rainey PB (2009) Experimental evolution of bet hedging. Nature 462:90–93

Bebber DP, Hynes J, Darrah PR, Boddy L, Fricker MD (2007) Biological solutions to transport network design. Proc R Soc B 274:2307–2315

Becker U, Dostal P, Jorritsma-Wienk LD, Matthies D (2008) The spatial scale of adaptive population differentiation in a wide-spread, well-dispersed plant species. Oikos 117:1865–1873

Beckman, CH (1987) The nature of wilt diseases of plants. Am Phytopathol Soc Press, St. Paul, Minnesota

Begon M, Harper JL, Townsend CR (1996) Ecology: individuals, populations and communities, 3rd edn. Blackwell Science, Oxford, UK

Béja O, Aravind L, Koonin EV, Suzuki MT, Hadd A et al (2000) Bacterial rhodopsin: evidence for a new type of phototrophy in the sea. Science 289:1902–1906

Bekker A, Holland HD, Lang PL, Rumble III D, Stein HJ et al (2004) Dating the rise of atmospheric oxygen. Nature 427:117–120

Bell AD (1979) The hexagonal branching pattern of rhizomes of *Alpinia speciosa* L. (Zingiberaceae). Ann Bot 43:209–223

Bell AD (1984) Dynamic morphology: a contribution to plant population ecology. In: Dirzo R, Sarukhan J (eds) Perspectives on plant population ecology. Sinauer Associates Inc., Sunderland, MA, pp 48–65

Bell G (1982) *The Masterpiece of Nature. The Evolution and Genetics of Sexuality*. University of California Press, Berkeley

Bell G (1984) Evolutionary and nonevolutionary theories of senescence. Am Nat 124:600–603

Bell G (1988) Recombination and the immortality of the germ line. J Evol Biol 1:67–82

Bell G (1988) Sex and death in Protozoa. The history of an obsession. Cambridge University Press, NY

Bell G (1992) Mid-life crisis. Evolution 46:854–856

Bell G (1994) The comparative biology of the alternation of generations. In: Kirkpatrick M (ed) Symposium on some mathematical questions in biology: the evolution of haploid-diploid life cycles, vol 25 (Am. Math. Soc. Providence, RI, pp 1–26

Bell G, Koufopanou VI (1991) The architecture of the life cycle in small organisms. Phil Trans R Soc Lond B 322:81–91

Bell-Petersen D, Cassone VM, Earnest DJ, Golden SS, Hardin PE et al (2005) Circadian rhythms from multiple oscillators: lessons from diverse organisms. Nature Rev Genet 6:544–556

Belovsky GE (1978) Diet optimization in a generalist herbivore: the moose. Theor Pop Biol 14:105–134

Belovsky GE (1984) Herbivore optimal foraging: a comparative test of three models. Am Nat 124:97–115

Bendel M, Kienast F, Rigling D (2006) Genetic population structure of three *Armillaria* species at the landscape scale: a case study from Swiss *Pinus mugo* forests. Mycol Res 110:705–712

Benedix JH Jr (1993) Area-restricted search by the plains pocket gopher (*Geomys bursarius*) in tallgrass prairie habitat. Behav Ecol 4:318–324

Bent AF, Mackey D (2007) Elicitors, effectors, and *R* genes: the new paradigm and a lifetime supply of questions. Annu Rev Phytopathol 45:399–436

Benveniste RE (1985) The contributions of retroviruses to the study of mammalian evolution. In: MacIntyre RJ (ed) Molecular evolutionary genetics. Plenum Press, NY, pp 359–417

Benveniste RE, Todaro GJ (1974) Evolution of type-C viral genes. Inheritance of exogenously acquired viral genes. Nature 252:456–459

Benveniste RE, Todaro GJ (1975) Segregation of RD-114 and FeLV-related sequences in crosses between domestic cat and leopard cat. Nature 257:506–508

Berbee ML (2001) The phylogeny of plant and animal pathogens in the Ascomycota. Physiol Mol Plant Pathol 59:165–187

Berbee ML, Taylor JW (1992) 18S ribosomal RNA gene sequence characters place the human pathogen Sporothrix schenckii in the genus Ophiostoma. 16:87–91

Berbee ML, Taylor JW (2010) Dating the molecular clock in fungi—how close are we? Fung Biol Rev 24:1–16

Berg HC (1993) Random walks in biology. Princeton University Press, Princeton, NJ

Berg, HC (2004). E. coli in motion. Springer, NY

Berg HC, Brown DA (1972) Chemotaxis in Escherichia coli analysed by three-dimensional tracking. Nature 239:500–504

Bergstrom CT, Lipsitch M, Levin BR (2000) Natural selection, infectious transfer and the existence conditions for bacterial plasmids. Genetics 155:1505–1519

Berkers CR, Maddocks ODK, Cheung EC, Mor I, Vousden KH (2013) Metabolic regulation by p53 family members. Cell Metab 18:617–633

Bernstein H, Byerly H, Hopf F, Michod RE (1985) DNA repair and complementation: the major factors in the origin and maintenance of sex. In: Halvorson HO, Monroy A (eds) The origin and evolution of sex. Liss, NY, pp 29–45

Bertrand H (2000) Role of mitochondrial DNA in the senescence and hypovirulence of fungi and potential for plant disease control. Annu Rev Phytopathol 38:397–422

Bertrand H, McDougall KJ, Pittenger TH (1968) Somatic cell variation during uninterrupted growth of Neurospora crassa in continuous growth tubes. J Gen Microbiol 50:337–350

Biller SJ, Berube PM, Lindell D, Chisholm SW (2015) Prochlorococcus: the structure and function of a collective diversity. Nature Rev Microbiol 13:13–27

Billiard S, López-Villavicencio M, Hood ME, Giraud T (2012) Sex, outcrossing and mating types: unsolved questions in fungi and beyond. J Evol Biol 25:1020–1038

Black BA, Sydeman WJ, Frank DC, Griffin D, Stable DW et al (2014) Six centuries of variability and extremes in a coupled marine-terrestrial ecosystem. Science 345:1498–1502

Blake WJ, Kaern M, Cantor CR, Collins JJ (2003) Noise in eukaryotic gene expression. Nature 422:633–637

Blonder B, Sloat L, Enquist BJ, McGill B (2014) Separating macroecological pattern and process: comparing ecological, economic, and geological systems. PLoS One 9(11):e112850. doi:10.1371/journal.pone.0112850

Blount ZD, Borland CZ, Lenski RE (2008) Historical contingency and the evolution of a key innovation in an experimental population of Escherichia coli. Proc Natl Acad Sci USA 105:7899–7906

Boardman RS, Cheetham AH, Oliver WA Jr (1973). Animal colonies: development and function through time. Dowden, Hutchinson & Ross, Stroudsburg, PA.

Bobay L-M, Traverse CC, Ochman H (2015) Impermanence of bacterial clones. Proc Natl Acad Sci 112:8893–8900

Bock R, Timmis JN (2008) Reconstructing evolution: gene transfer from plastids to the nucleus. BioEssays 30:556–566

Boddy L, Hynes J, Bebber DP, Fricker MD (2009) Saprotrophic cord systems: dispersal mechanisms in space and time. Mycoscience 50:9–19

Bonar L (1951) Two new fungi on Torreya. Mycologia 43:62–66

Bonito G, Smith ME, Nowak M, Healy RA, Guevara G et al (2013) Historical biogeography and diversification of truffles in the Tuberaceae and their newly identified Southern Hemisphere sister lineage. PLoS One 8:e52765. doi:10.1371/journal.pone.0052765

Bonner JT (1958) The relation of spore formation to recombination. Amer Nat 92:193–200

Bonner JT (1965) Size and cycle: an essay on the structure of biology. Princeton University Press, Princeton, NJ

Bonner JT (1974) On development. The biology of form. Harvard University Press, Cambridge, MA

Bonner JT (1982a) Evolutionary strategies and developmental constraints in the cellular slime molds. Am Nat 119:530–552

Bonner JT (ed) (1982b) Evolution and development. Springer-Verlag, NY

Bonner JT (1988) The evolution of complexity by means of natural selection. Princeton University Press, Princeton, NJ

Bonner JT (1993) Life cycles: reflections of an evolutionary biologist. Princeton University Press, Princeton, NJ

Bonner JT (1998) The origins of multicellularity. Integrat Biol 1:27–36

Bonner JT (2009) The social amoebae: the biology of cellular slime molds. Princeton University Press, Princeton, NJ

Bonner JT, Horn HS (1982) Selection for size, shape, and developmental timing. In: Bonner JT (ed) Evolution and development. Springer-Verlag, NY, pp 259–276

Bos J, Zhang Q, Vyawahare S, Rogers E, Rosenberg SM et al (2015) Emergence of antibiotic resistance from multinucleated bacterial filaments. Proc Natl Acad Sci USA 112:178–183

Boulding KE (1964). General systems as a point of view. In: Mesarovic MD (ed) Views on general systems theory. Proceedings of the 2nd Systems Symposium on Case Institute of Technology. Wiley, NY, pp 25–38

Box GEP, Draper NR (1987) Empirical model-building and response surfaces. John Wiley & Sons, NY

Boyce MS (1984) Restitution of *r*- and *K*-selection as a model of density-dependent selection. Annu Rev Ecol Syst 15:427–447

Braam J (2005) In touch: plant responses to mechanical stimuli. New Phytol 165:373–389

Braam J, Davis RW (1990) Rain-, wind-, and touch-induced expression of calmodulin and calmodulin-related genes in *Arabidopsis*. Cell 60:357–364

Bradshaw AD (1959) Population differentiation in *Agrostis tenuis* Sibth. I Morphological differentiation. New Phytol 58:208–227

Bradshaw AD (1965) Evolutionary significance of phenotypic plasticity in plants. Adv Genet 13:115–155

Bradshaw AD (1972) Some of the evolutionary consequences of being a plant. Evol Biol 5:25–47

Brandon RN (1984) The levels of selection. In: Brandon RN, Burian RM (eds) Genes, organisms and populations. MIT Press, Cambridge, MA, pp 133–141

Brasier M, McLoughlin N, Green O, Wacey D (2006) A fresh look at the fossil evidence for early Archaean cellular life. Phil Trans R Soc B 361:887–902

Brencic A, Winans SC (2005) Detection of and response to signals involved in host-microbe interactions by plant-associated bacteria. Microbiol Mol Biol Rev 69:155–194

Britten RJ, Davidson EH (1971) Repetitive and non-repetitive DNA sequences and a speculation on the origins of evolutionary novelty. Quart Rev Biol 46:111–138

Brock TD (1966) Principles of microbial ecology. Prentice-Hall, Englewood Cliffs, NJ

Brock TD (1971) Microbial growth rates in nature. Bacteriol Rev 35:39–58

Brock TD, Brock KM, Belly RT, Weiss RL (1972) *Sulfolobus:* a new genus of sulfur-oxidizing bacteria living at low pH and high temperature. Arch f Mikrobiol 84:54–68

Brower AVZ, DeSalle R, Vogler A (1996) Gene trees, species trees, and systematics: a cladistics perspective. Annu Rev Ecol Syst 27:423–450

Brown AJ, Casselton LA (2001) Mating in mushrooms: increasing the chances but prolonging the affair. Trends Genet 17:393–400

Brown CJ, Todd KM, Rosenzweig RF (1998) Multiple duplications of yeast hexose transport genes in response to selection in a glucose-limited environment. Mol Biol Evol 15:931–942

Brown GW, Flood MM (1947) Tumbler mortality. J Am Stat Assoc 42:562–574

Brown JH, West GB (eds) (2000) Scaling in biology. Oxford University Press, NY

Brown JH, Marquet PA, Taper ML (1993) Evolution of body size: consequences of an energetic definition of fitness. Am Nat 142:573–584

Brown JH, West GB, Enquist BJ (2000) Scaling in biology: patterns and processes, causes and consequences. In: Brown JH, West GB (eds) Scaling in biology. Oxford University Press, NY, pp 1–24

Brown JH, Gillooly JF, Allen AP, Savage VM, West GB (2004) Toward a metabolic theory of ecology. Ecology 85:1771–1789

Brown JKM, Hovmøller MS (2002) Aerial dispersal of pathogens on the global and continental scales and its impact on plant disease. Science 297:537–541

Brown JR (2003) Ancient horizontal gene transfer. Nature Rev Genet 4:121–132

Bruckner S, Mosch H-U (2012) Choosing the right lifestyle: adhesion and development in *Saccharomyces cerevisiae*. FEMS Microbiol Rev 36:25–58

Brun YV, Janakiraman R (2000) The dimorphic life cycle of *Caulobacter* and stalked bacteria. In: Brun YV, Shimkets LJ (eds) Prokaryotic development. American Society for Microbiology, Washington, DC, pp 297–317

Brun YV, Shimkets LJ (eds) (2000) Prokaryotic development. American Society for Microbiology Press, Washington, DC

Buchanan-Wollaston V, Passiatore JE, Cannon F (1987) The *moh* and *oriT* mobilization functions of a bacterial plasmid promote its transfer to plants. Nature 328:172–175

Burdon JJ (1987) Diseases and plant population biology. Cambridge University Press, Cambridge, UK

Burdon JJ (1987) Phenotypic and genetic patterns of resistance to the pathogen *Phakopsora pachyrhizi* in populations of *Glycine canescens*. Oecologia 73:257–267

Burkepile DE, Parker JD, Woodson CB et al (2006) Chemically mediated competition between microbes and animals: microbes as consumers in food webs. Ecology 87:2821–2831

Burr TJ, Otten L (1999) Crown gall of grape: biology and disease management. Annu Rev Phytopathol 37:53–80

Burstein D, Amaro F, Zusman T, Lifshitz Z, Cohen O et al (2016) Genomic analysis of 38 *Legionella* species identifies large and diverse effector repertoires. Nat Genet 48:167–175

Bushman F (2002) Lateral DNA transfer: mechanisms and consequences. Cold Spring Harbor Laboratory Press, Cold Spring Harbor, NY

Bushnell WR, Roelfs AP (eds) (1984) The cereal rusts, vol I. origins, specificity, structure, and physiology. Academic Press, NY

Buss L (1986) Competition and community organization on hard surfaces in the sea. In: Diamond J, Case TJ (eds) Community ecology. Harper & Row, New York, pp 517–536

Buss LW (1982) Somatic cell parasitism and the evolution of somatic tissue compatibility. Proc Nat Acad Sci 79:5337–5341

Buss LW (1983) Evolution, development, and the units of selection. Proc Natl Acad Sci (USA) 80:1387–1391

Buss LW (1985) The uniqueness of the individual revisited. In: Jackson JBC, Buss LW, Cook RE (eds) Population biology and evolution of clonal organisms. Yale University Press, New Haven, CT, pp 467–505

Buss LW (1987) The evolution of individuality. Princeton University Press, Princeton, NJ

Buss LW (1990) Competition within and between encrusting clonal invertebrates. Trends Ecol Evol 5:352–256

Buss LW, Blackstone NW (1991) An experimental exploration of Waddington's epigenetic landscape. Philos Trans R Soc Lond B 332:49–58

Buss LW, Grosberg RK (1990) Morphogenetic basis for phenotypic differences in hydroid competitive bahaviour. Nature 343:63–66

Butlin R (2002) The costs and benefits of sex: new insights from old asexual lineages. Nat Rev Genet 3:311–317

Butterfield NJ (2009) Modes of pre-Ediacaran multicellularity. Precambr Res 173:210–211

Butterfield NJ, Knoll AH, Swett K (1990) A bangiophyte red alga from the Proterozoic of Arctic Canada. Science 250:104–107

Cadavid LF (2004) Self-discrimination in colonial invertebrates: genetic control of allorecognition in the hydroid *Hydractinia*. Dev Comp Immunol 28:871–879

Calvino I (2014) The complete cosmicomics. Houghton Mifflin Harcourt, New York

Campbell A (1981) Evolutionary significance of accessory DNA elements in bacteria. Annu Rev Microbiol 35:55–83

Campbell CD, Eichler EE (2013) Properties and rates of germline mutations in humans. Trends Genet 29:575–584

Campillo F, Champagnat N (2012) Simulation and analysis of an individual-based model for graph-structured plant dynamics. Ecol Model 234:93–105

Campisi J (2001) From cells to organisms: can we learn about aging from cells in culture? Exp Gerontol 36:607–618

Carbone C, Gittleman JL (2002) A common rule for the scaling of carnivore density. Science 295:2273–2276

Carlile MJ (1980) From prokaryote to eukaryote: gains and losses. In: Gooday GW, Lloyd D, Trinci APJ (eds) The eukaryotic microbial cell. 13th symposium Society for General Microbiology. Cambridge University Press, Cambridge, UK, pp 1–40

Carlile MJ, Watkinson SC, Gooday GW (2001) The fungi, 2nd edn. Elsevier, New York

Carlisle PL, Banerjee MM, Lazzell A, Monteagudo C, Lopez-Ribot JL et al (2009) Expression levels of a filament-specific transcriptional regulator are sufficient to determine *Candida albicans* morphology and virulence. Proc Natl Acad Sci USA 106:599–604

Carroll SB (2001) Chance and necessity: the evolution of morphological complexity and diversity. Nature 409:1102–1109

Carroll SB (2008) Evo-devo and an expanding evolutionary synthesis: a genetic theory of morphological evolution. Cell 134:25–36

Carroll SB, Grenier JK, Weatherbee SD (2005) From DNA to diversity: molecular genetics and the evolution of animal design, 2nd edn. Blackwell, Malden, MA

Casadesus J, Low D (2006) Epigenetic regulation in the bacterial world. Microbiol Mol Biol Rev 70:830–856

Caspar DLD, Klug A (1962) Physical principles in the construction of regular viruses. Cold Spring Harb Symp Quant Biol 27:1–24

Caspi R, Foerster H, Fulcher CA, Hopkinson R, Ingraham J et al (2006) MetaCyc; a multiorganism database of metabolic pathways and enzymes. Nucleic Acids Res 34:D511–D516

Casselton LA, Economou A (1985) Dikaryon formation. In: Moore D, Casselton LA, Wood DA, Frankland JC (eds) Developmental biology of higher fungi. Cambridge University Press, NY, pp 213–229

Caswell H (1985) The evolutionary demography of clonal reproduction. In: Jackson JBC, Buss LW, Cook RE (eds) Population biology and evolution of clonal organisms. Yale University Press, New Haven, CT, pp 187–224

Caswell H, Salguero-Gomez R (2013) Age, stage and senescence in plants. J Ecol 101:585–595

Caten CE (1987) The genetic integration of fungal life styles. In: Rayner ADM, Brasier CM, Moore D (eds) Evolutionary biology of the fungi. Cambridge University Press, Cambridge, UK, pp 215–229

Cavalier-Smith T (2006) Cell evolution and Earth history: stasis and revolution. Philos Trans R Soc B 361:969–1006

Cavanaugh CM, McKiness ZP, Newton ILG, Stewart FJ (2013) Marine chemosynthetic symbioses. In: Rosenberg E, DeLong EF, Lory S, Stackebrandt E, Thompson F (eds) The prokaryotes, 4th edn. Springer, NY, pp 579–607

Ceccarelli M, Sarri V, Caceres ME, Cionini PG (2011) Intraspecific genotypic diversity in plants. Genome 54:701–709

Chae L, Kim T, Nilo-Poyanco R, Rhee SY (2014) Genomic signatures of specialized metabolism in plants. Science 344:510–513

Chambers HF, DeLeo FR (2009) Waves of resistance: Staphyloccus aureus in the antibiotic era. Nat Rev Microbiol 7:629–641

Chapman N, Miller AJ, Lindsey K, Whalley WR (2012) Roots, water, and nutrient acquisition: let's get physical. Trends Plant Sci 17:701–710

Chappell TM, Rausher MD (2016) Evolution of host range in Coleosporium ipomeae, a plant pathogen with multiple hosts. Proc Nat Acad Sci USA 113:5346–5351

Charlesworth B (1980) Evolution in age-structured populations, 1st edn. Cambridge University Press, Cambridge, UK

Charlesworth B (1985) The population genetics of transposable elements. In: Ohta T, Aoki K (eds) Population genetics and molecular evolution: papers marking the sixtieth birthday of Motoo Kimura. Sci. Soc. Press, Tokyo, pp 213–232

Charlesworth B (1994) Evolution in age-structured populations, 2nd edn. Cambridge University Press, Cambridge, UK

Charlesworth B (2000) Fisher, Medawar, Hamilton and the evolution of aging. Genetics 156:927–931

Chen H, Fink GR (2006) Feedback control of morphogenesis in fungi by aromatic alcohols. Genes Dev 20:1150–1161

Chen W, Koide RT, Adams TS, DeForest JL, Cheng L et al (2016) Root morphology and mycorrhizal symbioses together shape nutrient foraging strategies of temperate trees. Proc Nat Acad Sci USA 113:8741–8746

Chepurnov VA, Mann DG, Sabbe K, Vyverman W (2004) Experimental studies on sexual reproduction in diatoms. Int Rev Cytol 237:91–154

Cherry Vogt KS, Harmata KL, Coulombe HL et al (2011) Causes and consequences of stolon regression in a colonial hydroid. J Exp Biol 214:3197–3205

Cheverud JM (1984) Quantitative genetics and developmental constraints on evolution by selection. J Theor Biol 110:155–171

Chien M, Morozova I, Shi S, Sheng H, Chen J et al (2004) The genomic sequence of the accidental pathogen Legionella pneumophila. Science 305:1966–1968

Childs DZ, Metcalf CJE, Rees M (2010) Evolutionary bet-hedging in the real world: empirical evidence and challenges revealed by plants. Proc R Soc B 277:3055–3064

Chong S, Chen C, Ge H, Xie XS (2014) Mechanism of transcriptional bursting in bacteria. Cell 158:314–326

Claessen D, Rozen DE, Kuipers OP, Søgaard-Andersen L, van Wezel GP (2014) Bacterial solutions to multicellularity: a tale of biofilms, filaments and fruiting bodies. Nat Rev Microbiol 12:115–124

Clark TA, Anderson JB (2004) Dikaryons of the basidiomycete fungus Schizophyllum commune: evolution in long-term culture. Genetics 167:1663–1675

Clarke PH (1982) The metabolic versatility of pseudomonads. Antonie van Leeuwenhoek 48:105–130

Clatworthy JN, Harper JL (1962) The comparative biology of closely related species living in the same area. V. Inter- and intraspecific interference within cultures of Lemna spp. and Salvinia natans. J Exp Bot 13:307–324

Clausen J, Keck DD, Hiesey WM (1948) Experimental studies on the nature of species. III. Environmental responses of climatic races of Achillea. Carnegie Institute (Publ. no. 581), Washington, DC

Clay K, Kover P (1996) The red queen hypothesis and plant/pathogen interactions. Annu Rev Phytopathol 34:29–50

Cody ML (1966) A general theory of clutch size. Evolution 20:174–184

Cody ML (1986) Structural niches in plant communities. In: Diamond J, Case TJ (eds) Community ecology. Harper & Row, New York, pp 381–405

Coelho M, Dereli A, Haese A, Kuhn S, Malinovoska L et al (2013) Fission yeast does not age under favorable conditions, but does so after stress. Curr Biol 23:1844–1852

Coelho SM, Peters AF, Charrier B, Roze D, Destombe C et al (2007) Complex life cycles of multicellular eukaryotes: New approaches based on the use of model organisms. Gene 406:152–170

Cohen JB, Levinson AD (1988) A point mutation in the last intron responsible for increased expression and transforming activity of the c-Ha-ras oncogene. Nature 334:119–124

Cohen NR, Lobritz MA, Collins JJ (2013) Microbial persistence and the road to drug resistance. Cell Host Microbe 13:632–642

Cohen SE, Golden SS (2015) Circadian rythms in bacteria. Microbiol Mol Biol Rev 79:373–384

Colwell R, Grimes DJ (2000) Nonculturable microorganisms in the environment. ASM Press, Washington, DC

Comfort A (1979) The biology of senescence, 3rd edn. Elsevier, NY

Connell JH (1961) The influence of interspecific competition and other factors on the distribution of the barnacle *Chthamalus stellatus*. Ecology 42:710–723

Connell JH (1983) On the prevalence and relative importance of interspecific competition: evidence from field experiments. Am Nat 122:661–696

Cook RE (1979) Asexual reproduction: a further consideration. Am Nat 113:769–772

Cook RE (1983) Clonal plant populations. Am Sci 71:244–253

Cook RE (1985) Growth and development in clonal plant populations. In: Jackson JBC, Buss LW, Cook RE (eds) Population biology and evolution of clonal organisms. Yale University Press, New Haven, CT, pp 259–296

Cook RJ, Baker KF (1983) The nature and practice of biological control of plant pathogens. American Phytopathological Society Press, St. Paul, MN

Cooke RC, Rayner ADM (1984) Ecology of saprotrophic fungi. Longman, London

Cooper AF Jr, van Gundy SD (1971) Senescence, quiescence, and cryptobiosis. In: Zuckerman BM, Mai WF, Rhode RA (eds) Plant parasitic nematodes. vol. II. Cytogenetics, host-parasite interactions, and physiology. Academic Press, NY, pp 297–318.

Cooper VS, Lenski RE (2000) The population genetics of ecological specialization in evolving *Escherichia coli* populations. Nature 407:736–739

Corner EJH (1964) The life of plants. University of Chicago Press, Chicago

Costanzo M, Baryshnikova A, Bellay J, Kim Y, Spear ED et al (2010) The genetic landscape of a cell. Science 327:425–431

Costello EK, Stagaman K, Dethlefsen L, Bohannan BJM, Reiman DA (2012) The application of ecological theory toward an understanding of the human microbiome. Science 336:1255–1262

Costerton JW, Lewandowski Z, Caldwell DE, Korber DR, Lappin-Scott HM (1995) Microbial biofilms. Annu Rev Microbiol 49:711–745

Cox MM (1999) Recombinational DNA repair in bacteria and the RecA protein. Prog Nucleic Acids Res Mol Biol 63:310–366

Croll D, Sanders IR (2009) Recombination in *Glomus intraradices*, a supposed ancient asexual arbuscular mycorrhizal fungus. BMC Evol Biol 9:13. doi:10.1186/1471-2148-9-13.

Croucher NJ, Harris SR, Fraser C, Quail MA, Burton J et al (2011) Rapid pneumococcal evolution in response to clinical interventions. Science 331:430–434

Crow JF (1988) The importance of recombination. In: Michod RE, Levin BR (eds). The evolution of sex: an examination of current ideas. Sinauer Associates Inc., Sunderland, MA, pp 56–73

Crow JF (1994) Advantages of sexual reproduction. Dev Genet 15:205–213

Crow JF (2000) The origins, patterns and implications of human spontaneous mutation. Nat Rev Genet 1:40–47

Cullis CA (1983) Environmentally induced DNA changes in plants. CRC Crit Rev Plant Sci 1:117–131

Cullis CA (2005) Mechanisms and control of rapid genomic changes in flax. Ann Bot 95:201–206

Curtis PD, Brun YV (2010) Getting in the loop: regulation of development in *Caulobacter crescentus*. Microbiol Mol Biol Rev 74:13–41

D'Amato F (1985) Cytogenetics of plant cell and tissue cultures and their regenerates. CRC Crit Rev Plant Sci 3:73–112

Daan S, Dijkstra C, Tinbergen JM (1990) Family planning in the kestrel *(Falco tinnunculus):* the ultimate control of covariation of laying date and clutch size. Behaviour 114:1–4

Damuth J (1981) Population density and body size in mammals. Nature 290:699–700

Darrah PR, Fricker MD (2014) Foraging by a wood-decoomposing fungus is ecologically adaptive. Environ Microbiol 16:118–129

Darwin C (1859) On the origin of species by means of natural selection or the preservation of favoured races in the struggle for life. Murray, London

Davidson EH (2006) The regulatory genome: gene regulatory networks in development and evolution. Elsevier, Amsterdam

Dawkins R (1982) The extended phenotype. Oxford University Press, Oxford, UK

Dawkins R (1989) The selfish gene, 2nd edn. Oxford University Press, Oxford, UK

Dawkins R (2004) The ancestor's tale: a pilgrimage to the dawn of evolution. Houghton Mifflin, New York

Dawson TE (1998) Fog in the California redwood forest: ecosystem inputs and use by plants. Oecologia 117:476–485

de Bodt S, Maere S, Van de Peer Y (2005) Genome duplication and the origin of angiosperms. Trends Ecol Evol 20:591–597

de Jager M, Weissing FJ, Herman PMJ, Nolet BA, van de Koppel J (2011) Lévy walks evolve through interaction between movement and environmental complexity. Science 332:1551–1553

de Jong IG, Haccou P, Kuipers OP (2011) Bet hedging or not? A guide to proper classification of microbial survival strategies. Bioessays 33:215–223 de Kroon H, Hutchings MJ (1995) Morphological plasticity in clonal plants: the foraging concept reconsidered. J Ecol 83:143–152

de Kroon H, Huber H, Stuefer JF, van Groenendael JM (2005) A modular concept of phenotypic plasticity in plants. New Phytol 166:73–82

de la Mare W (1920) Miss T. In: The collected poems of Walter de la Mare 1901–1918. Vol. II. Originally published as part of Peacock Pie, 1913. Constable & Co, London

de Oliveira Dal'Molin CG, Nielsen LK (2013) Plant genome-scale metabolic reconstruction and modelling. Curr Opin Biotechnol 24:271–277

de Tomaso AW (2006) Allorecognition polymorphism versus parasitic stem cells. Trends Genet 22:485–490

de Vargas C, Audic S, Henry N, Decelle J, Mahe F et al (2015) Eukaryotic plankton diversity in the sunlit ocean. Science 348. doi:10.1126/science.1261605

de Wit R, Bouvier T (2006) 'Everything is everywhere, but, the environment selects': what did Baas Becking and Beijerinck really say? Environ Microbiol 8:755–758

de Witte LC, Stocklin J (2010) Longevity of clonal plants: why it matters and how to measure it. Ann Bot 106:859–870

Deamer DW (1997) The first living systems: a bioenergetic perspective. 61:239–261

Deering R, Dong F, Rambo D, Money NP (2001) Airflow patterns around mushrooms and their relationship to spore dispersal. Mycologia 93:732–736

Dekker J (2008) Gene regulation in the third dimension. Science 319:1793–1794

DeLong EF, Beja O (2010) The light-driven proton pump proteorhodopsin enhances bacterial survival during tough times. PLoS Biol 8:e1000359. doi:10.1371/journal.pbiol.l000359.

DeLong EF, Karl DM (2005) Genomic perspectives in microbial oceanography. Nature 437:336–342

DeLong JP, Okie JG, Moses ME, Sibly RM, Brown JH (2010) Shifts in metabolic scaling, production, and efficiency across major evolutionary transitions of life. Proc Nat Acad Sci USA 107:12941–12945

DeSilva HR (1938) Age and highway accidents. Sci Mon 47:536–545

Dethlefsen L, Eckburg PB, Bid EM, Relman DA (2006) Assembly of the human intestinal microbiota. Trends Ecol Evol 21:517–523

DeWitt TJ, Sih A, Wilson DS (1998) Costs and limits of phenotypic plasticity. Trends Ecol Evol 13:77–81

DeWitt TJ, Scheiner SM (eds) (2004) Phenotypic plasticity: functional and conceptual approaches. Oxford University Press, New York

DeWoody J, Rowe CA, Hipkins VD, Mock KE (2008) "Pando" lives: Molecular genetic evidence of a giant aspen clone in central Utah. Western North Am Nat 68:493–497

Dial KP, Marzluff JM (1988) Are the smallest organisms the most diverse? Ecology 69:1620–1624

Diamond J (1986) Overview: laboratory experiments, field experiments, and natural experiments. In: Diamond J, Case TJ (eds) Community ecology. Harper & Row, New York, pp 3–22

Dickman A, Cook S (1989) Fire and fungus in a mountain hemlock forest. Can J Bot 67:2005–2016

Dimond AE (1970) Biophysics and biochemistry of the vascular wilt syndrome. Annu Rev Phytopathol 8:301–322

Dix NJ, Webster J (1995) Fungal ecology. Chapman and Hall, London

Dixon DM, Salkin LF (1991) Association between the human pathogen Sporothrix schenckii and sphagnum moss. In: Andrews JH, Hirano SS (eds) Microbial ecology of leaves. Springer, NY, pp 237–249

Dobson AP, Lafferty KD, Kuris AM (2006) Parasites and food webs. In: Pascual M, Duime JA (eds) Ecological networks: linking structure to dynamics in food webs. Oxford University Press, NY, pp 119–135

Dobzhansky T (1956) What is an adaptive trait? Am Nat 90:337–347

Doebley J, Lukens L (1998) Transcriptional regulators and the evolution of plant form. Plant Cell 10:1075–1082

Dombroskie SL, Aarssen LW (2010) Within-genus size distributions in angiosperms: Small is better. Perspect Plant Ecol Evol Syst. 12:283–293

Domozych DS, Domozych CE (2014) Multicellularity in green algae: upsizing in a walled complex. Front Plant Sci 5. doi:10.3389/fpls.2014.00649

Donaldson GP, Lee SM, Mazmanian SK (2016) Gut biogeography of the bacterial microbiota. Nat Rev Microbiol 14:20–32

Doolittle RF, Feng D-F, Johnson MS, McClure MA (1989) Origins and evolutionary relationships of retroviruses. Q Rev Biol 64:1–30

Doolittle WF, Zhaxybayeva O (2013) What is a prokaryote? In: Rosenberg E et al (eds) The prokaryotes: prokaryotic biology and symbiotic associations, 4th edn. Springer, Berlin, pp 21–37

Drake JW (1974) The role of mutation in microbial evolution. In: Carlile MJ, Skehel JJ (eds) Evolution in the microbial world. 24th symposium of the society for general microbiology. Cambridge University Press, New York, pp 41–58

Drake JW (1991) A constant rate of spontaneous mutation in DNA-based microbes. Proc Nat Acad Sci USA 88:7160–7164

Drake JW, Charlesworth B, Charlesworth D, Crow JF (1998) Rates of spontaneous mutation. Genetics 148:1667–1686

Drescher K, Dunkel J, Nadel CP, van Teeffelen S, Grnja I et al (2016) Architectural transitions in *Vibrio cholerae* biofilms at single-cell resolution. Proc Nat Acad Sci USA 113:E2066–E2072

Dryja TP, McGee TL, Reichel E, Hahn LB, Cowley GS et al (1990) A point mutation of the rhodopsin gene in one form of retinitis pigmentosa. Nature 343:364–366

Dubnau D, Losick R (2006) Bistability in bacteria. Mol Microbiol 61:564–572

Dudley SA, Schmitt J (1996) Testing the adaptive plasticity hypothesis: Density-dependent selection on manipulated stem length in *Impatiens capensis*. Am Nat 147:445–465

Duncan KE, Istock CA, Graham JB, Ferguson N (1989) Genetic exchange between *Bacillus subtilis* and *Bacillus licheniformis*: variable hybrid stability and the nature of bacterial species. Evolution 43:1585–1609

Dunning Hotopp JC, Clark ME, Oliveira DCSG et al (2007) Widespread lateral gene transfer from intracellular bacteria to multicellular eukaryotes. Science 317:1753–1756

Duplessis S, Cuomo CA, Lin Y-C, Aerts A, Tisserant E et al (2011) Obligate biotrophy features unraveled by the genomic analysis of rust fungi. Proc Nat Acad Sci USA 108:9166–9171

Durrant A (1962) The environmental induction of heritable changes in *Linum*. Heredity 27:277–298

Dusenbery DB (1998) Fitness landscapes for effects of shape on chemotaxis and other behaviors of bacteria. J Bacteriol 180:5978–5983

Dutech C, Barres B, Bridier J, Robin C, Milgroom MG et al (2012) The chestnut blight fungus world tour: successive introduction events from diverse origins in an invasive plant fungal pathogen. Mol Ecol 21:3931–3946

Dworkin J, Shah IM (2010) Exit from dormancy in microbial organisms. Nat Rev Microbiol 8:890–896

Dworkin M (1985) Developmental biology of the bacteria. Benjamin/Cummings, Menlo Park, CA

Dworkin M (2006) Prokaryotic life cycles. In: Dworkin M et al (eds) The prokaryotes: a handbook on the biology of bacteria, Vol 2: Ecophysiology and biochemistry, 3rd edn. Springer, New York, pp 140–166

Dyall SD, Brown MT, Johnson PJ (2004) Ancient invasions: from endosymbionts to organelles. Science 304:253–257

Dykhuizen D (2016) Thoughts toward a theory of natural selection: The importance of microbial experimental evolution. Cold Spring Harb Perspect Biol 8. doi:10.1101/cshperspect.a018044.

Dykhuizen D, Davies M (1980) An experimental model: bacterial specialists and generalists competing in chemostats. Ecology 61:1213–1227

Dykhuizen DE (1990) Experimental studies of natural selection in bacteria. Annu Rev Ecol Syst 21:373–398

Dykhuizen DE, Hartl DL (1983) Selection in chemostats. Microbiol Rev 47:150–168

Eadie JM, Broekhoven L, Colgan P (1987) Size ratios and artifacts: Hutchinson's rule revisited. Am Nat 129:1–17

Eagle H (1955) Nutrition needs of mammalian cells in tissue culture. Science 122:501–504

Ehrensvärd G (1962) Life: origin and development. University of Chicago Press, Chicago

Ehrlich GD, Ahmed A, Earl J, Hiller NL, Costerton JW et al (2010) The distributed genome hypothesis as a rubric for understanding evolution *in situ* during chronic bacterial biofilm infectious processes. FEMS Immunol Med Microbiol 59:269–279

Eichten SR, Schmitz RJ, Springer NM (2014) Epigenetics: beyond chromatin modifications and complex genetic regulation. Plant Physiol 165:933–947

Eisenbach M (2004) Chemotaxis. Imperial College Press, London

El Albani A, Bengtson S, Canfield DE, Bekker A, Macchiarelli R et al (2010) Large colonial organisms with coordinated growth in oxygenated environments 2.1 Gyr ago. Nature 466:100–104

Elad Y, Williamson B, Tudzynski P, Delen N (eds) (2004) Botrytis: biology, pathology and control. Kluwer, Dordrecht, The Netherlands

Elbarbary RA, Lucas BA, Maquat LE (2016) Retrotransposons as regulators of gene expression. Science 351:aac7247. doi:10.1126/science.aac7247

Eldar A, Elowitz MB (2010) Functional roles for noise in genetic circuits. Nature 467:167–173

Elena SF, Lenski RE (2003) Evolution experiments with microorganisms: the dynamics and genetic bases of adaptation. Nature Rev Genet 4:457–469

Ellis J (1982) Promiscuous DNA—chloroplast genes inside plant mitochondria. Nature 299:678–679

Ellison AM, Bank MS, Clinton BD, Colburn EA, Elliott K et al (2005) Loss of foundation species: consequences for the structure and dynamics of forested ecosystems. Front Ecol Environ 3:379–486

Ellison CE, Hall C, Kowbel D, Welch J, Brem RB et al (2011) Population genomics and local adaptation in wild isolates of a model microbial eukaryote. Proc Nat Acad Sci USA 108:2831–2836

Elowitz MB, Levine AJ, Siggia ED, Swain PS (2002) Stochastic gene expression in a single cell. Science 297:1183–1186

Eltringham SK (1982) Elephants. Blanford Press, Poole, Dorset, UK

Embley TM, Williams TA (eds) (2015) Eukaryotic origins: progress and challenges. Philos Trans R Soc B 370(1678)

Embree M, Liu JK, Al-Bassam MM, Zengler K (2015) Networks of energetic and metabolic interactions define dynamics in microbial communities. Proc Nat Acad Sci USA 112:15, 450–15, 455

Emlen JM (1966) The role of time and energy in food preference. Am Nat 100:611–617

Ericsson M, Hanstorp D, Hagberg P, Enger J, Nystrom T (2000) Sorting out bacterial viability with optical tweezers. J Bacteriol 182:5551–5555

Esau K (1965) Plant anatomy, 2nd edn. Wiley, NY

Escoll P, Rolando M, Gomez-Valero L, Buchrieser C (2013) From amoeba to macrophages: exploring the molecular mechanisms of Legionella pneumophila infection in both hosts. Curr Top Microbiol Immunol 376:1–34

Estes JA, Terborgh J, Brashares JS, Power ME, Berger J et al (2011) Trophic downgrading of planet Earth. Science 333:310–306

Evert RF (2006) Esau's plant anatomy: meristems, cells and tissues of the plant body: their structure, function, and development, 3rd edn. Wiley, New York

Evert RF, Eichhorn SE (2013) Raven biology of plants, 8th edn. Freeman, NY

Fabrizio P, Longo VD (2003) The chronological life span of Saccharomyces cerevisiae. Aging Cell 2:73–81

Fassett NC (1957) A manual of aquatic plants. University of Wisconsin Press, Madison, WI

Fawcett HS (1925) Maintained growth rates in fungus cultures of long duration. Ann Appl Biol 12:191–198

Fedoroff NV (1983) Controlling elements in maize. In: Shapiro JA (ed) Mobile genetic elements. Academic Press, New York, pp 1–63

Fedoroff NV (1989) Maize transposable elements. In: Berg DE, Howe MM (eds) Mobile DNA. American Society for Microbiology Press, Washington, DC, pp 375–411

Feibleman JK (1954) Theory of integrative levels. Br J Philos Sci 5:59–66

Feil EJ, Spratt BG (2001) Recombination and the population structures of bacterial pathogens. Annu Rev Microbiol 55:561–590

Feldgarden M, Stoebel DM, Brisson D, Dykhuizen DE (2003) Size doesn't matter: Microbial selection experiments address ecological phenomena. Ecology 84:1679–1687

Fenchel T (1993) There are more small than large species? Oikos 68:375–378

Fenchel T (2002) Microbial behavior in a heterogeneous world. Science 296:1068–1071

Fenchel T (2008) Motility of bacteria in sediments. Aquat Microb Ecol 51:23–30

Feng S, Jacobsen SE, Reik W (2010) Epigenetic reprogramming in plant and animal development. Science 330:622–627

Ferrell DL (2008) Field fitness, phalanx-guerrilla morphological variation, and symmetry of colonial growth in the encrusting hydroid genus Hydractinia. J Mar Biol Assoc UK 88:1577–1587

Ferry JG, House CH (2006) The stepwise evolution of early life driven by energy conservation. Mol Biol Evol 23:1286–1292

Feschotte C (2008) Transposable elements and the evolution of regulatory networks. Nature Rev Genet 9:397–405

Figueiredo LM, Cross GAM, Janzen CJ (2009) Epigenetic regulation in African trypanosomes: a new kid on the block. Nature Rev Microbiol 7:504–513

Finch CE (1990) Longevity, senescence, and the genome. University of Chicago Press, Chicago, IL

Finch CE (2009) Update on slow aging and negligible senescence—a mini-review. Gerontology 55:307–313

Fincham JRS (1983) Genetics. Jones & Bartlett, Boston, MA

Finkel SE (2006) Long-term survival during stationary phase: evolution and the GASP phenotype. Nat Rev Microbiol 4:113–120

Finlay BJ, Esteban GF (2007) Body size and biogeography. In: Hildrew AG, Raffaelli DG, Edmonds-Brown R (eds) Body size: the structure and function of aquatic ecosystems. Cambridge University Press, Cambridge, UK, pp 167–185

Fiore-Donno AM, Nikolaev SI, Nelson M, Pawlowski J, Cavalier-Smith T et al (2010) Deep phylogeny and evolution of slime moulds (Mycetozoa). Protist 161:55–70

Fisher RA (1930) The genetical theory of natural selection. Clarendon Press, Oxford, UK

Flardh K, Buttner MJ (2009) Streptomyces morphogenetics: dissecting differentiation in a filamentous bacterium. Nat Rev Microbiol 7:36–49

Fliermans CB, Cherry WB, Orrison LH, Smith SJ, Tison DL, Pope DH (1981) Ecological distribution of Legionella pneumophila. Appl Environ Microbiol 41:9–16

Flor HH (1956) The complementary genic systems in flax and flax rust. Adv Genet 8:29–54

Flor HH (1971) Current status of the gene-for-gene concept. Annu Rev Phytopathol 9:275–296

Flores E, Herrero A (2010) Compartmentalized function through cell differentiation in filamentous cyanobacteria. Nature Rev Microbiol 8:39–49

Flot J-F, Hespeels B, Li X, Noel B, Arkhipova I et al (2013) Genomic evidence for ameiotic evolution in the bdelloid rotifer Adineta vaga. Nature 500:453–457

Flynn KJ, Hansen PJ (2013) Cutting the canopy to defeat the "selfish gene"; conflicting selection pressures for the integration pf phototrophy in mixotrophic protists. Protist 164:811–823

Flynn TM, O'Loughlin EJ, Mishra B, DiChristina TJ, Kemner KM (2014) Sulfur-mediated electron shuttling during bacterial iron reduction. Science 344:1039–1042

Fogg GE (1986) Picoplankton. Proc R Soc Lond B 228:1–30

Folse HJ III, Roughgarden J (2010) What is an individual organism? A multilevel selection perspective. Q Rev Biol 85:447–472

Folse HJ III, Roughgarden J (2011) Direct benefits of genetic mosaicism and intraorganismal selection: Modeling coevolution between a long-lived tree and a short-lived herbivore. Evolution 66:1091–1113

Forage RG, Harrison DEF, Pitt DE (1985) Effect of environment on microbial activity. In: Bull AT, Dalton H (eds) Comprehensive biotechnology, vol 1. Pergamon Press, NY, pp 251–280

Forche A, Alby K, Schaefer D, Johnson AD, Berman J et al (2008) The parasexual cycle in Candida albicans provides an alternative pathway to meiosis for the formation of recombinant strains. PLoS Biol 6:1084–1097

Foster PL (2000) Adaptive mutation: implications for evolution. BioEssays 22:1067–1074

Foster PL, Hanson AJ, Lee H, Popodi EM, Tang H (2013) On the mutational topology of the bacterial genome. G3 Genes Genom Genet 3:399–407

Fraenkel DG (2011) Yeast intermediary metabolism. Cold Spring Harbor Laboratory Press, Cold Spring Harbor, NY

Frank SA (2010) The trade-off between rate and yield in the design of microbial metabolism. J Evol Biol 23:609–613

Fraser C, Hanage WP, Spratt BG (2007) Recombination and the nature of bacterial speciation. Science 315:476–480

Fraser DW, McDade JE (1979) Legionellosis. Sci Am 241:82–99

Fraser DW, Tsai TR, Orenstein W, Parkin WE, Beecham HJ et al (1977) Legionnaires' disease: description of an epidemic of pneumonia. N Engl J Med 297:1189–1197

Fraser HB, Hirsh AE, Giaever G, Kumm J, Eisen MB (2004) Noise minimization in eukaryotic gene expression. PLoS Biol 2. doi: 10.1371/journal.pbio.0020137

Fraústo da Silva JJR, Williams RJP (1991) The biological chemistry of the elements: the inorganic chemistry of life. Clarendon Press of Oxford University Press, Oxford, UK

French DL, Laskov R, Scharff MD (1989) The role of somatic hypermutation in the generation of antibody diversity. Science 244:1152–1157

Freschotte C (2008) Transposable elements and the evolution of regulatory networks. Nat Rev Genet 9:397–405

Fricker MD, Boddy L, Nakagaki T, Bebber DP (2009) Adaptive biological networks. In: Gross T, Sayama H (eds) Adaptive networks: theory, models and applications. Springer, New York, pp 51–70

Fridman AL, Tainsky MA (2008) Critical pathways in cellular senescence and immortalization revealed by gene expression profiling. Oncogene 27:5975–5987

Friedman WE (2013) One genome, two ontogenies. Science 339:1045–1046

Funnell BE, Phillips GJ (eds) (2004) Plasmid biology. ASM Press, Washington, DC

Futuyma DJ (2009) Evolution, 2nd edn. Sinauer Associates Inc., Sunderland, Mass

Futuyma DJ, Moreno G (1988) The evolution of ecological specialization. Annu Rev Ecol Syst 19:201–233

Gadgil M (1971) Dispersal: population consequences and evolution. Ecology 52:253–261

Gadgil M, Bossert WH (1970) Life historical consequences of natural selection. Am Nat 104:1–24

Gagneux S, Davis Long C, Small PM, Van T, Schoolnik GK et al (2006) The competitive cost of antibiotic resistance in Mycobacterium tuberculosis. Science 312:1944–1946

Gaillard J-M, Allaine D, Pontier D, Yoccoz NG, Promislow DEL (1994) Senescence in natural populations of mammals: a reanalysis. Evolution 48:509–516

Gardner SN, Mangel M (1997) When can a clonal organism escape senescence? Am Nat 150:462–490

Garrett SD (1973) Deployment of reproductive resources by plant-pathogenic fungi: An application of E.J. Salisbury's generalization for flowering plants. Acta Bot Ind 1:1–9

Gaston KJ, Blackburn TM (2000) Pattern and process in macroecology. Blackwell Science, Oxford, UK

Gaston KJ, Blackburn TM, Lawton JH (1993) Comparing animals and automobiles: a vehicle for understanding body size and abundance relationships in species assemblages. Oikos 66:172–179

Gause GF (1932) Experimental studies on the struggle for existence. I. Mixed population of two species of yeast. J Exp Biol 9:389–402

Gelvin SB (2010) Plant proteins involved in Agrobacterium-mediated genetic transformation. Annu Rev Phytopathol 48:45–68

Geydan TD, Debets AJM, Verkley GJM, van Diepeningen AD (2012) Correlated evolution of senescence and ephemeral substrate use in the Sordariomycetes. Mol Ecol 21:2816–2828

Ghildiyal M, Zamore PD (2009) Small silencing RNAs: an expanding universe. Nat Rev Genet 10:94–108

Gibbs KA, Urbanowski ML, Greenberg EP (2008) Genetic determinants of self identity and social recognition in bacteria. Science 321:256–259

Gilbert C, Schaack S, Pace JK, Brindley PJ, Feschotte C (2010) A role for host-parasite interactions in the horizontal transfer of transposons across phyla. Nature 464:1347–1350

Gilbert SF, Epel D (2009) Ecological developmental biology: integrating epigenetics, medicine, and evolution. Sinauer Associates Inc., Sunderland, MA

Gill DE, Chao L, Perkins SL, Wolf JB (1995) Genetic mosaicism in plants and clonal animals. Annu Rev Ecol Syst 26:423–444

Gillespie JH (1974) Natural selection for within-generation variance in offspring number. Genetics 76:601–606

Gillie OJ (1968) Observations on the tube method of measuring growth rate in Neurospora crassa. J Gen Microbiol 51:185–194

Gimeno CJ, Ljungdahl PO, Styles CA, Fink GR (1992) Unipolar cell divisions in the yeast S. cerevisiae lead to filamentous growth: regulation by starvation and RAS. Cell 68:1077–1090

Gingerich PD (1983) Rates of evolution: effects of time and temporal scaling. Science 222:159–161

Givnish TJ (ed) (1986) On the economy of plant form and function. Cambridge University Press, Cambridge, UK

Givnish TJ (2002) Ecological constraints on the evolution of plasticity in plants. Evol Ecol 16:213–242

Givnish TJ, Sytsma KJ (eds) (1997) Molecular evolution and adaptive radiation. Cambridge University Press, New York

Givnish TJ, Wong C, Stuart-Williams H, Holloway-Phillips M, Farquhar GD (2014) Determinants of maximum tree height in Eucalyptus species along a rainfall gradient in Victoria, Australia. Ecology 95:2991–3007

Gjerde I, Blom HH, Lindblom L, Saetersdal M, Schei FH (2012) Community assembly in epiphytic lichens in early stages of colonization. Ecology 93:749–759

Gladfelter A, Berman J (2009) Dancing genomes: fungal nuclear positioning. Nature Rev Microbiol 7:875–886

Glass NL, Kaneko I (2003) Fatal attraction: nonself recognition and heterokaryon incompatibility in filamentous fungi. Eukaryot Cell 2:1–8

Glass NL, Jacobson DJ, Shiu PKT (2000) The genetics of hyphal fusion and vegetative incompatibility in filamentous ascomycete fungi. Annu Rev Genet 34:165–186

Glass NL, Rasmussen C, Roca MG, Read ND (2004) Hyphal homing, fusion and mycelial interconnectedness. Trends Microbiol 12:135–141

Gliddon CJ, Gouyon PH (1989) The units of selection. Trends Ecol Evol 4:204–209

Gogarten JP, Doolittle WF, Lawrence JG (2002) Prokaryotic evolution in light of gene transfer. Mol Biol Evol 19:2226–2238

Golden JW, Robinson SJ, Haselkorn R (1985) Rearrangement of nitrogen fixation genes during heterocyst differentiation in the cyanobacterium Anabaena. Nature 314:419–423

Goldman BS, Nierman WC, Kaiser D, Slater SC, Durkin AS et al (2006) Evolution of sensory complexity recorded in a myxobacterial genome. Proc Nat Acad Sci USA 103:15200–15205

Goldstein S (1974) Aging in vitro: growth of cultured cells from Galapagos tortoise. Exp Cell Res 83:297–362

Gomez-Alpizar L, Carbone I, Ristaino JB (2007) An Andean origin of Phytophthora infestans inferred from mitochondrial and nuclear gene genealogies. Proc Nat Acad Sci USA 104:3306–3311

Gomez-Consarnau L, Akram N, Lindell K, Pedersen A, Neutze R et al (2010) Proteorhodopsin phototrophy promotes survival of marine bacteria during starvation. PLoS Biol 8:e1000358. doi:10.1317/journal.pbio.1000358

Gomez-Valero L, Rusniok C, Buchrieser C (2009) Legionella pneumophila: population genetics, phylogeny and genomics. Infect Genet Evol 9:727–739

Goodwin SB, Cohen BA, Fry WE (1994) Panglobal distribution of a single clonal lineage of the Irish potato famine fungus. Proc Nat Acad Sci USA 91:11591–11595

Gordon DM (2013) The ecology of Escherichia coli. In: Donnenberg MS (ed) Escherichia coli: pathotypes and principles of pathogenesis, 2nd edn. Elsevier, London, UK, pp 3–20

Goto Y, Obata T, Kunisawa J, Sato S, Ivanov II et al (2014) Innate lymphoid cells regulate intestinal epithelial cell glycosylation. Science 345. doi:10.1126/science.1254009

Gottschalk G (1986) Bacterial metabolism, 2nd edn. Springer, New York

Gould SJ (1966) Allometry and size in ontogeny and phylogeny. Biol Rev 41:587–640

Gould SJ (1977) Ontogeny and phylogeny. Belknap Press of Harvard University Press, Cambridge, MA

Gould SJ (1980) The Panda's Thumb. Norton, New York

Gould SJ (1996) Full house: the spread of excellence from Plato to Darwin. Three Rivers Press, NY

Gould SJ, Lewontin RC (1979) The spandrels of San Marco and the Panglossian paradigm: a critique of the adaptationist programme. Proc R Soc B 205:581–598

Gow NAR, Gadd GM (1995) The growing fungus. Chapman & Hall, London, UK

Gow NAR, Robson GD, Gadd GM (eds) (1999) The fungal colony. In: Proceedings, symposium of

British Mycological Society, 1997. Cambridge University Press, Cambridge, UK

Graham LE (1993) Origin of land plants. Wiley, New York

Graham LE, Graham JM, Wilcox LW (2009) Algae, 2nd edn. Benjamin Cummings, San Francisco

Grant MC, Mitton JB, Linhart YB (1992) Even larger organisms. Nature 360:216

Graumann PL (2007) Cytoskeletal elements in bacteria. Annu Rev Microbiol 61:589–618

Graur D, Li W-H (2000) Fundamentals of molecular evolution, 2nd edn. Sinauer Associates Inc., Sunderland, MA

Greene E (1989) A diet-induced developmental polymorphism in a caterpillar. Science 243:643–646

Greene MT (1997) What cannot be said in science. Nature 388:619–620

Gregory PH (1961) The microbiology of the atmosphere. Leonard Hill, London

Gremer JR, Kimball S, Venable DL (2016) Within- and among-year germination in Sonoran Desert winter annuals: bet hedging and predictive germination in a variable environment. Ecol Lett 19:1209–1218

Gresham D, Hong J (2015) The functional basis of adaptive evolution in chemostats. FEMS Microbiol Rev 39:2–16

Griffiths AJF (1992) Fungal senescence. Annu Rev Genet 26:351–372

Griffiths AJF, Yang X (1993) Senescence in natural populations of Neurospora intermedia. Mycol Res 97:1379–1387

Griffiths AJF, Wessler SR, Carroll SB, Doebley J (2015) An introduction to genetic analysis, 11th edn. Freeman, NY

Grime JP, Crick JC, Rincon JE (1986) The ecological significance of plasticity. In: Jennings DH, Trewavas AJ (eds) Plasticity in plants. Proceedings of the 40th symposium Society for Experimental Biology. Company of Biologists Ltd., Cambridge, UK, pp 5–29

Grosberg RK (1988) The evolution of allorecognition specificity in clonal invertebrates. Q Rev Biol 63:377–412

Grosberg RK, Patterson MR (1989) Iterated ontogenies reiterated. Paleobiology 15:67–73

Grosberg RK, Quinn JF (1986) The genetic control and consequences of kin recognition by the larvae of a colonial marine invertebrate. Nature 322:456–459

Grosberg RK, Strathmann RR (1998) One cell, two cell, red cell, blue cell: the persistence of a unicellular stage in multicellular life histories. Trends Ecol Evol 13:112–116

Grosberg RK, Strathmann RR (2007) The evolution of multicellularity: a minor major transition? Annu Rev Ecol Evol Syst 38:621–654

Gross T, Sayana H (eds) (2009) Adaptive networks: theory, models and applications. Springer, NY

Guasto JS, Rusconi R, Stocker R (2012) Fluid mechanics of planktonic microorganisms. Annu Rev Fluid Mech 44:373–400

Guillaume F, Otto SP (2012) Gene functional trade-offs and the evolution of pleiotropy. Genetics 192:1389–1409

Gustafsson L, Part T (1990) Acceleration of senescence in the collared flycatcher Ficedula albicollis by reproductive costs. Nature 347:279–281

Hacker J, Kaper JB (2000) Pathogenicity islands and the evolution of microbes. Annu Rev Microbiol 54:641–679

Hadany L, SP Otto (2009) Condition-dependent sex and the rate of adaptation. Am Nat 174(suppl):S71–S78

Haigis MC, Yankner BA (2010) The aging stress response. Mol. Cell 40:333–344

Hairston NG Sr (1990) Ecological experiments: purpose, design, and execution. Cambridge University Press, New York

Haldane JBS (1953) Some animal life tables. J Inst Actuaries 79:83–89

Haldane JBS (1956) On being the right size. In: Newman JR (ed) The world of mathematics, vol II. Simon & Schuster, New York, pp 952–957

Hall-Stoodley L, Costerton JW, Stoodley P (2004) Bacterial biofilms: from the natural environment to infectious diseases. Nature Rev Microbiol 2:95–108

Hamilton WD (1966) The moulding of senescence by natural selection. J Theor Biol 12:12–45

Hamilton WD (1970) Review of the book "Evolution in Changing Environments". Science 167:1478–1480

Hamilton WD (1980) Sex versus non-sex versus parasite. Oikos 35:282–290

Hammerschmidt K, Rose CJ, Kerr B, Rainey PB (2014) Life cycles, fitness decoupling and the evolution of multicellularity. Nature 515:75–79

Hanski I, Gilpin ME (eds) (1997) Metapopulation biology: ecology, genetics, and evolution. Academic Press, New York

Hanson CA, Fuhrman JA, Horner-Devine MC, Martiny JBH (2012) Beyond biogeographic patterns: processes shaping the microbial landscape. Nat Rev Microbiol 10:497–506

Harder W, Dijkhuizen L, Veldkamp H (1984) Environmental regulation of microbial metabolism. In: Kelly DP, Carr NG (eds) The Microbe 1984. Part II. Prokaryotes and eukaryotes. Proceedings 36th symposium Society for General Microbiology. Cambridge University Press, Cambridge, UK, pp 51–95

Hardy NB, Peterson DA, von Dohlen CD (2015) The evolution of life cycle complexity in aphids:

Ecological optimization or historical constraint? Evolution 69:1423–1432

Harold FM (1990) To shape a cell: an inquiry into the causes of morphogenesis of microorganisms. Microbiol Rev 54:381–431

Harper JL (1977) Population biology of plants. Academic Press, New York

Harper JL (1978) The demography of plants with clonal growth. In: Freysen AHJ, Woldendorp JW (eds) Structure and functioning of plant populations. North-Holland Publishing Company, Amsterdam, pp 27–45

Harper JL (1981a) The concept of population in modular organisms. In: May RM (ed) Theoretical ecology: principles and applications. Blackwell, Oxford, UK, pp 53–77

Harper JL (1981b) The meanings of rarity. In: Synge H (ed) The biological aspects of rare plant conservation. Wiley, New York, pp 189–203

Harper JL (1982) After description. In: Newman EI (ed) The plant community as a working mechanism. Blackwell, Oxford, UK, pp 11–25

Harper JL (1984) Comments in the Forward to the book. In: Dirzo R, Sarukhan J (eds) Perspectives on plant population ecology. Sinauer Associates Inc., Sunderland, MA, pp xv–xviii

Harper JL (1985) Modules, branches, and the capture of resources. In: Jackson JBC, Buss LW, Cook RE (eds) Population biology and evolution of clonal organisms. Yale University Press, New Haven, CT, pp 1–33

Harper JL (1986) Preface to "Modular Organisms: Case Studies of Growth and Form. Papers relating to a discussion meeting on growth and form in modular organisms". Proc R Soc Lond B 228:111

Harper JL (1988) An apophasis of plant population biology. In: Davy AJ, Hutchings MJ, Watkinson AR (eds) Plant population ecology. Proceedings 28th symposium British Ecological Society. Blackwell Scientific Publishing, Oxford, UK, pp 435–452

Harper JL, Bell AD (1979) The population dynamics of growth form in organisms with modular construction. In: Anderson RM, Turner BD, Taylor LR (eds) Population dynamics. Blackwell, Oxford, UK, pp 29–52

Harper JL, White J (1974) The demography of plants. Annu Rev Ecol Syst 5:419–463

Harper JL, Rosen BR, White J (eds) (1986) The growth and form of modular organisms. Philos Trans R Soc Lond B 313:1–250

Hart JA (1988) Rust fungi and host plant coevolution: do primitive hosts harbor primitive parasites? Cladistics 4:339–366

Hartl DL, Dykhuizen DE (1984) The population genetics of Escherichia coli. Annu Rev Genet 18:31–68

Hartmann M, Grob C, Tarran GA, Martin AP, Burkill PH et al (2012) Mixotrophic basis of Atlantic oligotrophic ecosystems. Proc Nat Acad Sci USA 109:5756–5760

Haselkorn R, Golden JW, Lammers PJ, Mulligan ME (1987) Rearrangement of nif genes during cyanobacterial heterocyst differentiation. Philos Trans R Soc Lond B 317:173–181

Hashimoto M, Nozoe T, Nakaoka H, Okura R, Akiyoshi S et al (2016) Noise-driven growth rate gain in clonal cellular populations. Proc Nat Acad Sci USA 113:3251–3256

Hawkey PM, Jones AM (2009) The changing epidemiology of resistance. J Antimicrob Chemoth 64:i3–il0

Hayflick L (1965) The limited in vitro lifetime of human diploid cell strains. Exp Cell Res 37:614–636

He G, Elling AA, Deng XW (2011) The epigenome and plant development. Annu Rev Plant Biol 62:411–435

Heaton LLM, Jones NS, Fricker MD (2016) Energetic costs on fungal growth. Am Nat 187:E27–E40

Hedges SB (2002) The origin and evolution of model organisms. Nature Rev Genet 3:838–849

Heidinger BJ, Blount JD, Boner W, Griffiths K, Metcalfe NB et al (2012) Telomere length in early life predicts lifespan. Proc Natl Acad Sci USA 109:1743–1748

Heidstra R, Sabatini S (2014) Plant and animal stem cells: similar yet different. Nature Rev Mol Cell Biol 15:301–312

Hein AM, Carrara F, Brumley DR, Stocker R, Levin SA (2016) Natural search algorithms as a bridge between organisms, evolution, and ecology. Proc Nat Acad Sci USA 113:9413–9420

Heitman J (2006) Sexual reproduction and the evolution of microbial pathogens. Curr Biol 16:R711–R725

Heitman J (2010) Evolution of eukaryotic microbial pathogens via covert sexual reproduction. Cell Host Microbe 8:86–99

Helaine S, Kugelberg E (2014) Bacterial persisters: formation, eradication, and experimental systems. Trends Microbiol 22:417–424

Henderson KA, Gottschling DE (2008) A mother's sacrifice: what is she keeping herself for? Curr Opin Cell Biol 20:723–728

Henriques-Normark B, Blomberg C, Dagerhamn J, Battig P, Normark S (2008) The rise and fall of bacterial clones: Streptococcus pneumoniae. Nat Rev Microbiol 6:827–837

Herrmann RG, Maier RM, Schmitz-Linneweber C (2003) Eukaryotic genome evolution: rearrangement and coevolution of compartmentalized genetic information. Philos Trans R Soc Lond B 358:87–97

Herron MD, Michod RE (2008) Evolution of complexity in the volvocine algae: transitions in individuality through Darwin's eye. Evolution 62:436–451

Herron MD, Hackett JD, Aylward FO, Michod RE (2009) Triassic origin and early radiation of multicellular volvocine algae. Proc Nat Acad Sci USA 106:3254–3258

Hershberg R (2015) Mutation—the engine of evolution: studying mutation and its role in the evolution of bacteria. Cold Spring Harbor Perspect Biol doi:10.1101/cshperspect.a018077

Heslop-Harrison JS, Schwarzacher T (2011) Organisation of the plant genome in chromosomes. Plant J 66:18–33

Hickman MA, Zeng G, Forche A, Hirakawa MP, Abbey D et al (2013) The 'obligate diploid' Candida albicans forms mating-competent haploids. Nature 494:55–59

Hill AV (1950) The dimensions of animals and their muscular dynamics. Sci Prog (London) 38:209–230

Hiller NL, Ahmed A, Powell E, Martin DP, Eutsey R et al (2010) Generation of genic diversity among Streptococcus pneumoniae strains via horizontal gene transfer during a chronic polyclonal pediatric infection. PLoS Pathog 6:e1001108

Hillis DM, Moritz C, Mable BK (eds) (1996) Molecular systematics, 2nd edn. Sinauer Associates Inc., Sunderland, MA

Hinchliff CE, Smith SA, Allman JF, Burleigh JG, Chaudhary R et al (2015) Synthesis of phylogeny and taxonomy into a comprehensive tree of life. Proc Nat Acad Sci USA 112:12764–12769

Hirano SS, Upper CD (1989) Diel variation in population size and ice nucleation activity of Pseudomonas syringae on snap bean leaflets. Appl Environ Microbiol 55:623–630

Hirano SS, Upper CD (2000) Bacteria in the leaf ecosystem with emphasis on Pseudomonas syringae—a pathogen, ice nucleus, and epiphyte. Microbiol Mol Biol Rev 64:624–653

Hirano SS, Ostertag EM, Savage SA, Baker LS, Willis DK, Upper CD (1997) Contribution of the regulatory gene lemA to field fitness of Pseudomonas syringae pv. syringae. Appl Environ Microbiol 63:4304–4312

Hittinger CT, Carroll SB (2007) Gene duplication and the adaptive evolution of a classic genetic switch. Nature 449:677–681

Hoehler TM, Jorgensen BB (2013) Microbial life under extreme energy limitation. Nature Rev Microbiol 11:83–94

Hoffmann FG, Opazo JC, Hoogewijs D, Hankeln T, Ebner B et al (2012) Evolution of the globin gene family in Deuterostomes: Lineage-specific patterns of diversification and attrition. Mol Biol Evol 29:1735–1745

Holeski LM, Kearsley MJC, Whitham TG (2009) Separating ontogenetic and environmental determination of resistance to herbivory in cottonwood. Ecology 90:2969–2973

Horn HS (1971) The adaptive geometry of trees. Princeton University Press, Princeton, NJ

Horn HS (2000) Twigs, trees, and the dynamics of carbon in the landscape. In: Brown JH, West GB (eds) Scaling in biology. Oxford University Press, Oxford, UK, pp 199–220

Horn HS, May RM (1977) Limits to similarity among coexisting competitors. Nature 270:660–661

Horner-Devine MC, Lage M, Hughes JB, Bohannan BJM (2004) A taxa-area relationship for bacteria. Nature 432:750–753

Hu Y, Coates ARM (2003) Tuberculosis. In: Coates ARM (ed) Dormancy and Low-Growth States in Microbial Disease. Cambridge University Press, Cambridge, UK, pp 181–207

Huete-Ortega M, Cermeno P, Calvo-Diaz A, Maranon E (2012) Isometric size-scaling of metabolic rate and the size abundance distribution of phytoplankton. Proc R Soc B 279:1815–1823

Huey RB, Hertz PE (1984) Is a jack-of-all-temperatures a master of none? Evolution 38:441–444

Huffaker CB (1964) Fundamentals of biological weed control. In: DeBach P (ed) Biological control of insect pests and weeds. Reinhold, NY, pp 631–649

Hughes DJ, Hughes RN (1986) Metabolic implications of modularity: studies on the respiration and growth of Electrapilosa. Philos Trans R Soc Lond B 313:23–29

Hughes JS, Otto SP (1999) Ecology and the evolution of biphasic life cycles. Am Nat 154:306–320

Hughes RN (1989) A functional biology of clonal animals. Chapman and Hall, London

Hughes RN (2005) Lessons in modularity: the evolutionary ecology of colonial invertebrates. Sci Mar 60(Suppl 1):166–179

Hughes RN, Cancino JM (1985) An ecological overview of cloning in metazoa. In: Jackson JBC, Buss LW, Cook RE (eds) Population biology and ecology of clonal organisms. Yale University Press, New Haven, CT, pp 153–186

Hughes TP, Jackson JBC (1980) Do corals lie about their age? Some demographic consequences of partial mortality, fission, and fusion. Science 209:713–715

Hulbert SH, Webb CA, Smith SM, Sun Q (2001) Resistance gene complexes: evolution and utilization. Annu Rev Phytopathol 39:285–312

Humphries NE, Sims DW (2014) Optimal foraging strategies: Levy walks balance searching and patch exploitation under a very broad range of conditions. J Theor Biol 358:179–193

Humphries NE, Weimerskirch H, Queiroz N, Southall EJ, Sims DW (2012) Foraging success of biological Levy flights recorded in situ. Proc Nat Acad Sci USA 109:7169–7174

Hutchinson GE (1959) Homage to Santa Rosalia, or why are there so many kinds of animals? Am Nat 93:145–159

Hutchinson GE (1975) A treatise on limnology, vol III. Limnological Botany, Wiley, NY

Hutchinson GE, MacArthur RH (1959) A theoretical ecological model of size distributions among species of animals. Am Nat 93:117–125

Huxley JS (1958) Evolutionary processes and taxonomy with special reference to grades. Uppsala University Arsskr 21–38

Ibarra-Laclette E, Lyons E, Hernandez-Guzman G, Perez-Torres CA, Carretero-Paulet L et al (2013) Architecture and evolution of a minute plant genome. Nature 498:94–98

Inderjit DA, Wardle R Karban, Callaway RM (2011) The ecosystem and evolutionary contexts of allelopathy. Trends Ecol Evol 26:655–662

Ingold CT (1946) Size and form in agarics. Trans Br Mycol Soc 29:108–113

Ingold CT (1965) Spore Liberation. Clarendon Press of Oxford University Press, London

Ingold CT (1971) Fungal spores: their liberation and dispersal. Clarendon Press of Oxford University Press, London

Istock CA (1967) The evolution of complex life cycle phenomena: an ecological perspective. Evolution 21:592–605

Istock CA (1984) Boundaries to life history variation and evolution. In: Price PW, Slobodchikoff CN, Gaud WS (eds) A new ecology. Novel approaches to interactive systems. Wiley, NY, pp 143–168

Ivanova NN, Schweintek P, Tripp HJ, Rinke C, Pati A et al (2014) Stop codon reassignments in the wild. Science 344:909–913

Ives AR, Woody ST, Nordheim EV, Nelson C, Andrews JH (2004) The synergistic effects of stochasticity and dispersal on population densities. Am Nat 163:375–387

Jablonski D (1996) Body size and macroevolution. In: Jablonski D, Erwin DH, Lipps JH (eds) Evolutionary paleobiology. University of Chicago Press, Chicago, IL pp 256–289

Jackson HS (1931) Present evolutionary tendencies and the origin of life cycles in the Uredinales. Mem Torrey Bot Club 18:5–108

Jackson JBC (1977) Competition on marine hard substrata: the adaptive significance of solitary and colonial strategies. Am Nat 111:743–767

Jackson JBC (1979) Morphological strategies of sessile animals. In: Larwood G, Rosen BR (eds) Biology and systematics of colonial organisms. Academic Press, NY, pp 499–555

Jackson JBC (1985) Distribution and ecology of clonal and aclonal benthic invertebrates. In: Jackson JBC, Buss LW, Cook RE (eds) Population biology and evolution of clonal organisms. Yale University Press, New Haven, CT, pp 297–355

Jackson JBC, Coates AG (1986) Life cycles and evolution of clonal (modular) animals. Philos Trans R Soc B 313:7–22

Jackson JBC, Buss LW, Cook RE (eds) (1985) Population biology and evolution of clonal organisms. Yale University Press, New Haven, CT

Jackson RW, Johnson LJ, Clarke SR, Arnold DL (2011) Bacterial pathogen evolution: breaking news. Trends Genet 27:32–40

Jacob F (1982) The possible and the actual. Pantheon, NY

Jacob F, Wollman EL (1961) Sexuality and genetics of bacteria. Academic Press, NY

Jaenisch R (1988) Transgenic animals. Science 240:1468–1474

James A, Plank MJ, Edwards AM (2011) Assessing Lévy walks as models of animal foraging. J R Soc Interface 8:1233–1247

James TY, Kauff F, Schoch CL, Matheny PB, Hofstetter V et al (2006) Reconstructing the early evolution of Fungi using a six-gene phylogeny. Nature 443:818–822

James TY, Stenlid J, Olson A, Johannesson H (2008) Evolutionary significance of imbalanced nuclear ratios within heterokaryons of the basidiomycete fungus *Heterobasidion parviporum*. Evolution 62:2279–2296

Janzen DH (1968) Host plants as islands in evolutionary and contemporary time. Am Nat 102:592–595

Janzen DH (1977a) Why fruits rot, seeds mold, and meat spoils. Am Nat 111:691–713

Janzen DH (1977b) What are dandelions and aphids? Am Nat 111:586–589

Janzen DH, Wilson DE (1974) The cost of being dormant in the tropics. Biotropica 6:260–262

Jarosz DF, Lancaster AK, Brown JCS, Lindquist S (2014) An evolutionarily conserved prion-like element converts wild fungi from metabolic specialists to generalists. Cell 158:1072–1082

Jarosz DF, Brown JCS, Walker GA, Datta MS, Ung WL et al (2014) Cross-kingdom chemical communication drives a heritable, mutually beneficial prion-based transformation of metabolism. Cell 158:1083–1093

Jasmin J-N, Kassen R (2007) Evolution of a single niche specialist in variable environments. Proc R Soc B 274:2761–2767

Jasmin J-N, Kassen R (2007) On the experimental evolution of specialization and diversity in heterogeneous environments. Ecol Lett 10:272–281

Javaux EJ, Knoll AH, Walter MR (2001) Morphological complexity and ecological complexity in early eukaryotic systems. Nature 412:66–69

Jennings DH, Rayner ADM (eds) (1984) The ecology and physiology of the fungal mycelium. Proceedings symposium British Mycological Society. Cambridge University Press, Cambridge, UK.

Jern P, Coffin JM (2008) Effects of retroviruses on host genome function. Annu Rev Genet 42:709–732

Jiang C, Brown PJB, Ducret A, Brun YV (2014) Sequential evolution of bacterial morphology by co-option of a developmental regulator. Nature 506:489–493

Jinks JL (1952) Heterokaryosis: a system of adaptation in wild fungi. Proc R Soc Lond B 140:83–99

Johannes F, Colot V, Jansen RC (2008) Epigenome dynamics: a quantitative genetics perspective. Nature Rev Genet 9:883–890

Johnsen PJ, Dubnau D, Levin BR (2009) Episodic selection and the maintenance of competence and natural transformation in Bacillus subtilis. Genetics 181:1521–1533

Johnson ML, Gaines MS (1990) Evolution of dispersal: theoretical models and empirical tests using birds and mammals. Annu Rev Ecol Syst 21:449–480

Johnson MTJ, Agrawal AA (2005) Plant genotype and environment interact to shape a diverse arthropod community on evening primrose (Oenothera biennis). Ecology 86:874–885

Johnson, P.J.T. and B.R. Levin. 2013. Pharmacodynamics, population dynamics, and the evolution of persistence in Staphylococcus aureus. PLoS Genet 9. doi:10.1371/journal.pgen.1003123.

Johnston C, Martin B, Fichant G, Polard P, Claverys J-P (2014) Bacterial transformation: distribution, shared mechanisms and divergent control. Nat Rev Microbiol 12:181–196

Jones JS, Bingham ET (1995) Inbreeding depression in alfalfa and cross-pollinated crops. Plant Breed Rev 13:209–233

Jones JDG, Dangl JL (2006) The plant immune system. Nature 444:323–329

Jones OR, Scheuerlein A, Salguero-Gomez R, Giovanni Camarda C, Schaible R et al (2014) Diversity of ageing across the tree of life. Nature 505:169–173

Jorgensen BB (1977) Distribution of colorless sulfur bacteria (Beggiatoa spp.) in a coastal marine sediment. Mar Biol 41:19–28

Judson HF (1979) The eighth day of creation. Simon & Schuster, New York

Judson OP, Normark BB (1996) Ancient asexual scandals. Trends Ecol Evol 11:41–46

Jukes TH (1987) Transitions, transversions, and the molecular evolutionary clock. J Mol Evol 26:87–98

Junnila RK, List EO, Berryman DE, Murrey JW, Kopchick JJ (2013) The GH/IGF-1 axis in ageing and longevity. Nature Rev Endocrinol 9:366–376

Justice SS, Hunstad DA, Cegelski L, Hultgren SJ (2008) Morphological plasticity as a bacterial survival strategy. Nature Rev Microbiol 6:162–168

Kaern M, Elston TC, Blake WJ, Collins JJ (2005) Stochasticity in gene expression: from theories to phenotypes. Nature Rev Genet 6:451–464

Kahl MP Jr (1964) Food ecology of the wood stork (Mycteria americana) in Florida. Ecol Monogr 34:97–117

Kaiser D (1986) Control of multicellular development: Dictyostelium and Myxococcus. Annu Rev Genet 20:539–566

Kaiser D (2001) Building a multicellular organism. Annu Rev Genet 35:103–123

Kaiser D (2003) Coupling cell movement to multicellular development in myxobacteria. Nature Rev Microbiol 1:45–54

Karasov WH, Martinez del Rio C (2007) Physiological ecology: how animals process energy, nutrients and toxins. Princeton University Press, Princeton, NJ

Karl DM (2007) Microbial oceanography: paradigms, processes and promise. Nature Rev Microbiol 5:759–769

Kassen R (2002) The experimental evolution of specialists, generalists, and the maintenance of diversity. J Evol Biol 15:173–190

Kays S, Harper JL (1974) The regulation of plant and tiller density in a grass sward. J Ecol 62:97–105

Kazazian HHJ (2004) Mobile elements: drivers of genome evolution. Science 303:1626–1632

Keeling PJ, Palmer JD (2008) Horizontal gene transfer in eukaryotic evolution. Nature Rev Genet 9:605–618

Keen NT (1982) Specific recognition in gene-for-gene host-parasite systems. Adv Plant Pathol 1:35–82

Keen NT (1990) Gene-for-gene complementarity in plant-pathogen interactions. Annu Rev Genet 24:447–463

Kelley DS, Baross JA, Delaney JR (2002) Volcanoes, fluids, and life at mid-ocean ridge spreading centers. Annu Rev Earth Planet Sci 30:385–491

Kemperman JA, Barnes BV (1976) Clone size in American aspens. Can J Bot 54:2603–2607

Kessin RH (2001) Dictyostelium: evolution, cell biology, and the development of multicellularity. Cambridge University Prress, Cambridge, UK

Kester JC, Fortune SM (2014) Persisters and beyond: Mechanisms of phenotypic drug resistance and drug tolerance in bacteria. Crit Rev Biochem Mol Biol 49:91–101

Kidwell MG, Lisch DR (2001) Perspective: transposable elements, parasitic DNA, and genome evolution. Evolution 55:1–24

Kieser KJ, Rubin EJ (2014) How sisters grow apart: mycobacterial growth and division. Nat Rev Microbiol 12:550–562

Kimura M (1967) On the evolutionary adjustment of spontaneous mutation rates. Genet Res 9:23–34

Kimura M (1987) Molecular evolutionary clock and the neutral theory. J Mol Evol 26:24–33

King M-C, Wilson AC (1975) Evolution at two levels in humans and chimpanzees. Science 188:107–116

King N (2004) The unicellular ancestry of animal development. Dev Cell 7:313–325

King N, Westbrook MJ, Young SL, Kuo A, Abedin M et al (2008) The genome of the choanoflagellate *Monosiga brevicollis* and the origin of metazoans. Nature 451:783–788

Kinkel LL, Andrews JH, Nordheim EV (1989a) Fungal immigration dynamics and community development on apple leaves. Microb Ecol 18:45–58

Kinkel LL, Andrews JH, Nordheim EV (1989b) Microbial introductions to apple leaves: influences of altered immigration on fungal community dynamics. Microb Ecol 18:161–173

Kinkel LL, Andrews JH, Berbee FM, Nordheim EV (1987) Leaves as islands for microbes. Oecologia 71:405–408

Kirchman DL (2016) Growth rates of microbes in the oceans. Annu Rev Mar Sci 8:285–309

Kirk DL (2005) A twelve-step program for evolving multicellularity and a division of labor. BioEssays 27:299–310

Kirkwood TBL (1977) Evolution of ageing. Nature 279:301–304

Kirkwood TBL (1981) Repair and its evolution: Survival versus reproduction. In: Townsend CR, Calow P (eds) Physiological ecology: an evolutionary approach to resource use. Sinauer Associates Inc., Sunderland, MA, pp 165–189

Kirkwood TBL (2005) Understanding the odd science of aging. Cell 120:437–447

Kirkwood TBL, Holliday R (1979) The evolution of ageing and longevity. Proc R Soc Lond B 205:531–546

Kirkwood TBL, Holliday R (1986) Ageing as a consequence of natural selection. In: Bittles AH, Collins KJ (eds) The biology of human ageing. Cambridge University Press, Cambridge, UK, pp 1–16

Kirkwood TBL, Rose MR (1991) Evolution of senescence: late survival sacrificed for reproduction. Philos Trans R Soc Lond B 332:15–24

Klein BS, Tebbets B (2007) Dimorphism and virulence in fungi. Curr Opin Microbiol 10:314–319

Klein T, Randin C, Korner C (2015) Water availability predicts forest canopy height at the global scale. Ecol Lett 18:1311–1320

Klekowski EJ (2003) Plant clonality, mutation, diplontic selection and mutational meltdown. Biol J Linn Soc 79:61–67

Klekowski EJ Jr (1988) Mutation, developmental selection, and plant evolution. Columbia University Press, New York

Klekowski EJ Jr, Godfrey PJ (1989) Ageing and mutation in plants. Nature 340:389–391

Klekowski EJ Jr, Kazarinova-Fukshansky N, Mohr H (1985) Shoot apical meristems and mutation: Stratified meristems and angiosperm evolution. Am J Bot 72:1788–1800

Klekowski EJ Jr, Lowenfeld R, Klekowski EH (1996) Mangrove genetics. IV. Postzygotic mutations fixed as periclinal chimeras. Int J Plant Sci 157:398–405

Klimesová J, Nobis MP, Herben T (2015) Senescence, ageing and death of the whole plant: morphological prerequisites and constraints of plant immortality. New Phytol 206:14–18

Klingenberg CP (2008) Morphological integration and developmental modularity. Annu Rev Ecol Evol Syst 39:115–132

Kloosterman WP, Francioli LC, Hormozdiari F, Marschall T, Hehir-Kwa JY et al (2015) Characteristics of de novo structural changes in the human genome. Genome Res 25:792–801

Knoll AH (2011) The multiple origins of complex multicellularity. Annu Rev Earth Planet Sci 39:217–239

Knoll AH (2015) Paleobiological perspectives on early microbial evolution. Cold Spring Harb Perspect Biol. doi:10.1101/cshperspect.a018093

Knoll AH, Bambach RK (2000) Directionality in the history of life: diffusion from the left wall or repeated scaling of the right? Paleobiology 26:1–14

Koch AL (1971) The adaptive responses of *Escherichia coli* to a feast and famine existence. Adv Microb Physiol 6:147–217

Koch AL (1976) How bacteria face depression, recession and derepression. Perspect Biol Med 20:44–63

Koch AL (1981) Evolution of antibiotic resistance gene function. Microbiol Rev 45:355–378

Koch AL (1987) Why *Escherichia coli* should be renamed *Escherichia ilei*. In: Torriani-Gorini A et al (eds) Phosphate metabolism and cellular regulation in microorganisms. Am Soc Microbiol, Washington, DC, pp 300–305

Koch AL (1988) Why can't a cell grow infinitely fast? Can J Microbiol 34:421–426

Koch AL (1990) Diffusion: the crucial process in many stages of the biology of bacteria. Adv Microb Ecol 11:37–70

Koch AL (1995) Bacterial growth and form. Chapman and Hall, NY

Koch AL (1996) What size should a bacterium be? A question of scale. Annu Rev Microbiol 50:317–348

Koch AL, Schaechter M (1984) The world and ways of *E. coli*. In: Demain AL (ed) Biology of industrial microorganisms, vol 1. Addison-Wesley, Reading, MA, pp 1–25

Koch GW, Sillett SC, Jennings GM, Davis SD (2004) The limits to tree height. Nature 428:851–854

Koch GW, Sillett SC, Antoine ME, Williams CB (2015) Growth maximization trumps maintenance of leaf conductance in the tallest angiosperm. Oecologia 177:321–331

Koehl MAR (1986) Seaweeds in moving water: form and mechanical function. In: Givnish T (ed) On the Economy of Plant Form and Function. Cambridge University Press, Cambridge, UK, pp 603–634

Kohler A, Kuo A, Nagy LG, Morin E, Barry KW et al (2015) Convergent losses of decay mechanisms and rapid turnover of symbiosis genes in mycorrhizal mutualists. Nature Genet 47:410–415

Kohli Y, Morrall RAA, Anderson JB, Kohn LM (1992) Local and trans-Canadian clonal distribution of *Sclerotinia sclerotiorum* on canola. Phytopathology 82:875–880

Kohn LM (1995) The clonal dynamic in wild and agricultural plant-pathogen populations. Can J Bot 73(Suppl. 1):S1231–S1240

Koonin EV, Martin W (2005) On the origin of genomes and cells within inorganic compartments. Trends Genet 21:647–654

Koufopanou V (1994) The evolution of soma in the volvocales. Am Nat 143:907–931

Kozlowski J, Gawelczyk AT (2002) Why are species' body size distributions usually skewed to the right? Funct Ecol 16:417–432

Kubiena W (1932) Uber Fruchtkorperbildung und engere Standortwahl von Pilzen in Bodenhohlraumen. Arch f Mikrobiol 3:507–542

Kubiena WL (1938) Micropedology. Collegiate Press, Ames, Iowa

Kuhn G, Hijri M, Sanders IR (2001) Evidence for the evolution of multiple genomes in arbuscular mycorrhizal fungi. Nature 414:745–748

Kumar S (2005) Molecular clocks: four decades of evolution. Nature Rev Genet 6:654–662

Kussell E, Leibler S (2005) Phenotypic diversity, population growth, and information in fluctuating environments. Science 309:2075–2078

Kussell E, Kishony R, Balaban NQ, Leibler S (2005) Bacterial persistence: a model of survival in changing environments. Genetics 169:1807–1814

Kysela DT, Brown PJB, Huang KC, Brun YV (2013) Biological consequences and advantages of asymmetric bacterial growth. Annu Rev Microbiol 67:417–435

Lack D (1947) The significance of clutch size. Ibis 89:302–352

Lafferty KD, Dobson AP, Kuris AM (2006) Parasites dominate food web links. Proc Natl Acad Sci USA 103:11211–11216

Laird DJ, De Tomaso AW, Weissman IL (2005) Stem cells are units of natural selection in a colonial ascidian. Cell 123:1351–1360

Laney SR, Olson RJ, Sosik HM (2012) Diatoms favor their younger daughters. Limnol Oceanogr 57:1572–1578

Lang GI, Rice DP, Hickman MJ, Sodergren E, Weinstock GM et al (2013) Pervasive genetic hitchhiking and clonal interference in forty evolving yeast populations. Nature 500:571–574

Larson A, Kirk MM, Kirk DL (1992) Molecular phylogeny of the volvocine flagellates. Mol Biol Evol 9:85–105

Larson DW (2001) The paradox of great longevity in a short-lived tree species. Exp Gerontol 36:651–673

Larwood G, Rosen BR (eds) (1979) Biology and systematics of colonial organisms. Academic Press, NY

Lau HY, Ashbolt NJ (2009) The role of biofilms and protozoa in *Legionella* pathogenesis: implications for drinking water. J Appl Microbiol 107:368–378

Laub MT, McAdams HH, Feldblyum T, Fraser CM, Shapiro L (2000) Global analysis of the genetic network controlling a bacterial cell cycle. Science 290:2144–2148

Lauer AR (1952). Age and sex in relation to accidents. In: Baldock RH (ed.) Road-user characteristics. Highway Res Bd Bull, no. 60 Natl Acad Sci, Washington, DC, pp 25–35

Laurance WF, Nascimento HEM, Laurance SG, Andrade A, Ribeiro JELS et al (2006) Rapid decay of tree-community composition in Amazonian forest fragments. Proc Natl Acad Sci USA 103:19010–19014

Lauro FM, McDougald D, Thomas T, Williams TJ, Egan S et al (2009) The genomic basis for trophic strategy in marine bacteria. Proc Natl Acad Sci USA 106:15527–15533

Lawrence GJ, Dodds PN, Ellis JG (2007) Rust of flax and linseed caused by *Melampsora lini*. Mol Plant Pathol 8:349–364

Lawrence JG, Ochman H (1997) Amelioration of bacterial genomes: rates of change and exchange. J Mol Evol 44:383–397

Lehninger AL (1970) Biochemistry. Worth, New York, NY

Lengeler JW, Drews G, Schlegel HG (eds) (1999) Biology of the prokaryotes. Thieme, Stuttgart, Germany

Lennon JT, Jones SE (2011) Microbial seed banks: the ecological and evolutionary implications of dormancy. Nature Rev Microbiol 9:119–130

Leonard KJ, Szabo LJ (2005) Stem rust of small grains and grasses caused by *Puccinia graminis*. Mol Plant Pathol 6:99–111

Leppik EE (1953) Some viewpoints on the phylogeny of rust fungi. I. Coniferous rusts. Mycologia 45:46–74

Leppik EE (1959) Some viewpoints on the phylogeny of rust fungi. III. Origin of grass rusts. Mycologia 51:512–528

Leppik EE (1961) Some viewpoints on the phylogeny of rust fungi. IV. Stem rust genealogy. Mycologia 53:378–405

Lerat E, Daubin V, Ochman H, and Moran NA (2005) Evolutionary origins of genomic repertoires in bacteria. PLoS Biol 3:e30. doi:10.1371/journal. pbio.0030130

Levin BR (1988) The evolution of sex in bacteria. In: Michod RE, Levin BR (eds) The Evolution of sex: an examination of current ideas. Sinauer Associates Inc., Sunderland, MA, pp 194–211

Levin BR, Bergstrom CT (2000) Bacteria are different: observations, interpretations, speculations, and opinions about the mechanisms of adaptive evolution in prokaryotes. Proc Nat Acad Sci USA 97:6981–6985

Levin BR and Cornejo OE (2009) The population and evolutionary dynamics of homologous gene recombination in bacteria. PLoS Genet 5(8):e1000601.

Levin BR, Perrot V, Walker N (2000) Compensatory mutations, antibiotic resistance and the population genetics of adaptive evolution in bacteria. Genetics 154:985–997

Levin DA (1978) Some genetic consequences of being a plant. In: Brussard PF (ed) Ecological genetics: the interface. Springer-Verlag, NY, pp 189–212

Levin SA, Muller-Landau HC, Nathan R, Chave J (2003) The ecology and evolution of seed dispersal: a theoretical perspective. Annu Rev Ecol Evol Syst 34:575–604

Levins R (1965) Theory of fitness in a heterogeneous environment. V. Optimal genetic systems. Genetics 52:891–904.

Levins R (1966) The strategy of model building in population biology. Am Sci 54:421–431

Levins R (1968) Evolution in changing environments: some theoretical explorations. Princeton University Press, Princeton, NJ

Levins R (1969) Dormancy as an adaptive strategy. In: Woolhouse HW (ed) Dormancy and survival. Proceedings of 23rd Symposium Society for Experimental Biology. Cambridge University Press, Cambridge, UK, pp 1–10

Levitan DR (1989) Density-dependent size regulation in Diadema antillarium: effects on fecundity and survivorship. Ecology 70:1414–1424

Lewontin R (2000) The triple helix: gene, organism, and environment. Harvard University Press, Cambridge, MA

Lewontin RC (1974) The genetic basis of evolutionary change. Columbia University Press, NY

Lewontin RC (1978) Adaptation. Sci Am 239:212–230

Lewontin RC (1983) Gene, organism and environment. In: Bendall DS (ed) Evolution from molecules to men. Cambridge University Press, Cambridge, UK, pp 273–285

Leyk D, Erley O, Ridder D, Leurs M, Rüther T et al (2007) Age-related changes in marathon and half-marathon performances. Int J Sports Med 28:513–517

Li G-W, Xie XS (2011) Central dogma at the single-molecule level in living cells. Nature 475:308–315

Lima-Mendez G, Faust K, Henry N, Decelle J, Colin S, et al. (2015) Determinants of community structure in the global plankton interactome. Science 348. doi: 10.1126/science.1262073

Lindemann J, Constantinidou HA, Barchet WR, Upper CD (1982) Plants as sources of airborne bacteria, including ice nucleation-active bacteria. Appl Environ Microbiol 44:1059–1063

Lindstedt SL, Calder WA III (1981) Body size, physiological time, and longevity of homeothermic animals. Quart Rev Biol 56:1–16

Lindstedt SL, Swain SD (1988) Body size as a constraint of design and function. In: Boyce MS (ed) Evolution of life histories of mammals. Yale University Press, New Haven, CT, pp 93–105

Ling J, O'Donoghue P, Söll D (2015) Genetic code flexibility in microorganisms: novel mechanisms and impact on physiology. Nature Rev Microbiol 13:707–721

Linhart YB, Grant MC (1996) Evolutionary significance of local genetic differentiation in plants. Annu Rev Ecol Syst 27:237–277

Lively CM (1986) Predator-induced shell dimorphism in the acorn barnacle Chthamalus anisopoma. Evolution 40:232–242

Lively CM and Morran LT (2014) The ecology of sexual reproduction. J Evol Biol, pp 1292–1303

Locey KJ, Lennon JT (2016) Scaling laws predict global microbial diversity. Proc Natl Acad Sci USA 113:5970–5975

Locey KJ, Lennon JT (2016b). Reply to Willis: powerful predictions of biodiversity from ecological models and scaling laws. Proc Natl Acad Sci USA 113. doi:10.1073/pnas.l609635113

Lodato MA, Woodworth MB, Lee S, Evrony GD, Mehta BK et al (2015) Somatic mutation in single human neurons tracks developmental and transcriptional history. Science 350:94–98

Lonsdale D, Pautasso M, Holdenrieder O (2008) Wood-decaying fungi in the forest: conservation needs and management options. Eur J Forest Res 127:1–22

López-García P, Zivanovic Y, Deschamps P, Moreira D (2015) Bacterial gene import and mesophilic adaptation in archaea. Nature Rev Microbiol 13:447–456

Losick R, Desplan C (2008) Stochasticity and cell fate. Science 320:65–68

Lovett Doust L (1981a) Population dynamics and local specialization in a clonal perennial (*Ranunculus repens*). 1. The dynamics of ramets in contrasting habitats. J Ecol 69:743–755

Lovett Doust L (1981b) Intraclonal variation and competition in *Ranunculus repens*. New Phytol 89:495–502

Lubchenco J, Cubit J (1980) Heteromorphic life histories of certain marine algae as adaptations to variations in herbivory. Ecology 61:676–687

Lücking R, Huhndorf S, Pfister DH, Plata ER, Lumbsch HT (2009) Fungi evolved right on track. Mycologia 101:810–822

Lynch AJJ, Barnes RW, Cambecedes J, Vaillancourt RE (1998) Genetic evidence that *Lomatia tasmanica* (Proteaceae) is an ancient clone. Aust J Bot 46:25–33

Lynch M (2010) Evolution of the mutation rate. Trends Genet 26:345–352

Lynch M, Burger R, Butcher D, Gabriel W (1993) The mutational meltdown in asexual populations. J Heredity 84:339–344

MacArthur R and Levins R (1964) Competition, habitat selection, and character displacement in a patchy environment. Proc Natl Acad Sci USA 51:1207–1210

MacArthur RH (1972) Geographical ecology: patterns in the distribution of species. Princeton University Press, Princeton, NJ

MacArthur RH, Connell JH (1966) The biology of populations. Wiley, NY

MacArthur RH, Pianka ER (1966) On optimal use of a patchy environment. Am Nat 100:603–609

MacArthur RH, Wilson EO (1963) An equilibrium theory of insular zoogeography. Evolution 17:373–387

MacArthur RH, Wilson EO (1967) The theory of island biogeography. Princeton University Press, Princeton, NJ

Mace ME, Bell AA, Beckman CH (eds) (1981) Fungal wilt diseases of plants. Academic Press, NY

MacLean RC, Bell G (2002) Experimental adaptive radiation in *Pseudomonas*. Am Nat 160:569–581

MacLean RC, Bell G (2003) Divergent evolution during an experimental adaptive radiation. Proc R Soc Lond B 270:1645–1650

Macnab RM (1996) Flagella and motility. In: Neidhardt FC (ed) *Escherichia coli* and *Salmonella*: cellular and molecular biology. ASM Press. Washington, DC, pp 123–145

Madigan MT, Martinko JM, Stahl DA, and Clark DP (2012) Brock biology of microorganisms, 13th edn. Benjamin Cummings (Pearson), San Francisco, CA.

Madigan MT, Martinko JM, Bender KS, Buckley DH, Stahl DA, and Brock T (2015) Brock biology of microorganisms, 14th edn. Benjamin Cummings (Pearson), NY

Magor BG, De Tomaso A, Rinkevich B, Weissman IL (1999) Allorecognition in colonial tunicates: protection against pedatory cell lineages? Immunol Rev 167:69–79

Magyar G, Kun A, Oborny B, Stuefer JF (2007) Importance of plasticity and decision-making strategies for plant resource acquisition in spatio-temporally variable environments. New Phytol 174:182–193

Maharjan R, Seeto S, Notley-McRobb L, Ferenci T (2006) Clonal adaptive radiation in a constant environment. Science 313:514–517

Maheshri N, O'Shea EK (2007) Living with noisy genes: how cells function reliably with inherent variability in gene expression. Annu Rev Biophys Biomol Struct 36:413–434

Maier W, Begerow D, Weiss M, Oberwinkler F (2003) Phylogeny of the rust fungi: an approach using nuclear large subunit ribsomal DNA sequences. Can J Bot 81:12–23

Maier W, Wingfield BD, Mennicken M, Wingfield MJ (2007) Polyphyly and two emerging lineages in the rust genera *Puccinia* and *Uromyces*. Mycol Res 111:176–185

Maitra A, Dill KA (2015) Bacterial growth laws reflect the evolutionary importance of energy efficiency. Proc Natl Acad Sci USA 112:406–411

Makeyev EV, Maniatis T (2008) Multilevel regulation of gene expression by microRNAs. Science 319:1789–1790

Marcon E, Moens PB (2005) The evolution of meiosis: recruitment and modification of somatic DNA-repair proteins. BioEssays 27:795–808

Margulis L (1993) Symbiosis in cell evolution: microbial communities in the archean and proterozoic eons, 2nd edn. Freeman, New York

Mariscal C and Doolittle WF (2015) Eukaryotes first: how could that be? Phil Trans R Soc B 370. ▶http://dx.doi.org/10.1098/rstb.2014.0322

Marshall CR, Valentine JW (2010) The importance of preadapted genomes in the origin of the animal bodyplans and the Cambrian explosion. Evolution 64:1189–1201

Martin W, Baross J, Kelley D, Russell MJ (2008) Hydrothermal vents and the origin of life. Nature Rev Microbiol 6:805–814

Martinez DE, Levinton JS (1992) Asexual metazoans undergo senescence. Proc Natl Acad Sci USA 89:9920–9923

Martiny JBH, Bohannan BJM, Brown JH, Colwell RK, Fuhrman JA et al (2006) Microbial biogeography:

putting microorganisms on the map. Nature Rev Microbiol 4:102–112

Martone PT, Kost L, Boller M (2012) Drag reduction in wave-swept macroalgae: alternative strategies and new predictions. Am J Bot 99:806–815

Matsuo M, Ito Y, Yamaguchi R, Obokata J (2005) The rice nuclear genome continuously integrates, shuffles, and eliminates the chloroplast genome to cause chloroplast-nuclear DNA flux. Plant Cell 17:665–675

Maughan H, Birky CW Jr, Nicholson WL (2009) Transcriptome divergence and the loss of plasticity in Bacillus subtilis after 6,000 generations under relaxed selection for sporulation. J Bacteriol 191:428–433

May RM (1978) The dynamics and diversity of insect faunas In: Mound LA and Waloff N (eds) Diversity of insect faunas. Proceedings of 9th symposium Royal Entomological Society. Blackwell, Oxford, UK, pp. 188–204

May RM (1988) How many species are there on earth? Science 241:1441–1449

Maynard Smith J (1978a) Optimization theory in evolution. Annu Rev Ecol Syst 9:31–56

Maynard Smith J (1978b) The evolution of sex. Cambridge University Press, Cambridge, UK

Maynard Smith J (1986) Contemplating life without sex. Nature 324:300–301

Maynard Smith J (1988) Evolutionary progress and levels of selection. In: Nitecki MH (ed) Evolutionary progress. University of Chicago Press, Chicago, pp 219–230

Maynard Smith J, Szathmary E (1995) The major transitions in evolution. Freeman, Oxford, UK

Maynard Smith J, Dowson CG, Spratt BG (1991) Localized sex in bacteria. Nature 349:29–31

Maynard Smith J, Feil EJ, Smith NH (2000) Population structure and evolutionary dynamics of pathogenic bacteria. BioEssays 22:1115–1122

Maynard Smith J, Smith NH, O'Rourke M, Spratt BG (1993) How clonal are bacteria? Proc Natl Acad Sci USA 90:4384–4388

Mayr E (1970) Populations, species, and evolution. Harvard University Press, Cambridge, MA

Mayr E (1982) The growth of biological thought: diversity, evolution and inheritance. Belknap Press of Harvard University Press, Cambridge, MA

Mayr E (1990) A natural system of organisms. Nature 348:491

Mayr E (1997) This is biology. Belknap Press of Harvard University Press, Cambridge, MA

Mayr E (1998) Two empires or three? Proc Natl Acad Sci USA 95:9720–9723

Mazel D (2006) Integrons: agents of bacterial evolution. Nature Rev Microbiol 4:608–620

Mazel D, Marliere P (1989) Adaptive eradication of methionine and cysteine from cyanobacterial light-harvesting proteins. Nature 341:245–248

McClain CR, Allen AP, Tittensor DP, Rex MA (2012) Energetics of life on the deep seafloor. Proc Natl Acad Sci USA 109:15366–15371

McClintock B (1956) Controlling elements and the gene. Cold Spring Harbor Sympos Quant Biol 21:197–216

McCullen CA, Binns AN (2006) Agrobacterium tumefaciens and plant cell interactions and activities required for interkingdom macromolecular transfer. Annu Rev Cell Dev Biol 22:101–127

McDade JE, Shepard CC, Fraser DW, Tsai TR, Redus MA et al (1977) Legionnaires' disease: isolation of a bacterium and demonstration of its role in other respiratory disease. New Engl J Med 297:1197–1203

McDonald JF (1990) Macroevolution and retroviral elements. BioScience 40:183–191

McDonald JF, Strand DJ, Brown MR, Paskewitz SM, Csink AM et al (1988) Evidence of host-mediated regulation of retroviral element expression at the post-transcriptional level. In: Lambert ME, McDonald JF, Weinstein IB (eds) Eukaryotic transposable elements as mutagenic agents (Banbury Report no 30), Cold Spring Harbor Laboratory Press, Cold Spring Harbor, NY, pp 219–234

McDonald MJ, Rice DP, Desai MM (2016) Sex speeds adaptation by altering the dynamics of molecular evolution. Nature 531:233–236

McFadden CS (1986) Colony fission increases particle capture rates of a soft coral: advantages of being a small colony. J Exptl Mar Biol Ecol 103:1–20

McFall-Ngai M, Hadfield MG, Bosch TCG, Carey HV, Domazet-Loso T et al (2013) Animals in a bacterial world, a new imperative for the life sciences. Proc Natl Acad Sci USA 110:3229–3236

McIntosh RP (1985) The background of ecology: concept and theory. Cambridge University Press, Cambridge, UK

McKenney PT, Driks A, Eichenberger P (2013) The Bacillus subtilis endospore: assembly and functions of the multilayered coat. Nature Rev Microbiol 11:33–44

McKinney FK, Jackson JBC (1989) Bryozoan evolution. Unwin Hyman, Boston

McKinney ML (1990) Trends in body-size evolution. In: McNamara KJ (ed) Evolutionary trends. University of Arizona Press, Tucson, AZ, pp 75–118

McKitrick MC (1993) Phylogenetic constraint in evolutionary theory: has it any explanatory power? Annu Rev Ecol Syst 24:307–330

McMahon T (1973) Size and shape in biology. Science 179:1201–1204

McMahon TA, Bonner JT (1983) On size and life. Scientific American Books, NY

McPeek MA, Holt RD (1992) The evolution of dispersal in spatially and temporally varying environments. Am Nat 140:1010–1027

McShea DW (2001) The hierarchical structure of organisms: a scale and documentation of a trend in the maximum. Paleobiology 27:405–423

Medawar PB (1946) Old age and natural death. Modern Quart 1:30–56

Medawar PB (1952) An unsolved problem of biology. H.K. Lewis, London, UK

Meinhardt KA, Gehring CA (2012) Disrupting mycorrhizal mutualisms: a potential mechanism by which exotic tamarisk outcompetes native cottonwoods. Ecol Appl 22:532–549

Mencuccini M, Martinez-Vilalta J, Vanderklein D, Hamid HA, Korakaki E et al (2005) Size-mediated ageing reduces vigour in trees. Ecol Lett 8:1183–1190

Merchant SS, Helmann JD (2012) Elemental economy: microbial strategies for optimizing growth in the face of nutrient limitation. Adv Microb Physiol 60:91–210

Meyerowitz EM (2002) Plants compared to animals: the broadest comparative study of development. Science 295:1482–1485

Michod RE (2006) The group covariance effect and fitness trade-offs during evolutionary transitions in individuality. Proc Natl Acad Sci USA 103:9113–9117

Michod RE (2007) Evolution of individuality during the transition from unicellular to multicellular life. Proc Natl Acad Sci USA 104:8613–8618

Michod RE, Levin BR (eds) (1988) The evolution of sex: an examination of the current ideas. Sinauer Associates Inc., Sunderland, MA

Milkman R (1996). Recombinational exchange among clonal populations, In: Neidhardt FC (ed-in-chief) *Escherichia coli* and *Salmonella*: cellular and molecular biology, 2nd edn.vol 2. ASM Press, Washington, DC, pp 2663–2684

Milkman R and McKane M (1995) DNA sequence variation and recombination in *E. coli*, In: Baumberg S, et al. (ed) Population genetics of bacteria. 52nd Symposium Society for General Microbiology, Cambridge University Press, Cambridge, UK, pp 127–142

Miller MB, Bassler BL (2001) Quorum sensing in bacteria. Annu Rev Microbiol 55:165–199

Mills LS, Zimova M, Oyler J, Running S, Abatzoglou JT et al (2013) Camouflage mismatch in seasonal coat color due to decreased snow duration. Proc Natl Acad Sci USA 110:7360–7365

Mira A, Ochman H, Moran NA (2001) Deletional bias and the evolution of bacterial genomes. Trends Genet 17:589–596

Mitchell JG (2002) The energetics and scaling of search strategies in bacteria. Am Nat 160:727–740

Mitchell JG, Kogure K (2006) Bacterial motility: links to the environment and a driving force for microbial physics. FEMS Microbiol Ecol 55:3–16

Mitra A, Flynn KJ, Burkholder JM, Berge T, Calbet A, et al. (2014) The role of mixotrophic protists in the biological carbon pump. Biogeosciences 11:995–1005

Mitsuhashi S (ed) (1971) Transferable drug resistance factor R. University Park Press, Baltimore, MD

Mitton JB, Grant MC (1996) Genetic variation and the natural history of quaking aspen. BioScience 46:25–31

Mock KE, Rowe CA, Hooten MB, DeWoody J, Hipkins VD (2008) Clonal dynamics in western North American aspen (*Populus tremuloides*). Mol Ecol 17:4827–4844

Molenaar D, van Berlo R, de Ridder D, and Teusink B (2009). Shifts in growth strategies reflect tradeoffs in cellular economics. Mol Syst Biol 5. doi: 10.1038/msb.2009.82

Moles AT, Westoby M (2006) Seed size and plant strategy across the whole life cycle. Oikos 113:91–105

Monaghan P, Haussmann MF (2006) Do telomere dynamics link lifestyle and lifespan? Trends Ecol Evol 21:47–53

Monod J, Jacob F (1961) General conclusions: teleonomic mechanisms in cellular metabolism, growth, and differentiation. Cold Spring Harbor Sympos Quant Biol 26:389–401

Monro K, Poore AGB (2004) Selection in modular organisms: is intraclonal variation in macroalgae evolutionarily important? Am Nat 163:564–578

Monro K, Poore AGB (2009) The potential for evolutionary responses to cell-lineage selection on growth form and its plasticity in a red seaweed. Am Nat 173:151–163

Monro K, Poore AGB (2009) The evolvability of growth form in a clonal seaweed. Evolution 63:3147–3157

Monshausen GB, Haswell ES (2013) A force of nature: molecular mechanisms of mechano perception in plants. J Exp Bot 64:4663–4680

Montgomerie R, Lyon B, Holder K (2001) Dirty ptarmigan: behavioral modification of conspicuous male plumage. Behav Ecol 12:429–438

Mooney HA, Drake JA (eds) (1986) Ecology of biological invasions of North America and Hawaii. Springer-Verlag, NY

Moore CM, Mills MM, Arrigo KR, Berman-Frank I, Bopp L et al (2013) Processes and patterns of oceanic nutrient limitation. Nature Geosci 6:701–710

Moore RC, Purugganan MD (2005) The evolutionary dynamics of plant duplicate genes. Curr Opin Plant Biol 8:122–128

Mora C, Tittensor DP, Adl S, Simpson AGB, and Worm B (2011) How many species are there on Earth and in the ocean? PLoS Biol 9(8): el001127. doi:10.1371/journal.pbio.l001127

Morales M, Munne-Bosch S (2015) Secret of long life lies underground. New Phytol 205:463–467

Moran NA (1988) The evolution of host-plant alternation in aphids: evidence for specialization as a dead-end. Am Nat 132:681–706

Moran NA (1989) A 48-million-year-old aphid-host plant association and complex life cycle: biogeographic evidence. Science 245:173–177

Moran NA (1992) The evolutionary maintenance of alternative phenotypes. Am Nat 139:971–989

Moran NA (1994) Adaptation and constraint in the complex life cycles of animals. Annu Rev Ecol Syst 25:573–600

Moran NA, Whitham TG (1988) Evolutionary reduction of complex life cycles: loss of host-alternation in Pemphigus (Homoptera: Aphidae). Evolution 42:717–728

Morita RY (1997) Bacteria in oligotrophic environments: starvation-survival lifestyle. Chapman & Hall, NY

Morse DR, Lawton JH, Dodson MM, Williamson MH (1985) Fractal dimension of vegetation and the distribution of arthropod body lengths. Nature 314:731–733

Mortimer RK, Johnston JR (1959) Life span of individual yeast cells. Nature 183:1751–1752

Muller CH (1966) The role of chemical inhibition (allelopathy) in vegetational composition. Bull Torrey Bot Club 93:332–251

Muller HJ (1964) The relation of recombination to mutational advance. Mutation Res 1:2–9

Munne-Bosch S (2015) Senescence: is it universal or not? Trends Plant Sci 20:713–720

Munro RHM, Nielsen SE, Price MH, Stenhouse GB, Boyce MS (2006) Seasonal and diel patterns of grizzly bear diet and activity in West-Central Alberta. J Mammal 87:1112–1121

Murashige T (1974) Plant propagation through tissue cultures. Annu Rev Plant Physiol 25:135–166

Mydlarz LD, Jones LE, Harvell CD (2006) Innate immunity, environmental drivers, and disease ecology of marine and freshwater invertebrates. Annu Rev Ecol Evol Syst 37:251–288

Nakayama H, Nakayama N, Seiki S, Kojima M, Sakakibara H et al (2014) Regulation of the KNOX-GA gene module induces heterophyllic alteration in North American lake cress. The Plant Cell 26:4733–4748

Neidhardt FC, Ingraham JL, Schaechter M (1990) Physiology of the bacterial cell: a molecular approach. Sinauer Associates Inc., Sunderland, MA

Nelson CD, Spear RN, Andrews JH (2000) Automated image analysis of live/dead staining of the fungus Aureobasidium pullulans on microscope slides and leaf surfaces. BioTechniques 29:874–880

Nemecek JC, Wuthrich M, Klein BS (2006) Global control of dimorphism and virulence in fungi. Science 312:583–588

Nemergut DR, Schmidt SK, Fukami T, O'Neill SP, Bilinski TM et al (2013) Patterns and processes of microbial community assembly. Microbiol Mol Biol Rev 77:342–356

Ngugi HK, Scherm H (2006) Biology of flower-infecting fungi. Annu Rev Phytopathol 44:261–282

Nicotra AB, Atkin OK, Bonser SP, Davidson AM, Finnegan EJ et al (2010) Plant phenotypic plasticity in a changing climate. Trends Plant Sci 15:684–692

Nie V, Speakman JR, Wu Q, Zhang C, Hu Y et al (2015) Exceptionally low daily energy expenditure in the bamboo-eating giant panda. Science 349:171–174

Nikaido H (2009) Multidrug resistance in bacteria. Annu Rev Biochem 78:119–146

Niklas KJ (1994a) Plant allometry: the scaling of form and process. University Chicago Press, Chicago, IL

Niklas KJ (1994b) Morphological evolution through complex domains of fitness. Proc Natl Acad Sci USA 91:6772–6779

Niklas KJ (2000) The evolution of plant body plans—a biomechanical perspective. Ann Bot 85:411–438

Niklas KJ (2014) The evolutionary-developmental origins of multicellularity. Am J Bot 101:6–25

Niklas KJ, Kutschera U (2009) The evolutionary development of plant body plans. Funct Plant Biol 36:682–695

Niklas KJ, Kutschera U (2010) The evolution of the land plant life cycle. New Phytol 185:27–41

Niklas KJ, Spatz H-C (2004) Growth and hydraulic (not mechanical) constraints govern the scaling of tree height and mass. Proc Natl Acad Sci 101:15661–15663

Nisbet EG, Sleep NH (2001) The habitat and nature of early life. Nature 409:1083–1091

Noodén LD (ed) (2004) Plant cell death processes. Elsevier Academic Press, Amsterdam

Noodén LD, Penney JP (2001) Correlative controls of senescence and plant death in Arabidopsis thaliana (Brassicaceae). J Exp Bot 52:2151–2159

Norman TM, Lord ND, Paulsson J, Losick R (2013) Memory and modularity in cell-fate decision making. Nature 503:481–486

Normark BB, Judson OP, Moran NA (2003) Genomic signatures of ancient asexual lineages. Biol J Linn Soc 79:69–84

Noss RF (ed) (2000) The redwood forest: history, ecology and conservation of the coast redwoods. Island Press, Washington, DC

Nowell RW, Green S, Lane BF, Sharp PM (2014) The extent of genome flux and its role in the differentiation of bacterial lineages. Genome Biol Evol 6:1514–1529

Nussey DH, Froy H, Lemaitre J-F, Gaillard J-M, Austad SN (2013) Senescence in natural populations of animals: widespread evidence and its implications for bio-gerontology. Aging Res Rev 12:214–225

Nutman AP, Bennett VC, Friend CRL, Van Kranendonk MJ, Chivas AR (2016) Rapid emergence of life shown by discovery of 3,700-million-year-old microbial structures. Nature 537:535–538

Nystrom T (2002) Aging in bacteria. Curr Opin Microbiol 5:596–601

O'Connell LM, Ritland K (2004) Somatic mutations at microsatellite loci in Western redcedar (*Thuja plicata:* Cupressaceae). J Hered 95:172–176

Ochman H, Wilson AC (1987) Evolution in bacteria: evidence for a universal substitution rate in cellular genomes. J Mol Evol 26:74–86

Ochman H, Lawrence JG, Groisman EA (2000) Lateral gene transfer and the nature of bacterial innovation. Nature 405:299–304

Ochman H, Lerat E, and Daubin V (2005. Examining bacterial species under the specter of gene transfer and exchange. Proc Natl Acad Sci, USA, 102(Suppl. 1):6595–6599

Odum EP, Barrett GW (2005) Fundamentals of ecology, 5th edn. Thomson Brooks /Cole, Belmont, CA

Oinonen E (1967) Sporal regeneration of bracken in Finland in light of the dimensions and age of its clones. Acta For Fenn 83:3–96

Okamura B, Harmelin J-G, Jackson JBC (2001) Refuges revisited: enemies versus flow and feeding as determinants of sessile animal distribution and form. In: Jackson JBC, Lidgard S, McKinney FK (eds) Evolutionary patterns: growth, form, and tempo in the fossil record. University Chicago Press, Chicago, IL pp 61–93

Ono Y (2002) The diversity of nuclear cycle in microcyclic rust fungi (Uredinales) and its ecological and evolutionary implications. Mycoscience 43:421–439

Orive ME (1995) Senescence in organisms with clonal reproduction and life histories. Am Nat 145:90–108

Orive ME (2001) Somatic mutations in organisms with complex life histories. Theor Pop Biol 59:235–249

Orrock JL, Holt RD, Baskett ML (2010) Refuge-mediated apparent competition in plant-consumer interactions. Ecol Lett 13:11–20

Orshan G (1963) Seasonal dimorphism of desert and Mediterranean chamaephytes and its significance as a factor in their water economy. In: Rutter AJ, Whitehead FW (eds) The water relations of plants. Blackwell, Oxford, UK, pp 207–222

Orskov F, Orskov I (1983) Summary of a workshop on the clone concept in the epidemiology, taxonomy, and evolution of the enterobacteriaceae and other bacteria. J Infect Dis 148:346–457

Orth JD, Conrad TM, Na J, Lerman JA, Nam H, et al. (2011) A comprehensive genome-scale reconstruction of *Escherichia coli* metabolism—2011. Mol Syst Biol 7. doi:10.1038/msb.2011.65

Otto SP (2009) The evolutionary enigma of sex. Am Nat 174 (Suppl.):SI–S14

Otto SP, Hastings IM (1998) Mutation and selection within the individual. Genetica 102(103):507–524

Otto SP, Lenormand T (2002) Resolving the paradox of sex and recombination. Nature Rev Genet 3:252–261

Otto SP, Orive ME (1995) Evolutionary consequences of mutation and selection within an individual. Genetics 141:1173–1187

Otto SP, Whitton J (2000) Polyploid incidence and evolution. Annu Rev Genet 34:401–437

Otto SP, Scott ME, Immler S (2015) Evolution of haploid selection in predominantly diploid organisms. Proc Natl Acad Sci USA 112:15952–15957

Pace NR (2006) Time for a change. Nature 44:289

Pace NR (2008) The molecular tree of life changes how we see, teach microbial diversity. Microbe 3:15–20

Pace NR (2009) Mapping the tree of life: progress and prospects. Microbiol Mol Biol Rev 73:565–576

Pace NR, Saap J, Goldenfeld N (2012) Phylogeny and beyond: scientific, historical, and conceptual significance of the first tree of life. Proc Natl Acad Sci USA 109:1011–1018

Paillet FL (2002) Chestnut: history and ecology of a transformed species. J Biogeog 29:1517–1530

Paine RT (1969) A note on trophic complexity and community stability. Am Nat 103:91–93

Paine RT (1995) A conversation on refining the concept of keystone species. Conserv Biol 9:962–964

Palacio S, Millard P, Montserrat-Marti G (2006) Aboveground biomass allocation patterns within Mediterranean sub-shrubs: a quantitative analysis of seasonal dimorphism. Flora 201:612–622

Palkova Z, Wilkinson D, Vachova L (2014) Aging and differentiation in yeast populations: elders with different properties and functions. FEMS Yeast Res 14:96–108

Parker GA, Maynard Smith J (1990) Optimality theory in evolutionary biology. Nature 348:27–33

Partridge L, Barton NH (1996) On measuring the rate of ageing. Proc R Soc Lond B 263:1365–1371

Partridge L, Barton NH (1993) Optimality, mutation and the evolution of ageing. Nature 362:305–311

Pascual M, Dunne JA (eds) (2006) Ecological networks: linking structure to dynamics in food webs. Oxford University Press, New York

Patrick R (1967) The effect of invasion rate, species pool, and size of area on the structure of the diatom community. Proc Natl Acad Sci USA 58:1335–1342

Pawlowska TE, Taylor JW (2004) Organization of genetic variation in individuals of arbuscular mycorrhizal fungi. Nature 427:733–737

Payne JL, Boyer AG, Brown JH, Finnegan S, Kowalewski M et al (2009) Two-phase increase in the maximum size of life over 3.5 billion years reflects biological innovation and environmental opportunity. Proc Natl Acad Sci USA 106:24–27

Peabody RB, Peabody DC, Sicard KM (2000) A genetic mosaic in the fruiting stage of Armillaria gallica. Fungal Genet Biol 29:72–80

Pearl R (1940) Introduction to medical biometry and statistics, 3rd edn. Saunders, Philadelphia

Pedersen B (1995) An evolutionary theory of clonal senescence. Theor Pop Biol 47:292–320

Pedersen B (1999) Senescence in plants. In: Vuorisalo TO, Mutikainen PK (eds) Life history evolution in plants. Kluwer Academic Publishers, Boston, MA, pp 239–274

Pedersen B, Tuomi J (1995) Hierarchial selection and fitness in modular and clonal organisms. Oikos 73:167–180

Pegg GF (1985) Life in a black hole—the micro-environment of the vascular pathogen. Trans Br Mycol Soc 85:1–20

Pentecost A (1980) Aspects of competition in saxicolous lichen communities. Lichenologist 12:135–144

Peregrin-Alvarez JM, Sanford C, Parkinson J (2009). The conservation and evolutionary modularity of metabolism. Genome Biol 10:R63.1–R63.17

Perkins DD, Turner BC (1988) Neurospora from natural populations: toward the population biology of a haploid eukaryote. Exp Mycol 12:91–131

Perry G, Pianka ER (1997) Animal foraging: past, present and future. Trends Ecol Evol 12:360–364

Peter IS, Davidson EH (2015). Genomic control process: development and evolution. Academic Press (Elsevier), New York

Peters RH (1983) The ecological implications of body size. Cambridge University Press, Cambridge, UK

Petit RJ, Hampe A (2006) Some evolutionary consequences of being a tree. Annu Rev Ecol Syst 37:187–214

Petranka JW (2007) Evolution of complex life cycles of amphibians: bridging the gap between metapopulation dynamics and life history evolution. Evol Ecol 21:751–764

Pfeiffer T, Bonhoeffer S (2003) An evolutionary scenario for the transition to undifferentiated multicellularity. Proc Natl Acad Sci USA 100:1 095–1098

Pfeiffer T, Schuster S, Bonhoeffer S (2001) Cooperation and competition in the evolution of ATP-producing pathways. Science 292:504–507

Pfeiffer V, Lingner J (2013) Replication of telomeres and the regulation of telomerase. Cold Spring Harbor Perspect Biol 5:a010405

Pfennig DW, Wund MA, Snell-Rood EC, Cruickshank T, Schlichting CD et al (2010) Phenotypic plasticity's impacts on diversification and speciation. Trends Ecol Evol 25:459–467

Pfennig N (1989) Ecology of photosynthetic purple and green sulfur bacteria. In: Schlegel HG, Bowien B (eds) Autotrophic bacteria. Springer, New York, pp 97–116

Pianka ER (1970) On r- and K-selection. Am Nat 104:592–597

Pianka ER (1976) Natural selection of optimal reproductive tactics. Am Zool 16:775–784

Pianka ER (1978) Evolutionary ecology, 2nd edn. Harper & Row, New York

Pianka ER (2000) Evolutionary ecology, 6th edn. Addison Wesley Longman, San Francisco, CA

Pierik R, de Wit M (2014) Shade avoidance: phytochrome signalling and other aboveground neighbour detection cues. J Exp Bot 65:2815–2824

Piersma T, Drent J (2003) Phenotypic flexibility and the evolution of organismal design. Trends Ecol Evol 18:228–233

Pilcher JR, Baillie MGL, Schmidt B, Becker B (1984) A 7,272-year tree-ring chronology for western Europe. Nature 312:150–152

Pimm SL, Jones HL, Diamond J (1988) On the risk of extinction. Am Nat 132:757–785

Pineda-Krch M, Lehtila K (2004) Costs and benefits of genetic heterogeneity. J Evol Biol 17:1167–1177

Pires ND, Dolan L (2012) Morphological evolution in land plants: new design with old genes. Phil Trans R Soc B 367:508–518

Pirie NW (1973) "On being the right size". Annu Rev Microbiol 27:119–132

Plante S, Colchero F, Calme S (2014) Foraging strategy of a neotropical primate: how intrinsic and extrinsic factors influence destination and residence time. J Anim Ecol 83:116–125

Platt TG, Morton ER, Barton IS, Bever JD, Fuqua C (2014). Ecological dynamics and complex interactions of Agrobacterium megaplasmids. Front Plant Sci 5. doi: 10.3389/fpls.2014.00635

Poindexter JS (1981a) The caulobacters: ubiquitous unusual bacteria. Microbiol Rev 45:123–179

Poindexter JS (1981b) Oligotrophy: fast and famine existence. Adv Microb Ecol 5:63–89

Pomeroy LR (1974) The ocean's food web, a changing paradigm. BioScience 24:499–504

Pomeroy LR, leB Williams PJ, Azam F, Hobbie JE (2007) The microbial loop. Oceanography 20:28–33

Pontecorvo G (1946) Genetic systems based on heterocaryosis. Cold Spring Harbor Sympos Quant Biol 11:193–201

Pontecorvo G (1956) The parasexual cycle in fungi. Annu Rev Microbiol 10:393–400

Post DM (2002) The long and short of food-chain length. Trends Ecol Evol 17:269–277

Postgate JR, Calcott PH (1985) Ageing and death in microbes. In: Bull AT, Dalton H (eds) Comprehensive biotechnology: the principles, applications and regulations of biotechnology in industry agriculture and medicine. Pergamon Press, Oxford, UK, pp 239–249

Potoyan DA, Wolynes PG (2015) Dichotomous noise models of gene switches. J Chem Phys 143:195101. doi:10.1063/1.4935572

Pouchkina-Stantcheva NN, McGee BM, Boschetti C, Tolleter D, Chakrabortee S et al (2007) Functional divergence of former alleles in an ancient asexual invertebrate. Science 318:268–271

Poudyal M, Rosa S, Powell AE, Moreno M, Dellaporta SL et al (2007) Embryonic chimerism does not induce tolerance in an invertebrate model organism. Proc Natl Acad Sci USA 104:4559–4564

Pranting M, Andersson DI (2011) Escape from growth restriction in small colony variants of *Salmonella typhimurium* by gene amplification and mutation. Mol Microbiol 79:305–315

Preston FW (1948) The commonness and rarity of species. Ecology 29:254–283

Price CA, Weitz JS (2012) Allometric covariation: a hallmark behavior of plants and leaves. New Phytol 193:882–889

Price CA, Weitz JS, Savage VM, Stegen J, Clarke A et al (2012) Testing the metabolic theory of ecology. Ecol Lett 15:1465–1474

Price PB, Sowers T (2004) Temperature dependence of metabolic rates for microbial growth, maintenance, and survival. Proc Natl Acad Sci USA 101:4631–4636

Price PW (1980) Evolutionary Biology of Parasites. Princeton University Press, Princeton, NJ

Price PW (1984) Communities of specialists: vacant niches in ecological and evolutionary time. In: Strong DR, Simberloff D, Abele LG, Thistle AB (eds) Ecological communities: conceptual issues and the evidence. Princeton University Press, Princeton, NJ, pp 510–523

Prochnik SE, Umen J, Nedelcu AM, Hallmann A, Miller SM et al (2010) Genomic analysis of organismal complexity in the multicellular green alga *Volvox carteri*. Science 329:223–226

Prud'homme B, Gompel N, Carroll SB (2007) Emerging principles of regulatory evolution. Proc Natl Acad Sci USA 104:8605–8612

Prud'homme B, Gompel N, Rokas A, Kassner VA, Williams TM et al (2006) Repeated morphological evolution through *cis*-regulatory changes in a pleiotropic gene. Nature 440:1050–1053

Pu Y, Zhao Z, Li Y, Zou J, Ma Q et al (2016) Enhanced efflux activity facilitates drug tolerance in dormant bacterial cells. Mol Cell 62:284–294

Puhalla JE, Bell AE (1981) Genetics and biochemistry of wilt pathogens. In: Mace ME, Bell AA, Beckman CH (eds) Fungal wilt diseases of plants. Academic Press, NY, pp 145–192

Pulliam HR (1988) Sources, sinks, and population regulation. Am Nat 132:652–661

Purvis A, Orme CDL, Dolphin K (2003) Why are most species small-bodied? A phylogenetic view. In: Blackburn TM, Gaston KJ (eds) Macroecology: concepts and consequences. Blackwell Science, Oxford, UK, pp 155–173

Pyke GH (1983) Animal movements; an optimal foraging approach. In: Swingland IR, Greenwood PJ (eds) The ecology of animal movement. Clarendon Press, Oxford, UK, pp 7–31

Pyke GH (1984) Optimal foraging theory: a critical review. Annu Rev Ecol Syst 15:523–575

Pyke GH (2015) Understanding movements of organisms: it's time to abandon the Lévy foraging hypothesis. Meth Ecol Evol 6:1–16

Quaintenne G, van Gils JA, Bocher P, Dekinga A, Piersma T (2010) Diet selection in a molluscivore shorebird across Western Europe: does it show short- or long-term intake rate-maximization? J Anim Ecol 79:53–62

Raaijmakers JM, Mazzola M (2012) Diversity and natural functions of antibiotics produced by beneficial and plant pathogenic bacteria. Annu Rev Phytopathol 50:403–424

Raff RA (1987) Constraint, flexibility, and phylogenetic history in the evolution of direct development in sea urchins. Dev Biol 119:6–19

Raff RA (1996) The shape of life: genes, development, and the evolution of animal form. University Chicago Press, Chicago

Raff RA (2007) Written in stone: fossils, genes and evo-devo. Nature Rev Genet 8:911–920

Raffaele S, Kamoun S (2012) Genome evolution in filamentous plant pathogens: why bigger can be better. Nature Rev Microbiol 10:417–430

Rahman MH, Dayanandan S, Rajora OP (2000) Microsatellite DNA markers in *Populus tremuloides*. Genome 43:293–297

Rainey PB, De Monte S (2014) Resolving conflicts during the evolutionary transition to multicellular life. Annu Rev Ecol Evol Syst 45:599–620

Rainey PB, Travisano M (1998) Adaptive radiation in a heterogeneous environment. Nature 394:69–72

Rajon E, Desouhant E, Chevalier M, Debias F, Menu F (2014) The evolution of bet hedging in response to local ecological conditions. Am Nat 184:E1–E15

Rang CU, Peng AY, Chao L (2011) Temporal dynamics of bacterial aging and rejuvenation. Curr Biol 21:1813–1816

Raper JR, Flexer AS (1970) The road to diploidy with emphasis on a detour. In: Charles HP, Knight BCJG (eds) Organization and control in prokaryotic and eukaryotic cells. 20th Symposium of the Society for General Microbiology, Cambridge University Press, Cambridge, UK, pp 401–432

Ratcliff WC, Denison RF, Borrello M, Travisano M (2012) Experimental evolution of multicellularity. Proc Natl Acad Sci USA 109:1595–1600

Raup DM (1991) Extinction: bad genes or bad luck?. Norton, New York

Raven JA (1986) Physiological consequences of extremely small size for autotrophic organisms in the sea, pp 1–70. In: Platt T, Li WKW (eds) Photosynthetic picoplankton. (Proceedings of the NATO Advanced Study Institute). Canada Department of Fisheries & Oceans (Canadian Bulletin of Fisheries and Aquatic Sciences, vol 214), Ottawa, Canada

Raven JA (1998) The twelfth Tansley Lecture: small is beautiful: the picophytoplankton. Funct Ecol 12:503–513

Raymann K, Brochier-Armanet C, Gribaldo S (2015) The two-domain tree of life is linked to a new root for the Archaea. Proc Natl Acad Sci USA 112:6670–6675

Rayner ADM (1991) The challenge of the individualistic mycelium. Mycologia 83:48–71

Rayner ADM, Franks NR (1987) Evolutionary and ecological parallels between ants and fungi. Trends Ecol Evol 2:127–132

Rayner ADM, Watling R, Frankland JC (1985) Resource relations—an overview. In: Moore D, Casselton LA, Wood DA, Frankland JC (eds) Developmental biology of higher fungi. Cambridge University Press, Cambridge, UK, pp 1–40

Read J, Stokes A (2006) Plant biomechanics in an ecological context. Am J Bot 93:1546–1565

Reboud X, Bell G (1997) Experimental evolution in Chlamydomonas. III. Evolution of specialist and generalist types in environments that vary in space and time. Heredity 78:507–514

Redfield RJ (1988) Evolution of bacterial transformation: is sex with dead cells ever better than no sex at all? Genetics 119:213–221

Redfield RJ (2001) Do bacteria have sex? Nature Rev Genet 2:634–639

Rees M, Venable DL (2007) Why do big plants make big seeds? J Ecol 95:926–936

Reeves PR (1992) Variation in O-antigens, niche-specific selection and bacterial populations. FEMS Microbiol Lett 100:509–516

Reich PB (1995) Phenology of tropical forests: patterns, causes, and consequences. Can J Bot 164–174

Reich PB, Tjoelker MG, Machado J-L, Oleksyn J (2006) Universal scaling of respiratory metabolism, size and nitrogen in plants. Nature 439:457–461

Reich PB, Uhl C, Walters MB, Prugh L, Ellsworth DS (2004) Leaf demography and phenology in Amazonian rain forest: a census of 40,000 leaves of 23 tree species. Ecol Monogr 74:3–23

Reichenbach H (1993) Biology of the myxobacteria: ecology and taxonomy. In: Dworkin M, Kaiser D (eds) Myxobacteria II. American Society for Microbiology Press, Washington, DC, pp 13–63

Remold S (2012) Understanding specialism when the jack of all trades can be master of all. Proc R Soc B 279:4861–4869

Replansky T, Koufopanou V, Greig D, Bell G (2008) Saccharomyces sensu stricto as a model system for evolution and ecology. Trends Ecol Evol 23:494–501

Rescan M, Lenormand T, Roze D (2016) Interactions between genetic and ecological effects on the evolution of life cycles. Am Nat 187:19–34

Rettenmeyer CW (1963) Behavioral studies of army ants. Univ Kansas Sci Bull 44:281–465

Reynolds A (2015) Liberating Lévy walk research from the shackles of optimal foraging. Phys Life Rev 14:59–83

Rice WR (2002) Experimental tests of the adaptive significance of sexual recombination. Nature Rev Genet 3:241–251

Richardson K, Bennion OT, Tan S, Hoang AN, Cokol M et al (2016) Temporal and intrinsic factors of rifampicin tolerance in mycobacteria. Proc Natl Acad Sci 113:8302–8307

Richter DJ, King N (2013) The genomic and cellular foundations of animal origins. Annu Rev Genet 47:509–537

Ricklefs RE (1990) Ecology, 3rd edn. Freeman, New York

Ricklefs RE (2008) The evolution of senescence from a comparative perspective. Funct Ecol 22:379–392

Ricklefs RE, Scheuerlein A (2002) Biological implications of the Weibull and Gompertz models of aging. J Gerontol Biol Sci 57A:B69–B76

Riha K, McKnight TD, Griffing LR, Shippen DE (2001) Living with genome instability: plant responses to telomere dysfunction. Science 291:1797–1800

Riley M, Abe T, Arnand MB, Berlyn MKB, Blattner FR et al (2006) Escherichia coli K-12: a cooperatively developed annotation snapshot—2005. Nucleic Acids Res. 34:1–9

Riquelme M, Yarden O, Bartnicki-Garcia S, Bowman B, Castro-Longoria E et al (2011) Architecture and development of the Neurospora crassa hypha—a model for polarized growth. Fungal Biol 115:446–474

Roach DA (1993) Evolutionary senescence in plants. Genetica 91:53–64

Roach DA, Carey JR (2014) Population biology of aging in the wild. Annu Rev Ecol Evol Syst 45:421–443

Robertson E, Bradley A, Kuehn M, Evans M (1986) Germ-line transmission of genes introduced into cultured pluripotential cells by retroviral vector. Nature 323:445–448

Roche CM, Loros JJ, McCluskey K, Glass NL (2014) Neurospora crassa: looking back and looking forward at a model microbe. Am J Bot 101:2011–2035

Roelfs AP (1985) Wheat and rye stem rust. In: Roelfs AP, Bushnell WR (eds) The cereal rusts. Diseases, distribution, epidemiology, and control, vol II. Academic Press, New York, pp 3–37

Roelfs AP, Bushnell WR (eds) (1985) The cereal rusts. Diseases, distribution, epidemiology, and control, vol II. Academic Press, New York

Rohme D (1981) Evidence for a relationship between longevity of mammalian species and life spans of normal fibroblasts in vitro and erythrocytes in vivo. Proc Natl Acad Sci USA 78:5009–5013

Rokas A (2008) The molecular origins of multicellular transitions. Curr Opin Genet Dev 18:472–478

Rokas A (2008) The origins of multicellularity and the early history of the genetic toolkit for animal development. Annu Rev Genet 42:235–251

Ronce O (2007) How does it feel to be like a rolling stone? Ten questions about dispersal evolution. Annu Rev Ecol Evol Syst 38:231–253

Roper M, Ellison C, Taylor JW, Glass NL (2011) Nuclear and genome dynamics in multinucleate ascomycete fungi. Curr Biol 21:R786–R793

Roper M, Pepper RE, Brenner MP, Pringle A (2008) Explosively launched spores of ascomycete fungi have drag-minimizing shapes. Proc Natl Acad Sci USA 105:20583–20588

Roper M, Seminara A, Bandi MM, Cobb A, Dillard HR et al (2010) Dispersal of fungal spores on a cooperatively generated wind. Proc Natl Acad Sci USA 107:17474–17479

Roper M, Simonin A, Hickey PC, Leeder A, Glass NL (2013) Nuclear dynamics in a fungal chimera. Proc Natl Acad Sci USA 110:12875–12880

Rose MR (1991) Evolutionary biology of aging. Oxford University Press, NY

Rosen BR (1986) Modular growth and form of corals: a matter of metamers? Phil Trans R Soc Lond B 313:115–142

Rosen MJ, Davison M, Bhaya D, Fisher DS (2015) Fine-scale diversity and extensive recombination in a quasisexual bacterial population occupying a broad niche. Science 348:1019–1023

Rosengarten RD, Nicotra ML (2011) Model systems of invertebrate allorecognition. Curr Biol 21:R82–R92

Ross C (1988) The intrinsic rate of natural increase and reproductive effort in primates. J Zool Lond 214:199–219

Rossello-Mora R, Amann R (2001) The species concept for prokaryotes. FEMS Microbiol Rev 25:39–67

Rossetti V, Schirrmeister BE, Bernasconi MV, Bagheri HC (2010) The evolutionary path to terminal differentiation and division of labor in cyanobacteria. J Theor Biol 262:23–34

Roszak DB, Grimes DJ, Colwell RR (1984) Viable but non-recoverable stages of Salmonella enteritidis in aquatic systems. Can J Microbiol 30:334–338

Roth VL (1981) Constancy in the size ratios of sympatric species. Am Nat 118:394–404

Roughgarden J (1971) Density-dependent natural selection. Ecology 51:453–468

Roughgarden J, Gaines S, Possingham H (1988) Recruitment dynamics in complex life cycles. Science 241:1460–1466

Rubin H (2002) The disparity between human cell senescence in vitro and lifelong replication in vivo. Nature Biotechnol 20:675–681

Ruffel S, Krouk G, Ristova D, Shasha D, Birnbaum KD et al (2011) Nitrogen economics of root foraging: transitive closure of the nitrate-cytokinin relay and distinct systemic signaling for N supply vs. N demand. Proc Natl Acad Sci USA 108:18524–18529

Ruiz-Trillo I, Burger G, Holland PWH, King N, Lang BF et al (2007) The origins of multicellularity: a multitaxon genome initiative. Trends Genet 23:113–118

Rundle HD, Nagel L, Boughman JW, Schluter D (2000) Natural selection and parallel speciation in sympatric sticklebacks. Science 287:306–308

Ryan FJ, Beadle GW, Tatum EL (1943) The tube method of measuring the growth rate of Neurospora. Am J Bot 30:784–799

Saberi S, Emberly E (2013) Non-equilibrium polar localization of proteins in bacterial cells. PLoS One 8(5). doi: 10.1371/journal.pone.0064075

Sackville Hamilton NR, Harper JL (1989) The dynamics of Trifolium repens in a permanent pasture. I. The population dynamics of leaves and nodes per shoot axis. Proc R Soc Lond B 237:133–173

Sackville Hamilton NR, Schmid B, Harper JL (1987) Life-history concepts and the population biology of clonal organisms. Proc R Soc Lond B 232:35–57

Saito MA, McIlvin MR, Moran DM, Goepfert TJ, DiTullio GR et al (2014) Multiple nutrient stresses at intersecting Pacific Ocean biomes detected by protein biomarkers. Science 345:1173–1177

Sakakibara K, Ando S, Yip HK, Tamada Y, Hiwatashi Y et al (2013) KNOX2 genes regulate the haploid-to-diploid morphological transition in land plants. Science 339:1067–1070

Salama R, Sadaie M, Hoare M, Narita M (2014) Cellular senescence and its effector programs. Genes & Develop 28:99–114

Salisbury EJ (1942) The reproductive capacity of plants. Studies in quantitative biology. Bell & Sons, London

Salmond GPC, Fineran PC (2015) A century of the phage: past, present and future. Nature Rev Microbiol 13:777–786

Sanchez A, Choubey S, Kondev J (2013) Regulation of noise in gene expression. Annu Rev Biophys 42:469–491

Sanchez JA, Lasker HR, Nepomuceno EG et al (2004) Branching and self-organization in marine modular colonial organisms. Am Nat 163:E24–E39

Sandegren L, Andersson DI (2009) Bacterial gene amplification: implications for the evolution of antibiotic resistance. Nature Rev Microbiol 7:578–588

Sanders FE (1971) Effect of root and soil properties on the uptake of nutrients by competing roots. Thesis, DPhil Oxford University, Oxford, UK

Sarkar SF, Guttman DS (2004) Evolution of the core genome of Pseudomonas syringae, a highly clonal endemic plant pathogen. Appl Environ Microbiol 70:1999–2012

Satterwhite RS, Cooper TF (2015) Constraints on adaptation of Escherichia coli to mixed-resource environments increase over time. Evolution 69:2067–2078

Saupe SJ (2000) Molecular genetics of heterokaryon incompatibility in filamentous ascomycetes. Microbiol Mol Biol Rev 64:489–502

Savageau M (1983) Escherichia coli habitats, cell types, and molecular mechanisms of gene control. Am Nat 122:732–744

Savile DBO (1953) Short season adaptations in the rust fungi. Mycologia 45:75–87

Savile DBO (1955) A phylogeny of the basidiomycetes. Can J Bot 33:60–104

Savile DBO (1971) Coevolution of the rust fungi and their hosts. Quart Rev Biol 46:211–218

Savile DBO (1971) Co-ordinated studies of parasitic fungi and flowering plants. Le Naturaliste Canadien 98:535–552

Savile DBO (1976) Evolution of the rust fungi (Uredinales) as reflected by their ecological problems. Evol Biol 9:137–207

Scheiner SM (1993) Genetics and evolution of phenotypic plasticity. Annu Rev Ecol Syst 24:35–68

Schippers JHM, Schmidt R, Wagstaff C, Jing H-C (2015) Living to die and dying to live: the survival strategy behind leaf senescence. Plant Physiol 169:914–930

Schirrmeister BE, Antonelli A, Bagheri HC (2011) The origin of multicellularity in cyanobacteria. BMS Evol Biol II:45. doi:10.1186/1471-2148-11-45

Schlegel HG, Bowien B (eds) (1989) Autotrophic Bacteria. Springer, New York

Schlichting CD, Pigliucci M (eds) (1998) Phenotypic evolution: a reaction norm perspective. Sinauer Associates Inc., Sunderland, MA

Schmid B (1986) Spatial dynamics and integration within clones of grassland perennials with different growth form. Proc R Soc Lond B 228:173–186

Schmidt-Nielsen K (1984) Scaling: why is animal size so important?. Cambridge University Press, Cambridge, UK

Schmidt SK, Nemergut DR, Darcy JL, Lynch R (2014) Do bacterial and fungal communities assemble differently during primary succession? Mol Ecol 23:254–258

Schneider DJ, Collmer A (2010) Studying plant-pathogen interactions in the genomics era: beyond molecular Koch's postulates to systems biology. Annu Rev Phytopathol 48:457–479

Schon I, Martens K, van Dijk P (eds) (2009) Lost sex: the evolutionary biology of parthenogenesis. Springer, Heidelberg

Schopf JW, Kudryavtsev AB, Czaja AD, Tripathi AB (2007) Evidence of archaean life: stromatolites and microfossils. Precambrian Res 158:141–155

Schopf TJM, Raup DM, Gould SJ, Simberloff DS (1975) Genomic versus morphologic rates of evolution: influence of morphologic complexity. Paleobiology 1:63–70

Schoustra SE, Debets AJM, Slakhorst M, Hoekstra RF (2007) Mitotic recombination accelerates adaptation in the fungus Aspergillus nidulans. PLoS Genet 3:e68. doi:10.1371/journal.pgen.0030068

Schultz ST, Scofield DG (2009) Mutation accumulation in real branches: fitness assays for genomic deleterious mutation rate and effect in large-statured plants. Am Nat 174:163–175

Schulz HN, Jorgensen BB (2001) Big bacteria. Annu Rev Microbiol 55:105–137

Schulz HN, Brinkhoff T, Ferdelman TG, Hernandez Marina M, Teske A et al (1999) Dense populations of a giant sulfur bacterium in Namibian shelf sediments. Science 284:493–495

Scott MF, Otto SP (2014) Why wait? three mechanisms selecting for environment-dependent developmental delays. J Evol Biol 27:2219–2232

Sebens KP (1979) The energetics of asexual reproduction and colony formation in benthic marine invertebrates. Am Zool 19:683–697

Sebens KP (1982) The limits to indeterminate growth: an optimal size model applied to passive suspension feeders. Ecology 63:209–222

Sebens KP (1983) Asexual reproduction in *Anthopleura elegantissima* (Brandt) (Anthozoa: Actiniaria): seasonality and spatial extent of clones. Ecology 63:434–444

Sebens KP (1987) The ecology of indeterminate growth in animals. Annu Rev Ecol Syst 18:371–407

Sebens KP (2002) Energetic constraints, size gradients, and size limits in benthic marine invertebrates. Integr Compar Biol 42:853–861

Seilacher A (1967) Fossil behavior. Sci Am 217:72–80

Seilacher A, Buatois LA, Mangano MG (2005) Trace fossils in the Ediacaran-Cambrian transition: behavioral diversification, ecological turnover and environmental shift. Palaeogeogr Palaeoclimat Palaeoecol 227:323–356

Selander RK, Caugant DA, Whittam TS (1987) Genetic structure and variation in natural populations of *Escherichia coli*. In: Neidhardt FC (ed-in-chief) Escherichia coli and Salmonella typhimurium. Cellular and molecular biology, vol 2. American Society Microbiology Press, Washington, DC, pp 1625–1648

Selman C, Blount JD, Nussey DH, Speakman JR (2012) Oxidative damage, ageing, and life-history evolution: where now? Trends Ecol Evol 27:570–577

Sereno PC (1999) The evolution of dinosaurs. Science 284:2137–2147

Sewell GWF, Wilson JF (1964) Occurrence and dispersal of *Verticillium* conidia in xylem sap of the hop (*Humulus lupulus* L.). Nature 204:901

Shapiro BJ, Friedman J, Cordero OX, Preheim SP, Timberlake SC et al (2012) Population genomics of early events in the ecological differentiation of bacteria. Science 336:48–51

Shapiro JA (1985) Intercellular communication and genetic change in bacterial populations. In: Halvorson HO, Pramer D, Rogul M (eds) Engineered organisms in the environment: scientific issues. Washington, American Society Microbiology Press, Washington, DC, pp 63–69

Shapiro JA (1998) Thinking about bacterial populations as multicellular organisms. Annu Rev Microbiol 52:81–104

Shapiro JA (1999) Views about evolution are evolving. Am Soc Microbiol News 65:201–207

Shapiro JA (2002) Genome organization and reorganization in evolution: formatting for computation and function. Ann NY Acad Sci 981:111–134

Shapiro JA (2005) A 21st century view of evolution: genome system architecture, repetitive DNA, and natural genetic engineering. Gene 345:91–100

Shapiro JA (2011) Evolution: a view from the 21st Century. FT Press Science, Upper Saddle River, NJ

Shapiro JA, von Sternberg R (2005) Why repetitive DNA is essential to genome function. Biol Rev 80:1–24

Sharp PM, Shields DC, Wolfe KH, Li W-H (1989) Chromosomal location and evolutionary rate variation in eubacterial genes. Science 246:808–810

Shattock RC, Preece TF (2000) Tranzschel revisited: modern studies of the relatedness of different rust fungi confirm his law. Mycologist 14:113–117

Shendure J, Akey JM (2015) The origins, determinants, and consequences of human mutations. Science 349:1478–1483

Shoval O, Sheftel H, Shinar G, Hart Y, Ramote O et al (2012) Evolutionary trade-offs, pareto optimality, and the geometry of phenotype space. Science 336:1157–1160

Shuford WD, Timossi LC (1989) Plant communities of Marin country. Calif Native Plant Soc, Sacramento, CA

Sibly RM, Calow P (1986) Physiological ecology of animals: an evolutionary approach. Blackwell, Oxford, UK

Siefert JL, Fox GE (1998) Phylogenetic mapping of bacterial morphology. Microbiology 144:2803–2808

Signer RAJ, Morrison SJ (2013) Mechanisms that regulate stem cell aging and life span. Cell Stem Cell 12:152–165

Silander JA Jr (1985) Microevolution in clonal plants. In: Jackson JBC, Buss LW, Cook RE (eds) Population biology and evolution of clonal organisms. Yale University Press, New Haven, CT, pp 107–152

Silvertown J, Franco M, Perez-Ishiwara R (2001) Evolution of senescence in iteroparous perennial plants. Evol Ecol Res 3:393–412

Simons AM (2014) Playing smart vs playing safe: the joint expression of phenotypic plasticity and potential bet hedging across and within thermal environments. J Evol Biol 27:1047–1056

Sims DW, Reynolds AM, Humphries NE, Southail EJ, Wearmouth VJ et al (2014) Hierarchical random walks in trace fossils and the origin of optimal search behavior. Proc Natl Acad Sci USA 111:11073–11078

Singh RP, Hodson DP, Huerta-Espino J, Jin Y, Bhavani S et al (2011) The emergence of Ug99 races of the stem rust fungus is a threat to world wheat production. Annu Rev Phytopathol 49:465–481

Sistrom WR (1969). Microbial life, 2nd edn. Holt, Rinehart & Winston, NY

Sjamsuridzal W, Nishida H, Ogawa H, Kakishima M, Sugiyama J (1999) Phylogenetic positions of rust fungi parasitic on ferns: evidence from 18S rDNA sequence analysis. Mycoscience 40:21–27

Skold HN, Asplund ME, Wood CA, Bishop JDD (2011) Telomerase deficiency in a colonial ascidian after prolonged asexual propagation. J Exp Zool (Mol Dev Evol) 316:276–283

Slatkin M (1974) Hedging one's evolutionary bets. Nature 250:704–705

Slobodkin LB (1988) Intellectual problems of applied ecology. BioScience 38:337–342

Slotkin RK, Martienssen R (2007) Transposable elements and the epigenetic regulation of the genome. Nature Rev Genet 8:272–285

Smith AP, Palmer JO (1976) Vegetative reproduction and close packing in a successional plant species. Nature 261:232–233

Smith FE (1954) Quantitative aspects of population growth. In: Boell EJ (ed) Dynamics of growth processes. Proceedings of 11th Growth Symposium, Society Study Development and Growth. Princeton University Press, Princeton, NJ, pp 277–294

Smith JNM (1974) The food searching behaviour of two European thrushes. II. The adaptiveness of the search patterns. Behaviour 49:1–61

Smith ML, Bruhn JN, Anderson JB (1992) The fungus *Armillaria bulbosa* is among the largest and oldest living organisms. Nature 356:428–431

Smith SE, Read D (2008) Mycorrhizal symbiosis, 3rd ed. Academic Press, NY

Sniegowski PD (2005) Linking mutation to adaptation: overcoming stress at the spa. New Phytol 166:360–362

Sniegowski PD, Lenski RE (1995) Mutation and adaptation: the directed mutation controversy in evolutionary perspective. Annu Rev Ecol Syst 26:553–578

Sober E (1984) The nature of selection. Evolutionary theory in philosophical focus. MIT Press, Cambridge, MA

Solari CA, Kessler JO, Michod RE (2006) A hydrodynamics approach to the evolution of multicellularity: flagellar motility and germ-soma differentiation in volvocalean green algae. Am Nat 167:537–554

Solari CA, Ganguly S, Kessler JO, Michod RE, Goldstein RE (2006) Multicellularity and the functional interdependence of motility and molecular transport. Proc Natl Acad Sci USA 103:1353–1358

Son K, Brumley DR, Stocker R (2015) Live from under the lends: exploring microbial motility with dynamic imaging and microfluidics. Nature Rev Microbiol 13:761–775

Sonneborn TM (1930) Genetic studies on *Stentostomum incaudatum* (Nov. Spec.) I. The nature and origin of differences among individuals formed during vegetative reproduction. J Exp Zool Part A Ecol Genet Physiol 57:57–108

Soriano P, Jaenisch R (1986) Retroviruses as probes for mammalian development: allocation of cells to the somatic and germ lineages. Cell 46:19–29

Southwood TRE (1977) Habitat, the templet for ecological strategies? J Anim Ecol 46:337–365

Southwood TRE (1988) Tactics, strategies and templets. Oikos 52:3–18

Spang A, Saw JH, Jorgensen SL, Zaremba-Niedzwiedzka K, Martijn J et al (2015) Complex archaea that bridge the gap between prokaryotes and eukaryotes. Nature 521:173–179

Spanu PD, Abbott JC, Amselem J, Burgis TA, Soanes DM et al (2010) Genome expansion and gene loss in powdery mildew fungi reveal tradeoffs in extreme parasitism. Science 330:1543–1546

Speijer D, Lukes J, Elias M (2015) Sex is a ubiquitous, ancient, and inherent attribute of eukaryotic life. Proc Natl Acad Sci USA 112:8827–8834

Spence DHN (1982) The zonation of plants in freshwater lakes. Adv Ecol Res 12:37–125

Spratt BG, Maiden MCJ (1999) Bacterial population genetics, evolution and epidemiology. Phil Trans R Soc Lond B 354:701–710

Spratt BG, Hanage WP, Feil EJ (2001) The relative contributions of recombination and point mutation to the diversification of bacterial clones. Curr Opin Microbiol 4:602–606

Spudich JL, Koshland DE Jr (1976) Non-genetic individuality: chance in the single cell. Nature 262:467–471

Srivastava M, Simakov O, Chapman J, Fahey B, Gauthier MEA, Mitros T et al (2010) The *Amphimedon queenslandica* genome and the evolution of animal complexity. Nature 466:720–726

Staats M, van Baarlen P, van Kan JAL (2005) Molecular phylogeny of the plant pathogenic genus *Botrytis* and the evolution of host specificity. Mol Biol Evol 22:333–346

Stanley SM (1973) An explanation for Cope's rule. Evolution 27:1–26

Stanley SM (1975) Clades versus clones in evolution: why we have sex. Science 190:382–383

Stanley SM (1979) Macroevolution, pattern and process. Freeman, San Francisco

Staskawicz BJ, Mudgett MB, Dangl JL, Galan JE (2001) Common and contrasting themes of plant and animal diseases. Science 292:2285–2289

Staskawicz BJ, Ausubel FM, Baker BJ, Ellis JG, Jones JDG (1995) Molecular genetics of plant disease resistance. Science 268:661–667

Stearns SC (1977) The evolution of life history traits: a critique of the theory and a review of the data. Annu Rev Ecol Syst 8:145–171

Stearns SC (1982) The role of development in the evolution of life histories. In: Bonner JT (ed) Evolution and development. Springer-Verlag, NY, pp 237–258

Stearns SC (1983) The influence of size and phylogeny on patterns of covariation among life-history traits in the mammals. Oikos 41:173–187

Stearns SC (1989) The evolutionary significance of phenotypic plasticity. Bioscience 39:436–445

Stearns SC (1992) The evolution of life histories. Oxford University Press, NY

Stearns SC (2000) Life history evolution: successes, limitations, and prospects. Naturwissenschaften 87:476–486

Stearns SC, Magwene P (2003) The naturalist in a world of genomics. Am Nat 161:171–180

Stebbins GL (1968) Integration of development and evolutionary progress. In: Lewontin RC (ed) Population biology and evolution. Syracuse University Press, Syracuse, NY, pp 17–36

Steenkamp ET, Wright J, Baldauf SL (2006) The protistan origins of animals and fungi. Mol Biol Evol 23:93–106

Steffan SA, Chikaraishi Y, Currie CR, Horn H, Gaines-Day HR et al (2015) Microbes are trophic analogs of animals. Proc Natl Acad Sci USA 112:15119–15124

Steinert M, Hentschel U, Hacker J (2002) Legionella pneumophila, an aquatic microbe goes astray. FEMS Microbiol Rev 26:149–162

Steinkraus KA, Kaeberlein M, Kennedy BK (2008) Replicative aging in yeast: the means to the end. Annu Rev Cell Dev Biol 24:29–54

Stephanopoulos G, Aristidou AA, Nielsen J (1998) Metabolic engineering: principles and methodologies. Academic Press, San Diego

Stephens DW, Krebs JR (1986) Foraging theory. Princeton University Press, Princeton, NJ

Stephens DW, Brown JS, Ydenberg RC (eds) (2007) Foraging: behavior and ecology. University of Chicago Press, Chicago

Stephenson NL, Das AJ, Condit R, Russo SE, Baker PJ et al (2014) Rate of tree carbon accumulation increases continuously with tree size. Nature 507:90–93

Sterner RW, Elser JJ (2002) Ecological stoichiometry: the biology of elements from molecules to the biosphere. Princeton University Press, Princeton, NJ

Stevens PS (1974) Patterns in nature. Little, Brown & Co, Boston.

Stevens VM, Trochet A, Van Dyck H, Clobert J, Baguette M (2012) How is dispersal integrated in life histories: a quantitative analysis using butterflies. Ecol Lett 15:74–86

Stewart EJ, Madden R, Paul G, Taddei F (2005) Aging and death in an organism that reproduces by morphologically symmetric division. PLoS Biol 3(2):e45

Stewart I (1998) Life's other secret: the new mathematics of the living world. Wiley, NY

Stocker R, Seymour JR (2012) Ecology and physics of bacterial chemotaxis in the ocean. Microbiol Mol Biol Rev 76:792–812

Stoecker DK, Johnson MD, de Vargas C, Not F (2009) Acquired phototrophy in aquatic protists. Aquat Microb Ecol 57:279–310

Stomp M, Huisman J, de Jongh F, Veraart AJ, Gerla D et al (2004) Adaptive divergence in pigment composition promotes plankton biodiversity. Science 432:104–107

Stoner DS, Weissman IL (1996) Somatic and germ cell parasitism in a colonial ascidian: possible role for a highly polymorphic allorecognition system. Proc Natl Acad Sci USA 93:15254–15259

Stout JE, Yu VL, Best MG (1985) Ecology of Legionella pneumophila within water distribution systems. Appl Environ Microbiol 49:221–228

Stowe WC, Brodie-Kommit J, Stowe-Evans E (2011) Characterization of complementary chromatic adaptation in Gloeotrichia UTEX 583 and identification of a transposon-like insertion in the cpeBA operon. Plant Cell Physiol 52:553–562

Strathmann R (1974) The spread of sibling larvae of sedentary marine invertebrates. Am Nat 108:29–44

Straub CS, Ives AR, Gratton C (2011) Evidence for a trade-off between host-range breadth and host-use efficiency in aphid parasitoids. Am Nat 177:389–395

Strobeck C (1975) Selection in a fine-grained environment. Am Nat 109:419–425

Strong DRJ, Ray JTS (1975) Host tree location behavior of a tropical vine (Monstera gigantea) by skototropism. Science 190:804–806

Sudbery PE (2011) Growth of Candida albicans hyphae. Nature Rev Microbiol 9:737–748

Sulikowski D, Burke D (2011) Movement and memory: different cognitive strategies are used to search for resources with different natural distributions. Behav Ecol Sociobiol 65:621–631

Sunagawa S, Coelho LP, Chaffron S, Kultima JR, Labadie K et al (2015). Structure and function of the global ocean microbiome. Science 348. doi:10.1126/science1261359

Sung W, Ackerman MS, Miller SF, Doak TG, Lynch M (2012) Drift-barrier hypothesis and mutation-rate evolution. Proc Natl Acad Sci USA 109:18488–18492

Suter DM, Molina N, Gatfield D, Schneider K, Schibler U et al (2011) Mammalian genes are transcribed with widely different bursting kinetics. Science 332:472–474

Swanson MS, Hammer BK (2000) *Legionella pneumophila* pathogenesis: a fateful journey from amoebae to macrophages. Annu Rev Microbiol 54:567–613

Swedmark B (1964) The interstitial fauna of marine sand. Biol Rev 39:1–42

Swift MJ (1976) Species diversity and the structure of microbial communities in terrestrial habitats. In: Anderson JM, Macfadyen A (eds) The role of terrestrial and aquatic organisms in decomposition processes. Blackwell, Oxford, UK, pp 185–222

Szathmáry E (2015) Toward major evolutionary transitions theory 2.0. Proc Natl Acad Sci USA 112:10104–10111

Szollosi GJ, Derenyi I, Vellai T (2006) The maintenance of sex in bacteria is ensured by its potential to reload genes. Genetics 174:2173–2180

Szovenyi P, Ricca M, Hock Z, Shaw JA, Shimizu KK et al (2013) Selection is no more efficient in haploid than diploid life stages of an angiosperm and a moss. Mol Biol Evol 30:1929–1939

Tack AJM, Thrall PH, Barrett LG, Burdon JJ, Laine A-L (2012) Variation in infectivity and aggressiveness in space and time in wild host-pathogen systems: causes and consequences. J Evol Biol 25:1918–1936

Tagkopoulos I, Liu Y-C, Tavazoie S (2008) Predictive behavior within microbial genetic networks. Science 320:1313–1317

Takahashi M, Bienfang PK (1983) Size structure of phytoplankton biomass and photosynthesis in subtropical Hawaiian waters. Mar Biol 76:203–211

Takamatsu A, Takaba E, Takizawa G (2009) Environment-dependent morphology in plasmodium of true slime mold *Physarum polysephalum* and a network growth model. J Theor Biol 256:29–44

Tan J, Pu Z, Ryberg WA, Jiang L (2015) Resident-invader phylogenetic relatedness, not resident phylogenetic diversity, controls community invasibility. Am Nat 186:59–71

Tanaka MM, Valckenborgh F (2011) Escaping an evolutionary lobster trap: drug resistance and compensatory mutation in a fluctuating environment. Evolution 65:1376–1387

Tarnita CE, Washburne A, Martinez-Garcia R, Sgro AE, Levin SA (2015) Fitness tradeoffs between spores and nonaggregating cells can explain the coexistence of diverse genotypes in cellular slime molds. Proc Natl Acad Sci USA 112:2776–2781

Taylor JW, Jacobson DJ, Fisher MC (1999a) The evolution of asexual fungi: reproduction, speciation and classification. Annu Rev Phytopathol 37:197–246

Taylor JW, Geiser DM, Burt A, Koufopanou V (1999b) The evolutionary biology and population genetics underlying fungal strain typing. Clinical Microbiol Rev 12:126–146

Taylor JW, Hann-Soden C, Branco S, Sylvain I, Ellison CE (2015) Clonal reproduction in fungi. Proc Natl Acad Sci USA 112:8901–8908

Taylor PD, Wilson MA (2003) Palaeoecology and evolution of marine hard substrate communities. Earth Sci Rev 62:1–103

Taylor TB, Buckling A (2010) Competition and dispersal in *Pseudomonas aeruginosa*. Am Nat 176:83–89

Templeton AR, Rothman ED (1974) Evolution in heterogeneous environments. Am Nat 108:409–428

Tero A, Takagi S, Saigusa T, Ito K, Bebber DP et al (2010) Rules for biologically inspired adaptive network design. Science 327:439–442

Teschler JK, Zamorano-Sanchez D, Utada AS, Warner CJA, Wong GCL et al (2015) Living in the matrix: assembly and control of *Vibrio cholerae* biofilms. Nature Rev Microbiol 13:255–268

Thomas H (2013) Senescence, ageing and death of the whole plant. New Phytol 197:696–711

Thompson DW (1961) On growth and form (abridged edition edited by JT Bonner). Cambridge University Press, Cambridge, UK.

Thompson JN (1994) The coevolutionary process. Univ Chicago Press, Chicago

Thompson JN, Burdon JJ (1992) Gene-for-gene coevolution between plants and parasites. Nature 360:121–125

Thompson K, Rabinowitz D (1989) Do big plants have big seeds? Am Nat 133:722–728

Thompson W (1984) Distribution, development and functioning of mycelial cord systems of decomposer basidiomycetes of the deciduous woodland floor. In: Jennings DH, Rayner ADM (eds) The ecology and physiology of the fungal mycelium. Cambridge University Press, Cambridge, UK, pp 185–214

Thompson W, Rayner ADM (1982) Structure and development of mycelial cord systems of *Phanerochaete laevis* in soil. Trans Brit Mycol Soc 78:193–200

Tian D, Traw MB, Chen JQ, Kreitman M, Bergelson J (2003) Fitness costs of *R*-gene-mediated resistance in *Arabidopsis thaliana*. Nature 423:74–77

Tibayrenc M, Ayala FJ (2012) Reproductive clonality of pathogens: a perspective on pathogenic viruses, bacteria, fungi, and parasitic protozoa. Proc Natl Acad Sci USA 109:E3305–E3313

Tibayrenc M, Ayala FJ (2015) How clonal are *Neisseria* species? The epidemic clonality model revisited. Proc Natl Acad Sci USA 112:8909–8913

Tibayrenc M, Kjellberg F, Arnaud J, Oury B, Breniere SF et al (1991) Are eukaryotic microorganisms clonal or sexual? A population genetics vantage. Proc Natl Acad Sci USA 88:5129–5133

Tice MM, Lowe DR (2004) Photosynthetic microbial mats in the 3,416-Myr-old ocean. Nature 431:549–552

Timmis JN, Ayliffe MA, Huang CY, Martin W (2004) Endosymbiotic gene transfer: organelle genomes forge eukaryotic chromosomes. Nature Rev Genet 5:123–135

Ting CS, Rocap G, King J, Chisholm SW (2002) Cyanobacterial photosynthesis in the oceans: the origins and significance of divergent light-harvesting strategies. Trends Microbiol 10:134–142

To T-L, Maheshri N (2010) Noise can induce bimodality in positive transcriptional feedback loops without bistability. Science 327:1142–1145

Tomitani A, Knoll AH, Cavanaugh CM, Ohno T (2006) The evolutionary diversification of cyanobacteria: molecular-phylogenetic and paleontological perspectives. Proc Natl Acad Sci USA 5442–5447

Touchon M, Rocha EPC (2016) Coevolution of the organization and structure of prokaryotic genomes. Cold Spring Harbor Perspect Biol 8:a018168

Touchon M, Hoede C, Tenaillon O, Barbe V, Baeriswyl S et al (2009) Organised genome dynamics in Escherichia coli species results in highly diverse adaptive paths. PLoS Genet 5:e1000344. doi:10.1371/journal.pgen.1000344

Trail F (2007) Fungal cannons: explosive spore discharge in the Ascomycota. FEMS Microbiol Lett 276:12–18

Traub R, Wisseman JCL (1974) The ecology of chigger-borne rickettsiosis (scrub typhus). J Med Entomol 11:237–303

Travis JMJ, Dytham C (1999) Habitat persistence, habitat availability and the evolution of dispersal. Proc R Soc Lond B 266:723–728

Trinci APJ (1984) Regulation of hyphal branching and hyphal orientation. In: Jennings DH, Rayner ADM (eds) The ecology and physiology of the fungal mycelium. Cambridge University Press, Cambridge, UK, pp 23–52

Trinci APJ, Cutter EG (1986) Growth and form in lower plants and the occurrence of meristems. Phil Trans R Soc B 313:95–113

Trinci APJ, Thurston CF (1976) Transition to the non-growing state in eukaryotic micro-organisms. In: Gray TRG, Postgate JR (eds) The survival of vegetative microbes (Proceedings 24th Symposium of the Society for General Microbiology), pp 55–79. Cambridge University Press, Cambridge, UK.

Tsai IJ, Bensasson D, Burt A, Koufopanou V (2008) Population genomics of the wild yeast Saccharomyces paradoxus: quantifying the life cycle. Proc Natl Acad Sci USA 105:4957–4962

Tuch BB, Li H, Johnson AD (2008) Evolution of eukaryotic transcription circuits. Science 319:1797–1799

Tuomi J, Vuorisalo T (1989a) Hierarchical selection in modular organisms. Trends Ecol Evol 4:209–213

Tuomi J, Vuorisalo T (1989b) What are the units of selection in modular organisms? Oikos 54:227–233

Turkington R, Harper JL (1979a) The growth, distribution and neighbour relationships of Trifolium repens in a permanent pasture. I. Ordination, pattern and contact. J Ecol 67:201–218

Turkington R, Harper JL (1979b) The growth, distribution and neighbour relationships of Trifolium repens in a permanent pasture. IV. Fine-scale biotic differentiation. J Ecol 67:245–254

Turner MG, Gardner RH, O'Neill RV (2001) Landscape ecology in theory and practice: pattern and process. Springer, NY

Twain M (1874) Life on the Mississippi. Harper & Row, NY

Valero M, Richerd S, Perrot V, Destombe C (1992) Evolution of alternation of haploid and diploid phases in life cycles. Trends Ecol Evol 7:25–29

Valladares F, Niinemets U (2008) Shade tolerance, a key plant feature of complex nature and consequences. Annu Rev Ecol Evol Syst 39:237–257

Valladares F, Gianoli E, Gomez JM (2007) Ecological limits to plant phenotypic plasticity. New Phytol 176:749–763

van de Lagemaat LN, Landry J-R, Mager DL, Medstrand P (2003) Transposable elements in mammals promote regulatory variation and diversification of genes with specialized functions. Trends Genet 19:530–536

van der Merwe M, Ericson L, Walker J, Thrall PH, Burdon JJ (2007) Evolutionary relationships among species of Puccinia and Uromyces (Pucciniaceae, Uredinales) inferred from partial protein coding gene phylogenies. Mycol Res 111:163–175

van der Merwe MM, Walker J, Ericson L, Burdon JJ (2008) Coevolution with higher taxonomic host groups within the Puccinia/Uromyces rust lineage obscured by host jumps. Mycol Res 112:1387–1408

van Diepeningen AD, Engelmoer DJP, Sellem CH, Huberts DHEW, Slakhorst SM et al (2014) Does autophagy mediate age-dependent effect of dietary restriction responses in the filamenatous fungus Podospora anserina? Phil Trans R Soc

B 369:20130447. ▶http://dx.doi.org/10.1098/rtsb.2013.0447

van Dijk PJ (2003) Ecological and evolutionary opportunities of apomixis: insights from *Taraxacum* and *Chondrilla*. Phil Trans R Soc Lond B 358:1113–1121

van Dover CL (2000) The ecology of deep-sea hydrothermal vents. Princeton University Press, Princeton, NJ

van Durme M, Nowack MK (2016) Mechanisms of developmentally controlled cell death in plants. Curr Opin Plant Biol 29:29–37

van Gestel J, Vlamakis HH, Kolter R (2015) From cell differentiation to cell collectives: *Bacillus subtilis* uses division of labor to migrate. PLoS Biol 13. doi:10.1371/journal.pbio.1002141

van Hoek MJA, Merks RMH (2012) Redox balance is key to explaining full vs. partial switching to low-yield metabolism. BMC Syst Biol 2. doi:▶biomedcentral.com/1752-0509/6/22

van Kleunen M, Fischer M (2001) Adaptive evolution of plastic foraging responses in a clonal plant. Ecology 82:3309–3319

van Schaik CP, Terborgh JW, Wright SJ (1993) The phenology of tropical forests: Adaptive significance and consequences for primary consumers. Annu Rev Ecol Syst 24:353–377

van Valen L (1973) Body size and numbers of plants and animals. Evolution 27:27–35

van Valen L (1978) Arborescent animals and other colonids. Nature 276:318

Varmus H (1988) Retroviruses. Science 240:1427–1435

Varmus H, Brown P (1989) Retroviruses. In: Berg DE, Howe MM (eds) Mobile DNA. Am Soc Microbiol Press, Washington, DC, pp 53–108

Varmus HE (1983) Retroviruses. In: Shapiro DA (ed) Mobile genetic elements. Academic Press, NY, pp 411–503

Vasek FC (1980) Creosote bush: long-lived clones in the mojave desert. Am J Bot 67:246–255

Vaupel JW, Baudisch A, Dolling M, Roach DA, Gampe J (2004) The case for negative senescence. Theor Pop Biol 65:339–351

Veening J-W, Smits WK, Kuipers OP (2008) Bistability, epigenetics, and bet-hedging in bacteria. Annu Rev Microbiol 62:193–210

Veening J-W, Stewart EJ, Berngruber TW, Taddei F, Kuipers OP et al (2008) Bet-hedging and epigenetic inheritance in bacterial cell development. Proc Natl Acad Sci USA 105:4393–4398

Velicer GJ, Lenski RE (1999) Evolutionary trade-offs under conditions of resource abundance and scarcity: experiments with bacteria. Ecology 80:1168–1179

Vellend M (2010) Conceptual synthesis in community ecology. Quart Rev Ecol 85:183–206

Venter JC, Remington K, Heidelberg JF, Halpern AL, Rusch D et al (2004) Environmental genome shotgun sequencing of the Sargasso Sea. Science 304:66–74

Verhulst PF (1838) Notice sur la loi que la population suit dans son accroissement. Corresp Math Phys 10:113–121

Vermeij GJ (1999) Inequality and the directionality of history. Am Nat 153:243–253

Verstrepen KJ, Jansen A, Lewitter F, Fink GR (2005) Intragenic tandem repeats generate functional variability. Nat Genet 37:986–990

Viswanathan GM, da Luz MGE, Raposo EP, Stanley HE (2011) The physics of foraging: an introduction to random searches and biological encounters. Cambridge University Press, Cambridge, UK

Vlamakis H, Chai Y, Beauregard P, Losick R, Kolter R (2013) Sticking together: building a biofilm the *Bacillus subtilis* way. Nature Rev Microbiol 11:157–168

Vogel S (1988) Life's devices: the physical world of animals and plants. Princeton University Press, Princeton, NJ

Vogel S (2003) Comparative biomechanics: Life's physical world. Princeton University Press, Princeton, NJ

Vogelstein B, Papadopoulos N, Velculescu VE, Zhou S, Diaz LA Jr et al (2013) Cancer genome landscapes. Science 339:1546–1558

Von Uexküll J (1957) (orig. 1934) A stroll through the worlds of animals and men: a picture book of invisible worlds. In: Schiller CH (ed and trans) Instinctive behavior: the development of a modern concept. International Universities Press, New York, pp 5–80

Wächtershäuser G (2006) From volcanic origins of chemoautotrophic life to bacteria, archaea and eukarya. Phil Trans R Soc B 361:1787–1808

Wagner JK, Brun YV (2007) Out on a limb: how the *Caulobacter* stalk can boost the study of bacterial cell shape. Mol Microbiol 64:28–33

Wainwright SA, Biggs WD, Currey JD, Gosline JM (1976) Mechanical design in organisms. Princeton University Press, Princeton, NJ

Wajnberg E, Fauvergue X, Pons O (2000) Patch leaving decision rales and the Marginal Value Theorem: an experimental analysis and a simulation model. Behav Ecol 11:577–586

Wakamoto Y, Dhar N, Chait R, Schneider K, Signorino-Gelo F et al (2013) Dynamic persistence of antibiotic-stressed mycobacteria. Science 339:91–95

Walbot V, Cullis CA (1983) The plasticity of the plant genome - is it a requirement for success? Plant Mol Biol Report 1:3–11

Walbot V, Cullis CA (1985) Rapid genomic change in higher plants. Annu Rev Plant Physiol 36:367–396

Walbot V, Evans MMS (2003) Unique features of the plant life cycle and their consequences. Nature Rev Genet 4:369–379

Walker JJ, Pace NR (2007) Endolithic microbial ecosystems. Annu Rev Microbiol 61:331–347

Waller DM, Steingraeber DA (1985) Branching and modular growth: theoretical models and empirical patterns. In: Jackson JBC, Buss LW, Cook RE (eds) Population biology and evolution of clonal organisms. Yale University Press, New Haven, CT, pp 225–257

Wang P, Robert L, Pelletier J, Dang WL, Taddei F et al (2010) Robust growth of Escherichia coli. Curr Biol 20:1099–1103

Ward BA, Follows MJ (2016) Marine mixotrophy increases trophic transfer efficiency, mean organism size, and vertical carbon flux. Proc Natl Acad Sci USA 113:2958–2963

Warner DA, Miller DAW, Bronikowski AM, Janzen FJ (2016) Decades of field data reveal that turtles senesce in the wild. Proc Natl Acad Sci USA 113:6502–6507

Watkinson AR, White J (1985) Some life history consequences of modular construction in plants. Phil Trans R Soc Lond B 313:31–51

Watson JD, Baker TA, Bell SP, Gann A, Levine M, Losick R (2008) Molecular biology of the gene, 6th edn. Cold Spring Harbor Laboratory Press, Woodbury, NY

Webber HJ (1903) New horticultural and agricultural terms. Science 18:501–503

Webster J, Weber RWS (2007) Introduction to fungi. Cambridge Univ Press, New York

Wegener HC (2003) Antibiotics in animal feed and their role in resistance development. Curr Opin Microbiol 6:439–445

Wei Y, Wang X, Liu J, Nememan I, Singh AH et al (2011) The population dynamics of bacteria in physically structured habitats and the adaptive virtue of random motility. Proc Natl Acad Sci USA 108:4047–4052

Weigel D, Jurgens G (2002) Stem cells that make stems. Nature 415:751–754

Weiher E, Keddy P (eds) (1999) Ecological assembly rules: perspectives, advances, retreats. Cambridge University Press, Cambridge, UK

Weismann A (1892) Die Continuitat des Keimplasmas als Grundlage einer Theorie der Vererbung. Gustav Fischer, Jena, Germany

Welch DM, Meselson M (2000) Evidence for the evolution of bdelloid rotifers without sexual reproduction or genetic exchange. Science 288:1211–1215

West GB, Brown JH, Enquist BJ (1997) A general model for the origin of allometric scaling laws in biology. Science 276:122–126

West GB, Brown JH, Enquist BJ (2000) The origin of universal scaling laws in biology. In: Brown JH, West GB (eds) Scaling in biology. Oxford University Press, NY, pp 87–112

West RR, McKinney FK, Fagerstrom JA, Vacelet J (2011) Biological interactions among extant and fossil clonal organisms. Facies 57:351–374

West-Eberhard MJ (1989) Phenotypic plasticity and the origins of diversity. Annu Rev Ecol Syst 20:249–278

Westman WE (1981) Seasonal dimorphism of foliage in California coastal sage scrub. Oecologia 51:385–388

White J (1979) The plant as a metapopulation. Annu Rev Ecol Syst 10:109–145

Whitham TG, Schweitzer JA (2002) Leaves as islands of spatial and temporal variation: consequences for plant herbivores, pathogens, communities and ecosystems. In: Lindow SE, Hecht-Poinar EI, Elliott VJ (eds) Phyllosphere microbiology. APS Press, St. Paul, MN, pp 279–298

Whitham TG, Slobodchikoff CN (1981) Evolution by individuals, plant-herbivore interactions, and mosaics of genetic variability: The adaptive significance of somatic mutations in plants. Oecologia 49:287–292

Whitham TG, Bailey JK, Schweitzer JA, Shuster SM, Bangert RK, LeRoy CJ et al (2006) A framework for community and ecosystem genetics: from genes to ecosystems. Nature Rev Genet 7:511–523

Whittingham WF, Raper KB (1960) Nonviability of stalk cells in Dictyostelium. Proc Natl Acad Sci USA 46:642–649

Whittle CA, Otto SP, Johnston MO, Krochko JE (2009) Adaptive epigenetic memory of ancestral temperature regime in Arabidopsis thaliana. Botany 87:650–657

Wilbur HM (1980) Complex life cycles. Annu Rev Ecol Syst 11:67–93

Wilbur HM, Collins JP (1973) Ecological aspects of amphibian metamorphosis. Science 182:1305–1314

Wilbur HM, Tinkle DW, Collins JP (1974) Environmental uncertainty, trophic level and resource availability in life history evolution. Am Nat 108:805–817

Wilkins AS, Holliday R (2009) The evolution of meiosis from mitosis. Genetics 181:3–12

Wilkinson HT, Miller RD, Miller RL (1981) Infiltration of fungal and bacterial propagules into soil. Soil Sci Soc Amer J 45:1034–1039

Williams CB, Sillett SC (2007) Epiphyte communities on redwood (*Sequoia sempervirens*) in northwestern California. The Bryologist 110:420–452

Williams GC (1957) Pleiotropy, natural selection, and the evolution of senescence. Evolution 11:398–411

Williams GC (1966) Adaptation and natural selection. Princeton University Press, Princeton, NJ

Williams GC (1975) Sex and evolution. Princeton University Press, Princeton, NJ

Williams GC (1999) The Tithonus error in modem gerontology. Quart Rev Biol 74:405–415

Williams TA, Foster PG, Cox CJ, Embley TM (2013) An archaeal origin of eukaryotes supports only two primary domains of life. Nature 504:231–236

Williams TM, Wolfe L, Davis T, Kendall T, Richter B et al (2014) Instantaneous energetics of puma kills reveal advantage of felid sneak attacks. Science 346:81–85

Willis A (2016) Extrapolating abundance curves has no predictive power for estimating microbial diversity. Proc Natl Acad Sci USA 113. doi:10.1073/pnas.1608281113

Wilson AC, Carlson SS, White TJ (1977) Biochemical evolution. Annu Rev Biochem 46:573–639

Wilson CG, Sherman PW (2013) Spatial and temporal escape from fungal parasitism in natural communities of anciently asexual bdelloid rotifers. Proc R Soc B 280:20131255

Wilson EO (1975) Sociobiology. Belknap Press of Harvard University Press, Cambridge, Mass

Wingfield BD, Ericson L, Szaro T, Burdon JJ (2004) Phylogenetic patterns in the Uredinales. Aust Plant Pathol 327–335

Winogradsky S (1949) Microbiologie du Sol, Problemes et Methodes. Masson, Paris

Wiser MJ, Ribeck N, Lenski RE (2013) Long-term dynamics of adaptation in asexual populations. Science 342:1364–1367

Woese CR (1987) Bacterial evolution. Microbiol Rev 51:221–271

Woese CR (1998a) The universal ancestor. Proc Natl Acad Sci USA 95:6854–6859

Woese CR (1998b) Default taxonomy: Ernst Mayr's view of the microbial world. Proc Natl Acad Sci USA 95:11043–11046

Woese CR (2002) On the evolution of cells. Proc Natl Acad Sci 99:8742–8747

Woese CR, Fox GE (1977) Phylogenetic structure of the prokaryotic domain: the primary kingdoms. Proc Natl Acad Sci USA 74:5088–5090

Wolda H (1987) Seasonality and the community. In: Gee JHR, Giller PS (eds) Organization of communities: past and present (27th Symposium British Ecological Society). Blackwell, London, UK, pp 69–95

Wolda H (1989) Seasonal cues in tropical organisms. Rainfall? Not necessarily! Oecologia 80:437–442

Wong S, Anand M, Bauch CT (2011) Agent-based modelling of clonal plant propagation across space: Recapturing fairy rings, power laws and other phenomena. Ecol Informatics 6:127–135

Woodruff LL (1926) Eleven thousand generations of *Paramecium*. Quart Rev Biol 1:436–438

Woolhouse MEJ, Taylor LH, Hayden DT (2001) Population biology of multihost pathogens. Science 292:1109–1112

Worrall JJ (1997) Somatic incompatibihty in basidiomycetes. Mycologia 89:24–36

Wozniak RAF, Waldor MK (2010) Integrative and conjugative elements: mosaic mobile genetic elements enabling dynamic lateral gene flow. Nature Rev Microbiol 8:552–563

Wright FL (1953) The future of architecture. Horizon Press, NY

Wright S (1932) The roles of mutation, inbreeding, crossbreeding and selection in evolution. In: Jones DF (ed) Proceedings of the sixth international congress of genetics, vol I. Brooklyn Bot. Garden, NY, pp 356–366

Wright S (1978) Evolution and the genetics of populations. In: Variability within and among natural populations, vol 4. University of Chicago Press, Chicago, pp 356–366

Wuest SE, Philipp MA, Guthörl D, Schmid B, Grossniklaus U (2016) Seed production affects maternal growth and senescence in Arabidopsis. Plant Physiol 171:392–404

Xie XS, Choi PJ, Li G-W, Lee NK, Lia G (2008) Single-molecule approach to molecular biology in living bacterial cells. Annu Rev Biophys 37:417–444

Yanagita T (1977) Cellular age in micro-organisms. In: Ishikawa T, Maruyama T, Matsumiya H (eds) Growth and differentiation in micro-organisms. University Park Press, Baltimore, Maryland, pp 1–36

Yip V, Beekman M, Latty T (2014) Foraging strategies of the acellular slime moulds *Didymium iridis* and *Didymium bahiense*. Fungal Ecol 11:29–36

Yoon HS, Hackett J, Ciniglia C, Pinto G, Bhattacharya D (2004) A molecular timeline for the origin of photosynthetic eukaryotes. Mol Biol Evol 21:809–818

Young KD (2006) The selective value of bacterial shape. Microbiol Mol Biol Rev 70:660–703

Zambryski P (1989) *Agrobacterium*-plant cell DNA transfer. In: Berg DE, Howe MM (eds) Mobile DNA. Am Soc Microbiol Press, Washington, DC, pp 309–333

Zamenhof S, Eichhorn HH (1967) Study of microbial evolution through loss of biosynthetic functions:

establishment of "defective" mutants. Nature 216:456–458

Zamer WE, Scheiner SM (2014) A conceptual framework for organismal biology: linking theories, models, and data. Integr Compar Biol 54:736–756

Zeyl C, Mizesko M, de Visser JAGM (2001) Mutational meltdown in laboratory yeast populations. Evolution 55:909–917

Zeyl C, Vanderford T, Carter M (2003) An evolutionary advantage of haploidy in large yeast populations. Science 299:555–558

Zhang J, Khan SA, Hasse C, Ruf S, Heckel DG et al (2015) Full crop protection from an insect pest by expression of long double-stranded RNAs in plastids. Science 347:991–994

Zhang Y (2014) Persisters, persistent infections and the Yin-Yang model. Emerg Microbes Infections 3. doi:10.1038/emi.20014.3

Zhong S, Khodursky A, Dykhuizen DE, Dean AM (2004) Evolutionary genomics of ecological specialization. Proc Natl Acad Sci USA 101:11719–11724

Zhong S, Miller SP, Dykhuizen DE, Dean AM (2009) Transcription, translation, and the evolution of specialists and generalists. Mol Biol Evol 26:2261–2678

Zinder SH, Dworkin M (2013) Morphological and physiological diversity. In: Rosenberg E, DeLong EF, Lory S, Stackebrandt E, Thompson F (eds) The prokaryotes: prokaryotic biology and symbiotic associations, 4th edn. Springer, NY, pp 89–122

Zinger L, Gobet A, Pommier T (2012) Two decades of describing the unseen majority of aquatic microbial diversity. Mol Ecol 21:1878–1896

Zinser ER, Lindell D, Johnson ZI, Futschik ME, Steglich C et al (2009) Choreography of the transcriptome, photophysiology, and cell cycle of a minimal photoautotroph. Prochlorococcus PLoS ONE 4(4):e5135. doi:10.1371/journal.pone.0005135

Zuckerkandl E, Pauling L (1965) Evolutionary divergence and convergence in proteins. In: Bryson V, Kogel HJ (eds) Evolving genes and proteins. Academic Press, NY, pp 97–166

Index

Note: Page numbers followed by *f*, *t* and *n* refer to figures, tables, and footnotes, respectively

© Springer Science+Business Media LLC 2017
J.H. Andrews, *Comparative Ecology of Microorganisms and Macroorganisms*,
DOI 10.1007/978-1-4939-6897-8

Printed in the United States
By Bookmasters